TABLE OF CONTENTS

VOLUME II

Page

Table of contents (cont'd)

ECE/TRANS/190 (Vol. II)

ECONOMIC COMMISSION FOR EUROPE

Committee on Inland Transport

European Agreement concerning the International Carriage of Dangerous Goods by Inland Waterways (ADN)

- Annexed Regulations, as revised as of 1 January 2007 (cont'd)

UNITED NATIONS
New York and Geneva, 2006

NOTE

The designations employed and the presentation of the material in this publication do not imply the expression of any opinion whatsoever on the part of the Secretariat of the United Nations concerning the legal status of any country, territory city or area, or of its authorities, or concerning the delimitation of its frontiers or boundaries.

ECE/TRANS/190 (Vol. II)

PUBLICATION DES NATIONS UNIES
Sales No.: E.06.VIII.2
ISBN 92-1-139118-0 *(Complete set of 2 volumes)* ISBN 92-1-139117-2 (Vol. II)

Volumes I end II not to be sold separately.

PART 2

Classification

CHAPTER 2.1

GENERAL PROVISIONS

2.1.1   **Introduction**

2.1.1.1   The classes of dangerous goods according to ADN are the following:

Class 1   Explosive substances and articles
Class 2   Gases
Class 3   Flammable liquids
Class 4.1   Flammable solids, self-reactive substances and solid desensitized explosives
Class 4.2   Substances liable to spontaneous combustion
Class 4.3   Substances which, in contact with water, emit flammable gases
Class 5.1   Oxidizing substances
Class 5.2   Organic peroxides
Class 6.1   Toxic substances
Class 6.2   Infectious substances
Class 7   Radioactive material
Class 8   Corrosive substances
Class 9   Miscellaneous dangerous substances and articles

2.1.1.2   Each entry in the different classes has been assigned a UN number. The following types of entries are used:

A.   Single entries for well defined substances or articles including entries for substances covering several isomers, e.g.:

UN No. 1090   ACETONE
UN No. 1104   AMYL ACETATES
UN No. 1194   ETHYL NITRITE SOLUTION

B.   Generic entries for a well defined group of substances or articles, which are not n.o.s. entries, e.g.:

UN No. 1133   ADHESIVES
UN No. 1266   PERFUMERY PRODUCTS
UN No. 2757   CARBAMATE PESTICIDE, SOLID, TOXIC
UN No. 3101   ORGANIC PEROXIDE TYPE B, LIQUID

C.   Specific n.o.s. entries covering a group of substances or articles of a particular chemical or technical nature, not otherwise specified, e.g.:

UN No. 1477   NITRATES, INORGANIC, N.O.S.
UN No. 1987   ALCOHOLS, N.O.S.

D.   General n.o.s. entries covering a group of substances or articles having one or more dangerous properties, not otherwise specified, e.g.:

UN No. 1325   FLAMMABLE SOLID, ORGANIC, N.O.S.
UN No. 1993   FLAMMABLE LIQUID, N.O.S.

The entries defined under B., C. and D. are defined as collective entries.

2.1.1.3   For packing purposes, substances other than those of Classes 1, 2, 5.2, 6.2 and 7, and other than self-reactive substances of Class 4.1 are assigned to packing groups in accordance with the degree of danger they present:

Packing group I: Substances presenting high danger;
Packing group II: Substances presenting medium danger;
Packing group III: Substances presenting low danger.

The packing group(s) to which a substance is assigned is (are) indicated in Table A of Chapter 3.2.

2.1.1.4   For the purpose of carriage in tank vessels, some substances may be further subdivided.

2.1.2   **Principles of classification**

2.1.2.1   The dangerous goods covered by the heading of a class are defined on the basis of their properties according to sub-section 2.2.x.1 of the relevant class. Assignment of dangerous goods to a class and a packing group is made according to the criteria mentioned in the same sub-section 2.2.x.1. Assignment of one or several subsidiary risk(s) to a dangerous substance or article is made according to the criteria of the class or classes corresponding to those risks, as mentioned in the appropriate sub-section(s) 2.2.x.1.

2.1.2.2   All dangerous goods entries are listed in Table A of Chapter 3.2 in the numerical order of their UN Number. This table contains relevant information on the goods listed, such as name, class, packing group(s), label(s) to be affixed, packing and carriage provisions[a].

NOTE: An alphabetical list of these entries can be found in Chapter 3.2, Table B.

2.1.2.3   Dangerous goods which are listed or defined in sub-section 2.2.x.2 of each class are not to be accepted for carriage.

2.1.2.4   Goods not mentioned by name, i.e. goods not listed as single entries in Table A of Chapter 3.2 and not listed or defined in one of the above-mentioned sub-sections 2.2.x.2 shall be assigned to the relevant class in accordance with the procedure of section 2.1.3. In addition, the subsidiary risk (if any) and the packing group (if any) shall be determined. Once the class, subsidiary risk (if any) and packing group (if any) have been established the relevant UN number shall be determined. The decision trees in sub-sections 2.2.x.3 (list of collective entries) at the end of each class indicate the relevant parameters for selecting the relevant collective entry (UN number). In all cases the most specific collective entry covering the properties of the substance or article shall be selected, according to the hierarchy indicated in 2.1.1.2 by the letters B, C and D respectively. If the substance or article cannot be classified under entries of type B or C according to 2.1.1.2, then, and only then shall it be classified under an entry of type D.

2.1.2.5   On the basis of the test procedures of Chapter 2.3 and the criteria set out in sub-sections 2.2.x.1 of classes when it is so specified, it may be determined that a substance, solution or mixture of a certain class, mentioned by name in Table A of Chapter 3.2, does not meet the criteria of that class. In such a case, the substance, solution or mixture is deemed not to belong to that class.

2.1.2.6   For the purposes of classification, substances with a melting point or initial melting point of 20 °C or lower at a pressure of 101.3 kPa shall be considered to be liquids. A viscous substance for which a specific melting point cannot be determined shall be subjected to the

---

[a]   *Note by the Secretariat: An alphabetic list of these entries has been prepared by the secretariat and is reproduced in Table B of Chapter 3.2. This table is not an official part of the ADN.*

ASTM D 4359-90 test or to the test for determining fluidity (penetrometer test) prescribed in 2.3.4.

2.1.3   **Classification of substances, including solutions and mixtures (such as preparations and wastes), not mentioned by name**

2.1.3.1   Substances including solutions and mixtures not mentioned by name shall be classified according to their degree of danger on the basis of the criteria mentioned in sub-section 2.2.x.1 of the various classes. The danger(s) presented by a substance shall be determined on the basis of its physical and chemical characteristics and physiological properties. Such characteristics and properties shall also be taken into account when such experience leads to a more stringent assignment.

2.1.3.2   A substance not mentioned by name in Table A of Chapter 3.2 presenting a single hazard shall be classified in the relevant class under a collective entry listed in sub-section 2.2.x.3 of that class.

2.1.3.3   A solution or mixture containing only one dangerous substance mentioned by name in Table A of Chapter 3.2, together with one or more non-dangerous substance(s), shall be regarded as the dangerous substance listed by name, unless:

     (a)   the solution or mixture is specifically mentioned by name in Table A of Chapter 3.2; or

     (b)   it is quite clear from the entry for the dangerous substance that it is applicable only to the pure or technically pure substance; or

     (c)   the class, physical state or packing group of the solution or mixture is different from that of the dangerous substance.

In the cases referred to under (b) or (c) above, the solution or mixture shall be classified as a substance not mentioned by name in the relevant class under a collective entry listed in sub-section 2.2.x.3 of that class taking account of the subsidiary risks presented by that solution or mixture, if any, unless the solution or mixture do not meet the criteria of any class, in which case they are not subject to ADN.

2.1.3.4   Solutions and mixtures containing a substance belonging to one of the entries mentioned in 2.1.3.4.1 or 2.1.3.4.2 shall be classified in accordance with the provisions of these paragraphs.

2.1.3.4.1   Solutions and mixtures containing one of the following substances mentioned by name shall always be classified under the same entry as the substance they contain, provided they do not have the hazard characteristics as indicated in 2.1.3.5.3:

     –   Class 3

         UN No. 1921 PROPYLENEIMINE, STABILIZED;

         UN No. 2481 ETHYL ISOCYANATE;

         UN No. 3064 NITROGLYCERIN SOLUTION IN ALCOHOL with more than 1% but not more than 5% nitroglycerin;

– Class 6.1

UN No. 1051 HYDROGEN CYANIDE, STABILIZED, containing less than 3% water;

UN No. 1185 ETHYLENEIMINE, STABILIZED;

UN No. 1259 NICKEL CARBONYL;

UN No. 1613 HYDROCYANIC ACID, AQUEOUS SOLUTION (HYDROGEN CYANIDE, AQUEOUS SOLUTION);

UN No. 1614 HYDROGEN CYANIDE, STABILIZED, containing not more than 3% water and absorbed in a porous inert material;

UN No. 1994 IRON PENTACARBONYL;

UN No. 2480 METHYL ISOCYANATE;

UN No. 3294 HYDROGEN CYANIDE, SOLUTION IN ALCOHOL, with not more than 45% hydrogen cyanide;

– Class 8

UN No. 1052 HYDROGEN FLUORIDE, ANHYDROUS;

UN No. 1744 BROMINE or UN No. 1744 BROMINE SOLUTION;

UN No. 1790 HYDROFLUORIC ACID with more than 85% hydrogen fluoride;

UN No. 2576 PHOSPHORUS OXYBROMIDE, MOLTEN.

2.1.3.4.2   Solutions and mixtures containing a substance belonging to one of the following entries of Class 9:

UN No. 2315 POLYCHLORINATED BIPHENYLS, LIQUID;

UN No. 3151 POLYHALOGENATED BIPHENYLS, LIQUID;

UN No. 3151 POLYHALOGENATED TERPHENYLS, LIQUID;

UN No. 3152 POLYHALOGENATED BIPHENYLS, SOLID;

UN No. 3152 POLYHALOGENATED TERPHENYLS, SOLID; or

UN No. 3432 POLYCHLORINATED BIPHENYLS, SOLID

shall always be classified under the same entry of Class 9 provided that:

- they do not contain any additional dangerous component other than components of packing group III of classes 3, 4.1, 4.2, 4.3, 5.1, 6.1 or 8; and

- they do not have the hazard characteristics as indicated in 2.1.3.5.3.

2.1.3.5   Substances not mentioned by name in Table A of Chapter 3.2, having more than one hazard characteristic and solutions or mixtures containing several dangerous substances shall be classified under a collective entry (see 2.1.2.4) and packing group of the appropriate class in accordance with their hazard characteristics. Such classification according to the hazard characteristics shall be carried out as follows:

2.1.3.5.1 The physical and chemical characteristics and physiological properties shall be determined by measurement or calculation and the substance, solution or mixture shall be classified according to the criteria mentioned in sub-section 2.2.x.1 of the various classes.

2.1.3.5.2 If this determination is not possible without disproportionate cost or effort (as for some kinds of wastes), the substance, solution or mixture shall be classified in the class of the component presenting the major hazard.

2.1.3.5.3 If the hazard characteristics of the substance, solution or mixture fall within more than one class or group of substances listed below then the substance, solution or mixture shall be classified in the class or group of substances corresponding to the major hazard on the basis of the following order of precedence:

(a) Material of Class 7 (apart from radioactive material in excepted packages where the other hazardous properties take precedence);

(b) Substances of Class 1;

(c) Substances of Class 2;

(d) Liquid desensitized explosives of Class 3;

(e) Self-reactive substances and solid desensitized explosives of Class 4.1;

(f) Pyrophoric substances of Class 4.2;

(g) Substances of Class 5.2;

(h) Substances of Class 6.1 or Class 3 which, on the basis of their inhalation toxicity, are to be classified under Packing group I (Substances meeting the classification criteria of Class 8 and having an inhalation toxicity of dust and mist (LC_{50}) in the range of Packing group I and a toxicity through oral ingestion or dermal contact only in the range of Packing group III or less, shall be allocated to Class 8);

(i) Infectious substances of Class 6.2.

2.1.3.5.4 If the hazard characteristics of the substance fall within more than one class or group of substances not listed in 2.1.3.5.3 above, the substance shall be classified in accordance with the same procedure but the relevant class shall be selected according to the precedence of hazards table in 2.1.3.10.

2.1.3.6 The most specific applicable collective entry (see 2.1.2.4) shall always be used, i.e. a general n.o.s. entry shall only be used if a generic entry or a specific n.o.s. entry cannot be used.

2.1.3.7 Solutions and mixtures of oxidizing substances or substances with an oxidizing subsidiary risk may have explosive properties. In such a case they are not to be accepted for carriage unless they meet the requirements for Class 1.

2.1.3.8 For the purposes of ADN, substances, solutions and mixtures (such as preparations and wastes) which cannot be assigned to Classes 1 to 8 or Class 9 entries other than UN Nos 3077 and 3082, but which may be assigned to UN Nos. 3077 or 3082 on the basis of the test methods and criteria of section 2.3.5 shall be considered to be pollutant to the aquatic environment.

2.1.3.9   Wastes which do not meet the criteria for classification in classes 1 to 9 but are covered by the *Basel Convention on the Control of Transboundary Movements of Hazardous Wastes and their Disposal* may be carried under UN Nos. 3077 or 3082.

2.1.3.10 Table of precedence of hazards

Class and packing group	4.1,II	4.1,III	4.2,II	4.2,III	4.3,I	4.3,II	4.3,III	5.1,I	5.1,II	5.1,III	6.1,I DERMAL	6.1,I ORAL	6.1,II	6.1,III	8,I	8,II	8,III	9
3,I	SOL 4.1 / LIQ 3.1	SOL 4.1 / LIQ 3.1	SOL 4.2 / LIQ 3.1	SOL 4.2 / LIQ 3.1	4.3,I	4.3,I	4.3,I	SOL 5.1,I / LIQ 3.1	SOL 5.1,I / LIQ 3.1	SOL 5.1,I / LIQ 3.1	3,I	3,I	3,I	3,I	3,I	3,I	3,I	3,I
3,II	SOL 4.1 / LIQ 3.II	SOL 4.1 / LIQ 3.II	SOL 4.2 / LIQ 3.II	SOL 4.2 / LIQ 3.II	4.3,I	4.3,I	4.3,I	SOL 5.1,I / LIQ 3.I	SOL 5.1,II / LIQ 3.II	SOL 5.1,II / LIQ 3.II	3,I	3,I	3,II	3,II	8,I	3,II	3,II	3,II
3,III	SOL 4.1 / LIQ 3.III	SOL 4.1 / LIQ 3.III	SOL 4.2 / LIQ 3.III	SOL 4.2 / LIQ 3.III	4.3,I	4.3,II	4.3,III	SOL 5.1,I / LIQ 3.I	SOL 5.1,II / LIQ 3.II	SOL 5.1,III / LIQ 3.III	6.1,I	6.1,I	6.1,II	3,III */	8,I	8,II	3,III	3,III
4.1,II			4.2,II	4.2,II	4.3,I	4.3,II	4.3,II	5.1,I	4.1,II	4.1,II	6.1,I	6.1,I	SOL 4.1,II / LIQ 6.1,II	SOL 4.1,II / LIQ 6.1,II	8,I	SOL 4.1,II / LIQ 8,II	SOL 4.1,II / LIQ 8,II	4.1,II
4.1,III			4.2,II	4.2,III	4.3,I	4.3,II	4.3,III	5.1,I	4.1,III	4.1,III	6.1,I	6.1,I	6.1,II	SOL 4.1,III / LIQ 6.1,III	8,I	8,II	SOL 4.1,III / LIQ 8,III	4.1,III
4.2,II					4.3,I	4.3,II	4.3,II	5.1,I	4.2,II	4.2,II	6.1,I	6.1,I	4.2,II	4.2,II	8,I	4.2,II	4.2,II	4.2,II
4.2,III					4.3,I	4.3,II	4.3,III	5.1,I	5.1,II	4.2,III	6.1,I	6.1,I	6.1,II	4.2,III	8,I	8,II	4.2,III	4.2,III
4.3,I								5.1,I	4.3,I	4.3,I	6.1,I	4.3,I	4.3,I	4.3,I	4.3,I	4.3,I	4.3,I	4.3,I
4.3,II								5.1,I	4.3,II	4.3,II	6.1,I	4.3,II	4.3,II	4.3,II	8,I	4.3,II	4.3,II	4.3,II
4.3,III								5.1,I	5.1,II	4.3,III	6.1,I	6.1,I	6.1,II	4.3,III	8,I	8,II	4.3,III	4.3,III
5.1,I											5.1,I	5.1,I	5.1,I	5.1,I	5.1,I	5.1,I	5.1,I	5.1,I
5.1,II											6.1,I	5.1,II	5.1,II	5.1,II	8,I	5.1,II	5.1,II	5.1,II
5.1,III											6.1,I	6.1,I	6.1,II	5.1,III	8,I	8,II	5.1,III	5.1,III
6.1,I DERMA															SOL 6.1,I / LIQ 8,I	6.1,I	6.1,I	6.1,I
6.1,I ORAL															SOL 6.1,I / LIQ 8,I	6.1,I	6.1,I	6.1,I
6.1,II INHAL															SOL 6.1,I / LIQ 8,I	6.1,I	6.1,I	6.1,I
6.1,II DERMA															SOL 6.1,I / LIQ 8,I	SOL 6.1,II / LIQ 8,II	6.1,II	6.1,II
6.1,II ORAL															8,I	SOL 6.1,II / LIQ 8,II	6.1,II	6.1,II
6.1,III															8,I	8,II	8,III	6.1,II
8,I																		8,I
8,II																		8,II
8,III																		8,III

SOL = Solid substances and mixtures
LIQ = Liquid substances, mixtures and solutions
DERMAL = Dermal toxicity
ORAL = Oral toxicity
INHAL = Inhalation toxicity
*/ Class 6.1 for pesticides

NOTE 1: Examples to explain the use of the table

Classification of a single substance

Description of the substance to be classified:

An amine not mentioned by name meeting the criteria for Class 3, packing group II as well as those for Class 8, packing group I.

Procedure:

The intersection of line 3 II with column 8 I gives 8 I.
This amine has therefore to be classified in Class 8 under:

UN No. 2734 AMINES LIQUID, CORROSIVE, FLAMMABLE, N.O.S. or UN No. 2734 POLYAMINES, LIQUID, CORROSIVE, FLAMMABLE, N.O.S.
packing group I

Classification of a mixture

Description of the mixture to be classified:

Mixture consisting of a flammable liquid classified in Class 3, packing group III, a toxic substance in Class 6.1, packing group II and a corrosive substance in Class 8, packing group I.

Procedure

The intersection of line 3 III with column 6.1 II gives 6.1 II.
The intersection of line 6.1 II with column 8 I gives 8 I LIQ.
This mixture not further defined has therefore to be classified in Class 8 under:

UN No. 2922 CORROSIVE LIQUID, TOXIC, N.O.S.
packing group I.

NOTE 2: Examples for the classification of mixtures and solutions under a class and a packing group:

A phenol solution of Class 6.1, (II), in benzene of Class 3, (II) is to be classified in Class 3, (II); this solution is to be classified under UN No. 1992 FLAMMABLE LIQUID, TOXIC, N.O.S., Class 3, (II), by virtue of the toxicity of the phenol.

A solid mixture of sodium arsenate of Class 6.1, (II) and sodium hydroxide of Class 8, (II) is to be classified under UN No. 3290 TOXIC SOLID, CORROSIVE, INORGANIC, N.O.S., in Class 6.1 (II).

A solution of crude or refined naphthalene of Class 4.1, (III) in petrol of Class 3, (II), is to be classified under UN No. 3295 HYDROCARBONS, LIQUID, N.O.S. in Class 3, (II).

A mixture of hydrocarbons of Class 3, (III), and of polychlorinated biphenyls (PCB) of Class 9, (II), is to be classified under UN No. 2315 POLYCHLORINATED BIPHENYLS, LIQUID or UN No. 3432 POLYCHLORINATED BIPHENYLS, SOLID in Class 9, (II).

A mixture of propyleneimine of Class 3, and polychlorinated biphenyls (PCB) of Class 9, (II), is to be classified under UN No. 1921 PROPYLENEIMINE, INHIBITED in Class 3.

2.1.4 Classification of samples

2.1.4.1 When the class of a substance is uncertain and it is being carried for further testing, a tentative class, proper shipping name and UN number shall be assigned on the basis of the consignor's knowledge of the substance and application of:

(a) the classification criteria of Chapter 2.2; and

(b) the requirements of this Chapter.

The most severe packing group possible for the proper shipping name chosen shall be used.

Where this provision is used the proper shipping name shall be supplemented with the word "SAMPLE" (e.g., "FLAMMABLE LIQUID, N.O.S., SAMPLE"). In certain instances, where a specific proper shipping name is provided for a sample of a substance considered to meet certain classification criteria (e.g., GAS SAMPLE, NON-PRESSURIZED, FLAMMABLE, UN No. 3167) that proper shipping name shall be used. When an N.O.S. entry is used to carry the sample, the proper shipping name need not be supplemented with the technical name as required by special provision 274 of Chapter 3.3.

2.1.4.2 Samples of the substance shall be carried in accordance with the requirements applicable to the tentative assigned proper shipping name provided:

(a) the substance is not considered to be a substance not accepted for carriage by sub-sections 2.2.x.2 of Chapter 2.2 or by Chapter 3.2;

(b) the substance is not considered to meet the criteria for Class 1 or considered to be an infectious substance or a radioactive material;

(c) the substance is in compliance with 2.2.41.1.15 or 2.2.52.1.9 if it is a self-reactive substance or an organic peroxide, respectively;

(d) the sample is carried in a combination packaging with a net mass per package not exceeding 2.5 kg; and

(e) the sample is not packed together with other goods.

CHAPTER 2.2

CLASS SPECIFIC PROVISIONS

2.2.1   **Class 1   Explosive substances and articles**

2.2.1.1   *Criteria*

2.2.1.1.1   The heading of Class 1 covers:

(a)   Explosive substances: solid or liquid substances (or mixtures of substances) capable by chemical reaction of producing gases at such a temperature and pressure and at such a speed as to cause damage to the surroundings.

Pyrotechnic substances: substances or mixtures of substances designed to produce an effect by heat, light, sound, gas or smoke or a combination of these as the result of non-detonating self-sustaining exothermic chemical reactions.

NOTE 1: Substances which are not themselves explosive but which may form an explosive mixture of gas, vapour or dust are not substances of Class 1.

NOTE 2: Also excluded from Class 1 are: water- or alcohol-wetted explosives of which the water or alcohol content exceeds the limits specified and those containing plasticizers - these explosives are assigned to Class 3 or Class 4.1 - and those explosives which, on the basis of their predominant hazard, are assigned to Class 5.2.

(b)   Explosive articles: articles containing one or more explosive or pyrotechnic substances.

NOTE: Devices containing explosive or pyrotechnic substances in such small quantity or of such a character that their inadvertent or accidental ignition or initiation during carriage would not cause any manifestation external to the device by projection, fire, smoke, heat or loud noise are not subject to the requirements of Class 1.

(c)   Substances and articles not mentioned above which are manufactured with a view to producing a practical effect by explosion or a pyrotechnic effect.

2.2.1.1.2   Any substance or article having or suspected of having explosive properties shall be considered for assignment to Class 1 in accordance with the tests, procedures and criteria prescribed in Part I, Manual of Tests and Criteria.

A substance or article assigned to Class 1 can only be accepted for carriage when it has been assigned to a name or n.o.s. entry listed in Table A of Chapter 3.2 and meets the criteria of the Manual of Tests and Criteria.

2.2.1.1.3   The substances and articles of Class 1 shall be assigned to a UN Number and a name or n.o.s. entry listed in Table A of Chapter 3.2. Interpretation of the names of substances and articles in Table A of Chapter 3.2 shall be based upon the glossary in 2.2.1.1.8.

Samples of new or existing explosive substances or articles carried for purposes including: testing, classification, research and development quality control, or as a commercial sample, other than initiating explosive, may be assigned to UN No. 0190 SAMPLES, EXPLOSIVE.

The assignment of explosive substances and articles not mentioned by name as such in Table A of Chapter 3.2 to an n.o.s entry of Class 1 or UN No. 0190 SAMPLES, EXPLOSIVE as well as the assignment of certain substances the carriage of which is subject to a specific authorization by the competent authority according to the special provisions referred to in Column (6) of Table A of Chapter 3.2 shall be made by the competent authority of the country of origin. This competent authority shall also approve in writing the conditions of carriage of these substances and articles. If the country of origin is not a Contracting Party to ADN, the classification and the conditions of carriage shall be recognized by the competent authority of the first country Contracting Party to ADN reached by the consignment.

2.2.1.1.4   Substances and articles of Class 1 shall have been assigned to a division in accordance with 2.2.1.1.5 and to a compatibility group in accordance with 2.2.1.1.6. The division shall be based on the results of the tests described in section 2.3.1 applying the definitions in 2.2.1.1.5. The compatibility group shall be determined in accordance with the definitions in 2.2.1.1.6. The classification code shall consist of the division number and the compatibility group letter.

2.2.1.1.5   *Definition of divisions*

Division 1.1   Substances and articles which have a mass explosion hazard (a mass explosion is an explosion which affects almost the entire load virtually instantaneously).

Division 1.2   Substances and articles which have a projection hazard but not a mass explosion hazard.

Division 1.3   Substances and articles which have a fire hazard and either a minor blast hazard or a minor projection hazard or both, but not a mass explosion hazard:

(a)   combustion of which gives rise to considerable radiant heat; or

(b)   which burn one after another, producing minor blast or projection effects or both.

Division 1.4   Substances and articles which present only a slight risk of explosion in the event of ignition or initiation during carriage. The effects are largely confined to the package and no projection of fragments of appreciable size or range is to be expected. An external fire shall not cause virtually instantaneous explosion of almost the entire contents of the package.

Division 1.5   Very insensitive substances having a mass explosion hazard which are so insensitive that there is very little probability of initiation or of transition from burning to detonation under normal conditions of carriage. As a minimum requirement they must not explode in the external fire test.

Division 1.6   Extremely insensitive articles which do not have a mass explosion hazard. The articles contain only extremely insensitive detonating substances and demonstrate a negligible probability of accidental initiation or propagation.

NOTE: The risk from articles of Division 1.6 is limited to the explosion of a single article.

2.2.1.1.6   *Definition of compatibility groups of substances and articles*

A   Primary explosive substance.

B   Article containing a primary explosive substance and not having two or more effective protective features. Some articles, such as detonators for blasting, detonator assemblies for blasting and primers, cap-type, are included, even though they do not contain primary explosives.

C   Propellant explosive substance or other deflagrating explosive substance or article containing such explosive substance.

D   Secondary detonating explosive substance or black powder or article containing a secondary detonating explosive substance, in each case without means of initiation and without a propelling charge, or article containing a primary explosive substance and having two or more effective protective features.

E   Article containing a secondary detonating explosive substance, without means of initiation, with a propelling charge (other than one containing a flammable liquid or gel or hypergolic liquids).

F   Article containing a secondary detonating explosive substance with its own means of initiation, with a propelling charge (other than one containing a flammable liquid or gel or hypergolic liquids) or without a propelling charge.

G   Pyrotechnic substance, or article containing a pyrotechnic substance, or article containing both an explosive substance and an illuminating, incendiary, tear- or smoke-producing substance (other than a water-activated article or one which contains white phosphorus, phosphides, a pyrophoric substance, a flammable liquid or gel or hypergolic liquids).

H   Article containing both an explosive substance and white phosphorus.

J   Article containing both an explosive substance and a flammable liquid or gel.

K   Article containing both an explosive substance and a toxic chemical agent.

L   Explosive substance or article containing an explosive substance and presenting a special risk (e.g. due to water activation or the presence of hypergolic liquids, phosphides or a pyrophoric substance) necessitating isolation of each type.

N   Articles containing only extremely insensitive detonating substances.

S   Substance or article so packed or designed that any hazardous effects arising from accidental functioning are confined within the package unless the package has been degraded by fire, in which case all blast or projection effects are limited to the extent that they do not significantly hinder or prevent fire-fighting or other emergency response efforts in the immediate vicinity of the package.

NOTE 1*: Each substance or article, packed in a specified packaging, may be assigned to one compatibility group only. Since the criterion of compatibility group S is empirical, assignment to this group is necessarily linked to the tests for assignment of a classification code.*

NOTE 2*: Articles of compatibility groups D and E may be fitted or packed together with their own means of initiation provided that such means have at least two effective protective* *features designed to prevent an explosion in the event of accidental functioning of the means of initiation. Such packages shall be assigned to compatibility groups D or E.*

NOTE 3: Articles of compatibility groups D and E may be packed together with their own means of initiation, which do not have two effective protective features (i.e. means of initiation assigned to compatibility group B), provided that they comply with mixed packing provision MP 21 of Section 4.1.10 of ADR.. Such packages shall be assigned to compatibility groups D or E.

NOTE 4: Articles may be fitted or packed together with their own means of ignition provided that the means of ignition cannot function during normal conditions of carriage.

NOTE 5: Articles of compatibility groups C, D and E may be packed together. Such packages shall be assigned to compatibility group E.

2.2.1.1.7	*Assignment of fireworks to divisions*

2.2.1.1.7.1   Fireworks shall normally be assigned to divisions 1.1, 1.2, 1.3, and 1.4 on the basis of test data derived from Test Series 6 of the Manual of Tests and Criteria. However, since the range of such articles is very extensive and the availability of test facilities may be limited, assignment to divisions may also be made in accordance with the procedure in 2.2.1.1.7.2.

2.2.1.1.7.2   Assignment of fireworks to UN No. 0333, 0334, 0335 or 0336 may be made on the basis of analogy, without the need for Test Series 6 testing, in accordance with the default fireworks classification table in 2.2.1.1.7.5. Such assignment shall be made with the agreement of the competent authority. Items not specified in the table shall be classified on the basis of test data derived from Test Series 6.

NOTE 1: The addition of other types of fireworks to column 1 of the table in 2.2.1.1.7.5 shall only be made on the basis of full test data submitted to the UN Sub-Committee of Experts on the Transport of Dangerous Goods for consideration.

NOTE 2: Test data derived by competent authorities which validates, or contradicts the assignment of fireworks specified in column 4 of the table in 2.2.1.1.7.5 to divisions in column 5 should be submitted to the UN Sub-Committee of Experts on the Transport of Dangerous Goods for information.

2.2.1.1.7.3   Where fireworks of more than one division are packed in the same package they shall be classified on the basis of the highest division unless test data derived from Test Series 6 indicate otherwise.

2.2.1.1.7.4   The classification shown in the table in 2.2.1.1.7.5 applies only for articles packed in fibreboard boxes (4G).

2.2.1.1.7.5   *Default fireworks classification table* [1]

NOTE 1: References to percentages in the table, unless otherwise stated, are to the mass of all pyrotechnic composition (e.g. rocket motors, lifting charge, bursting charge and effect charge).

NOTE 2: "Flash composition" in this table refers to pyrotechnic compositions containing an oxidizing substance, or black powder, and a metal powder fuel that are used to produce an aural report effect or used as a bursting charge in fireworks devices.

---

[1]   *This table contains a list of firework classifications which may be used in the absence of Test Series 6 data (see 2.2.1.1.7.2).*

NOTE 3:   *Dimensions in mm refer to:*

- *for spherical and peanut shells the diameter of the sphere of the shell;*
- *for cylinder shells the length of the shell;*
- *for a shell in mortar, Roman candle, shot tube firework or mine the inside diameter of the tube comprising or containing the firework;*
- *for a bag mine or cylinder mine, the inside diameter of the mortar intended to contain the mine.*

Type	Includes: / Synonym:	Definition	Specification	Classification
Shell, spherical or cylindrical	Spherical display shell: aerial shell, colour shell. dye shell, multi-break shell. multi-effect shell, nautical shell. parachute shell, smoke shell. star shell: report shell: maroon, salute. sound shell. thunderclap, aerial shell kit	Device with or without propellant charge, with delay fuse and bursting charge, pyrotechnic unit(s) or loose pyrotechnic composition and designed to be projected from a mortar	All report shells	1.1G
			Colour shell: ≥ 180 mm	1.1G
			Colour shell: < 180 mm with > 25% flash composition, as loose powder and/or report effects	1.1G
			Colour shell: < 180 mm with ≤ 25% flash composition, as loose powder and/or report effects	1.3G
			Colour shell: ≤ 50 mm, or ≤ 60 g pyrotechnic composition, with ≤ 2% flash composition as loose powder and/or report effects	1.4G
	Peanut shell	Device with two or more spherical aerial shells in a common wrapper propelled by the same propellant charge with separate external delay fuses	The most hazardous spherical aerial shell determines the classification	
	Preloaded mortar, shell in mortar	Assembly comprising a spherical or cylindrical shell inside a mortar from which the shell is designed to be projected	All report shells	1.1G
			Colour shell: ≥ 180 mm	1.1G
			Colour shell: > 50 mm and < 180 mm	1.2G
			Colour shell: ≤ 50 mm, or ≤ 60 g pyrotechnic composition, with ≤ 25% flash composition as loose powder and/or report effects	1.3G

Type	Includes: / Synonym:	Definition	Specification	Classification
Shell, spherical or cylindrical (cont'd)	Shell of shells (spherical) (*Reference to percentages for shell of shells are to the gross mass of the fireworks article*)	Device without propellant charge, with delay fuse and bursting charge, containing report shells and inert materials and designed to be projected from a mortar	> 120 mm	1.1G
		Device without propellant charge, with delay fuse and bursting charge, containing report shells ≤ 25g flash composition per report unit, with ≤ 33% flash composition and ≥ 60% inert materials and designed to be projected from a mortar	≤ 120 mm	1.3G
		Device without propellant charge, with delay fuse and bursting charge, containing colour shells and/or pyrotechnic units and designed to be projected from a mortar	> 300 mm	1.1G
		Device without propellant charge, with delay fuse and bursting charge, containing colour shells ≤ 70mm and/or pyrotechnic units, with ≤ 25% flash composition and ≤ 60% pyrotechnic composition and designed to be projected from a mortar	> 200 mm and ≤ 300 mm	1.3G
		Device with propellant charge, with delay fuse and bursting charge, containing colour shells ≤ 70 mm and/or pyrotechnic units, with ≤ 25% flash composition and ≤ 60% pyrotechnic composition and designed to be projected from a mortar	≤ 200 mm	1.3G
Battery/ combination	Barrage, bombardos, cakes, finale box, flowerbed, hybrid, multiple tubes, shell cakes, banger batteries, flash banger batteries	Assembly including several elements either containing the same type or several types each corresponding to one of the types of fireworks listed in this table, with one or two points of ignition	The most hazardous firework type determines the classification	

Type	Includes: / Synonym:	Definition	Specification	Classification
Roman candle	Exhibition candle, candle, bombettes	Tube containing a series of pyrotechnic units consisting of alternate pyrotechnic composition, propellant charge, and transmitting fuse	≥ 50 mm inner diameter, containing flash composition, or < 50 mm with > 25% flash composition	1.1G
			≥ 50 mm inner diameter, containing no flash composition	1.2G
			< 50 mm inner diameter and ≤ 25% flash composition	1.3G
			≤ 30 mm inner diameter, each pyrotechnic unit ≤ 25 g and ≤ 5% flash composition	1.4G
Shot tube	Single shot Roman candle, small preloaded mortar	Tube containing a pyrotechnic unit consisting of pyrotechnic composition, propellant charge with or without transmitting fuse	≤ 30 mm inner diameter and pyrotechnic unit > 25 g, or > 5% and ≤ 25% flash composition	1.3G
			≤ 30 mm inner diameter, pyrotechnic unit ≤ 25 g and ≤ 5% flash composition	1.4G
Rocket	Avalanche rocket, signal rocket, whistling rocket, bottle rocket, sky rocket, missile type rocket, table rocket	Tube containing pyrotechnic composition and/or pyrotechnic units, equipped with stick(s) or other means for stabilization of flight, and designed to be propelled into the air	Flash composition effects only	1.1G
			Flash composition > 25% of the pyrotechnic composition	1.1G
			> 20 g pyrotechnic composition and flash composition ≤ 25%	1.3G
			≤ 20 g pyrotechnic composition, black powder bursting charge and ≤ 0.13 g flash composition per report and ≤ 1 g in total	1.4G

Type	Includes: / Synonym:	Definition	Specification	Classification
Mine	Pot-a-feu, ground mine, bag mine, cylinder mine	Tube containing propellant charge and pyrotechnic units and designed to be placed on the ground or to be fixed in the ground. The principal effect is ejection of all the pyrotechnic units in a single burst producing a widely dispersed visual and/or aural effect in the air or:	> 25% flash composition, as loose powder and/ or report effects	1.1G
			≥ 180 mm and ≤ 25% flash composition, as loose powder and/ or report effects	1.1G
		Cloth or paper bag or cloth or paper cylinder containing propellant charge and pyrotechnic units, designed to be placed in a mortar and to function as a mine	< 180 mm and ≤ 25% flash composition, as loose powder and/ or report effects	1.3G
			≤ 150 g pyrotechnic composition, containing ≤ 5% flash composition as loose powder and/ or report effects. Each pyrotechnic unit ≤ 25 g, each report effect < 2g; each whistle, if any, ≤ 3 g	1.4G
Fountain	Volcanos, gerbs, showers, lances, Bengal fire, flitter sparkle, cylindrical fountains, cone fountains, illuminating torch	Non-metallic case containing pressed or consolidated pyrotechnic composition producing sparks and flame	≥ 1 kg pyrotechnic composition	1.3G
			< 1 kg pyrotechnic composition	1.4G
Sparkler	Handheld sparklers, non-handheld sparklers, wire sparklers	Rigid wire partially coated (along one end) with slow burning pyrotechnic composition with or without an ignition tip	Perchlorate based sparklers: > 5 g per item or > 10 items per pack	1.3G
			Perchlorate based sparklers: ≤ 5 g per item and ≤ 10 items per pack; Nitrate based sparklers: ≤ 30 g per item	1.4G
Bengal stick	Dipped stick	Non-metallic stick partially coated (along one end) with slow-burning pyrotechnic composition and designed to be held in the hand	Perchlorate based items: > 5 g per item or > 10 items per pack	1.3 G
			Perchlorate based items: ≤ 5 g per item and ≤ 10 items per pack; nitrate based items: ≤ 30 g per item	1.4G

Type	Includes: / Synonym:	Definition	Specification	Classification
Low hazard fireworks and novelties	Table bombs, throwdowns, crackling granules, smokes, fog, snakes, glow worm, serpents, snaps, party poppers	Device designed to produce very limited visible and/ or audible effect which contains small amounts of pyrotechnic and/or explosive composition.	Throwdowns and snaps may contain up to 1.6 mg of silver fulminate; snaps and party poppers may contain up to 16 mg of potassium chlorate/red phosphorous mixture; other articles may contain up to 5 g of pyrotechnic composition, but no flash composition	1.4G
Spinner	Aerial spinner, helicopter, chaser, ground spinner	Non-metallic tube or tubes containing gas- or spark-producing pyrotechnic composition, with or without noise producing composition, with or without aerofoils attached	Pyrotechnic composition per item > 20 g, containing ≤ 3% flash composition as report effects, or whistle composition ≤ 5 g	1.3G
			Pyrotechnic composition per item ≤ 20 g, containing ≤ 3% flash composition as report effects, or whistle composition ≤ 5 g	1.4G
Wheels	Catherine wheels, Saxon	Assembly including drivers containing pyrotechnic composition and provided with a means of attaching it to a support so that it can rotate	≥ 1 kg total pyrotechnic composition, no report effect, each whistle (if any) ≤ 25 g and ≤ 50 g whistle composition per wheel	1.3G
			< 1 kg total pyrotechnic composition, no report effect, each whistle (if any) ≤ 5 g and ≤ 10 g whistle composition per wheel	1.4G
Aerial wheel	Flying Saxon, UFO's, rising crown	Tubes containing propellant charges and sparks-flame- and/or noise producing pyrotechnic compositions, the tubes being fixed to a supporting ring	> 200 g total pyrotechnic composition or > 60 g pyrotechnic composition per driver, ≤ 3% flash composition as report effects, each whistle (if any) ≤ 25 g and ≤ 50 g whistle composition per wheel	1.3G

Type	Includes: / Synonym:	Definition	Specification	Classification
			≤ 200 g total pyrotechnic composition and ≤ 60 g pyrotechnic composition per driver, ≤ 3% flash composition as report effects, each whistle (if any) ≤ 5 g and ≤ 10 g whistle composition per wheel	1.4G
Selection pack	Display selection box, display selection pack, garden selection box, indoor selection box; assortment	A pack of more than one type each corresponding to one of the types of fireworks listed in this table	The most hazardous firework type determines the classification	
Firecracker	Celebration cracker, celebration roll, string cracker	Assembly of tubes (paper or cardboard) linked by a pyrotechnic fuse, each tube intended to produce an aural effect	Each tube ≤ 140 mg of flash composition or ≤ 1 g black powder	1.4G
Banger	Salute, flash banger, lady cracker	Non-metallic tube containing report composition intended to produce an aural effect	> 2 g flash composition per item	1.1G
			≤ 2 g flash composition per item and ≤ 10 g per inner packaging	1.3G
			≤ 1 g flash composition per item and ≤ 10 g per inner packaging or ≤ 10 g black powder per item	1.4G

2.2.1.1.8   *Glossary of names*

NOTE 1: The descriptions in the glossary are not intended to replace the test procedures, nor to determine the hazard classification of a substance or article of Class 1. Assignment to the correct division and a decision on whether Compatibility Group S is appropriate shall be based on testing of the product in accordance with the Manual of Tests and Criteria, Part 1 or by analogy with similar products which have already been tested and assigned in accordance with the procedures of the Manual of Tests and Criteria.

NOTE 2: The figures given after the names refer to the relevant UN numbers (Column (2) of Table A of Chapter 3.2). For the classification code, see 2.2.1.1.4.

AIR BAG INFLATORS or AIR BAG MODULES or SEAT-BELT PRETENSIONERS: UN No. 0503

Articles which contain pyrotechnic substances and are used as life-saving vehicle airbags or seat-belts.

AMMUNITION, ILLUMINATING, with or without burster, expelling charge or propelling charge: UN Nos. 0171, 0254, 0297

Ammunition designed to produce a single source of intense light for lighting up an area. The term includes illuminating cartridges, grenades and projectiles; and illuminating and target identification bombs.

NOTE: The following articles: CARTRIDGES, SIGNAL; SIGNAL DEVICES HAND; SIGNALS, DISTRESS; FLARES, AERIAL; FLARES, SURFACE are not included in this definition. They are listed separately.

AMMUNITION, INCENDIARY, liquid or gel, with burster, expelling charge or propelling charge: UN No. 0247

Ammunition containing liquid or gelatinous incendiary substance. Except when the incendiary substance is an explosive per se, it also contains one or more of the following: a propelling charge with primer and igniter charge; a fuze with burster or expelling charge.

AMMUNITION, INCENDIARY, WHITE PHOSPHORUS with burster, expelling charge or propelling charge: UN Nos. 0243, 0244

Ammunition containing white phosphorus as incendiary substance. It also contains one or more of the following: a propelling charge with primer and igniter charge; a fuze with burster or expelling charge.

AMMUNITION, INCENDIARY with or without burster, expelling charge or propelling charge: UN Nos. 0009, 0010, 0300

Ammunition containing incendiary composition. Except when the composition is an explosive per se, it also contains one or more of the following: a propelling charge with primer and igniter charge; a fuze with burster or expelling charge.

AMMUNITION, PRACTICE: UN Nos. 0362, 0488

Ammunition without a main bursting charge, containing a burster or expelling charge. Normally it also contains a fuze and a propelling charge.

AMMUNITION, PROOF: UN No. 0363

Ammunition containing pyrotechnic substances, used to test the performance or strength of new ammunition, weapon components or assemblies.

AMMUNITION, SMOKE, WHITE PHOSPHORUS, with burster, expelling charge or propelling charge: UN Nos. 0245, 0246

Ammunition containing white phosphorus as a smoke-producing substance. It also contains one or more of the following: a propelling charge with primer and igniter charge; a fuze with burster or expelling charge. The term includes grenades, smoke.

AMMUNITION, SMOKE with or without burster, expelling charge or propelling charge: UN Nos. 0015, 0016, 0303

Ammunition containing a smoke-producing substance such as chlorosulphonic acid mixture or titanium tetrachloride; or a smoke-producing pyrotechnic composition based on hexachloroethane or red phosphorus. Except when the substance is an explosive per se, the ammunition also contains one or more of the following: a propelling charge with primer and igniter charge; a fuze with burster or expelling charge. The term includes grenades, smoke.

NOTE: SIGNALS, SMOKE are not included in this definition. They are listed separately.

AMMUNITION, TEAR-PRODUCING, with burster, expelling charge or propelling charge: UN Nos. 0018, 0019, 0301

Ammunition containing a tear-producing substance. It also contains one or more of the following: a pyrotechnic substance; a propelling charge with primer and igniter charge; a fuze with burster or expelling charge.

ARTICLES, EXPLOSIVE, EXTREMELY INSENSITIVE (ARTICLES EEI): UN No. 0486

Articles containing only extremely insensitive detonating substances (EIDS) which demonstrate a negligible probability of accidental initiation or propagation under normal conditions of transport, and which have passed Test Series 7.

ARTICLES, PYROPHORIC: UN No. 0380

Articles which contain a pyrophoric substance (capable of spontaneous ignition when exposed to air) and an explosive substance or component. The term excludes articles containing white phosphorus.

ARTICLES, PYROTECHNIC, for technical purposes: UN Nos. 0428, 0429, 0430, 0431, 0432

Articles which contain pyrotechnic substances and are used for technical purposes such as heat generation, gas generation, theatrical effects, etc.

NOTE: The following articles: all ammunition; CARTRIDGES, SIGNAL; CUTTERS, CABLE, EXPLOSIVE; FIREWORKS; FLARES, AERIAL; FLARES, SURFACE; RELEASE DEVICES, EXPLOSIVE; RIVETS, EXPLOSIVE; SIGNAL DEVICES, HAND; SIGNALS,

DISTRESS; SIGNALS, RAILWAY TRACK, EXPLOSIVES; SIGNALS, SMOKE are not included in this definition. They are listed separately.

BLACK POWDER (GUNPOWDER), COMPRESSED or BLACK POWDER (GUNPOWDER), IN PELLETS: UN No. 0028

Substance consisting of a pelletized form of black powder.

BLACK POWDER (GUNPOWDER), granular or as meal: UN No. 0027

Substance consisting of an intimate mixture of charcoal or other carbon and either potassium nitrate or sodium nitrate, with or without sulphur.

BOMBS, WITH FLAMMABLE LIQUID, with bursting charge: UN Nos. 0399, 0400

Articles which are dropped from aircraft, consisting of a tank filled with inflammable liquid and bursting charge.

BOMBS, PHOTO-FLASH: UN No. 0038

Explosive articles which are dropped from aircraft to provide brief, intense illumination for photography. They contain a charge of detonating explosive without means of initiation or with means of initiation containing two or more effective protective features.

BOMBS, PHOTO-FLASH: UN No. 0037

Explosive articles which are dropped from aircraft to provide brief, intense illumination for photography. They contain a charge of detonating explosive with means of initiation not containing two or more effective protective features.

BOMBS, PHOTO-FLASH: UN Nos. 0039, 0299

Explosive articles which are dropped from aircraft to provide brief, intense illumination for photography. They contain a photo-flash composition.

BOMBS with bursting charge: UN Nos. 0034; 0035

Explosive articles which are dropped from aircraft, without means of initiation or with means of initiation containing two or more effective protective features.

BOMBS with bursting charge: UN Nos. 0033, 0291

Explosive articles which are dropped from aircraft, with means of initiation not containing two or more effective protective features.

BOOSTERS WITH DETONATOR: UN Nos. 0225, 0268

Articles consisting of a charge of detonating explosive with means of initiation. They are used to increase the initiating power of detonators or detonating cord.

BOOSTERS without detonator: UN Nos. 0042, 0283

Articles consisting of a charge of detonating explosive without means of initiation. They are used to increase the initiating power of detonators or detonating cord.

BURSTERS, explosive: UN No. 0043

Articles consisting of a small charge of explosive used to open projectiles or other ammunition in order to disperse their contents.

CARTRIDGES, FLASH: UN Nos. 0049, 0050

Articles consisting of a casing, a primer and flash powder, all assembled in one piece ready for firing.

CARTRIDGES FOR WEAPONS, BLANK: UN Nos. 0326, 0413, 0327, 0338, 0014

Ammunition consisting of a closed cartridge case with a centre or rim fire primer and a charge of smokeless or black powder but no projectile. It produces a loud noise and is used for training, saluting, propelling charge, starter pistols, etc. The term includes ammunition, blank.

CARTRIDGES FOR WEAPONS, INERT PROJECTILE: UN Nos. 0328, 0417, 0339, 0012

Ammunition consisting of a projectile without bursting charge but with a propelling charge with or without a primer. The articles may include a tracer, provided that the predominant hazard is that of the propelling charge.

CARTRIDGES FOR WEAPONS with bursting charge: UN Nos. 0006, 0321, 0412

Ammunition consisting of a projectile with a bursting charge without means of initiation or with means of initiation containing two or more effective protective features; and a propelling charge with or without a primer. The term includes fixed (assembled) ammunition, semi-fixed (partially assembled) ammunition and separate loading ammunition when the components are packed together.

CARTRIDGES FOR WEAPONS with bursting charge: UN Nos. 0005, 0007, 0348

Ammunition consisting of a projectile with a bursting charge with means of initiation not containing two or more effective protective features; and a propelling charge with or without a primer. The term includes fixed (assembled) ammunition, semi-fixed (partially assembled) ammunition and separate loading ammunition when the components are packed together.

CARTRIDGES, OIL WELL: UN Nos. 0277, 0278

Articles consisting of a thin casing of fibreboard, metal or other material containing only propellant powder which projects a hardened projectile to perforate an oil well casing.

NOTE: CHARGES, SHAPED are not included in this definition. They are listed separately.

CARTRIDGES, POWER DEVICE: UN Nos. 0275, 0276, 0323, 0381

Articles designed to accomplish mechanical actions. They consist of a casing with a charge of deflagrating explosive and a means of ignition. The gaseous products of the deflagration produce inflation, linear or rotary motion or activate diaphragms, valves or switches or project fastening devices or extinguishing agents.

CARTRIDGES, SIGNAL: UN Nos. 0054, 0312, 0405

Articles designed to fire coloured flares or other signals from signal pistols, etc.

CARTRIDGES, SMALL ARMS: UN Nos. 0417, 0339, 0012

Ammunition consisting of a cartridge case fitted with a centre or rim fire primer and containing both a propelling charge and solid projectile. They are designed to be fired in weapons of calibre not larger than 19.1 mm. Shot-gun cartridges of any calibre are included in this description.

NOTE: CARTRIDGES, SMALL ARMS, BLANK, are not included in this definition. They are listed separately. Some military small arms cartridges are not included in this definition. They are listed under CARTRIDGES FOR WEAPONS, INERT PROJECTILE.

CARTRIDGES, SMALL ARMS, BLANK: UN Nos. 0014, 0327, 0338

Ammunition consisting of a closed cartridge case with a centre or rim fire primer and a charge of smokeless or black powder. The cartridge cases contain no projectiles. The cartridges are designed to be fired from weapons with a calibre of at most 19.1 mm and serve to produce a loud noise and are used for training, saluting, propelling charge, starter pistols, etc.

CASES, CARTRIDGE, EMPTY, WITH PRIMER: UN Nos. 0379; 0055

Articles consisting of a cartridge case made from metal, plastics or other non-inflammable material, in which the only explosive component is the primer.

CASES, COMBUSTIBLE, EMPTY, WITHOUT PRIMER: UN Nos. 0447, 0446

Articles consisting of a cartridge case made partly or entirely from nitrocellulose.

CHARGES, BURSTING, PLASTICS BONDED: UN Nos. 0457, 0458, 0459, 0460

Articles consisting of a charge of detonating explosive, plastics bonded, manufactured in a specific form without a casing and without means of initiation. They are designed as components of ammunition such as warheads.

CHARGES, DEMOLITION: UN No. 0048

Articles containing a charge of a detonating explosive in a casing of fibreboard, plastics, metal or other material. The articles are without means of initiation or with means of initiation containing two or more effective protective features.

NOTE: The following articles: BOMBS; MINES; PROJECTILES are not included in this definition. They are listed separately.

CHARGES, DEPTH: UN No. 0056

Articles consisting of a charge of detonating explosive contained in a drum or projectile without means of initiation or with means of initiation containing two or more effective protective features. They are designed to detonate under water.

CHARGES, EXPLOSIVE, COMMERCIAL without detonator: UN Nos. 0442, 0443, 0444, 0445

Articles consisting of a charge of detonating explosive without means of initiation, used for explosive welding, jointing, forming and other metallurgical processes.

CHARGES, PROPELLING, FOR CANNON: UN Nos. 0242, 0279, 0414

Charges of propellant in any physical form for separate-loading ammunition for cannon.

CHARGES, PROPELLING: UN Nos. 0271, 0272, 0415, 0491

Articles consisting of a charge of a propellant charge in any physical form, with or without a casing, as a component of rocket motors or for reducing the drag of projectiles.

CHARGES, SHAPED, without detonator: UN Nos. 0059, 0439, 0440, UN 0441

Articles consisting of a casing containing a charge of detonating explosive with a cavity lined with rigid material, without means of initiation. They are designed to produce a powerful, penetrating jet effect.

CHARGES, SHAPED, FLEXIBLE, LINEAR: UN Nos. 0237, 0288

Articles consisting of a V-shaped core of a detonating explosive clad by a flexible sheath.

CHARGES, SUPPLEMENTARY, EXPLOSIVE: UN No. 0060

Articles consisting of a small removable booster placed in the cavity of a projectile between the fuze and the bursting charge.

COMPONENTS, EXPLOSIVE TRAIN, N.O.S.: UN Nos. 0382, 0383, 0384, 0461

Articles containing an explosive designed to transmit detonation or deflagration within an explosive train.

CONTRIVANCES, WATER-ACTIVATED with burster, expelling charge or propelling charge: UN Nos. 0248, 0249

Articles whose functioning depends upon physico-chemical reaction of their contents with water.

CORD, DETONATING, flexible: UN Nos. 0065, 0289

Article consisting of a core of detonating explosive enclosed in spun fabric and a plastics or other covering. The covering is not necessary if the spun fabric is sift-proof.

CORD (FUSE) DETONATING, metal clad: UN Nos. 0102, 0290

Article consisting of a core of detonating explosive clad by a soft metal tube with or without protective covering.

CORD (FUSE) DETONATING, MILD EFFECT, metal clad: UN No. 0104

Article consisting of a core of detonating explosive clad by a soft metal tube with or without a protective covering. The quantity of explosive substance is so small that only a mild effect is manifested outside the cord.

CORD, IGNITER: UN No. 0066

Article consisting of textile yarns covered with black powder or another fast burning pyrotechnic composition and of a flexible protective covering; or it consists of a core of black powder surrounded by a flexible woven fabric. It burns progressively along its length with an external flame and is used to transmit ignition from a device to a charge or primer.

CUTTERS, CABLE, EXPLOSIVE: UN No. 0070

Articles consisting of a knife-edged device which is driven by a small charge of deflagrating explosive into an anvil.

DETONATOR ASSEMBLIES, NON-ELECTRIC for blasting: UN Nos. 0360, 0361, 0500

Non-electric detonators assembled with and activated by such means as safety fuse, shock tube, flash tube or detonating cord. They may be of instantaneous design or incorporate delay elements. Detonating relays incorporating detonating cord are included.

DETONATORS, ELECTRIC for blasting: UN Nos. 0030, 0255, 0456

Articles specially designed for the initiation of blasting explosives. These detonators may be constructed to detonate instantaneously or may contain a delay element. Electric detonators are activated by an electric current.

DETONATORS FOR AMMUNITION: UN Nos. 0073, 0364, 0365, 0366

Articles consisting of a small metal or plastics tube containing explosives such as lead azide, PETN or combinations of explosives. They are designed to start a detonation train.

DETONATORS, NON-ELECTRIC for blasting: UN Nos. 0029, 0267, 0455

Articles specially designed for the initiation of blasting explosives. These detonators may be constructed to detonate instantaneously or may contain a delay element. Non-electric detonators are activated by such means as shock tube, flash tube, safety fuse, other igniferous device or flexible detonating cord. Detonating relays without detonating cord are included.

EXPLOSIVE, BLASTING, TYPE A: UN No. 0081

Substances consisting of liquid organic nitrates such as nitroglycerine or a mixture of such ingredients with one or more of the following: nitrocellulose; ammonium nitrate or other inorganic nitrates; aromatic nitro-derivatives, or combustible materials, such as wood-meal and aluminium powder. They may contain inert components such as kieselguhr, and additives such as colouring agents and stabilizers. Such explosives shall be in powdery, gelatinous or elastic form. The term includes dynamite; gelatine, blasting and gelatine dynamites.

EXPLOSIVE, BLASTING, TYPE B: UN Nos. 0082, 0331

Substances consisting of

(a)   a mixture of ammonium nitrate or other inorganic nitrates with an explosive such as trinitrotoluene, with or without other substances such as wood-meal and aluminium powder; or

(b)   a mixture of ammonium nitrate or other inorganic nitrates with other combustible substances which are not explosive ingredients. In both cases they may contain inert components such as kieselguhr, and additives such as colouring agents and stabilizers. Such explosives must not contain nitroglycerine, similar liquid organic nitrates or chlorates.

EXPLOSIVE, BLASTING, TYPE C: UN No. 0083

Substances consisting of a mixture of either potassium or sodium chlorate or potassium, sodium or ammonium perchlorate with organic nitro-derivatives or combustible materials such as wood-meal or aluminium powder or a hydrocarbon. They may contain inert components such as kieselguhr and additives such as colouring agents and stabilizers. Such explosives must not contain nitroglycerine or similar liquid organic nitrates.

EXPLOSIVE, BLASTING, TYPE D: UN No. 0084

Substances consisting of a mixture of organic nitrated compounds and combustible materials such as hydrocarbons and aluminium powder. They may contain inert components such as kieselguhr and additives such as colouring agents and stabilizers. Such explosives must not contain nitroglycerine, similar liquid organic nitrates, chlorates and ammonium nitrate. The term generally includes plastic explosives.

EXPLOSIVES, BLASTING, TYPE E: UN Nos. 0241, 0332

Substances consisting of water as an essential ingredient and high proportions of ammonium nitrate or other oxidizers, some or all of which are in solution. The other constituents may include nitro-derivatives such as trinitrotoluene, hydrocarbons or aluminium powder. They may contain inert components such as kieselguhr and additives such as colouring agents and stabilizers. The term includes explosives, emulsion, explosives, slurry and explosives, watergel.

FIREWORKS: UN Nos. 0333, 0334, 0335, 0336, 0337

Pyrotechnic articles designed for entertainment.

FLARES, AERIAL: UN Nos. 0093, 0403, 0404, 0420, 0421;

Articles containing pyrotechnic substances which are designed to be dropped from an aircraft to illuminate, identify, signal or warn.

FLARES, SURFACE: UN Nos. 0092, 0418, 0419

Articles containing pyrotechnic substances which are designed for use on the surface to illuminate, identify, signal or warn.

FLASH POWDER: UN Nos. 0094, 0305

Pyrotechnic substance which, when ignited, produces an intense light.

FRACTURING DEVICES, EXPLOSIVE without detonator, for oil wells: UN No. 0099

Articles consisting of a charge of detonating explosive contained in a casing without means of initiation. They are used to fracture the rock around a drill shaft to assist the flow of crude oil from the rock.

FUSE, IGNITER, tubular, metal clad: UN No. 0103

Article consisting of a metal tube with a core of deflagrating explosive.

FUSE, NON-DETONATING: UN No. 0101

Article consisting of cotton yarns impregnated with fine black powder (quickmatch). It burns with an external flame and is used in ignition trains for fireworks, etc.

FUSE, SAFETY: UN No. 0105

Article consisting of a core of fine grained black powder surrounded by a flexible woven fabric with one or more protective outer coverings. When ignited, it burns at a predetermined rate without any external explosive effect.

FUZES, DETONATING: UN Nos. 0106, 0107, 0257, 0367

Articles with explosive components designed to produce a detonation in ammunition. They incorporate mechanical, electrical, chemical or hydrostatic components to initiate the detonation. They generally incorporate protective features.

FUZES, DETONATING with protective features: UN Nos. 0408, 0409, 0410

Articles with explosive components designed to produce a detonation in ammunition. They incorporate mechanical, electrical, chemical or hydrostatic components to initiate the detonation. The detonating fuze must incorporate two or more effective protective features.

FUZES, IGNITING: UN Nos. 0316, 0317, 0368

Articles with primary explosive components designed to produce a deflagration in ammunition. They incorporate mechanical, electrical, chemical or hydrostatic components to start the deflagration. They generally incorporate protective features.

GRENADES, hand or rifle, with bursting charge: UN Nos. 0284, 0285

Articles which are designed to be thrown by hand or to be projected by a rifle. They are without means of initiation or with means of initiation containing two or more effective protective features.

GRENADES, hand or rifle, with bursting charge: UN Nos. 0292, 0293

Articles which are designed to be thrown by hand or to be projected by a rifle. They are with means of initiation not containing two or more effective protective features.

GRENADES, PRACTICE, hand or rifle: UN Nos. 0110, 0372, 0318, 0452

Articles without a main bursting charge which are designed to be thrown by hand or to be projected by a rifle. They contain the priming device and may contain a spotting charge.

HEXOTONAL: UN No. 0393

Substance consisting of an intimate mixture of cyclotrimethylenetrinitramine (RDX), trinitrotoluene (TNT) and aluminium.

HEXOLITE (HEXOTOL), dry or wetted with less than 15 % water, by mass: UN No. 0118

Substance consisting of an intimate mixture of cyclotrimethylenetrinitramine (RDX) and trinitrotoluene (TNT). The term includes "Composition B".

IGNITERS: UN Nos. 0121, 0314, 0315, 0325, 0454

Articles containing one or more explosive substances designed to produce a deflagration in an explosive train. They may be actuated chemically, electrically or mechanically.

NOTE: *The following articles: CORD, IGNITER; FUSE, IGNITER; FUSE, NON-DETONATING; FUZES, IGNITING; LIGHTERS, FUSE; PRIMERS, CAP TYPE; PRIMERS, TUBULAR are not included in this definition. They are listed separately.*

JET PERFORATING GUNS, CHARGED, oil well, without detonator: UN Nos. 0124, 0494

Articles consisting of a steel tube or metallic strip, into which are inserted shaped charges connected by detonating cord, without means of initiation.
LIGHTERS, FUSE: UN No. 0131

Articles of various design actuated by friction, percussion or electricity and used to ignite safety fuse.

MINES with bursting charge: UN Nos. 0137, 0138

Articles consisting normally of metal or composition receptacles filled with a detonating explosive, without means of initiation or with means of initiation containing two or more effective protective features. They are designed to be operated by the passage of ships, vehicles or personnel. The term includes "Bangalore torpedoes".

MINES with bursting charge: UN Nos. 0136, 0294

Articles consisting normally of metal or composition receptacles filled with a detonating explosive, with means of initiation not containing two or more effective protective features. They are designed to be operated by the passage of ships, vehicles or personnel. The term includes "Bangalore torpedoes".

OCTOLITE (OCTOL), dry or wetted with less than 15 % water, by mass: UN No. 0266

Substance consisting of an intimate mixture of cyclotetramethylenetetranitramine (HMX) and trinitrotoluene (TNT).

OCTONAL: UN No. 0496

Substance consisting of an intimate mixture of cyclotetramethylenetetranitramine (HMX), trinitrotoluene (TNT) and aluminium.

PENTOLITE, dry or wetted with less than 15 % water, by mass: UN No. 0151

Substance consisting of an intimate mixture of pentaerythrite tetranitrate (PETN) and trinitrotoluene (TNT).

POWDER CAKE (POWDER PASTE), WETTED with not less than 17 % alcohol, by mass; POWDER CAKE (POWDER PASTE), WETTED with not less than 25 % water, by mass: UN Nos. 0433, 0159

Substance consisting of nitrocellulose impregnated with not more than 60 % of nitroglycerine or other liquid organic nitrates or a mixture of these.

POWDER, SMOKELESS: UN Nos. 0160, 0161

Substance based on nitrocellulose used as propellant. The term includes propellants with a single base (nitrocellulose (NC) alone), those with a double base (such as NC and nitroglycerine (NG)) and those with a triple base (such as NC/NG/nitroguanidine).

NOTE: Cast, pressed or bag-charges of smokeless powder are listed under CHARGES, PROPELLING or CHARGES, PROPELLING, FOR CANON.

PRIMERS, CAP TYPE: UN Nos. 0044, 0377, 0378

Articles consisting of a metal or plastics cap containing a small amount of primary explosive mixture that is readily ignited by impact. They serve as igniting elements in small arms cartridges and in percussion primers for propelling charges.
PRIMERS, TUBULAR: UN Nos. 0319, 0320, 0376

Articles consisting of a primer for ignition and an auxiliary charge of deflagrating explosive such as black powder used to ignite the propelling charge in a cartridge case for cannon, etc.

PROJECTILES, inert with tracer: UN Nos. 0345, 0424, 0425

Articles such as a shell or bullet, which are projected from a cannon or other gun, rifle or other small arm.

PROJECTILES with burster or expelling charge: UN Nos. 0346, 0347

Articles such as a shell or bullet, which are projected from a cannon or other gun. They are without means of initiation or with means of initiation containing two or more effective protective features. They are used to scatter dyes for spotting or other inert materials.

PROJECTILES with burster or expelling charge: UN Nos. 0426, 0427

Articles such as a shell or bullet, which are projected from a cannon or other gun. They are with means of initiation not containing two or more effective protective features. They are used to scatter dyes for spotting or other inert materials.

PROJECTILES with burster or expelling charge: UN Nos. 0434, 0435

Articles such as a shell or bullet, which are projected from a cannon or other gun, rifle or other small arm. They are used to scatter dyes for spotting or other inert materials.

PROJECTILES with bursting charge: UN Nos. 0168, 0169, 0344

Articles such as a shell or bullet, which are projected from a cannon or other gun. They are without means of initiation or with means of initiation containing two or more effective protective features.

PROJECTILES with bursting charge: UN Nos. 0167, 0324

Articles such as a shell or bullet, which are projected from a cannon or other gun. They are with means of initiation not containing two or more effective protective features.

PROPELLANT, LIQUID: UN Nos. 0495, 0497

Substance consisting of a deflagrating liquid explosive, used for propulsion.

PROPELLANT, SOLID: UN Nos. 0498, 0499, 0501

Substance consisting of a deflagrating solid explosive, used for propulsion.

RELEASE DEVICES, EXPLOSIVE: UN No. 0173

Articles consisting of a small charge of explosive with means of initiation and rods or links. They sever the rods or links to release equipment quickly.

RIVETS, EXPLOSIVE: UN No. 0174

Articles consisting of a small charge of explosive inside a metallic rivet.

ROCKET MOTORS: UN Nos. 0186, 0280, 0281

Articles consisting of a charge of explosive, generally a solid propellant, contained in a cylinder fitted with one or more nozzles. They are designed to propel a rocket or a guided missile.

ROCKET MOTORS, LIQUID FUELLED: UN Nos. 0395, 0396

Articles consisting of a liquid fuel within a cylinder fitted with one or more nozzles. They are designed to propel a rocket or a guided missile.

ROCKET MOTORS WITH HYPERGOLIC LIQUIDS with or without expelling charge: UN Nos. 0322, 0250

Articles consisting of a hypergolic fuel contained in a cylinder fitted with one or more nozzles. They are designed to propel a rocket or a guided missile.

ROCKETS, LINE THROWING: UN Nos. 0238, 0240, 0453

Articles consisting of a rocket motor which is designed to extend a line.

ROCKETS, LIQUID FUELLED with bursting charge: UN Nos. 0397, 0398

Articles consisting of a liquid fuel within a cylinder fitted with one or more nozzles and fitted with a warhead. The term includes guided missiles.

ROCKETS with bursting charge: UN Nos. 0181, 0182

Articles consisting of a rocket motor and a warhead without means of initiation or with means of initiation containing two or more effective protective features. The term includes guided missiles.

ROCKETS with bursting charge: UN Nos. 0180, 0295

Articles consisting of a rocket motor and a warhead with means of initiation not containing two or more effective protective features. The term includes guided missiles.

ROCKETS with expelling charge: UN Nos. 0436, 0437, 0438

Articles consisting of a rocket motor and a charge to expel the payload from a rocket head. The term includes guided missiles.

ROCKETS with inert head: UN Nos. 0183, 0502

Articles consisting of a rocket motor and an inert head. The term includes guided missiles.

SAMPLES, EXPLOSIVE, other than initiating explosive UN No. 0190

New or existing explosive substances or articles, not yet assigned to a name in Table A of Chapter 3.2 and carried in conformity with the instructions of the competent authority and generally in small quantities, inter alia, for the purposes of testing, classification, research and development, or quality control, or as commercial samples.

NOTE: Explosive substances or articles already assigned to another name in Table A of Chapter 3.2 are not included in this definition.

SIGNAL DEVICES, HAND: UN Nos. 0191, 0373

Portable articles containing pyrotechnic substances which produce visual signals or warnings. The term includes small surface flares such as highway or railway flares and small distress flares.

SIGNALS, DISTRESS, ship: UN Nos. 0194, 0195

Articles containing pyrotechnic substances designed to produce signals by means of sound, flame or smoke or any combination thereof.

SIGNALS, RAILWAY TRACK, EXPLOSIVE: UN Nos. 0192, 0193, 0492, 0493

Articles containing a pyrotechnic substance which explodes with a loud report when the article is crushed. They are designed to be placed on a rail.

SIGNALS, SMOKE: UN Nos. 0196, 0197, 0313, 0487

Articles containing pyrotechnic substances which emit smoke. In addition they may contain devices for emitting audible signals.

SOUNDING DEVICES, EXPLOSIVE: UN Nos. 0374, 0375

Articles consisting of a charge of detonating explosive, without means of initiation or with means of initiation containing two or more effective protective features. They are dropped from ships and function when they reach a predetermined depth or the sea bed.

SOUNDING DEVICES, EXPLOSIVE: UN Nos. 0204, 0296

Articles consisting of a charge of detonating explosive with means of initiation not containing two or more effective protective features. They are dropped from ships and function when they reach a predetermined depth or the sea bed.

SUBSTANCES, EXPLOSIVE, VERY INSENSITIVE (Substances, EVI), N.O.S.: UN No. 0482

Substances presenting a mass explosion hazard but which are so insensitive that there is very little probability of initiation or of transition from burning to detonation under normal conditions of transport, and which have passed Test Series 5.

TORPEDOES, LIQUID FUELLED with inert head: UN No. 0450

Articles consisting of a liquid explosive system to propel the torpedo through the water, with an inert head.

TORPEDOES, LIQUID FUELLED with or without bursting charge: UN No. 0449

Articles consisting of either a liquid explosive system to propel the torpedo through the water, with or without a warhead; or a liquid non-explosive system to propel the torpedo through the water, with a warhead.

TORPEDOES with bursting charge: UN No. 0451

Articles consisting of a non-explosive system to propel the torpedo through the water, and a warhead without means of initiation or with means of initiation containing two or more effective protective features.

TORPEDOES with bursting charge: UN No. 0329

Articles consisting of an explosive system to propel the torpedo through the water, and a warhead without means of initiation or with means of initiation containing two or more effective protective features.

TORPEDOES with bursting charge: UN No. 0330

Articles consisting of an explosive or non-explosive system to propel the torpedo through the water, and a warhead with means of initiation not containing two or more effective protective features.

TRACERS FOR AMMUNITION: UN Nos. 0212, 0306

Sealed articles containing pyrotechnic substances, designed to reveal the trajectory of a projectile.

TRITONAL: UN No. 0390

Substance consisting of trinitrotoluene (TNT) mixed with aluminium.

WARHEADS, ROCKET with burster or expelling charge: UN No. 0370

Articles consisting of an inert payload and a small charge of detonating or deflagrating explosive, without means of initiation or with means of initiation containing two or more effective protective features. They are designed to be fitted to a rocket motor to scatter inert material. The term includes warheads for guided missiles.

WARHEADS, ROCKET with burster or expelling charge: UN No. 0371

Articles consisting of an inert payload and a small charge of detonating or deflagrating explosive, with means of initiation not containing two or more effective protective features. They are designed to be fitted to a rocket motor to scatter inert material. The term includes warheads for guided missiles.

WARHEADS, ROCKET with bursting charge: UN Nos. 0286, 0287

Articles consisting of a detonating explosive, without means of initiation or with means of initiation containing two or more effective protective features. They are designed to be fitted to a rocket. The term includes warheads for guided missiles.

WARHEADS, ROCKET with bursting charge: UN No. 0369

Articles consisting of a detonating explosive, with means of initiation not containing two or more effective protective features. They are designed to be fitted to a rocket. The term includes warheads for guided missiles.

WARHEADS, TORPEDO with bursting charge: UN No. 0221

Articles consisting of a detonating explosive, without means of initiation or with means of initiation containing two or more effective protective features. They are designed to be fitted to a torpedo.

2.2.1.2 *Substances and articles not accepted for carriage*

2.2.1.2.1   Explosive substances which are unduly sensitive according to the criteria of the Manual of Tests and Criteria, Part I, or are liable to spontaneous reaction, as well as explosive substances and articles which cannot be assigned to a name or n.o.s. entry listed in Table A of Chapter 3.2, shall not be accepted for carriage.

2.2.1.2.2   Articles of compatibility group K shall not be accepted for carriage (1.2K, UN No. 0020 and 1.3K, UN No. 0021).

2.2.1.3 *List of collective entries*

Classification code (see 2.2.1.1.4)	UN No	Name of the substance or article
1.1A	0473	SUBSTANCES, EXPLOSIVE, N.O.S.
1.1B	0461	COMPONENTS, EXPLOSIVE TRAIN, N.O.S.
1.1C	0474	SUBSTANCES, EXPLOSIVE, N.O.S.
	0497	PROPELLANT, LIQUID
	0498	PROPELLANT, SOLID
	0462	ARTICLES, EXPLOSIVE, N.O.S.
1.1D	0475	SUBSTANCES, EXPLOSIVE, N.O.S.
	0463	ARTICLES, EXPLOSIVE, N.O.S.
1.1E	0464	ARTICLES, EXPLOSIVE, N.O.S.
1.1F	0465	ARTICLES, EXPLOSIVE, N.O.S.
1.1G	0476	SUBSTANCES, EXPLOSIVE, N.O.S.
1.1L	0357	SUBSTANCES, EXPLOSIVE, N.O.S.
	0354	ARTICLES, EXPLOSIVE, N.O.S.
1.2B	0382	COMPONENTS, EXPLOSIVE TRAIN, N.O.S.
1.2C	0466	ARTICLES, EXPLOSIVE, N.O.S.
1.2D	0467	ARTICLES, EXPLOSIVE, N.O.S.
1.2E	0468	ARTICLES, EXPLOSIVE, N.O.S.
1.2F	0469	ARTICLES, EXPLOSIVE, N.O.S.
1.2L	0358	SUBSTANCES, EXPLOSIVE, N.O.S.
	0248	CONTRIVANCES, WATER-ACTIVATED with burster, expelling charge or propelling charge
	0355	ARTICLES, EXPLOSIVE, N.O.S.
1.3C	0132	DEFLAGRATING METAL SALTS OF AROMATIC NITRO-DERIVATIVES, N.O.S.
	0477	SUBSTANCES, EXPLOSIVE, N.O.S.
	0495	PROPELLANT, LIQUID
	0499	PROPELLANT, SOLID
	0470	ARTICLES, EXPLOSIVE, N.O.S.
1.3G	0478	SUBSTANCES, EXPLOSIVE, N.O.S.
1.3L	0359	SUBSTANCES, EXPLOSIVE, N.O.S.
	0249	CONTRIVANCES, WATER-ACTIVATED with burster, expelling charge or propelling charge
	0356	ARTICLES, EXPLOSIVE, N.O.S.
1.4B	0350	ARTICLES, EXPLOSIVE, N.O.S.
	0383	COMPONENTS, EXPLOSIVE TRAIN, N.O.S.
1.4C	0479	SUBSTANCES, EXPLOSIVE, N.O.S.
	0351	ARTICLES, EXPLOSIVE, N.O.S.
	0501	PROPELLANT, SOLID
1.4D	0480	SUBSTANCES, EXPLOSIVE, N.O.S.
	0352	ARTICLES, EXPLOSIVE, N.O.S.
1.4E	0471	ARTICLES, EXPLOSIVE, N.O.S.
1.4F	0472	ARTICLES, EXPLOSIVE, N.O.S.
1.4G	0485	SUBSTANCES, EXPLOSIVE, N.O.S.
	0353	ARTICLES, EXPLOSIVE, N.O.S.
1.4S	0481	SUBSTANCES, EXPLOSIVE, N.O.S.
	0349	ARTICLES, EXPLOSIVE, N.O.S.
	0384	COMPONENTS, EXPLOSIVE TRAIN, N.O.S.
1.5D	0482	SUBSTANCES, EXPLOSIVE, VERY INSENSITIVE (SUBSTANCES, EVI) N.O.S.

Classification code (see 2.2.1.1.4)	UN No	Name of the substance or article
1.6N	0486	ARTICLES, EXPLOSIVE, EXTREMELY INSENSITIVE (ARTICLES, EEI)
	0190	SAMPLES, EXPLOSIVE other than initiating explosive ***NOTE:*** *Division and Compatibility Group shall be defined as directed by the competent authority and according to the principles in 2.2.1.1.4.*

2.2.2   **Class 2   Gases**

2.2.2.1   *Criteria*

2.2.2.1.1   The heading of Class 2 covers pure gases, mixtures of gases, mixtures of one or more gases with one or more other substances and articles containing such substances.

A gas is a substance which:

(a)   at 50 °C has a vapour pressure greater than 300 kPa (3 bar); or

(b)   is completely gaseous at 20° C at the standard pressure of 101.3 kPa .

NOTE 1: *UN No. 1052 HYDROGEN FLUORIDE ANHYDROUS is nevertheless classified in Class 8.*

NOTE 2: *A pure gas may contain other components deriving from its production process or added to preserve the stability of the product, provided that the level of these components does not change its classification or its conditions of carriage, such as filling ratio, filling pressure, test pressure.*

NOTE 3: *N.O.S. entries in 2.2.2.3 may cover pure gases as well as mixtures.*

NOTE 4: *Carbonated beverages are not subject to the provisions of ADN.*

2.2.2.1.2   The substances and articles of Class 2 are subdivided as follows:

1.   *Compressed gas:* a gas which when packaged under pressure for carriage is entirely gaseous at -50 °C; this category includes all gases with a critical temperature less than or equal to -50 °C;

2.   *Liquefied gas:* a gas which when packaged under pressure for carriage is partially liquid at temperatures above -50 °C. A distinction is made between:

> *High pressure liquefied gas*: a gas with a critical temperature above -50 °C and equal to or below +65 °C; and

> *Low pressure liquefied gas:* a gas with a critical temperature above +65 °C;

3.   *Refrigerated liquefied gas:* a gas which when packaged for carriage is made partially liquid because of its low temperature;

4.   *Dissolved gas:* a gas which when packaged under pressure for carriage is dissolved in a liquid phase solvent;

5.   Aerosol dispensers and receptacles, small, containing gas (gas cartridges);

6.   Other articles containing gas under pressure;

7.   Non-pressurized gases subject to special requirements (gas samples).

2.2.2.1.3   Substances and articles (except aerosols) of Class 2 are assigned to one of the following groups according to their hazardous properties, as follows:

A   asphyxiant;

O oxidizing;

F flammable;

T toxic;

TF toxic, flammable;

TC toxic, corrosive;

TO toxic, oxidizing;

TFC toxic, flammable, corrosive;

TOC toxic, oxidizing, corrosive.

For gases and gas mixtures presenting hazardous properties associated with more than one group according to the criteria, the groups designated by letter T take precedence over all other groups. The groups designated by letter F take precedence over the groups designated by letters A or O.

NOTE 1: In the UN Model Regulations, the IMDG Code and the ICAO Technical Instructions, gases are assigned to one of the following three divisions, based on the primary hazard:

Division 2.1: *flammable gases (corresponding to the groups designated by the capital letter F);*

Division 2.2: *non-flammable, non-toxic gases (corresponding to the groups designated by the capital letters A or O);*

Division 2.3: *toxic gases (corresponding to the groups designated by the capital letter T (i.e. T, TF, TC, TO, TFC and TOC).*

NOTE 2: Receptacles, small containing gas (UN No. 2037) shall be assigned to the groups A to TOC according to the hazard of the contents. For aerosols (UN No. 1950), see 2.2.2.1.6.

NOTE 3: Corrosive gases are considered to be toxic, and are therefore assigned to the group TC, TFC or TOC.

NOTE 4: Mixtures containing more than 21% oxygen by volume shall be classified as oxidizing.

2.2.2.1.4 If a mixture of Class 2 mentioned by name in Table A of Chapter 3.2 meets different criteria as mentioned in 2.2.2.1.2 and 2.2.2.1.5, this mixture shall be classified according to the criteria and assigned to an appropriate N.O.S. entry.

2.2.2.1.5 Substances and articles (except aerosols) of Class 2 which are not mentioned by name in Table A of Chapter 3.2 shall be classified under a collective entry listed in 2.2.2.3 in accordance with 2.2.2.1.2 and 2.2.2.1.3. The following criteria shall apply:

Asphyxiant gases

Gases which are non-oxidizing, non-flammable and non-toxic and which dilute or replace oxygen normally in the atmosphere.

Flammable gases

Gases which at 20 °C and a standard pressure of 101.3 kPa:

(a)   are ignitable when in a mixture of 13% or less by volume with air; or

(b)   have a flammable range with air of at least 12 percentage points regardless of the lower flammable limit.

Flammability shall be determined by tests or by calculation, in accordance with methods adopted by ISO (see ISO 10156:1996).

Where insufficient data are available to use these methods, tests by a comparable method recognized by the competent authority of the country of origin may be used.

If the country of origin is not a Contracting Party to ADN these methods shall be recognized by the competent authority of the first country Contracting Party to ADN reached by the consignment.

Oxidizing gases

Gases, which may, generally by providing oxygen, cause or contribute to the combustion of other material more than air does. Oxidizing ability is determined either by tests or by calculation methods adopted by ISO (see ISO 10156:1996 and ISO 10156-2:2005).

Toxic gases

NOTE: Gases meeting the criteria for toxicity in part or completely owing to their corrosivity are to be classified as toxic. See also the criteria under the heading "Corrosive gases" for a possible subsidiary corrosivity risk.

Gases which:

(a)   are known to be so toxic or corrosive to humans as to pose a hazard to health; or

(b)   are presumed to be toxic or corrosive to humans because they have a LC_{50} value for acute toxicity equal to or less than 5 000 ml/m³ (ppm) when tested in accordance with 2.2.61.1.

In the case of gas mixtures (including vapours of substances from other classes) the following formula may be used:

$$LC_{50} \text{ Toxic (mixture)} = \frac{1}{\sum_{i=1}^{n} \frac{f_i}{T_i}}$$

where f_i =   mole fraction of the ith component substance of the mixture;

          T_i =   toxicity index of the ith component substance of the mixture.
                          The T_i equals the LC_{50} value as found in packing instruction P200 of 4.1.4.1 of ADR.
                          When no LC_{50} value is listed in packing instruction P200 of 4.1.4.1 of ADR, a LC_{50} value available in scientific literature shall be used.

When the LC_{50} value is unknown, the toxicity index is determined by using the lowest LC_{50} value of substances of similar physiological and chemical effects, or through testing if this is the only practical possibility.

Corrosive gases

Gases or gas mixtures meeting the criteria for toxicity completely owing to their corrosivity are to be classified as toxic with a subsidiary corrosivity risk.

A gas mixture that is considered to be toxic due to the combined effects of corrosivity and toxicity has a subsidiary risk of corrosivity when the mixture is known by human experience to be destructive to the skin, eyes or mucous membranes or when the LC_{50} value of the corrosive components of the mixture is equal to or less than 5 000 ml/m^3 (ppm) when the LC_{50} is calculated by the formula:

$$LC_{50} \text{ Corrosive (mixture)} = \frac{1}{\displaystyle\sum_{i=1}^{n} \frac{f_{ci}}{T_{ci}}}$$

where f_{ci} =   mole fraction of the ith corrosive component substance of the mixture;

           T_{ci} =   toxicity index of the ith corrosive component substance of the mixture.
The T_{ci} equals the LC50 value as found in packing instruction P200 of 4.1.4.1 of ADR.
When no LC_{50} value is listed in packing instruction P200 of 4.1.4.1 of ADR, a LC_{50} value available in scientific literature shall be used.
When the LC_{50} value is unknown the toxicity index is determined by using the lowest LC_{50} value of substances of similar physiological and chemical effects, or through testing if this is the only practical possibility.

2.2.2.1.6   *Aerosols*

Aerosols (UN No. 1950) are assigned to one of the following groups according to their hazardous properties, as follows:

A   asphyxiant;

O   oxidizing;

F   flammable;

T   toxic;

C   corrosive;

CO   corrosive, oxidizing;

FC   flammable, corrosive;

TF   toxic, flammable;

TC   toxic, corrosive;

TO   toxic, oxidizing;

TFC   toxic, flammable, corrosive

TOC   toxic, oxidizing, corrosive.

The classification depends on the nature of the contents of the aerosol dispenser.

NOTE: Gases, which meet the definition of toxic gases according to 2.2.2.1.5 or of pyrophoric gases according to packing instruction P200 in 4.1.4.1 of ADR, shall not be used as a propellant in an aerosol dispenser. Aerosols with contents meeting the criteria for packing group 1 for toxicity or corrosivity shall not be accepted for carriage (see also 2.2.2.2.2).

The following criteria shall apply:

(a)   Assignment to group A shall apply when the contents do not meet the criteria for any other group according to sub-paragraphs (b) to (f) below;

(b)   Assignment to group O shall apply when the aerosol contains an oxidizing gas according to 2.2.2.1.5;

(c)   Assignment to group F shall apply if the contents include 85% by mass or more flammable components and the chemical heat of combustion is 30 kJ/g or more.

It shall not apply if the contents contain 1% by mass or less flammable components and the heat of combustion is less than 20 kJ/g.

Otherwise the aerosol shall be tested for flammability in accordance with the tests described in the *Manual of Tests and Criteria*, Part III, section 31. Extremely flammable and flammable aerosols shall be assigned to group F;

NOTE: Flammable components are flammable liquids, flammable solids or flammable gases and gas mixtures as defined in Notes 1 to 3 of sub-section 31.1.3 of Part III of the Manual of Tests and Criteria. This designation does not cover pyrophoric, self-heating or water-reactive substances. The chemical heat of combustion shall be determined by one of the following methods ASTM D 240, ISO/FDIS 13943: 1999 (E/F) 86.1 to 86.3 or NFPA 30B.

(d)   Assignment to group T shall apply when the contents, other than the propellant of aerosol dispensers to be ejected, are classified as Class 6.1, packing groups II or III;

(e)   Assignment to group C shall apply when the contents, other than the propellant of aerosol dispensers to be ejected, meet the criteria for Class 8, packing groups II or III;

(f)   When the criteria for more than one group amongst groups O, F, T, and C are met, assignment to groups CO, FC, TF, TC TO, TFC or TOC shall apply, as relevant.

2.2.2.2   *Gases not accepted for carriage*

2.2.2.2.1   Chemically unstable substances of Class 2 shall not be accepted for carriage, unless the necessary steps have been taken to prevent all possibility of a dangerous reaction e.g. decomposition, dismutation or polymerisation under normal conditions during transport.

To this end particular care shall be taken to ensure that receptacles and tanks do not contain any substances liable to promote these reactions.

2.2.2.2.2   The following substances and mixtures shall not be accepted for carriage:

- UN No. 2186 HYDROGEN CHLORIDE, REFRIGERATED LIQUID;

- UN No. 2421 NITROGEN TRIOXIDE;

- UN No. 2455 METHYL NITRITE;

- Refrigerated liquefied gases which cannot be assigned to classification codes 3A, 3O or 3F;

- Dissolved gases which cannot be classified under UN Nos. 1001, 2073 or 3318;

- Aerosols where gases which are toxic according to 2.2.2.1.5 or pyrophoric according to packing instruction P200 in 4.1.4.1 of ADR are used as propellants;

- Aerosols with contents meeting the criteria for packing group I for toxicity or corrosivity (see 2.2.61 and 2.2.8);

- Receptacles, small, containing gases which are very toxic (LC_{50} lower than 200 ppm) or pyrophoric according to packing instruction P200 in 4.1.4.1 of ADR.

Compressed gases		
Classification code	**UN No**	**Name and description**
1 A	1956	COMPRESSED GAS, N.O.S.
1 O	3156	COMPRESSED GAS, OXIDIZING, N.O.S.
1 F	1964	HYDROCARBON GAS MIXTURE, COMPRESSED, N.O.S.
	1954	COMPRESSED GAS, FLAMMABLE, N.O.S.
1T	1955	COMPRESSED GAS, TOXIC, N.O.S.
1 TF	1953	COMPRESSED GAS, TOXIC, FLAMMABLE, N.O.S.
1 TC	3304	COMPRESSED GAS, TOXIC, CORROSIVE, N.O.S.
1 TO	3303	COMPRESSED GAS, TOXIC, OXIDIZING, N.O.S.
1 TFC	3305	COMPRESSED GAS, TOXIC, FLAMMABLE, CORROSIVE, N.O.S.
1 TOC	3306	COMPRESSED GAS, TOXIC, OXIDIZING, CORROSIVE, N.O.S.

Liquefied gases		
Classification code	**UN No**	**Name and description**
2 A	1058	LIQUEFIED GASES, non-flammable, charged with nitrogen, carbon dioxide or air
	1078	REFRIGERANT GAS, N.O.S. such as mixtures of gases, indicated by the letter R, which as: Mixture F1, have a vapour pressure at 70 °C not exceeding 1.3 MPa (13 bar) and a density at 50 °C not lower than that of dichlorofluoromethane (1.30 kg/l); Mixture F2, have a vapour pressure at 70 °C not exceeding 1.9 MPa (19 bar) and a density at 50 °C not lower than that of dichlorodifluoromethane (1.21 kg/l); Mixture F3, have a vapour pressure at 70 °C not exceeding 3 MPa (30 bar) and a density at 50 °C not lower than that of chlorodifluoromethane (1.09 kg/l). ***NOTE:*** *Trichlorofluoromethane (Refrigerant R 11), 1,1,2-trichloro-1,2,2-trifluoroethane (Refrigerant R 113), 1,1,1-trichloro-2,2,2-trifluoroethane (Refrigerant R 113a), 1-chloro-1,2,2-trifluoroethane (Refrigerant R 133) and 1-chloro-1,1,2-trifluoroethane (Refrigerant R 133b) are not substances of Class 2. They may, however, enter into the composition of mixtures F1 to F3.*
	1968	INSECTICIDE GAS, N.O.S.
	3163	LIQUEFIED GAS, N.O.S.
2 O	3157	LIQUEFIED GAS, OXIDIZING, N.O.S.

		Liquefied gases (cont'd)
Classification code	UN No	Name and description
2 F	1010	BUTADIENES AND HYDROCARBON MIXTURE, STABILIZED, having a vapour pressure at 70 °C not exceeding 1.1 MPa (11 bar) and a density at 50 °C not lower than 0.525 kg/l. *NOTE: Butadienes, stabilized are also classified under UN No. 1010, see Table A of Chapter 3.2.*
	1060	METHYLACETYLENE AND PROPADIENE MIXTURE, STABILIZED such as mixtures of methylacetylene and propadiene with hydrocarbons, which as: Mixture P1, contain not more than 63% methylacetylene and propadiene by volume and not more than 24% propane and propylene by volume, the percentage of C_4- saturated hydrocarbons being not less than 14% by volume; and as Mixture P2, contain not more than 48% methylacetylene and propadiene by volume and not more than 50% propane and propylene by volume, the percentage of C_4- saturated hydrocarbons being not less than 5% by volume, as well as mixtures of propadiene with 1 to 4% methylacetylene.
	1965	HYDROCARBON GAS MIXTURE, LIQUEFIED, N.O.S such as mixtures, which as: Mixture A, have a vapour pressure at 70 °C not exceeding 1.1 MPa (11 bar) and a density at 50 °C not lower than 0.525 kg/l; Mixture A01, have a vapour pressure at 70 °C not exceeding 1.6 MPa (16 bar) and a relative density at 50 °C not lower than 0.516 kg/l; Mixture A02, have a vapour pressure at 70 °C not exceeding 1.6 MPa (16 bar) and a relative density at 50 °C not lower than 0.505 kg/l; Mixture A0, have a vapour pressure at 70 °C not exceeding 1.6 MPa (16 bar) and a density at 50 °C not lower than 0.495 kg/l; Mixture A1, have a vapour pressure at 70 °C not exceeding 2.1 MPa (21 bar) and a density at 50 °C not lower than 0.485 kg/l; Mixture B1 have a vapour pressure at 70 °C not exceeding 2.6 MPa (26 bar) and a relative density at 50 °C not lower than 0.474 kg/l; Mixture B2 have a vapour pressure at 70 °C not exceeding 2.6 MPa (26 bar) and a relative density at 50 °C not lower than 0.463 kg/l; Mixture B, have a vapour pressure at 70 °C not exceeding 2.6 MPa (26 bar) and a density at 50 °C not lower than 0.450 kg/l; Mixture C, have a vapour pressure at 70 °C not exceeding 3.1 MPa (31 bar) and a relative density at 50 °C not lower than 0.440 kg/l; *NOTE 1: In the case of the foregoing mixtures, the use of the following names customary in the trade is permitted for describing these substances: for mixture A01, A02 and A0: BUTANE; for mixture C: PROPANE.* *NOTE 2: UN No. 1075 PETROLEUM GASES, LIQUEFIED may be used as an alternative entry for UN No. 1965 HYDROCARBON GAS MIXTURE LIQUEFIED, N.O.S. for carriage prior to or following maritime or air carriage.*
	3354	INSECTICIDE GAS, FLAMMABLE, N.O.S.
	3161	LIQUEFIED GAS, FLAMMABLE, N.O.S.

Liquefied gases (cont'd)

Classification code	UN No	Name and description
2 T	1967	INSECTICIDE GAS, TOXIC, N.O.S.
	3162	LIQUEFIED GAS, TOXIC, N.O.S.
2 TF	3355	INSECTICIDE GAS, TOXIC, FLAMMABLE, N.O.S.
	3160	LIQUEFIED GAS, TOXIC, FLAMMABLE, N.O.S.
2 TC	3308	LIQUEFIED GAS, TOXIC, CORROSIVE, N.O.S.
2 TO	3307	LIQUEFIED GAS, TOXIC, OXIDIZING, N.O.S.
2 TFC	3309	LIQUEFIED GAS, TOXIC, FLAMMABLE, CORROSIVE, N.O.S.
2 TOC	3310	LIQUEFIED GAS, TOXIC, OXIDIZING, CORROSIVE, N.O.S.

Refrigerated liquefied gases

Classification code	UN No	Name and description
3 A	3158	GAS, REFRIGERATED LIQUID, N.O.S.
3 O	3311	GAS, REFRIGERATED LIQUID, OXIDIZING, N.O.S.
3 F	3312	GAS, REFRIGERATED LIQUID, FLAMMABLE, N.O.S.

Dissolved gases

Classification code	UN No	Name and description
4		Only substances listed in Table A of Chapter 3.2 are to be accepted for carriage.

Aerosols and receptacles, small, containing gas

Classification code	UN No	Name and description
5	1950	AEROSOLS
	2037	RECEPTACLES, SMALL CONTAINING GAS (GAS CARTRIDGES) without a release device, non-refillable

Other articles containing gas under pressure

Classification code	UN No	Name and description
6A	2857	REFRIGERATING MACHINES containing non-flammable, non-toxic gases or ammonia solutions (UN 2672)
	3164	ARTICLES, PRESSURIZED, PNEUMATIC (containing non-flammable gas) or
	3164	ARTICLES, PRESSURIZED, HYDRAULIC (containing non-flammable gas)
6F	3150	DEVICES, SMALL, HYDROCARBON GAS POWERED or
	3150	HYDROCARBON GAS REFILLS FOR SMALL DEVICES, with release device

Gas samples

Classification code	UN No	Name and description
7 F	3167	GAS SAMPLE, NON-PRESSURIZED, FLAMMABLE, N.O.S., not refrigerated liquid
7 T	3169	GAS SAMPLE, NON-PRESSURIZED, TOXIC, N.O.S., not refrigerated liquid
7 TF	3168	GAS SAMPLE, NON-PRESSURIZED, TOXIC, FLAMMABLE, N.O.S., not refrigerated liquid

2.2.3   **Class 3   Flammable liquids**

2.2.3.1   *Criteria*

2.2.3.1.1   The heading of Class 3 covers substances and articles containing substances of this Class which:

–   are liquids according to subparagraph (a) of the definition for "liquid" in 1.2.1;

–   have at 50 °C a vapour pressure of not more than 300 kPa (3 bar) and are not completely gaseous at 20 °C and at standard pressure of 101.3 kPa; and

–   have a flash-point of not more than 60 °C (see 2.3.3.1 for the relevant test).

The heading of Class 3 also covers liquid substances and molten solid substances with a flash-point of more than 60°C and which are carried or handed over for carriage whilst heated at temperatures equal to or higher than their flash-point. These substances are assigned to UN No. 3256.

The heading of Class 3 also covers liquid desensitized explosives. Liquid desensitized explosives are explosive substances which are dissolved or suspended in water or other liquid substances, to form an homogeneous liquid mixture to suppress their explosive properties. Such entries in Table A of Chapter 3.2 are UN Nos. 1204, 2059, 3064, 3343, 3357 and 3379.

For the purpose of carriage in tank vessels, the heading of Class 3 also covers the following substances which:

–   have a flash-point above 60° C and which are carried or handed over for carriage at a temperature within a range of 15 K below the flash-point;

–   have an auto-ignition temperature of 200° C or below and which are not mentioned elsewhere.

NOTE 1: Substances having a flash-point above 35 °C, non-toxic and non-corrosive, which, do not sustain combustion according to the criteria of 32.2.5 of Part III of the Manual of Tests and Criteria are not substances of Class 3; if, however, these substances are handed over for carriage and carried whilst heated at temperatures equal to or higher than their flash-point, they are substances of Class 3.

NOTE 2: By derogation from paragraph 2.2.3.1.1 above, diesel fuel, gasoil, heating oil (light) having a flash-point above 60 °C and not more than 100 °C shall be deemed substances of Class 3, UN No. 1202.

NOTE 3: Liquids which are highly toxic on inhalation, having a flash-point below 23 °C and toxic substances, having a flash-point of 23 °C or above are substances of Class 6.1 (see 2.2.61.1).

NOTE 4: Flammable liquid substances and preparations used as pesticides, which are highly toxic, toxic or slightly toxic and have a flash-point of 23 °C or above are substances of Class 6.1 (see 2.2.61.1).

NOTE 5: For the purpose of carriage in tank vessels, substances having a flash-point above 60° C and not more than 100° C are substances of Class 9 (identification number 9003).

2.2.3.1.2   The substances and articles of Class 3 are subdivided as follows:

F   Flammable liquids, without subsidiary risk:

F1   Flammable liquids having a flash-point of or below 60 °C;
F2   Flammable liquids having a flash-point above 60 °C which are carried or handed over for carriage at or above their flash-point (elevated temperature substances);
F3   Substances having a flash-point above 60° C which are carried or handed over for carriage at a temperature within a range of 15 K below the flash-point;
F4   Substances having an auto-ignition temperature of 200° C or below and which are not mentioned elsewhere.

FT   Flammable liquids, toxic:

FT1   Flammable liquids, toxic;
FT2   Pesticides;

FC   Flammable liquids, corrosive;

FTC   Flammable liquids, toxic, corrosive;

D   Liquid desensitized explosives.

2.2.3.1.3   Substances and articles classified in Class 3 are listed in Table A of Chapter 3.2. Substances not mentioned by name in Table A of Chapter 3.2 shall be assigned to the relevant entry of 2.2.3.3 and the relevant packing group in accordance with the provisions of this section. Flammable liquids shall be assigned to one of the following packing groups according to the degree of danger they present for carriage:

Packing Group	Flash-point (closed cup)	Initial boiling point
I	--	≤ 35°C
II [a]	< 23°C	> 35°C
III [a]	≥ 23°C and ≤ 60°C	> 35°C

[a]   See also 2.2.3.1.4

For a liquid with (a) subsidiary risk(s), the packing group determined in accordance with the table above and the packing group based on the severity of the subsidiary risk(s) shall be considered; the classification and packing group shall then be determined in accordance with the table of precedence of hazards in 2.1.3.10.

2.2.3.1.4   Liquid or viscous mixtures and preparations, including those containing no more than 20% nitrocellulose with a nitrogen content not exceeding 12.6% (by dry mass), shall be assigned to packing group III only if the following requirements are met:

(a)   the height of the separated layer of solvent is less than 3 % of the total height of the sample in the solvent-separation test (see Manual of Tests and Criteria, Part III, sub-section 32.5.1); and

(b)   the viscosity [2] and flash-point are in accordance with the following table:

---

[2]   *Viscosity determination: Where the substance concerned is non-Newtonian, or where a flow cup method of viscosity determination is otherwise unsuitable, a variable shear-rate viscometer shall be used to determine the dynamic viscosity coefficient of the substance, at 23 °C, at a number of shear rates. The values obtained are plotted against shear rate and then extrapolated to zero shear rate. The dynamic viscosity thus obtained, divided by the density, gives the apparent kinematic viscosity at near-zero shear rate.*

Kinematic viscosity (extrapolated) ν (at near-zero shear rate) mm²/s at 23 °C	Flow time t in accordance with ISO 2431:1993		Flash-point in °C
	in s	Jet diameter in mm	
20 < ν ≤ 80	20 < t ≤ 60	4	above 17
80 < ν ≤ 135	60 < t ≤ 100	4	above 10
135 < ν ≤ 220	20 < t ≤ 32	6	above 5
220 < ν ≤ 300	32 < t ≤ 44	6	above -1
300 < ν ≤ 700	44 < t ≤ 100	6	above -5
700 < <	100 < t	6	-5 and below

NOTE: Mixtures containing more than 20% but not more than 55% nitrocellulose with a nitrogen content not exceeding 12.6% by dry mass are substances assigned to UN No. 2059.

Mixtures having a flash-point below 23 °C and containing:

– *more than 55 % nitrocellulose, whatever their nitrogen content; or*

– *not more than 55 % nitrocellulose with a nitrogen content above 12.6 % by dry mass,*

are substances of Class 1 (UN Nos. 0340 or 0342) or of Class 4.1 (UN Nos. 2555, 2556 or 2557).

2.2.3.1.5   Non-toxic and non-corrosive solutions and homogeneous mixtures having a flash-point of 23 °C or above (viscous substances, such as paints or varnishes, excluding substances containing more than 20 % nitrocellulose) packed in receptacles of less than 450 litres capacity, are not subject to ADN if, in the solvent-separation test (see Manual of Tests and Criteria, Part III, sub-section 32.5.1), the height of the separated layer of solvent is less than 3 % of the total height, and if the substances at 23 °C have, in the flow cup conforming to ISO 2431:1993 having a jet 6 mm in diameter, a flow time of:

(a)   not less than 60 seconds, or

(b)   not less than 40 seconds and contain not more than 60 % of substances of Class 3.

2.2.3.1.6   If substances of Class 3, as a result of admixtures, come into categories of risk different from those to which the substances mentioned by name in Table A of Chapter 3.2 belong, these mixtures or solutions shall be assigned to the entries to which they belong on the basis of their actual degree of danger.

NOTE: For the classification of solutions and mixtures (such as preparations and wastes) see also 2.1.3.

2.2.3.1.7   On the basis of the test procedures in accordance with 2.3.3.1 and 2.3.4, and the criteria set out in 2.2.3.1.1, it may also be determined whether the nature of a solution or a mixture mentioned by name or containing a substance mentioned by name is such that the solution or mixture is not subject to the provisions for this Class (see also 2.1.3).

2.2.3.2   ***Substances not accepted for carriage***

2.2.3.2.1   Substances of Class 3 which are liable to form peroxides easily (as happens with ethers or with certain heterocyclic oxygenated substances) shall not be accepted for carriage if their peroxide content, calculated as hydrogen peroxide (H_2O_2), exceeds 0.3%. The peroxide content shall be determined as indicated in 2.3.3.2.

2.2.3.2.2   The chemically unstable substances of Class 3 shall not be accepted for carriage unless the necessary steps have been taken to prevent their dangerous decomposition or polymerization during carriage. To this end, it shall be ensured in particular that receptacles and tanks do not contain any substance liable to promote these reactions.

2.2.3.2.3   Liquid desensitized explosives other than those listed in Table A of Chapter 3.2 shall not be accepted for carriage as substances of Class 3.

2.2.3.3 *List of collective entries*

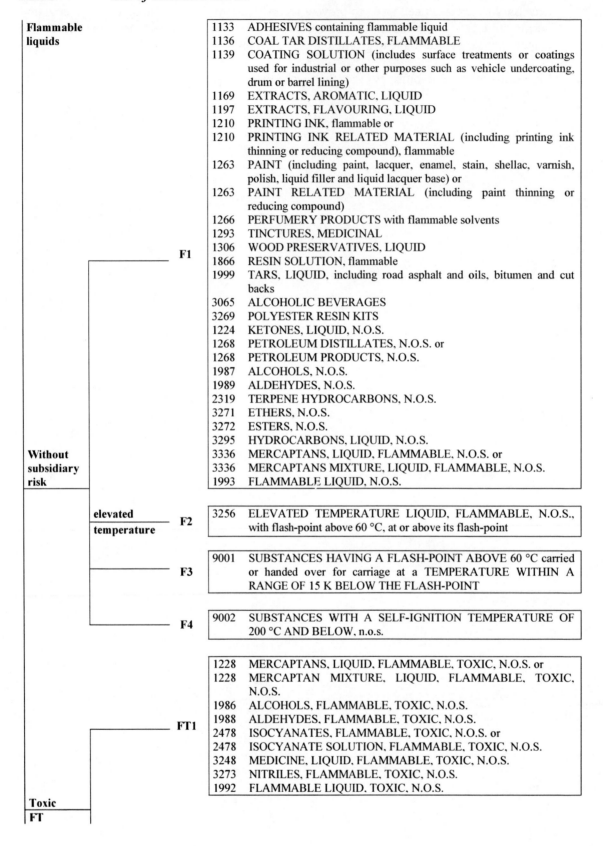

Flammable liquids			
		F1	1133 ADHESIVES containing flammable liquid
			1136 COAL TAR DISTILLATES, FLAMMABLE
			1139 COATING SOLUTION (includes surface treatments or coatings used for industrial or other purposes such as vehicle undercoating, drum or barrel lining)
			1169 EXTRACTS, AROMATIC, LIQUID
			1197 EXTRACTS, FLAVOURING, LIQUID
			1210 PRINTING INK, flammable or
			1210 PRINTING INK RELATED MATERIAL (including printing ink thinning or reducing compound), flammable
			1263 PAINT (including paint, lacquer, enamel, stain, shellac, varnish, polish, liquid filler and liquid lacquer base) or
			1263 PAINT RELATED MATERIAL (including paint thinning or reducing compound)
			1266 PERFUMERY PRODUCTS with flammable solvents
			1293 TINCTURES, MEDICINAL
			1306 WOOD PRESERVATIVES, LIQUID
			1866 RESIN SOLUTION, flammable
			1999 TARS, LIQUID, including road asphalt and oils, bitumen and cut backs
			3065 ALCOHOLIC BEVERAGES
			3269 POLYESTER RESIN KITS
			1224 KETONES, LIQUID, N.O.S.
			1268 PETROLEUM DISTILLATES, N.O.S. or
			1268 PETROLEUM PRODUCTS, N.O.S.
			1987 ALCOHOLS, N.O.S.
			1989 ALDEHYDES, N.O.S.
			2319 TERPENE HYDROCARBONS, N.O.S.
			3271 ETHERS, N.O.S.
			3272 ESTERS, N.O.S.
			3295 HYDROCARBONS, LIQUID, N.O.S.
			3336 MERCAPTANS, LIQUID, FLAMMABLE, N.O.S. or
Without subsidiary risk			3336 MERCAPTANS MIXTURE, LIQUID, FLAMMABLE, N.O.S.
			1993 FLAMMABLE LIQUID, N.O.S.
	elevated temperature	F2	3256 ELEVATED TEMPERATURE LIQUID, FLAMMABLE, N.O.S., with flash-point above 60 °C, at or above its flash-point
		F3	9001 SUBSTANCES HAVING A FLASH-POINT ABOVE 60 °C carried or handed over for carriage at a TEMPERATURE WITHIN A RANGE OF 15 K BELOW THE FLASH-POINT
		F4	9002 SUBSTANCES WITH A SELF-IGNITION TEMPERATURE OF 200 °C AND BELOW, n.o.s.
		FT1	1228 MERCAPTANS, LIQUID, FLAMMABLE, TOXIC, N.O.S. or
			1228 MERCAPTAN MIXTURE, LIQUID, FLAMMABLE, TOXIC, N.O.S.
			1986 ALCOHOLS, FLAMMABLE, TOXIC, N.O.S.
			1988 ALDEHYDES, FLAMMABLE, TOXIC, N.O.S.
			2478 ISOCYANATES, FLAMMABLE, TOXIC, N.O.S. or
			2478 ISOCYANATE SOLUTION, FLAMMABLE, TOXIC, N.O.S.
			3248 MEDICINE, LIQUID, FLAMMABLE, TOXIC, N.O.S.
			3273 NITRILES, FLAMMABLE, TOXIC, N.O.S.
			1992 FLAMMABLE LIQUID, TOXIC, N.O.S.
Toxic FT			

(cont'd)

Toxic				
FT			2758	CARBAMATE PESTICIDE, LIQUID, FLAMMABLE, TOXIC
			2760	ARSENICAL PESTICIDE, LIQUID, FLAMMABLE, TOXIC
			2762	ORGANOCHLORINE PESTICIDE, LIQUID, FLAMMABLE, TOXIC
			2764	TRIAZINE PESTICIDE, LIQUID, FLAMMABLE, TOXIC
			2772	THIOCARBAMATE PESTICIDE, LIQUID, FLAMMABLE, TOXIC
			2776	COPPER BASED PESTICIDE, LIQUID, FLAMMABLE, TOXIC
			2778	MERCURY BASED PESTICIDE, LIQUID, FLAMMABLE, TOXIC
	Pesticide **(f.p<23 °C)**	**FT2**	2780	SUBSTITUTED NITROPHENOL PESTICIDE, LIQUID, FLAMMABLE, TOXIC
			2782	BIPYRIDILIUM PESTICIDE, LIQUID, FLAMMABLE, TOXIC
			2784	ORGANOPHOSPHORUS PESTICIDE, LIQUID, FLAMMABLE, TOXIC
			2787	ORGANOTIN PESTICIDE, LIQUID, FLAMMABLE, TOXIC
			3024	COUMARIN DERIVATIVE PESTICIDE, LIQUID, FLAMMABLE, TOXIC
			3346	PHENOXYACETIC ACID DERIVATIVE PESTICIDE, LIQUID, FLAMMABLE, TOXIC
			3350	PYRETHROID PESTICIDE, LIQUID, FLAMMABLE TOXIC
			3021	PESTICIDE, LIQUID, FLAMMABLE, TOXIC, N.O.S.

NOTE : The classification of a pesticide under an entry shall be effected on the basis of the active ingredient, of the physical state of the pesticide and any subsidiary risks it may exhibit.

Corrosive			3469	PAINT, FLAMMABLE, CORROSIVE (including paint, lacquer, enamel, stain, shellac, varnish, polish, liquid filler and liquid lacquer base) or
		FC	3469	PAINT RELATED MATERIAL, FLAMMABLE, CORROSIVE (including paint thinning or reducing compound)
			2733	AMINES, FLAMMABLE, CORROSIVE, N.O.S. or
			2733	POLYAMINES, FLAMMABLE, CORROSIVE, N.O.S.
			2985	CHLOROSILANES, FLAMMABLE, CORROSIVE, N.O.S.
			3274	ALCOHOLATES SOLUTION, N.O.S., in alcohol
			2924	FLAMMABLE LIQUID, CORROSIVE, N.O.S.

Toxic, **corrosive**	**FTC**	3286	FLAMMABLE LIQUID, TOXIC, CORROSIVE, N.O.S.

Liquid **desensitised**	**D**	3343	NITROGLYCERIN MIXTURE, DESENSITIZED, LIQUID, FLAMMABLE, N.O.S. with not more than 30% nitroglycerin by mass
Explosive		3357	NITROGLYCERIN MIXTURE, DESENSITIZED, LIQUID, N.O.S. with not more than 30% nitroglycerin by mass
		3379	DESENSITIZED EXPLOSIVE, LIQUID, N.O.S.

2.2.41 **Class 4.1** **Flammable solids, self-reactive substance and solid desensitized explosives**

2.2.41.1 *Criteria*

2.2.41.1.1 The heading of Class 4.1 covers flammable substances and articles, desensitized explosives which are solids according to subparagraph (a) of the definition "solid" in 1.2.1 and self-reactive liquids or solids.

The following are assigned to Class 4.1:

– readily flammable solid substances and articles (see paragraphs 2.2.41.1.3 to 2.2.41.1.8);

– self-reactive solids or liquids (see paragraphs 2.2.41.1.9 to 2.2.41.1.17);

– solid desensitized explosives (see 2.2.41.1.18);

– substances related to self-reactive substances (see 2.2.41.1.19).

2.2.41.1.2 The substances and articles of Class 4.1 are subdivided as follows:

F Flammable solids, without subsidiary risk:

F1 Organic;
F2 Organic, molten;
F3 Inorganic;

FO Flammable solids, oxidizing;

FT Flammable solids, toxic:

FT1 Organic, toxic;
FT2 Inorganic, toxic;

FC Flammable solids, corrosive:

FC1 Organic, corrosive;
FC2 Inorganic, corrosive;

D Solid desensitized explosives without subsidiary risk;

DT Solid desensitized explosives, toxic;

SR Self-reactive substances:

SR1 Not requiring temperature control;
SR2 Requiring temperature control.

Flammable solids

Definition and properties

2.2.41.1.3   *Flammable solids* are readily combustible solids and solids which may cause fire through friction.

Readily combustible solids are powdered, granular, or pasty substances which are dangerous if they can be easily ignited by brief contact with an ignition source, such as a burning match, and if the flame spreads rapidly. The danger may come not only from the fire but also from toxic combustion products. Metal powders are especially dangerous because of the difficulty of extinguishing a fire since normal extinguishing agents such as carbon dioxide or water can increase the hazard.

Classification

2.2.41.1.4   Substances and articles classified as flammable solids of Class 4.1 are listed in Table A of Chapter 3.2. The assignment of organic substances and articles not mentioned by name in Table A of Chapter 3.2 to the relevant entry of sub-section 2.2.41.3 in accordance with the provisions of Chapter 2.1 can be based on experience or on the results of the test procedures in accordance with Part III, sub-section 33.2.1 of the Manual of Tests and Criteria. The assignment of inorganic substances not mentioned by name shall be based on the results of the test procedures in accordance with Part III, sub-section 33.2.1 of the Manual of Tests and Criteria; experience shall also be taken into account when it leads to a more stringent assignment.

2.2.41.1.5   When substances not mentioned by name are assigned to one of the entries listed in 2.2.41.3 on the basis of the test procedures in accordance with the Manual of Tests and Criteria, Part III, sub-section 33.2.1, the following criteria apply:

(a)   With the exception of metal powders or powders of metal alloys, powdery, granular or pasty substances shall be classified as readily flammable substances of Class 4.1 if they can be easily ignited by brief contact with an ignition source (e.g. a burning match), or if, in the event of ignition, the flame spreads rapidly, the burning time is less than 45 seconds for a measured distance of 100 mm or the rate of burning is greater than 2.2 mm/s.

(b)   Metal powders or powders of metal alloys shall be assigned to Class 4.1 if they can be ignited by a flame and the reaction spreads over the whole length of the sample in 10 minutes or less.

Solids which may cause fire through friction shall be classified in Class 4.1 by analogy with existing entries (e.g. matches) or in accordance with any appropriate special provision.

2.2.41.1.6   On the basis of the test procedure in accordance with the Manual of Tests and Criteria, Part III, Section 33.2.1 and the criteria set out in 2.2.41.1.4 and 2.2.41.1.5, it may also be determined whether the nature of a substance mentioned by name is such that the substance is not subject to the provisions for this Class.

2.2.41.1.7   If substances of Class 4.1, as a result of admixtures, come into different categories of risk from those to which the substances mentioned by name in Table A of Chapter 3.2 belong, these mixtures shall be assigned to the entries to which they belong on the basis of their actual degree of danger.

NOTE: *For the classification of solutions and mixtures (such as preparations and wastes), see also 2.1.3.*

Assignment of packing groups

2.2.41.1.8   Flammable solids classified under the various entries in Table A of Chapter 3.2 shall be assigned to packing groups II or III on the basis of test procedures of the Manual of Tests and Criteria, Part III, sub-section 33.2.1, in accordance with the following criteria:

(a)   Readily flammable solids which, when tested, have a burning time of less than 45 seconds over a measured distance of 100 mm shall be assigned to:

Packing group II: if the flame passes the wetted zone;

Packing group III: if the wetted zone stops the flame for at least four minutes;

(b)   Metal powders or powders of metal alloys shall be assigned to:

Packing group II: if, when tested, the reaction spreads over the whole length of the sample in five minutes or less;

Packing group III: if, when tested, the reaction spreads over the whole length of the sample in more than five minutes.

For solids which may cause fire through friction, the packing group shall be assigned by analogy with existing entries or in accordance with any special provision.

Self-reactive substances

Definitions

2.2.41.1.9   For the purposes of ADN, *self-reactive substances* are thermally unstable substances liable to undergo a strongly exothermic decomposition even without participation of oxygen (air). Substances are not considered to be self-reactive substances of Class 4.1, if:

(a)   they are explosives according to the criteria of Class 1;

(b)   they are oxidizing substances according to the classification procedure for Class 5.1 (see 2.2.51.1) except that mixtures of oxidizing substances which contain 5.0% or more of combustible organic substances shall be subjected to the classification procedure defined in Note 2;

(c)   they are organic peroxides according to the criteria of Class 5.2 (see 2.2.52.1);

(d)   their heat of decomposition is less than 300 J/g; or

(e)   their self-accelerating decomposition temperature (SADT) (see NOTE 2 below) is greater than 75 °C for a 50 kg package.

NOTE 1: *The heat of decomposition can be determined using any internationally recognised method e.g. differential scanning calorimetry and adiabatic calorimetry.*

NOTE 2:   *Mixtures of oxidizing substances meeting the criteria of Class 5.1 which contain 5.0% or more of combustible organic substances, which do not meet the criteria mentioned in (a), (c), (d) or (e) above, shall be subjected to the self-reactive substance classification procedure.*

A mixture showing the properties of a self-reactive substance, type B to F, shall be classified as a self-reactive substance of Class 4.1.

A mixture showing the properties of a self-reactive substance, type G, according to the principle given in 20.4.3 (g) of Part II of the Manual of Tests and Criteria shall be considered for classification as a substance of Class 5.1 (see 2.2.51.1).

NOTE 3: *The self-accelerating decomposition temperature (SADT) is the lowest temperature at which self-accelerating decomposition may occur with a substance in the packaging as used during carriage. Requirements for the determination of the SADT are given in the Manual of Tests and Criteria, Part II, Chapter 20 and section 28.4.*

NOTE 4: *Any substance which shows the properties of a self-reactive substance shall be classified as such, even if this substance gives a positive test result according to 2.2.42.1.5 for inclusion in Class 4.2.*

Properties

2.2.41.1.10   The decomposition of self-reactive substances can be initiated by heat, contact with catalytic impurities (e.g. acids, heavy-metal compounds, bases), friction or impact. The rate of decomposition increases with temperature and varies with the substance. Decomposition, particularly if no ignition occurs, may result in the evolution of toxic gases or vapours. For certain self-reactive substances, the temperature shall be controlled. Some self-reactive substances may decompose explosively, particularly if confined. This characteristic may be modified by the addition of diluents or by the use of appropriate packagings. Certain self-reactive substances burn vigorously. Self-reactive substances are, for example, some compounds of the types listed below:

aliphatic azo compounds (-C-N=N-C-);
organic azides (-C-N_3);
diazonium salts (-CN_2^+ Z$^-$);
N-nitroso compounds (-N-N=O); and
aromatic sulphohydrazides (-SO_2-NH-NH_2).

This list is not exhaustive and substances with other reactive groups and some mixtures of substances may have similar properties.

Classification

2.2.41.1.11   Self-reactive substances are classified into seven types according to the degree of danger they present. The types of self-reactive substances range from type A, which is not accepted for carriage in the packaging in which it is tested, to type G, which is not subject to the provisions for self-reactive substances of Class 4.1. The classification of types B to F is directly related to the maximum quantity allowed in one packaging. The principles to be applied for classification as well as the applicable classification procedures, test methods and criteria and an example of a suitable test report are given in Part II of the Manual of Tests and Criteria.

2.2.41.1.12   Self-reactive substances which have already been classified and are already permitted for carriage in packagings are listed in 2.2.41.4, those already permitted for carriage in IBCs are listed in 4.1.4.2 of ADR, packing instruction IBC520 and those already permitted for carriage in tanks according to Chapter 4.2 of ADR are listed in 4.2.5.2 of ADR, portable tank instruction T23. Each permitted substance listed is assigned to a generic entry of Table A of Chapter 3.2 (UN Nos. 3221 to 3240), and appropriate subsidiary risks and remarks providing relevant transport information are given.

The collective entries specify:

–   self-reactive substances types B to F, see 2.2.41.1.11 above;

–   physical state (liquid/solid); and

–   temperature control (when required), see 2.2.41.1.17 below.

The classification of the self-reactive substances listed in 2.2.41.4 is based on the technically pure substance (except where a concentration of less than 100% is specified).

2.2.41.1.13   Classification of self-reactive substances not listed in 2.2.41.4, 4.1.4.2 of ADR, packing instruction IBC520 or 4.2.5.2 of ADR, portable tank instruction T23 and assignment to a collective entry shall be made by the competent authority of the country of origin on the basis of a test report. The statement of approval shall contain the classification and the relevant conditions of carriage. If the country of origin is not a Contracting Party to ADN, the classification and the conditions of carriage shall be recognized by the competent authority of the first country Contracting Party to ADN reached by the consignment.

2.2.41.1.14   Activators, such as zinc compounds, may be added to some self-reactive substances to change their reactivity. Depending on both the type and the concentration of the activator, this may result in a decrease in thermal stability and a change in explosive properties. If either of these properties is altered, the new formulation shall be assessed in accordance with the classification procedure.

2.2.41.1.15   Samples of self-reactive substances or formulations of self-reactive substances not listed in 2.2.41.4, for which a complete set of test results is not available and which are to be carried for further testing or evaluation, shall be assigned to one of the appropriate entries for self-reactive substances type C provided the following conditions are met:

–   the available data indicate that the sample would be no more dangerous than self-reactive substances type B;

–   the sample is packaged in accordance with packing method OP2 of 4.1.4.1 of ADR and the quantity per transport unit is limited to 10 kg;

–   the available data indicate that the control temperature, if any, is sufficiently low to prevent any dangerous decomposition and sufficiently high to prevent any dangerous phase separation.

Desensitization

2.2.41.1.16   In order to ensure safety during carriage, self-reactive substances are in many cases desensitized by use of a diluent. Where a percentage of a substance is stipulated, this refers to the percentage by mass, rounded to the nearest whole number. If a diluent is used, the self-reactive substance shall be tested with the diluent present in the concentration and form used in carriage. Diluents which may allow a self-reactive substance to concentrate to a dangerous extent in the event of leakage from a packaging shall not be used. Any diluent shall be compatible with the self-reactive substance. In this regard, compatible diluents are those solids or liquids which have no detrimental influence on the thermal stability and hazard type of the self-reactive substance. Liquid diluents in formulations requiring temperature control (see 2.2.41.1.14) shall have a boiling point of at least 60 °C and a flash-point not less than 5 °C. The boiling point of the liquid shall be at least 50 °C higher than the control temperature of the self-reactive substance.

Temperature control requirements

2.2.41.1.17   Certain self-reactive substances may only be carried under temperature controlled conditions. The control temperature is the maximum temperature at which the self-reactive substance can be safely carried. It is assumed that the temperature of the immediate surroundings of a package only exceeds 55 °C during carriage for a relatively short time in a 24 hour period. In the event of loss of temperature control, it may be necessary to implement emergency procedures. The emergency temperature is the temperature at which such procedures shall be implemented.

The control and emergency temperatures are derived from the SADT (see table 1). The SADT shall be determined in order to decide whether a substance shall be subjected to temperature control during carriage. Provisions for the determination of the SADT are given in the Manual of Tests and Criteria, Part II, Chapter 20 and Section 28.4.

Table 1   Derivation of control and emergency temperatures

Type of receptacle	SADT [a]	Control temperature	Emergency temperature
Single packagings and IBCs	20 °C or less	20 °C below SADT	10 °C below SADT
	over 20 °C to 35 °C	15 °C below SADT	10 °C below SADT
	over 35 °C	10 °C below SADT	5 °C below SADT
Tanks	not greater than 50 °C	10 °C below SADT	5 °C below SADT

[a]   *SADT of the substance as packaged for carriage.*

Self-reactive substances with an SADT not greater than 55 °C shall be subject to temperature control during carriage. Where applicable, control and emergency temperatures are listed in 2.2.41.4. The actual temperature during carriage may be lower than the control temperature but shall be selected so as to avoid dangerous separation of phases.

Solid desensitized explosives

2.2.41.1.18   Solid desensitized explosives are substances which are wetted with water or alcohols or are diluted with other substances to suppress their explosive properties. Such entries in Table A of Chapter 3.2 are: UN Nos. 1310, 1320, 1321, 1322, 1336, 1337, 1344, 1347, 1348, 1349, 1354, 1355, 1356, 1357, 1517, 1571, 2555, 2556, 2557, 2852, 2907, 3317, 3319, 3344, 3364, 3365, 3366, 3367, 3368, 3369, 3370, 3376 and 3380.

Substances related to self-reactive substances

2.2.41.1.19   Substances that:

(a)   have been provisionally accepted into Class 1 according to Test Series 1 and 2 but exempted from Class 1 by Test Series 6;

(b)   are not self-reactive substances of Class 4.1; and

(c)   are not substances of Classes 5.1 or 5.2

are also assigned to Class 4.1. UN Nos. 2956, 3241, 3242 and 3251 are such entries.

2.2.41.2   Substances not accepted for carriage

2.2.41.2.1   The chemically unstable substances of Class 4.1 shall not be accepted for carriage unless the necessary steps have been taken to prevent their dangerous decomposition or polymerization during carriage. To this end, it shall in particular be ensured that receptacles and tanks do not contain any substance liable to promote these reactions.

2.2.41.2.2   Flammable solids, oxidizing, assigned to UN No. 3097 shall not be accepted for carriage unless they meet the requirements for Class 1 (see also 2.1.3.7).

2.2.41.2.3   The following substances shall not be accepted for carriage:

- Self-reactive substances of type A (see Manual of Tests and Criteria, Part II, paragraph 20.4.2 (a));

- Phosphorus sulphides which are not free from yellow and white phosphorus;

- Solid densitized explosives other than those listed in Table A of Chapter 3.2;

- Inorganic flammable substances in the molten form other than UN No. 2448 SULPHUR, MOLTEN;

- Barium azide with a water content less than 50% (mass).

2.2.41.3 List of collective entries

Flammable solids	**without subsidiary risk**	organic	**F1**	3175 SOLIDS CONTAINING FLAMMABLE LIQUID, N.O.S. 1353 FIBRES IMPREGNATED WITH WEAKLY NITRATED NITROCELLULOSE, N.O.S. or 1353 FABRICS IMPREGNATED WITH WEAKLY NITRATED NITROCELLULOSE, N.O.S. 1325 FLAMMABLE SOLID, ORGANIC, N.O.S.
		organic molten	**F2**	3176 FLAMMABLE SOLID, ORGANIC, MOLTEN, N.O.S.
		inorganic	**F3**	3089 METAL POWDER, FLAMMABLE, N.O.S. [a] [b] 3181 METAL SALTS OF ORGANIC COMPOUNDS, FLAMMABLE, N.O.S. 3182 METAL HYDRIDES, FLAMMABLE, N.O.S. [c] 3178 FLAMMABLE SOLID, INORGANIC, N.O.S.
	oxidizing		**FO**	3097 FLAMMABLE SOLID, OXIDIZING, N.O.S. (not allowed, see Para. 2.2.41.2.2)
	toxic **FT**	organic	**FT1**	2926 FLAMMABLE SOLID, TOXIC, ORGANIC, N.O.S.
		inorganic	**FT2**	3179 FLAMMABLE SOLID, TOXIC, INORGANIC, N.O.S.
	corrosive **FC**	organic	**FC1**	2925 FLAMMABLE SOLID, CORROSIVE, ORGANIC, N.O.S.
		inorganic	**FC2**	3180 FLAMMABLE SOLID, CORROSIVE, INORGANIC, N.O.S.
Solid desensitized explosives	without subsidiary risk		**D**	3319 NITROGLYCERIN MIXTURE, DESENSITIZED, SOLID, N.O.S. with more than 2% but not more than 10% nitroglycerin by mass 3344 PENTAERYTHRITE TETRANITRATE MIXTURE, DESENSITIZED, SOLID, N.O.S. with more than 10% but not more than 20% PETN by mass 3380 DESENSITIZED EXPLOSIVE, SOLID, N.O.S.
	toxic		**DT**	Only those listed in Table A of Chapter 3.2 are to be accepted for carriage as substances of Class 4.1
Self-reactive substances SR	not requiring temperature control		**SR1**	SELF-REACTIVE LIQUID TYPE A ⎫ Not accepted for carriage, SELF-REACTIVE SOLID TYPE A ⎬ see 2.2.41.2.3 3221 SELF-REACTIVE LIQUID TYPE B 3222 SELF-REACTIVE SOLID TYPE B 3223 SELF-REACTIVE LIQUID TYPE C 3224 SELF-REACTIVE SOLID TYPE C 3225 SELF-REACTIVE LIQUID TYPE D 3226 SELF-REACTIVE SOLID TYPE D 3227 SELF-REACTIVE LIQUID TYPE E 3228 SELF-REACTIVE SOLID TYPE E 3229 SELF-REACTIVE LIQUID TYPE F 3230 SELF-REACTIVE SOLID TYPE F SELF-REACTIVE LIQUID TYPE G ⎫ Not subject to the provisions applicable to SELF-REACTIVE SOLID TYPE G ⎬ Class 4.1, see 2.2.41.1.11
	requiring temperature control		**SR2**	3231 SELF-REACTIVE LIQUID TYPE B, TEMPERATURE CONTROLLED 3232 SELF-REACTIVE SOLID TYPE B, TEMPERATURE CONTROLLED 3233 SELF-REACTIVE LIQUID TYPE C, TEMPERATURE CONTROLLED 3234 SELF-REACTIVE SOLID TYPE C, TEMPERATURE CONTROLLED 3235 SELF-REACTIVE LIQUID TYPE D, TEMPERATURE CONTROLLED 3236 SELF-REACTIVE SOLID TYPE D, TEMPERATURE CONTROLLED 3237 SELF-REACTIVE LIQUID TYPE E, TEMPERATURE CONTROLLED 3238 SELF-REACTIVE SOLID TYPE E, TEMPERATURE CONTROLLED 3239 SELF-REACTIVE LIQUID TYPE F, TEMPERATURE CONTROLLED 3240 SELF-REACTIVE SOLID TYPE F, TEMPERATURE CONTROLLED

---

[a]   *Metals and metal alloys in powdered or other flammable form, liable to spontaneous combustion, are substances of Class 4.2.*

[b]   *Metals and metal alloys in powdered or other flammable form, which in contact with water, emit flammable gases, are substances of Class 4.3.*

[c]   *Metals hydrides which, in contact with water, emit flammable gases, are substances of Class 4.3. Aluminium borohydride or aluminium borohydride in devices are substances of Class 4.2, UN No. 2870.*

2.2.41.4 **List of currently assigned self-reactive substances in packagings**

In the column "Packing Method" codes "OP1" to "OP8" refer to packing methods in 4.1.4.1 of ADR, packing instruction P520 (see also 4.1.7.1 of ADR). Self-reactive substances to be carried shall fulfil the classification and the control and emergency temperatures (derived from the SADT) as listed. For substances permitted in IBCs, see 4.1.4.2 of ADR, packing instruction IBC520 and, for those permitted in tanks according Chapter 4.2 of ADR, see 4.2.5.2 of ADR, portable tank instruction T23.

NOTE: The classification given in this table is based on the technically pure substance (except where a concentration of less than 100 % is specified). For other concentrations, the substance may be classified differently following the procedures given in Part II of the Manual of Tests and Criteria and in 2.2.41.1.17.

SELF-REACTIVE SUBSTANCE	Concen-tration (%)	Packing method	Control temperature (°C)	Emergency temperature (°C)	UN generic entry	Remarks
ACETONE-PYROGALLOL COPOLYMER 2-DIAZO-1-NAPHTHOL-5-SULPHONATE	100	OP8			3228	
AZODICARBONAMIDE FORMULATION TYPE B, TEMPERATURE CONTROLLED	< 100	OP5			3232	(1) (2)
AZODICARBONAMIDE FORMULATION TYPE C	< 100	OP6			3224	(3)
AZODICARBONAMIDE FORMULATION TYPE C, TEMPERATURE CONTROLLED	< 100	OP6			3234	(4)
AZODICARBONAMIDE FORMULATION TYPE D	< 100	OP7			3226	(5)
AZODICARBONAMIDE FORMULATION TYPE D, TEMPERATURE CONTROLLED	< 100	OP7			3236	(6)
2,2' -AZODI(2,4-DIMETHYL-4-METHOXYVALERONITRILE)	100	OP7	-5	+5	3236	
2,2' -AZODI(2,4-DIMETHYL-VALERONITRILE)	100	OP7	+10	+15	3236	
2,2' -AZODI(ETHYL-2-METHYLPROPIONATE)	100	OP7	+20	+25	3235	
1,1-AZODI(HEXAHYDROBENZONITRILE)	100	OP7			3226	
2,2' -AZODI(ISOBUTYRONITRILE	100	OP6	+40	+45	3234	
2,2' -AZODI(ISOBUTYRONITRILE) as a water based paste	≤ 50	OP6			3224	
2,2' -AZODI(2-METHYLBUTYRO-NITRILE)	100	OP7	+35	+40	3236	
BENZENE-1,3-DISULPHONYL HYDRAZIDE, as a paste	52	OP7			3226	
BENZENE SULPHONYL HYDRAZIDE	100	OP7			3226	
4-(BENZYL(ETHYL)AMINO)-3-ETHOXY-BENZENEDIAZONIUM ZINC CHLORIDE	100	OP7			3226	
4-(BENZYL(METHYL)AMINO)-3-ETHOXYBENZENEDIAZONIUM ZINC CHLORIDE	100	OP7	+40	+45	3236	
3-CHLORO-4-DIETHYLAMINOBENZENE-DIAZONIUM ZINC CHLORIDE	100	OP7			3226	

SELF-REACTIVE SUBSTANCE	Concen-tration (%)	Packing method	Control temperature (°C)	Emergency temperature (°C)	UN generic entry	Remarks
2-DIAZO-1-NAPHTHOL-4-SULPHONYL CHLORIDE	100	OP5			3222	(2)
2-DIAZO-1-NAPHTHOL-5-SULPHONYL CHLORIDE	100	OP5			3222	(2)
2-DIAZO-1-NAPHTHOL SULPHONIC ACID ESTER MIXTURE, TYPE D	< 100	OP7			3226	(9)
2,5-DIBUTOXY-4-(4-MORPHOLINYL)-BENZENEDIAZONIUM, TETRACHLOROZINCATE (2:1)	100	OP8			3228	
2,5-DIETHOXY-4-MORPHOLINO-BENZENEDIAZONIUM ZINC CHLORIDE	67-100	OP7	+35	+40	3236	
2,5-DIETHOXY-4-MORPHOLINO-BENZENEDIAZONIUM ZINC CHLORIDE	66	OP7	+40	+45	3236	
2,5-DIETHOXY-4-MORPHOLINO-BENZENEDIAZONIUM TETRAFLUOROBORATE	100	OP7	+30	+35	3236	
2,5-DIETHOXY-4-(4-MORPHOLINYL)-BENZENEDIAZONIUM SULPHATE	100	OP7			3226	
2,5-DIETHOXY-4-(PHENYLSULPHONYL)-BENZENEDIAZONIUM ZINC CHLORIDE	67	OP7	+40	+45	3236	
DIETHYLENEGLYCOL BIS (ALLYL CARBONATE) + DI-ISOPROPYLPEROXYDICARBONATE	≥ 88 + ≤ 12	OP8	-10	0	3237	
2,5-DIMETHOXY-4-(4-METHYL-PHENYLSULPHONYL)BENZENE-DIAZONIUM ZINC CHLORIDE	79	OP7	+40	+45	3236	
4-(DIMETHYLAMINO)-BENZENEDIAZONIUM TRICHLOROZINCATE (-1)	100	OP8			3228	
4-DIMETHYLAMINO-6-(2-DIMETHYL-AMINOETHOXY) TOLUENE-2-DIAZONIUM ZINC CHLORIDE	100	OP7	+40	+45	3236	
N,N'-DINITROSO-N,N'- DIMETHYL TEREPHTHALAMIDE, as a paste	72	OP6			3224	
N,N'-DINITROSOPENTAMETHYLENE-TETRAMINE	82	OP6			3224	(7)
DIPHENYLOXIDE-4,4'-DISULPHONYL HYDRAZIDE	100	OP7			3226	
4-DIPROPYLAMINOBENZENE-DIAZONIUM ZINC CHLORIDE	100	OP7			3226	
2-(N,N-ETHOXYCARBONYL-PHENYLAMINO)-3-METHOXY-4-(N-METHYL-N-CYCLOHEXYLAMINO) BENZENEDIAZONIUM ZINC CHLORIDE	63-92	OP7	+ 40	+ 45	3236	
2-(N,N-ETHOXYCARBONYL-PHENYLAMINO)-3-METHOXY-4-(N-METHYL-N-CYCLOHEXYLAMINO) BENZENEDIAZONIUM ZINC CHLORIDE	62	OP7	+ 35	+ 40	3236	

SELF-REACTIVE SUBSTANCE	Concen-tration (%)	Packing method	Control temperature (°C)	Emergency temperature (°C)	UN generic entry	Remarks
N-FORMYL-2-(NITROMETHYLENE)-1,3-PERHYDROTHIAZINE	100	OP7	+45	+50	3236	
2-(2-HYDROXYETHOXY)-1-(PYRROLIDIN-1-YL)BENZENE-4-DIAZONIUM ZINC CHLORIDE	100	OP7	+ 45	+ 50	3236	
3-(2-HYDROXYETHOXY)-4-(PYRROLIDIN-1-YL)BENZENE DIAZONIUM ZINC CHLORIDE	100	OP7	+40	+45	3236	
2-(N,N-METHYLAMINOETHYL-CARBONYL)-4-(3,4-DIMETHYL-PHENYLSULPHONYL)BENZENE-DIAZONIUM HYDROGEN SULPHATE	96	OP7	+45	+50	3236	
4-METHYLBENZENESULPHONYL-HYDRAZIDE	100	OP7			3226	
3-METHYL-4-(PYRROLIDIN-1-YL)BENZENEDIAZONIUM TETRAFLUOROBORATE	95	OP6	+45	+50	3234	
4-NITROSOPHENOL	100	OP7	+35	+40	3236	
SELF-REACTIVE LIQUID, SAMPLE		OP2			3223	(8)
SELF-REACTIVE LIQUID, SAMPLE, TEMPERATURE CONTROLLED		OP2			3233	(8)
SELF-REACTIVE SOLID, SAMPLE		OP2			3224	(8)
SELF-REACTIVE SOLID, SAMPLE, TEMPERATURE CONTROLLED		OP2			3234	(8)
SODIUM 2-DIAZO-1-NAPHTHOL-4-SULPHONATE	100	OP7			3226	
SODIUM 2-DIAZO-1-NAPHTHOL-5-SULPHONATE	100	OP7			3226	
TETRAMINE PALLADIUM (II) NITRATE	100	OP6	+30	+35	3234	

Remarks

(1)   Azodicarbonamide formulations which fulfil the criteria of paragraph 20.4.2 (b) of the Manual of Tests and Criteria. The control and emergency temperatures shall be determined by the procedure given in 2.2.41.1.17.

(2)   "EXPLOSIVE" subsidiary risk label required (Model No. 1, see 5.2.2.2.2).

(3)   Azodicarbonamide formulations which fulfil the criteria of paragraph 20.4.2 (c) of the Manual of Tests and Criteria.

(4)   Azodicarbonamide formulations which fulfil the criteria of paragraph 20.4.2 (c) of the Manual of Tests and Criteria. The control and emergency temperatures shall be determined by the procedure given in 2.2.41.1.17.

(5)   Azodicarbonamide formulations which fulfil the criteria of paragraph 20.4.2 (d) of the Manual of Tests and Criteria.

(6)   Azodicarbonamide formulations which fulfil the criteria of paragraph 20.4.2 (d) of the Manual of Tests and Criteria. The control and emergency temperatures shall be determined by the procedure given in 2.2.41.1.17.

(7)   With a compatible diluent having a boiling point of not less than 150 °C.

(8)   See 2.2.41.1.15.

(9)   This entry applies to mixtures of esters of 2-diazo-1-naphthol-4-sulphonic acid and 2-diazo-1-naphthol-5-sulphonic acid which fulfil the criteria of paragraph 20.4.2 (d) of the *Manual of Test and Criteria.*

2.2.42   **Class 4.2   Substances liable to spontaneous combustion**

2.2.42.1   *Criteria*

2.2.42.1.1   The heading of Class 4.2 covers:

–   *Pyrophoric substances* which are substances, including mixtures and solutions (liquid or solid), which even in small quantities ignite on contact with air within five minutes. These are the Class 4.2 substances the most liable to spontaneous combustion; and

–   *Self-heating substances and articles* which are substances and articles, including mixtures and solutions, which, on contact with air, without energy supply, are liable to self-heating. These substances will ignite only in large amounts (kilogrammes) and after long periods of time (hours or days).

2.2.42.1.2   The substances and articles of Class 4.2 are subdivided as follows:

S   Substances liable to spontaneous combustion, without subsidiary risk:

S1   Organic, liquid;
S2   Organic, solid;
S3   Inorganic, liquid;
S4   Inorganic, solid;
S5   Organometallic;

SW   Substances liable to spontaneous combustion, which, in contact with water, emit flammable gases;

SO   Substances liable to spontaneous combustion, oxidizing;

ST   Substances liable to spontaneous combustion, toxic:

ST1   Organic, toxic, liquid;
ST2   Organic, toxic, solid;
ST3   Inorganic, toxic, liquid;
ST4   Inorganic, toxic, solid;

SC   Substances liable to spontaneous combustion, corrosive:

SC1   Organic, corrosive, liquid;
SC2   Organic, corrosive, solid;
SC3   Inorganic, corrosive, liquid;
SC4   Inorganic, corrosive, solid.

Properties

2.2.42.1.3   Self-heating of these substances, leading to spontaneous combustion, is caused by reaction of the substance with oxygen (in the air) and the heat developed not being conducted away rapidly enough to the surroundings. Spontaneous combustion occurs when the rate of heat production exceeds the rate of heat loss and the auto-ignition temperature is reached.

Classification

2.2.42.1.4   Substances and articles classified in Class 4.2 are listed in Table A of Chapter 3.2. The assignment of substances and articles not mentioned by name in Table A of Chapter 3.2 to the relevant specific N.O.S. entry of 2.2.42.3 in accordance with the provisions of Chapter 2.1 can be based on experience or the results of the test procedures in accordance with the Manual of Tests and Criteria, Part III, Section 33.3. Assignment to general N.O.S. entries of Class 4.2 shall be based on the results of the test procedures in accordance with the Manual of Tests and Criteria, Part III, section 33.3; experience shall also be taken into account when it leads to a more stringent assignment.

2.2.42.1.5   When substances or articles not mentioned by name are assigned to one of the entries listed in 2.2.42.3 on the basis of the test procedures in accordance with the Manual of Tests and Criteria, Part III, section 33.3, the following criteria shall apply:

(a)   Solids liable to spontaneous combustion (pyrophoric) shall be assigned to Class 4.2 when they ignite on falling from a height of 1 m or within five minutes;

(b)   Liquids liable to spontaneous combustion (pyrophoric) shall be assigned to Class 4.2 when:

(i)   on being poured on an inert carrier, they ignite within five minutes, or

(ii)   in the event of a negative result of the test according to (i), when poured on a dry, indented filter paper (Whatman No. 3 filter), they ignite or carbonize it within five minutes;

(c)   Substances in which, in a 10 cm sample cube, at 140 °C test temperature, spontaneous combustion or a rise in temperature to over 200 °C is observed within 24 hours shall be assigned to Class 4.2. This criterion is based on the temperature of the spontaneous combustion of charcoal, which is at 50 °C for a sample cube of 27 m^3. Substances with a temperature of spontaneous combustion higher than 50 °C for a volume of 27 m^3 are not to be assigned to Class 4.2.

NOTE 1: Substances carried in packages with a volume of not more than 3 m^3 are exempted from Class 4.2 if, tested with a 10 cm sample cube at 120 °C, no spontaneous combustion nor a rise in temperature to over 180 °C is observed within 24 hours.

NOTE 2: Substances carried in packages with a volume of not more than 450 litres are exempted from Class 4.2 if, tested with a 10 cm sample cube at 100 °C, no spontaneous combustion nor a rise in temperature to over 160 °C is observed within 24 hours.

NOTE 3: Since organometallic substances can be classified in Class 4.2 or 4.3 with additional subsidiary risks, depending on their properties, a specific classification flow chart for these substances is given in 2.3.6.

2.2.42.1.6   If substances of Class 4.2, as a result of admixtures, come into different categories of risk from those to which the substances mentioned by name in Table A of Chapter 3.2 belong, these mixtures shall be assigned to the entries to which they belong on the basis of their actual degree of danger.

NOTE: For the classification of solutions and mixtures (such as preparations and wastes), see also 2.1.3.

2.2.42.1.7   On the basis of the test procedure in the Manual of Tests and Criteria, Part III, section 33.3 and the criteria set out in 2.2.42.1.5, it may also be determined whether the nature of a substance mentioned by name is such that the substance is not subject to the provisions for this Class.

Assignment of packing groups

2.2.42.1.8   Substances and articles classified under the various entries in Table A of Chapter 3.2 shall be assigned to packing groups I, II or III on the basis of test procedures of the Manual of Tests and Criteria, Part III, section 33.3, in accordance with the following criteria:

(a)   Substances liable to spontaneous combustion (pyrophoric) shall be assigned to packing group I;

(b)   Self-heating substances and articles in which, in a 2.5 cm sample cube, at 140 °C test temperature, spontaneous combustion or a rise in temperature to over 200 °C is observed within 24 hours, shall be assigned to packing group II;
Substances with a temperature of spontaneous combustion higher than 50 °C for a volume of 450 litres are not to be assigned to packing group II;

(c)   Slightly self-heating substances in which, in a 2.5 cm sample cube, the phenomena referred to under (b) are not observed, in the given conditions, but in which in a 10 cm sample cube at 140 °C test temperature spontaneous combustion or a rise in temperature to over 200 °C is observed within 24 hours, shall be assigned to packing group III.

2.2.42.2   ***Substances not accepted for carriage***

The following substances shall not be accepted for carriage:

–   UN No. 3255 tert-BUTYL HYPOCHLORITE; and

–   Self-heating solids, oxidizing, assigned to UN No. 3127 unless they meet the requirements for Class 1 (see 2.1.3.7).

2.2.42.3 List of collective entries

Substances liable to spontaneous combustion		**liquid**	**S1**	2845 3183	PYROPHORIC LIQUID, ORGANIC, N.O.S. SELF-HEATING LIQUID, ORGANIC, N.O.S.
Without subsidiary risk **S**	**organic**	**solid**	**S2**	1373 2006 3313 2846 3088	FIBRES or FABRICS, ANIMAL or VEGETABLE or SYNTHETIC, N.O.S. with oil PLASTICS, NITROCELLULOSE-BASED, SELF-HEATING, N.O.S. ORGANIC PIGMENTS, SELF HEATING PYROPHORIC SOLID, ORGANIC, N.O.S. SELF-HEATING SOLID, ORGANIC, N.O.S.
		liquid	**S3**	3194 3186	PYROPHORIC LIQUID, INORGANIC, N.O.S. SELF-HEATING LIQUID, INORGANIC, N.O.S.
	inorganic	**solid**	**S4**	1383 1383 1378 2881 3189[a] 3205 3200 3190	PYROPHORIC METAL, N.O.S. or PYROPHORIC ALLOY, N.O.S. METAL CATALYST, WETTED with a visible excess of liquid METAL CATALYST, DRY METAL POWDER, SELF-HEATING, N.O.S. ALKALINE EARTH METAL ALCOHOLATES, N.O.S. PYROPHORIC SOLID, INORGANIC, N.O.S. SELF-HEATING SOLID, INORGANIC, N.O.S.
	organo-metallic		**S5**	3392 3391 3400	ORGANOMETALLIC SUBSTANCE, LIQUID, PYROPHORIC ORGANOMETALLIC SUBSTANCE, SOLID, PYROPHORIC ORGANOMETALLIC SUBSTANCE, SOLID, SELF-HEATING
Water-reactive			**SW**	3394 3393	ORGANOMETALLIC SUBSTANCE, LIQUID, PYROPHORIC, WATER-REACTIVE ORGANOMETALLIC SUBSTANCE, SOLID, PYROPHORIC, WATER-REACTIVE
Oxidizing			**SO**	3127	SELF-HEATING SOLID, OXIDIZING, N.O.S. (not allowed, see 2.2.42.2)
Toxic **ST**	**organic**	**liquid**	**ST1**	3184	SELF-HEATING LIQUID, TOXIC, ORGANIC, N.O.S.
		solid	**ST2**	3128	SELF-HEATING SOLID, TOXIC, ORGANIC, N.O.S.
	inorganic	**liquid**	**ST3**	3187	SELF-HEATING LIQUID, TOXIC, INORGANIC, N.O.S.
		solid	**ST4**	3191	SELF-HEATING SOLID, TOXIC, INORGANIC, N.O.S.
Corrosive **SC**	**organic**	**liquid**	**SC1**	3185	SELF-HEATING LIQUID, CORROSIVE, ORGANIC, N.O.S.
		solid	**SC2**	3126	SELF-HEATING SOLID, CORROSIVE, ORGANIC, N.O.S.
	inorganic	**liquid**	**SC3**	3188	SELF-HEATING LIQUID, CORROSIVE, INORGANIC, N.O.S.
		solid	**SC4**	3206 3192	ALKALI METAL ALCOHOLATES, SELF-HEATING, CORROSIVE, N.O.S. SELF-HEATING SOLID, CORROSIVE, INORGANIC, N.O.S.

---

[a] Dust and powder of metals, non toxic in a non-spontaneous combustible form which nevertheless, in contact with water, emit flammable gases, are substances of Class 4.3.

2.2.43   **Class 4.3   Substances which, in contact with water, emit flammable gases**

2.2.43.1   *Criteria*

2.2.43.1.1   The heading of Class 4.3 covers substances which react with water to emit flammable gases liable to form explosive mixtures with air, and articles containing such substances.

2.2.43.1.2   Substances and articles of Class 4.3 are subdivided as follows:

W   Substances which, in contact with water, emit flammable gases, without subsidiary risk, and articles containing such substances:

   W1   Liquid;
   W2   Solid;
   W3   Articles;

WF1   Substances which, in contact with water, emit flammable gases, liquid, flammable;

WF2   Substances which, in contact with water, emit flammable gases, solid, flammable;

WS   Substances which, in contact with water, emit flammable gases, solid, self-heating;

WO   Substances which, in contact with water, emit flammable gases, oxidizing, solid;

WT   Substances which, in contact with water, emit flammable gases, toxic:

   WT1   Liquid;
   WT2   Solid;

WC   Substances which, in contact with water, emit flammable gases, corrosive:

   WC1   Liquid;
   WC2   Solid;

WFC   Substances which, in contact with water, emit flammable gases, flammable, corrosive.

Properties

2.2.43.1.3   Certain substances in contact with water may emit flammable gases that can form explosive mixtures with air. Such mixtures are easily ignited by all ordinary sources of ignition, for example naked lights, sparking handtools or unprotected light bulbs. The resulting blast wave and flames may endanger people and the environment. The test method referred to in 2.2.43.1.4 below is used to determine whether the reaction of a substance with water leads to the development of a dangerous amount of gases which may be flammable. This test method shall not be applied to pyrophoric substances.

Classification

2.2.43.1.4   Substances and articles classified in Class 4.3 are listed in Table A of Chapter 3.2. The assignment of substances and articles not mentioned by name in Table A of Chapter 3.2 to the relevant entry of 2.2.43.3 in accordance with the provisions of Chapter 2.1 shall be based on the results of the test procedure in accordance with the Manual of Tests and Criteria, Part III, Section 33.4; experience shall also be taken into account when it leads to a more stringent assignment.

2.2.43.1.5   When substances not mentioned by name are assigned to one of the entries listed in 2.2.43.3 on the basis of the test procedure in accordance with the Manual of Tests and Criteria, Part III, Section 33.4, the following criteria shall apply:

A substance shall be assigned to Class 4.3 if:

(a)   spontaneous ignition of the gas emitted takes place in any step of the test procedure; or

(b)   there is an evolution of flammable gas at a rate greater than 1 litre per kilogram of the substance to be tested per hour.

NOTE: Since organometallic substances can be classified in Classes 4.2 or 4.3 with additional subsidiary risks, depending on their properties, a specific classification flow chart for these substances is given in 2.3.6.

2.2.43.1.6   If substances of Class 4.3, as a result of admixtures, come into different categories of risk from those to which the substances mentioned by name in Table A of Chapter 3.2 belong, these mixtures shall be assigned to the entries to which they belong on the basis of their actual degree of danger.

NOTE: For the classification of solutions and mixtures (such as preparations and wastes) see also 2.1.3.

2.2.43.1.7   On the basis of the test procedures in accordance with the Manual of Tests and Criteria, Part III, Section 33.4, and the criteria set out in paragraph 2.2.43.1.5, it may also be determined whether the nature of a substance mentioned by name is such that the substance is not subject to the provisions for this Class.

Assignment of packing groups

2.2.43.1.8   Substances and articles classified under the various entries in Table A of Chapter 3.2 shall be assigned to packing groups I, II or III on the basis of test procedures of the Manual of Tests and Criteria, Part III, section 33.4, in accordance with the following criteria:

(a)   Packing group I shall be assigned to any substance which reacts vigorously with water at ambient temperature and generally demonstrates a tendency for the gas produced to ignite spontaneously, or one which reacts readily with water at ambient temperatures such that the rate of evolution of flammable gas is equal to or greater than 10 litres per kilogram of substance over any one minute period;

(b)   Packing group II shall be assigned to any substance which reacts readily with water at ambient temperature such that the maximum rate of evolution of flammable gas is equal to or greater than 20 litres per kilogram of substance per hour, and which does not meet the criteria of packing group I;

(c)   Packing group III shall be assigned to any substance which reacts slowly with water at ambient temperature such that the maximum rate of evolution of flammable gas is greater than 1 litre per kilogram of substance per hour, and which does not meet the criteria of packing groups I or II.

2.2.43.2   *Substances not accepted for carriage*

Water-reactive solids, flammable, assigned to UN No. 3132, water-reactive solids, oxidizing, assigned to UN No. 3133 and water-reactive solids, self-heating, assigned to UN No.3135 shall not be accepted for carriage unless they meet the requirements for Class 1 (see also 2.1.3.7).

2.2.43.3   *List of collective entries*

Substances which, in contact with water, emit flammable gases				

				1389 ALKALI METAL AMALGAM, LIQUID 1391 ALKALI METAL DISPERSION having a flash-point above 60 °C or 1391 ALKALINE EARTH METAL DISPERSION having a flash-point above 60 °C 1392 ALKALINE EARTH METAL AMALGAM, LIQUID 1420 POTASSIUM METAL ALLOYS, LIQUID 1421 ALKALI METAL ALLOY, LIQUID, N.O.S. 1422 POTASSIUM SODIUM ALLOYS, LIQUID 3398 ORGANOMETALLIC SUBSTANCE, LIQUID, WATER-REACTIVE 3148 WATER-REACTIVE LIQUID, N.O.S.
		liquid	W1	
Without subsidiary risk		solid	W2 ^a	1390 ALKALI METAL AMIDES 3401 ALKALI METAL AMALGAM, SOLID
W				3402 ALKALINE EARTH METAL AMALGAM, SOLID 3170 ALUMINIUM SMELTING BY-PRODUCTS or 3170 ALUMINIUM REMELTING BY-PRODUCTS 3403 POTASSIUM METAL ALLOYS, SOLID 3404 POTASSIUM SODIUM ALLOYS, SOLID 1393 ALKALINE EARTH METAL ALLOY, N.O.S. 1409 METAL HYDRIDES, WATER-REACTIVE, N.O.S. 3208 METALLIC SUBSTANCE, WATER-REACTIVE, N.O.S. 3395 ORGANOMETALLIC SUBSTANCE, SOLID, WATER-REACTIVE 2813 WATER-REACTIVE SOLID, N.O.S.
		articles	W3	3292 BATTERIES, CONTAINING SODIUM or 3292 CELLS, CONTAINING SODIUM
Liquid, flammable			WF1	1391 ALKALI METAL DISPERSION having a flash-point of not more than 60 °C or 1391 ALKALINE EARTH METAL DISPERSION having a flash-point of not more than 60 °C 3399 ORGANOMETALLIC SUBSTANCE, LIQUID, WATER-REACTIVE, FLAMMABLE
Solid, flammable			WF2	3396 ORGANOMETALLIC SUBSTANCE, SOLID, WATER-REACTIVE, FLAMMABLE 3132 WATER-REACTIVE SOLID, FLAMMABLE, N.O.S. (not allowed, see 2.2.43.2)
Solid, self-heating			WS ^b	3397 ORGANOMETALLIC SUBSTANCE, SOLID, WATER-REACTIVE, SELF-HEATING 3209 METALLIC SUBSTANCE, WATER-REACTIVE, SELF-HEATING, N.O.S. 3135 WATER-REACTIVE SOLID, SELF-HEATING, N.O.S. (not allowed, see 2.2.43.2)
Solid, oxidizing			WO	3133 WATER-REACTIVE SOLID, OXIDIZING, N.O.S. (not allowed, see 2.2.43.2)
Toxic		liquid	WT1	3130 WATER-REACTIVE LIQUID, TOXIC, N.O.S.
WT		solid	WT2	3134 WATER-REACTIVE SOLID, TOXIC, N.O.S.
Corrosive		liquid	WC1	3129 WATER-REACTIVE LIQUID, CORROSIVE, N.O.S.
WC		solid	WC2	3131 WATER-REACTIVE SOLID, CORROSIVE, N.O.S.
Flammable, corrosive			WFC ^c	2988 CHLOROSILANES, WATER-REACTIVE, FLAMMABLE, CORROSIVE, NO.S. (No other collective entry with this classification code available, if need be, classification under a collective entry with a classification code to be determined according to the table of precedence of hazard in 2.1.3.10.)

---

^a   *Metals and metal alloys which, in contact with water, do not emit flammable gases and are not pyrophoric or self-heating, but which are readily flammable, are substances of Class 4.1. Alkaline-earth metals and alkaline-earth metal alloys in pyrophoric form are substances of Class 4.2. Dust and powders of metals in pyrophoric form are substances of Class 4.2. Metals and metal alloys in pyrophoric form are substances of Class 4.2. Compounds of phosphorus with heavy metals such as iron, copper, etc. are not subject to the provisions of ADN.*

^b   *Metals and metal alloys in pyrophoric form are substances of Class 4.2.*

^c   *Chlorosilanes, having a flash-point of less than 23 °C and which, in contact with water, do not emit flammable gases, are substances of Class 3. Chlorosilanes, having a flash-point equal to or greater than 23 °C and which, in contact with water, do not emit flammable gases, are substances of Class 8.*

2.2.51   **Class 5.1 Oxidizing substances**

2.2.51.1   *Criteria*

2.2.51.1.1   The heading of Class 5.1 covers substances which, while in themselves not necessarily combustible, may, generally by yielding oxygen, cause or contribute to the combustion of other material, and articles containing such substances.

2.2.51.1.2   The substances of Class 5.1 and articles containing such substances are subdivided as follows:

        O   Oxidizing substances without subsidiary risk or articles containing such substances:

            O1   Liquid;
            O2   Solid;
            O3   Articles;

        OF   Oxidizing substances, solid, flammable;

        OS   Oxidizing substances, solid, self-heating;

        OW Oxidizing substances, solid which, in contact with water, emit flammable gases;

        OT   Oxidizing substances, toxic:

            OT1 Liquid;
            OT2 Solid;

        OC   Oxidizing substances, corrosive:

            OC1 Liquid;
            OC2 Solid;

        OTC Oxidizing substances, toxic, corrosive.

2.2.51.1.3   Substances and articles classified in Class 5.1 are listed in Table A of Chapter 3.2. The assignment of substances and articles not mentioned by name in Table A of Chapter 3.2 to the relevant entry of 2.2.51.3 in accordance with the provisions of Chapter 2.1 can be based on the tests, methods and criteria in paragraphs 2.2.51.1.6 to 2.2.51.1.9 below and the Manual of Tests and Criteria, Part III, Section 34.4. In the event of divergence between test results and known experience, judgement based on known experience shall take precedence over test results.

2.2.51.1.4   If substances of Class 5.1, as a result of admixtures, come into different categories of risk from those to which the substances mentioned by name in Table A of Chapter 3.2 belong, these mixtures or solutions shall be assigned to the entries to which they belong on the basis of their actual degree of danger.

          NOTE: For the classification of solutions and mixtures (such as preparations and wastes), see also Section 2.1.3.

2.2.51.1.5   On the basis of the test procedures in the Manual of Tests and Criteria, Part III, Section 34.4 and the criteria set out in 2.2.51.1.6-2.2.51.1.9 it may also be determined whether the nature of a substance mentioned by name in Table A of Chapter 3.2 is such that the substance is not subject to the provisions for this class.

Oxidizing solids

Classification

2.2.51.1.6   When oxidizing solid substances not mentioned by name in Table A of Chapter 3.2 are assigned to one of the entries listed in 2.2.51.3 on the basis of the test procedure in accordance with the Manual of Tests and Criteria, Part III, sub-section 34.4.1, the following criteria shall apply:

A solid substance shall be assigned to Class 5.1 if, in the 4:1 or the 1:1 sample-to-cellulose ratio (by mass) tested, it ignites or burns or exhibits mean burning times equal to or less than that of a 3:7 mixture (by mass) of potassium bromate and cellulose.

Assignment of packing groups

2.2.51.1.7   Oxidizing solids classified under the various entries in Table A of Chapter 3.2 shall be assigned to packing groups I, II or III on the basis of test procedures of the Manual of Tests and Criteria, Part III, sub-section 34.4.1, in accordance with the following criteria:

(a)   Packing group I: any substance which, in the 4:1 or 1:1 sample-to-cellulose ratio (by mass) tested, exhibits a mean burning time less than the mean burning time of a 3:2 mixture, by mass, of potassium bromate and cellulose;

(b)   Packing group II: any substance which, in the 4:1 or 1:1 sample-to-cellulose ratio (by mass) tested, exhibits a mean burning time equal to or less than the mean burning time of a 2:3 mixture (by mass) of potassium bromate and cellulose and the criteria for packing group I are not met;

(c)   Packing group III: any substance which, in the 4:1 or 1:1 sample-to-cellulose ratio (by mass) tested, exhibits a mean burning time equal to or less than the mean burning time of a 3:7 mixture (by mass) of potassium bromate and cellulose and the criteria for packing groups I and II are not met.

Oxidizing liquids

Classification

2.2.51.1.8   When oxidizing liquid substances not mentioned by name in Table A of Chapter 3.2 are assigned to one of the entries listed in sub-section 2.2.51.3 on the basis of the test procedure in accordance with the Manual of Tests and Criteria, Part III, sub-section 34.4.2, the following criteria shall apply:

A liquid substance shall be assigned to Class 5.1 if, in the 1:1 mixture, by mass, of substance and cellulose tested, it exhibits a pressure rise of 2070 kPa gauge or more and a mean pressure rise time equal to or less than the mean pressure rise time of a 1:1 mixture, by mass, of 65% aqueous nitric acid and cellulose.

Assignment of packing groups

2.2.51.1.9   Oxidizing liquids classified under the various entries in Table A of Chapter 3.2 shall be assigned to packing groups I, II or III on the basis of test procedures of the Manual of Tests and Criteria, Part III, section 34.4.2, in accordance with the following criteria:

(a)   Packing group I: any substance which, in the 1:1 mixture, by mass, of substance and cellulose tested, spontaneously ignites; or the mean pressure rise time of a 1:1 mixture, by mass, of substance and cellulose is less than that of a 1:1 mixture, by mass, of 50% perchloric acid and cellulose;

(b)   Packing group II: any substance which, in the 1:1 mixture, by mass, of substance and cellulose tested, exhibits a mean pressure rise time less than or equal to the mean pressure rise time of a 1:1 mixture, by mass, of 40% aqueous sodium chlorate solution and cellulose; and the criteria for packing group I are not met;

(c)   Packing group III: any substance which, in the 1:1 mixture, by mass, of substance and cellulose tested, exhibits a mean pressure rise time less than or equal to the mean pressure rise time of a 1:1 mixture, by mass, of 65% aqueous nitric acid and cellulose; and the criteria for packing groups I and II are not met.

2.2.51.2   *Substances not accepted for carriage*

2.2.51.2.1   The chemically unstable substances of Class 5.1 shall not be accepted for carriage unless the necessary steps have been taken to prevent their dangerous decomposition or polymerization during carriage. To this end it shall in particular be ensured that receptacles and tanks do not contain any material liable to promote these reactions.

2.2.51.2.2   The following substances and mixtures shall not be accepted for carriage:

–   oxidizing solids, self-heating, assigned to UN No. 3100, oxidizing solids, water-reactive, assigned to UN No. 3121 and oxidizing solids, flammable, assigned to UN No. 3137, unless they meet the requirements for Class 1 (see also 2.1.3.7);

–   hydrogen peroxide, not stabilized or hydrogen peroxide, aqueous solutions, not stabilized containing more than 60 % hydrogen peroxide;

–   tetranitromethane not free from combustible impurities;

–   perchloric acid solutions containing more than 72 % (mass) acid, or mixtures of perchloric acid with any liquid other than water;

–   chloric acid solution containing more than 10 % chloric acid or mixtures of chloric acid with any liquid other than water;

–   halogenated fluor compounds other than UN Nos. 1745 BROMINE PENTAFLUORIDE; 1746 BROMINE TRIFLUORIDE and 2495 IODINE PENTAFLUORIDE of Class 5.1 as well as UN Nos. 1749 CHLORINE TRIFLUORIDE and 2548 CHLORINE PENTAFLUORIDE of Class 2;

–   ammonium chlorate and its aqueous solutions and mixtures of a chlorate with an ammonium salt;

–   ammonium chlorite and its aqueous solutions and mixtures of a chlorite with an ammonium salt;

–   mixtures of a hypochlorite with an ammonium salt;

–   ammonium bromate and its aqueous solutions and mixtures of a bromate with an ammonium salt;

–　ammonium permanganate and its aqueous solutions and mixtures of a permanganate with an ammonium salt;

–　ammonium nitrate containing more than 0.2 % combustible substances (including any organic substance calculated as carbon) unless it is a constituent of a substance or article of Class 1;

–　fertilizers having an ammonium nitrate content (in determining the ammonium nitrate content, all nitrate ions for which a molecular equivalent of ammonium ions is present in the mixture shall be calculated as ammonium nitrate) or a content in combustible substances exceeding the values specified in special provision 307 except under the conditions applicable to Class 1;

–　ammonium nitrite and its aqueous solutions and mixtures of an inorganic nitrite with an ammonium salt;

–　mixtures of potassium nitrate, sodium nitrite and an ammonium salt.

2.2.51.3   *List of collective entries*

Oxidizing substances	liquid	O1	3210 CHLORATES, INORGANIC, AQUEOUS SOLUTION, N.O.S.
			3211 PERCHLORATES, INORGANIC, AQUEOUS SOLUTION, N.O.S.
			3213 BROMATES, INORGANIC, AQUEOUS SOLUTION, N.O.S.
			3214 PERMANGANATES, INORGANIC, AQUEOUS SOLUTION, N.O.S.
			3216 PERSULPHATES, INORGANIC, AQUEOUS SOLUTION, N.O.S.
			3218 NITRATES, INORGANIC, AQUEOUS SOLUTION, N.O.S.
			3219 NITRITES, INORGANIC, AQUEOUS SOLUTION, N.O.S.
			3139 OXIDIZING LIQUID, N.O.S.

Without subsidiary risk **O**	solid	O2	1450 BROMATES, INORGANIC, N.O.S
			1461 CHLORATES, INORGANIC, N.O.S.
			1462 CHLORITES, INORGANIC, N.O.S.
			1477 NITRATES, INORGANIC, N.O.S
			1481 PERCHLORATES, INORGANIC, N.O.S.
			1482 PERMANGANATES, INORGANIC, N.O.S.
			1483 PEROXIDES, INORGANIC, N.O.S
			2627 NITRITES, INORGANIC, N.O.S.
			3212 HYPOCHLORITES, INORGANIC, N.O.S.
			3215 PERSULPHATES, INORGANIC, N.O.S.
			1479 OXIDIZING SOLID, N.O.S.

	articles	O3	3356 OXYGEN GENERATOR, CHEMICAL
Solid, flammable		OF	3137 OXIDIZING SOLID, FLAMMABLE, N.O.S. (not allowed, see 2.2.51.2)
Solid, self-heating		OS	3100 OXIDIZING SOLID, SELF-HEATING, N.O.S. (not allowed, see 2.2.51.2)
Solid, water reactive		OW	3121 OXIDIZING SOLID, WATER REACTIVE, N.O.S. (not allowed, see 2.2.51.2)
Toxic **OT**	liquid	OT1	3099 OXIDIZING LIQUID, TOXIC, N.O.S.
	solid	OT2	3087 OXIDIZING SOLID, TOXIC, N.O.S.
Corrosive **OC**	liquid	OC1	3098 OXIDIZING LIQUID, CORROSIVE, N.O.S.
	solid	OC2	3085 OXIDIZING SOLID, CORROSIVE, N.O.S.
Toxic, corrosive		OTC	(No collective entry with this classification code available; if need be, classification under a collective entry with a classification code to be determined according to the table of precedence of hazard in 2.1.3.10.)

2.2.52 **Class 5.2** **Organic peroxides**

2.2.52.1 *Criteria*

2.2.52.1.1   The heading of Class 5.2 covers organic peroxides and formulations of organic peroxides.

2.2.52.1.2   The substances of Class 5.2 are subdivided as follows:

P1   Organic peroxides, not requiring temperature control;
P2   Organic peroxides, requiring temperature control.

Definition

2.2.52.1.3   *Organic peroxides* are organic substances which contain the bivalent -O-O- structure and may be considered derivatives of hydrogen peroxide, where one or both of the hydrogen atoms have been replaced by organic radicals.

Properties

2.2.52.1.4   Organic peroxides are liable to exothermic decomposition at normal or elevated temperatures. The decomposition can be initiated by heat, contact with impurities (e.g. acids, heavy-metal compounds, amines), friction or impact. The rate of decomposition increases with temperature and varies with the organic peroxide formulation. Decomposition may result in the evolution of harmful, or flammable, gases or vapours. For certain organic peroxides the temperature shall be controlled during carriage. Some organic peroxides may decompose explosively, particularly if confined. This characteristic may be modified by the addition of diluents or by the use of appropriate packagings. Many organic peroxides burn vigorously. Contact of organic peroxides with the eyes is to be avoided. Some organic peroxides will cause serious injury to the cornea, even after brief contact, or will be corrosive to the skin.

NOTE: Test methods for determining the flammability of organic peroxides are set out in the Manual of Tests and Criteria, Part III, sub-section 32.4. Because organic peroxides may react vigorously when heated, it is recommended to determine their flash-point using small sample sizes such as described in ISO 3679:1983.

Classification

2.2.52.1.5   Any organic peroxide shall be considered for classification in Class 5.2 unless the organic peroxide formulation contains:

(a)   not more than 1.0 % available oxygen from the organic peroxides when containing not more than 1.0 % hydrogen peroxide;

(b)   not more than 0.5 % available oxygen from the organic peroxides when containing more than 1.0 % but not more than 7.0 % hydrogen peroxide.

NOTE: The available oxygen content (%) of an organic peroxide formulation is given by the formula

$$16 \times 3 \, (n_i \times c_i / m_i)$$

where:

n_i =   *number of peroxygen groups per molecule of organic peroxide i;*
c_i =   *concentration (mass %) of organic peroxide i; and*
m_i =   *molecular mass of organic peroxide i.*

2.2.52.1.6   Organic peroxides are classified into seven types according to the degree of danger they present. The types of organic peroxide range from type A, which is not accepted for carriage in the packaging in which it is tested, to type G, which is not subject to the provisions of Class 5.2. The classification of types B to F is directly related to the maximum quantity allowed in one packaging. The principles to be applied to the classification of substances not listed in 2.2.52.4 are set out in the Manual of Tests and Criteria, Part II.

2.2.52.1.7   Organic peroxides which have already been classified and are already permitted for carriage in packagings are listed in 2.2.52.4, those already permitted for carriage in IBCs are listed in 4.1.4.2 of ADR, packing instruction IBC520 and those already permitted for carriage in tanks in accordance with Chapters 4.2 and 4.3 are listed in 4.2.5.2 of ADR, portable tank instruction T23. Each permitted substance listed is assigned to a generic entry of Table A of Chapter 3.2 (UN Nos. 3101 to 3120) and appropriate subsidiary risks and remarks providing relevant transport information are given.

These generic entries specify:

–   the type (B to F) of organic peroxide (see 2.2.52.1.6 above);

–   physical state (liquid/solid); and

–   temperature control (when required), see 2.2.52.1.15 to 2.2.52.1.18.

Mixtures of these formulations may be classified as the same type of organic peroxide as that of the most dangerous component and be carried under the conditions of carriage given for this type. However, as two stable components can form a thermally less stable mixture, the self-accelerating decomposition temperature (SADT) of the mixture shall be determined and, if necessary, the control and emergency temperatures derived from the SADT in accordance paragraph 2.2.52.1.16.

2.2.52.1.8   Classification of organic peroxides, formulations or mixtures of organic peroxides not listed in 2.2.52.4, 4.1.4.2 of ADR, packing instruction IBC520 or 4.2.5.2 of ADR, portable tank instruction T23, and assignment to a collective entry shall be made by the competent authority of the country of origin. The statement of approval shall contain the classification and the relevant conditions of carriage. If the country of origin is not a Contracting Party to ADN, the classification and conditions of carriage shall be recognized by the competent authority of the first country Contracting Party to ADN reached by the consignment.

2.2.52.1.9   Samples of organic peroxides or formulations of organic peroxides not listed in 2.2.52.4, for which a complete set of test results is not available and which are to be carried for further testing or evaluation, shall be assigned to one of the appropriate entries for organic peroxides type C provided the following conditions are met:

–   the available data indicate that the sample would be no more dangerous than organic peroxides type B;

–   the sample is packaged in accordance with packing method OP2 of 4.1.4.1 of ADR and the quantity per transport unit is limited to 10 kg;

–   the available data indicate that the control temperature, if any, is sufficiently low to prevent any dangerous decomposition and sufficiently high to prevent any dangerous phase separation.

Desensitization of organic peroxides

2.2.52.1.10   In order to ensure safety during carriage, organic peroxides are in many cases desensitized by organic liquids or solids, inorganic solids or water. Where a percentage of a substance is stipulated, this refers to the percentage by mass, rounded to the nearest whole number. In general, desensitization shall be such that, in case of spillage, the organic peroxide will not concentrate to a dangerous extent.

2.2.52.1.11   Unless otherwise stated for the individual organic peroxide formulation, the following definition(s) shall apply to diluents used for desensitization:

–   diluents type A are organic liquids which are compatible with the organic peroxide and which have a boiling point of not less than 150 °C. Type A diluents may be used for desensitizing all organic peroxides.

–   diluents type B are organic liquids which are compatible with the organic peroxide and which have a boiling point of less than 150 °C but not less than 60 °C and a flash-point of not less than 5 °C.

Type B diluents may be used for desensitization of all organic peroxides provided that the boiling point of the liquid is at least 60 °C higher than the SADT in a 50 kg package.

2.1.52.1.12   Diluents, other than type A or type B, may be added to organic peroxide formulations as listed in 2.2.52.4 provided that they are compatible. However, replacement of all or part of a type A or type B diluent by another diluent with differing properties requires that the organic peroxide formulation be re-assessed in accordance with the normal acceptance procedure for Class 5.2.

2.2.52.1.13   Water may only be used for the desensitization of organic peroxides which are listed in 2.2.52.4 or in the competent authority decision according to 2.2.52.1.8 as being "with water" or "as a stable dispersion in water". Samples of organic peroxides or formulations of organic peroxides not listed in 2.2.52.4 may also be desensitized with water provided the requirements of 2.2.52.1.9 are met.

2.2.52.1.14   Organic and inorganic solids may be used for desensitization of organic peroxides provided that they are compatible. Compatible liquids and solids are those which have no detrimental influence on the thermal stability and hazard type of the organic peroxide formulation.

Temperature control requirements

2.2.52.1.15   Certain organic peroxides may only be carried under temperature-controlled conditions. The control temperature is the maximum temperature at which the organic peroxide can be safely carried. It is assumed that the temperature of the immediate surroundings of a package only exceeds 55 °C during carriage for a relatively short time in a 24 hour period. In the event of loss of temperature control, it may be necessary to implement emergency procedures. The emergency temperature is the temperature at which such procedures shall be implemented.

2.2.52.1.16   The control and emergency temperatures are derived from the SADT which is defined as the lowest temperature at which self-accelerating decomposition may occur with a substance in the packaging as used during carriage (see Table 1). The SADT shall be determined in order to decide whether a substance shall be subjected to temperature control during carriage. Provisions for the determination of the SADT are given in the Manual of Tests and Criteria, Part II, Sections 20 and 28.4.

Table 1: Derivation of control and emergency temperatures

Type of receptacle	SADT [a]	Control temperature	Emergency temperature
Single packagings and IBCs	20 °C or less	20 °C below SADT	10 °C below SADT
	over 20 °C to 35 °C	15 °C below SADT	10 °C below SADT
	over 35 °C	10 °C below SADT	5 °C below SADT
Tanks	Not greater than 50 °C	10 °C below SADT	5 °C below SADT

[a]   *SADT of the substance as packaged for carriage.*

2.2.52.1.17   The following organic peroxides shall be subject to temperature control during carriage:

–   organic peroxides types B and C with an SADT ≤ 50 °C;

–   organic peroxides type D showing a medium effect when heated under confinement with an SADT ≤ 50 °C or showing a low or no effect when heated under confinement with an SADT ≤ 45 °C; and

–   organic peroxides types E and F with an SADT ≤ 45 °C.

NOTE: Provisions for the determination of the effects of heating under confinement are given in the Manual of Tests and Criteria, Part II, Section 20 and Sub-section 28.4.

2.2.52.1.18   Where applicable, control and emergency temperatures are listed in 2.2.52.4. The actual temperature during carriage may be lower than the control temperature but shall be selected so as to avoid dangerous separation of phases.

2.2.52.2   *Substances not accepted for carriage*

Organic peroxides, type A, shall not be accepted for carriage under the provisions of Class 5.2 (see Manual of Tests and Criteria, Part II, paragraph 20.4.3 (a)).

2.2.52.3 *List of collective entries*

Organic peroxides			ORGANIC PEROXIDE TYPE A, LIQUID	} Not accepted for carriage,
			ORGANIC PEROXIDE TYPE A, SOLID	} see 2.2.52.2
		3101	ORGANIC PEROXIDE TYPE B, LIQUID	
		3102	ORGANIC PEROXIDE TYPE B, SOLID	
		3103	ORGANIC PEROXIDE TYPE C, LIQUID	
		3104	ORGANIC PEROXIDE TYPE C, SOLID	
Not requiring temperature control	**P1**	3105	ORGANIC PEROXIDE TYPE D, LIQUID	
		3106	ORGANIC PEROXIDE TYPE D, SOLID	
		3107	ORGANIC PEROXIDE TYPE E, LIQUID	
		3108	ORGANIC PEROXIDE TYPE E, SOLID	
		3109	ORGANIC PEROXIDE TYPE F, LIQUID	
		3110	ORGANIC PEROXIDE TYPE F, SOLID	
			ORGANIC PEROXIDE TYPE G, LIQUID	} Not subject to the provisions
			ORGANIC PEROXIDE TYPE G, SOLID	} applicable to Class 5.2, see 2.2.52.1.6

		3111	ORGANIC PEROXIDE TYPE B, LIQUID, TEMPERATURE CONTROLLED
		3112	ORGANIC PEROXIDE TYPE B, SOLID, TEMPERATURE CONTROLLED
		3113	ORGANIC PEROXIDE TYPE C, LIQUID, TEMPERATURE CONTROLLED
		3114	ORGANIC PEROXIDE TYPE C, SOLID, TEMPERATURE CONTROLLED
Requiring temperature control	**P2**	3115	ORGANIC PEROXIDE TYPE D, LIQUID, TEMPERATURE CONTROLLED
		3116	ORGANIC PEROXIDE TYPE D, SOLID, TEMPERATURE CONTROLLED
		3117	ORGANIC PEROXIDE TYPE E, LIQUID, TEMPERATURE CONTROLLED
		3118	ORGANIC PEROXIDE TYPE E, SOLID, TEMPERATURE CONTROLLED
		3119	ORGANIC PEROXIDE TYPE F, LIQUID, TEMPERATURE CONTROLLED
		3120	ORGANIC PEROXIDE TYPE F, SOLID, TEMPERATURE CONTROLLED

2.2.52.4 *List of currently assigned organic peroxides in packagings*

In the column "Packing Method", codes "OP1" to "OP8" refer to packing methods in 4.1.4.1 of ADR packing instruction P520 (see also 4.1.7.1 of ADR). Organic peroxides to be carried shall fulfill the classification and he control and emergency temperatures (derived from the SADT) as listed. For substances permitted in IBCs, see 4.1.4.2 of ADR, packing instruction IBC520 and, for those permitted in tanks according to Chapters 4.2 and 4.3 of ADR, see 4.2.5.2 of ADR, portable tank instruction T23.

ORGANIC PEROXIDE	Concentration (%)	Diluent type A (%)	Diluent type B 1) (%)	Inert solid (%)	Water	Packing Method	Control temperature (°C)	Emergency temperature (°C)	Number (Generic entry)	Subsidiary risks and remarks
ACETYL ACETONE PEROXIDE	≤ 42	≥ 48			≥ 8	OP7			3105	2)
"	≤ 32 as a paste					OP7			3106	20)
ACETYL CYCLOHEXANESULPHONYL PEROXIDE	≤ 82				≥ 12	OP4	-10	0	3112	3)
"	≤ 32		≥ 68			OP7	-10	0	3115	
tert-AMYL HYDROPEROXIDE	≤ 88	≥ 6			≥ 6	OP8			3107	
tert-AMYL PEROXYACETATE	≤ 62	≥ 38				OP8			3107	
tert-AMYL PEROXYBENZOATE	≤ 100					OP5			3103	
tert-AMYL PEROXY-2-ETHYLHEXANOATE	≤ 100					OP7	+20	+25	3115	
tert-AMYL PEROXY-2-ETHYLHEXYL CARBONATE	≤ 100					OP7			3105	
tert-AMYLPEROXY ISOPROPYL CARBONATE	≤ 77	≥ 23				OP5			3103	
tert-AMYL PEROXYNEODECANOATE	≤ 77		≥ 23			OP7	0	+10	3115	
tert-AMYL PEROXYPIVALATE	≤ 77		≥ 23			OP5	+10	+15	3113	
tert-AMYLPEROXY-3,5,5-TRIMETHYLHEXANOATE	≤ 100					OP5			3101	3)
tert-BUTYL CUMYL PEROXIDE	> 42 - 100					OP8			3107	
"	≤ 52			≥ 48		OP8			3108	
n-BUTYL-4,4-DI-(tert-BUTYLPEROXY)VALERATE	> 52 - 100					OP5			3103	
"	≤ 52			≥ 48		OP8			3108	
tert-BUTYL HYDROPEROXIDE	>79 - 90				≥ 10	OP5			3103	13)
"	≤ 80	≥ 20				OP7			3105	4) 13)
"	≤ 79				> 14	OP8			3107	13) 23)
"	≤ 72				≥ 28	OP8			3109	13)
tert-BUTYL HYDROPEROXIDE + DI-tert-BUTYLPEROXIDE	< 82 + > 9				≥ 7	OP5			3103	13)
tert-BUTYL MONOPEROXYMALEATE	> 52 - 100					OP5			3102	3)
"	≤ 52	≥ 48				OP6			3103	
"	≤ 52			≥ 48		OP8			3108	
"	≤ 52 as a paste					OP8			3108	
tert-BUTYL PEROXYACETATE	> 52 - 77	≥ 23				OP5			3101	3)
"	> 32 - 52	≥ 48				OP6			3103	
"	≤ 32		≥ 68			OP8			3109	

ORGANIC PEROXIDE	Concentration (%)	Diluent type A (%)	Diluent type B 1) (%)	Inert solid (%)	Water	Packing Method	Control temperature (°C)	Emergency temperature (°C)	Number (Generic entry)	Subsidiary risks and remarks
tert-BUTYL PEROXYBENZOATE	> 77 - 100					OP5			3103	
"	> 52 - 77	≥ 23				OP7			3105	
"	≤ 52			≥ 48		OP7			3106	
tert-BUTYL PEROXYBUTYL FUMARATE	≤ 52	≥ 48				OP7			3105	
1,6-Di-(tert-BUTYLPEROXY-CARBONYLOXY) HEXANE	≤ 72	≥ 28				OP5			3103	
tert-BUTYL PEROXYCROTONATE	≤ 77	≥ 23				OP7			3105	
tert-BUTYL PEROXYDIETHYLACETATE	≤ 100					OP5	+20	+25	3113	
tert-BUTYL PEROXY-2-ETHYLHEXANOATE	> 52 - 100					OP6	+20	+25	3113	
"	> 32 - 52		≥ 48			OP8	+30	+35	3117	
"	≤ 52			≥ 48		OP8	+20	+25	3118	
"	≤ 32		≥ 68			OP8	+40	+45	3119	
tert-BUTYL PEROXY-2-ETHYLHEXANOATE + 2,2-DI-(tert-BUTYLPEROXY)BUTANE	≥ 12 + ≤ 14	≥ 14		≥ 60		OP7			3106	
"	≤ 31 + ≤ 36		≥ 33			OP7	+35	+40	3115	
tert-BUTYL PEROXY-2-ETHYLHEXYLCARBONATE	≤ 100					OP7			3105	
tert-BUTYL PEROXYISOBUTYRATE	> 52 - 77		≥ 23			OP5	+15	+20	3111	3)
"	≤ 52		≥ 48			OP7	+15	+20	3115	
tert-BUTYLPEROXY ISOPROPYLCARBONATE	≤ 77	≥ 23				OP5			3103	
1-(2-tert-BUTYLPEROXY ISOPROPYL)-3-ISOPROPENYLBENZENE	≤ 77	≥ 23				OP7			3105	
"	≤ 42			≥ 58		OP8			3108	
"	≤ 100					OP5			3103	
tert-BUTYL PEROXY-2-METHYLBENZOATE	> 77 - 100					OP7	-5	+5	3115	
tert-BUTYL PEROXYNEODECANOATE	≤ 77	≥ 23				OP7	0	+10	3115	
"	≤ 52 as a stable dispersion in water					OP8	0	+10	3119	
"	≤ 42 as a stable dispersion in water					OP8	0	+10	3118	
"	≤ 32	≥ 68				OP8	0	+10	3119	
tert-BUTYL PEROXYNEOHEPTANOATE	≤ 77	≥ 23				OP7	0	+10	3115	
"	≤ 42 as a stable dispersion in water					OP8	0	+10	3117	

ORGANIC PEROXIDE	Concentration (%)	Diluent type A (%)	Diluent type B 1) (%)	Inert solid (%)	Water	Packing Method	Control temperature (°C)	Emergency temperature (°C)	Number (Generic entry)	Subsidiary risks and remarks
tert-BUTYL PEROXYPIVALATE	> 67 - 77	≥ 23				OP5	0	+10	3113	
"	> 27 - 67		≥ 33			OP7	0	+10	3115	
"	≤ 27		≥ 73			OP8	+30	+35	3119	
tert-BUTYLPEROXY STEARYLCARBONATE	≤ 100					OP7			3106	
tert-BUTYL PEROXY-3,5,5-TRIMETHYLHEXANOATE	> 32 - 100					OP7			3105	
"	≤ 32		≥ 68			OP8			3109	
3-CHLOROPEROXYBENZOIC ACID	> 57 - 86			≥ 14		OP1			3102	3)
"	≤ 57			≥ 3	≥ 40	OP7			3106	
"	≤ 77			≥ 6	≥ 17	OP7			3106	
CUMYL HYDROPEROXIDE	> 90 - 98	≤ 10				OP8			3107	13)
"	≤ 90	≥ 10				OP8			3109	13) 18)
CUMYL PEROXYNEODECANOATE	≤ 77		≥ 23			OP7	-10	0	3115	
"	≤ 52 as a stable dispersion in water					OP8	-10	0	3119	
CUMYL PEROXYNEOHEPTANOATE	≤ 77	≥ 23				OP7	-10	0	3115	
CUMYL PEROXYPIVALATE	≤ 77		≥ 23			OP7	-5	+5	3115	
CYCLOHEXANONE PEROXIDE(S)	≤ 91				≥ 9	OP6			3104	13)
"	≤ 72	≥ 28				OP7			3105	5)
"	≤ 72 as a paste					OP7			3106	5) 20)
"	≤ 32			≥ 68					Exempt	29)
DIACETONE ALCOHOL PEROXIDES	≤ 57		≥ 26		≥ 8	OP7	+40	+45	3115	6)
DIACETYL PEROXIDE	≤ 27		≥ 73			OP7	+20	+25	3115	7) 13)
DI-tert-AMYL PEROXIDE	≤ 100					OP8			3107	
1,1-DI-(tert-AMYLPEROXY)CYCLOHEXANE	≤ 82	≥ 18				OP6			3103	

ORGANIC PEROXIDE	Concentration (%)	Diluent type A (%)	Diluent type B 1) (%)	Inert solid (%)	Water (%)	Packing Method	Control temperature (°C)	Emergency temperature (°C)	Number (Generic entry)	Subsidiary risks and remarks
DIBENZOYL PEROXIDE	> 51 - 100			≤ 48		OP2			3102	3)
"	> 77 - 94				≥ 6	OP4			3102	3)
"	≤ 77				≥ 23	OP6			3104	
"	≤ 62			≥ 28	≥ 10	OP7			3106	20)
"	> 52 – 62 as a paste					OP7			3106	
"	> 35 - 52			≥ 48		OP7			3106	
"	> 36 - 42	≥ 18			≤ 40	OP8			3107	
"	≤ 56.5 as a paste				≥ 15	OP8			3108	20)
"	≤ 52 as a paste					OP8			3108	20)
"	≤ 42 as a stable dispersion in water					OP8			3109	
"	≤ 35			≥ 65					Exempt	29)
DI-(4-tert-BUTYLCYCLOHEXYL) PEROXYDICARBONATE	≤ 100					OP6	+30	+35	3114	
"	≤ 42 as a stable dispersion in water					OP8	+30	+35	3119	25)
DI-tert-BUTYL PEROXIDE	> 52 - 100					OP8			3107	
"	≤ 52		≥ 48			OP8			3109	
DI-tert-BUTYL PEROXYAZELATE	≤ 52	≥ 48				OP7			3105	
2,2-DI-(tert-BUTYLPEROXY)BUTANE	≤ 52	≥ 48				OP6			3103	
1,1-DI-(tert-BUTYLPEROXY) CYCLOHEXANE	> 80 - 100					OP5			3101	3)
"	> 52 - 80	≥ 20				OP5			3103	
"	> 42 - 52	≥ 48				OP7			3105	
"	≤ 42	≥ 13		≥ 45		OP7			3106	
"	≤ 27	≥ 25				OP8			3107	21)
"	≤ 42	≥ 58				OP8			3109	
"	≤ 13	≥ 13				OP8			3109	
DI-n-BUTYL PEROXYDICARBONATE	> 27 - 52		≥ 74			OP7	-15	-5	3115	
"	≤ 27		≥ 48			OP8	-10	0	3117	
"	≤ 42 as a stable dispersion in water (frozen)		≥ 73			OP8	-15	-5	3118	

ORGANIC PEROXIDE	Concentration (%)	Diluent type A (%)	Diluent type B 1) (%)	Inert solid (%)	Water	Packing Method	Control temperature (°C)	Emergency temperature (°C)	Number (Generic entry)	Subsidiary risks and remarks
DI-sec-BUTYL PEROXYDICARBONATE	> 52 - 100					OP4	-20	-10	3113	
"	≤ 52		≥ 48			OP7	-15	-5	3115	
DI-(2-tert-BUTYLPEROXYISOPROPYL)BENZENE(S)	> 42 - 100			≤ 57		OP7			3106	29)
"	≤ 42			≥ 58					Exempt	
DI-(tert-BUTYLPEROXY) PHTHALATE	> 42 - 52	≥ 48				OP7			3105	
"	≤ 52 as a paste					OP7			3106	20)
2,2-DI-(tert-BUTYLPEROXY)PROPANE	≤ 42	≥ 58				OP8			3107	
"	≤ 52	≥ 48				OP7			3105	
"	≤ 42	≥ 13		≥ 45		OP7			3106	
1,1-DI-(tert-BUTYLPEROXY)-3,3,5-TRIMETHYLCYCLOHEXANE	> 90 - 100					OP5			3101	3)
"	> 57 - 90	≥ 10				OP5			3103	
"	≤ 77		≥ 23			OP5			3103	
"	≤ 57			≥ 43		OP8			3110	
"	≤ 57	≥ 43				OP8			3107	
"	≤ 32	≥ 26	≥ 42			OP8			3107	
DICETYL PEROXYDICARBONATE	≤ 100					OP7	+30	+35	3116	
"	≤ 42 as a stable dispersion in water					OP8	+30	+35	3119	
DI-4-CHLOROBENZOYL PEROXIDE	≤ 77				≥ 23	OP5			3102	3)
"	≤ 52 as a paste					OP7			3106	20)
"	≤ 32			≥ 68					Exempt	29)
DICUMYL PEROXIDE	> 52 - 100			≤ 57		OP8			3110	12)
"	≤ 52			≥ 48					Exempt	29)
DICYCLOHEXYL PEROXYDICARBONATE	> 91 - 100					OP3	+10	+15	3112	3)
"	≤ 91					OP5	+5	+10	3114	
"	≤ 42 as a stable dispersion in water				≥ 9	OP8	+15	+20	3119	
DIDECANOYL PEROXIDE	≤ 100					OP6	+30	+35	3114	
2,2-DI-(4,4-DI (tert-BUTYLPEROXY) CYCLOHEXYL) PROPANE	≤ 42			≥ 58		OP7			3106	
DI-2,4-DICHLOROBENZOYL PEROXIDE	≤ 22		≥ 78			OP8			3107	
"	≤ 77				≥ 23	OP5			3102	3)
"	≤ 52 as a paste with silicon oil					OP7			3106	
DI-(2-ETHOXYETHYL) PEROXYDICARBONATE	≤ 52		≥ 48			OP7	-10	0	3115	
1-(2-ETHYLHEXANOYL-PEROXY)-1,3-DIMETHYLBUTYL PEROXYPIVALATE	≤ 52	≥ 45	≥ 10			OP7	-20	-10	3115	

ORGANIC PEROXIDE	Concentration (%)	Diluent type A (%)	Diluent type B 1) (%)	Inert solid (%)	Water	Packing Method	Control temperature (°C)	Emergency temperature (°C)	Number (Generic entry)	Subsidiary risks and remarks
DI-(2-ETHYLHEXYL) PEROXYDICARBONATE	> 77 - 100					OP5	-20	-10	3113	
"	≤ 77		≥ 23			OP7	-15	-5	3115	
"	≤ 62 as a stable dispersion in water					OP8	-15	-5	3117	
"	≤ 52 as a stable dispersion in water					OP8	-15	-5	3119	
"	≤ 52 as a stable dispersion in water (frozen)					OP8	-15	-5	3120	
2,2-DIHYDROPEROXYPROPANE	≤ 27			≥ 73		OP5			3102	3)
DI-(1-HYDROXYCYCLOHEXYL) PEROXIDE	≤ 100					OP7			3106	
DIISOBUTYRYL PEROXIDE	> 32 - 52		≥ 48			OP5	-20	-10	3111	3)
"	≤ 32		≥ 68			OP7	-20	-10	3115	
DI-ISOPROPYLBENZENE DIHYDROPEROXIDE	≤ 82	≥ 5			≥ 5	OP7			3106	24)
DIISOPROPYL PEROXYDICARBONATE	> 52-100					OP2	-15	-5	3112	3)
"	≤ 52		≥ 48			OP7	-20	-10	3115	
"	≤ 28	≥ 72				OP7	-15	-5	3115	
DILAUROYL PEROXIDE	≤ 100					OP7			3106	
"	≤ 42 as a stable dispersion in water					OP8			3109	
DI-(3-METHOXYBUTYL) PEROXYDICARBONATE	≤ 52		≥ 48			OP7	-5	+5	3115	
DI-(2-METHYLBENZOYL) PEROXIDE	≤ 87				≥ 13	OP7	+30	+35	3112	3)
DI-(3-METHYLBENZOYL) PEROXIDE + BENZOYL (3-METHYLBENZOYL) PEROXIDE + DIBENZOYL PEROXIDE	≤ 20 + ≤ 18 + ≤ 4		≥ 58			OP7	+35	+40	3115	
DI-(4-METHYLBENZOYL) PEROXIDE	≤ 52 as a paste with silicon oil					OP7			3106	
2,5-DIMETHYL-2,5-DI-(BENZOYLPEROXY)HEXANE	> 82-100			≥ 18		OP5			3102	3)
"	≤ 82					OP7			3106	
"	≤ 82				≥ 18	OP5			3104	

ORGANIC PEROXIDE	Concentration (%)	Diluent type A (%)	Diluent type B 1) (%)	Inert solid (%)	Water	Packing Method	Control temperature (°C)	Emergency temperature (°C)	Number (Generic entry)	Subsidiary risks and remarks
2,5-DIMETHYL-2,5-DI-(tert-BUTYLPEROXY)HEXANE	> 52 – 100					OP7			3105	
"	≤ 47 as a paste					OP8			3108	
"	≤ 52	≥ 48				OP8			3109	
"	≤ 77			≥ 23		OP8			3108	
2,5-DIMETHYL-2,5-DI-(tert-BUTYLPEROXY)HEXYNE-3	>52-86	≥ 14				OP5			3103	26)
"	≤ 52			≥ 48		OP7			3106	
"	> 86-100					OP5			3101	3)
2,5-DIMETHYL-2,5-DI-(2-ETHYLHEXANOYLPEROXY)HEXANE	≤ 100					OP5	+20	+25	3113	
2,5-DIMETHYL-2,5-DIHYDROPEROXYHEXANE	≤ 82				≥ 18	OP6			3104	
2,5-DIMETHYL-2,5-DI-(3,5,5-TRIMETHYLHEXANOYLPEROXY)HEXANE	≤ 77	≥ 23				OP7			3105	
1,1-DIMETHYL-3-HYDROXYBUTYL PEROXYNEOHEPTANOATE	≤ 52	≥ 48				OP8	0	+10	3117	
DIMYRISTYL PEROXYDICARBONATE	≤ 100					OP7	+20	+25	3116	
"	≤ 42 as a stable dispersion in water					OP8	+20	+25	3119	
DI-(2-NEODECANOYLPEROXYISOPROPYL) BENZENE	≤ 52	≥ 48				OP7	-10	0	3115	
DI-n-NONANOYL PEROXIDE	≤ 100					OP7	0	+10	3116	
DI-n-OCTANOYL PEROXIDE	≤ 100					OP5	+10	+15	3114	
DI-(2-PHENOXYETHYL) PEROXYDICARBONATE	>85-100				≥ 15	OP5			3102	3)
"	≤ 85					OP7			3106	
DIPROPIONYL PEROXIDE	≤ 27		≥ 73			OP8	+15	+20	3117	
DI-n-PROPYL PEROXYDICARBONATE	≤ 100					OP3	-25	-15	3113	
"	≤ 77		≥ 23			OP5	-20	-10	3113	

ORGANIC PEROXIDE	Concentration (%)	Diluent type A (%)	Diluent type B 1) (%)	Inert solid (%)	Water	Packing Method	Control temperature (°C)	Emergency temperature (°C)	Number (Generic entry)	Subsidiary risks and remarks
DISUCCINIC ACID PEROXIDE	> 72-100					OP4			3102	3) 17)
"	≤ 72				≥ 28	OP7	+10	+15	3116	
DI-(3,5,5-TRIMETHYLHEXANOYL) PEROXIDE	> 38-82	≥ 18				OP7	0	+10	3115	
"	≤ 52 as a stable dispersion in water					OP8	+10	+15	3119	
"	≤ 38	≥ 62				OP8	+20	+25	3119	
ETHYL 3.3-DI-(tert-AMYLPEROXY)BUTYRATE	≤ 67	≥ 33				OP7			3105	
ETHYL 3.3-DI-(tert-BUTYLPEROXY)BUTYRATE	> 77 - 100					OP5			3103	
"	≤ 77	≥ 23				OP7			3105	
"	≤ 52			≥ 48		OP7			3106	
"	≤ 52	≥ 48				OP7			3105	
"	≤ 52			≥ 48		OP7			3106	
tert-HEXYL PEROXYNEODECANOATE	≤ 71	≥ 29				OP7	0	+10	3115	
tert-HEXYL PEROXYPIVALATE	≤ 72		≥ 28			OP7	+10	+15	3115	
ISOPROPYL sec-BUTYL PEROXYDICARBONATE +DI-sec-BUTYL PEROXYDICARBONATE +DI-ISOPROPYL PEROXYDICARBONATE	≤ 32 + ≤ 15 - 18 ≤ 12 - 15	≥ 38				OP7	-20	-10	3115	
ISOPROPYL sec-BUTYL PEROXYDICARBONATE + DI-sec-BUTYL PEROXYDICARBONATE + DI-ISOPROPYL PEROXYDICARBONATE	≤ 52 + ≤ 28 + ≤ 22					OP5	-20	-10	3111	3)
ISOPROPYLCUMYL HYDROPEROXIDE	≤ 72	≥ 28				OP8			3109	13)
p-MENTHYL HYDROPEROXIDE	> 72 - 100					OP7			3105	13)
"	≤ 72	≥ 28				OP8			3109	27)
METHYLCYCLOHEXANONE PEROXIDE(S)	≤ 67		≥ 33			OP7	+35	+40	3115	
METHYL ETHYL KETONE PEROXIDE(S)	(see remark 8)	≥ 48				OP5			3101	3) 8) 13)
"	(see remark 9)	≥ 55				OP7			3105	9)
"	(see remark 10)	≥ 60				OP8			3107	10)
METHYL ISOBUTYL KETONE PEROXIDE(S)	≤ 62	≥ 19				OP7			3105	22)
ORGANIC PEROXIDE. LIQUID. SAMPLE						OP2			3102	11)
ORGANIC PEROXIDE. LIQUID. SAMPLE, TEMPERATURE CONTROLLED						OP2			3113	11)
ORGANIC PEROXIDE. SOLID. SAMPLE						OP2			3104	11)
ORGANIC PEROXIDE. SOLID. SAMPLE, TEMPERATURE CONTROLLED						OP2			3114	11)
PEROXYACETIC ACID. DISTILLED. TYPE F, stabilized	≤ 41					M	+30	+35	3119	13) 30)
PEROXYACETIC ACID. TYPE D, stabilized	≤ 43					OP7			3105	13) 14) 19)

ORGANIC PEROXIDE	Concentration (%)	Diluent type A (%)	Diluent type B 1) (%)	Inert solid (%)	Water	Packing Method	Control temperature (°C)	Emergency temperature (°C)	Number (Generic entry)	Subsidiary risks and remarks
PEROXYACETIC ACID. TYPE E, stabilized	≤ 43					OP8			3107	13) 14) 19)
PEROXYACETIC ACID. TYPE F. stabilized	≤ 43					OP8			3109	13) 14) 19)
PEROXYLAURIC ACID	≤ 100					OP8	+35	+40	3118	
PINANYL HYDROPEROXIDE	56 - 100					OP7			3105	13)
"	≤ 56	≥ 44				OP8			3109	
POLYETHER POLY-tert-BUTYLPEROXY-CARBONATE	≤ 52		≥ 23			OP8			3107	
1,1,3,3-TETRAMETHYLBUTYL HYDROPEROXIDE	≤ 100					OP7			3105	
1,1,3,3-TETRAMETHYLBUTYL PEROXY-2 ETHYLHEXANOATE	≤ 100					OP7	+15	+20	3115	
1,1,3,3- TETRAMETHYLBUTYL PEROXYNEODECANOATE	≤ 72		≥ 28			OP7	-5	+5	3115	
"	≤ 52 as a stable dispersion in water					OP8, N	-5	+5	3119	
1,1,3,3-TETRAMETHYL-BUTYL PEROXYPIVALATE	≤ 77	≥ 23				OP7	0	+10	3115	
3,6,9-TRIETHYL-3,6,9-TRIMETHYL-1,4,7 TRIPEROXONANE	≤ 42	≥ 58				OP7			3105	28)

Remarks (refer to the last column of the Table in 2.2.52.4):

1)   *Diluent type B may always be replaced by diluent type A. The boiling point of diluent type B shall be at least 60°C higher than the SADT of the organic peroxide.*

2)   *Available oxygen ≤ 4.7%.*

3)   *"EXPLOSIVE" subsidiary risk label required (Model No.1, see 5.2.2.2.2).*

4)   *Diluent may be replaced by di-tert-butyl peroxide.*

5)   *Available oxygen ≤ 9%.*

6)   *With ≤ 9% hydrogen peroxide; available oxygen ≤ 10%.*

7)   *Only non-metallic packagings allowed.*

8)   *Available oxygen > 10% and ≤ 10.7%, with or without water.*

9)   *Available oxygen ≤ 10%, with or without water.*

10)   *Available oxygen ≤ 8.2%, with or without water.*

11)   *See 2.2.52.1.9.*

12)   *Up to 2000 kg per receptacle assigned to ORGANIC PEROXIDE TYPE F on the basis of large scale trials.*

13)   *"CORROSIVE" subsidiary risk label required (Model No.8, see 5.2.2.2.2).*

14)   *Peroxyacetic acid formulations which fulfil the criteria of the Manual of Tests and Criteria, paragraph 20.4.3 (d).*

15)   *Peroxyacetic acid formulations which fulfil the criteria of the Manual of Tests and Criteria, paragraph 20.4.3 (e).*

16)   *Peroxyacetic acid formulations which fulfil the criteria of the Manual of Tests and Criteria, paragraph 20.4.3 (f).*

17)   *Addition of water to this organic peroxide will decrease its thermal stability.*

18)   *No "CORROSIVE" subsidiary risk label (Model No.8, see 5.2.2.2.2) required for concentrations below 80%.*

19)   *Mixtures with hydrogen peroxide, water and acid(s).*

20)   *With diluent type A, with or without water.*

21)   *With ≥ 25% diluent type A by mass, and in addition ethylbenzene.*

22)   *With ≥ 19% diluent type A by mass, and in addition methyl isobutyl ketone.*

23)   *With < 6% di-tert-butyl peroxide.*

24)   *With ≤ 8% 1-isopropylhydroperoxy-4-isopropylhydroxybenzene.*

25)   *Diluent type B with boiling point > 110 °C.*

26)   *With < 0.5% hydroperoxides content.*

27)   *For concentrations more than 56%, "CORROSIVE" subsidiary risk label required (Model No.8, see 5.2.2.2.2).*

28)   *Available active oxygen ≤ 7.6% in diluent type A having a 95% boil-off point in the range of 200 - 260 °C.*

29)   *Not subject to the requirements of ADN for Class 5.2.*

2.2.61   **Class 6.1**   **Toxic substances**

2.2.61.1   *Criteria*

2.2.61.1.1   The heading of Class 6.1 covers substances of which it is known by experience or regarding which it is presumed from experiments on animals that in relatively small quantities they are able by a single action or by action of short duration to cause damage to human health, or death, by inhalation, by cutaneous absorption or by ingestion.

2.2.61.1.2   Substances of Class 6.1 are subdivided as follows:

   T   Toxic substances without subsidiary risk:

       T1   Organic, liquid;
       T2   Organic, solid;
       T3   Organometallic substances;
       T4   Inorganic, liquid;
       T5   Inorganic, solid;
       T6   Liquid, used as pesticides;
       T7   Solid, used as pesticides;
       T8   Samples;
       T9   Other toxic substances;

   TF   Toxic substances, flammable:

       TF1   Liquid;
       TF2   Liquid, used as pesticides;
       TF3   Solid;

   TS   Toxic substances, self-heating, solid;

   TW   Toxic substances, which, in contact with water, emit flammable gases:

       TW1   Liquid;
       TW2   Solid;

   TO   Toxic substances, oxidizing:

       TO1   Liquid;
       TO2   Solid;

   TC   Toxic substances, corrosive:

       TC1   Organic, liquid;
       TC2   Organic, solid;
       TC3   Inorganic, liquid;
       TC4   Inorganic, solid;

   TFC   Toxic substances, flammable, corrosive.

   Definitions

2.2.61.1.3   For the purposes of ADN:

   LD_{50} *(median lethal dose) for acute oral toxicity* is the statistically derived single dose of a substance that can be expected to cause death within 14 days in 50 per cent of young adult albino rats when administered by the oral route. The LD$_{50}$ value is expressed in terms of mass of test substance per mass of test animal (mg/kg);

LD$_{50}$ for acute dermal toxicity is that dose of the substance which, administered by continuous contact for 24 hours with the bare skin of albino rabbits, is most likely to cause death within 14 days in one half of the animals tested. The number of animals tested shall be sufficient to give a statistically significant result and be in conformity with good pharmacological practice. The result is expressed in milligrams per kg body mass;

LC$_{50}$ for acute toxicity on inhalation is that concentration of vapour, mist or dust which, administered by continuous inhalation to both male and female young adult albino rats for one hour, is most likely to cause death within 14 days in one half of the animals tested. A solid substance shall be tested if at least 10% (by mass) of its total mass is likely to be dust in a respirable range, e.g. the aerodynamic diameter of that particle-fraction is 10 μm or less. A liquid substance shall be tested if a mist is likely to be generated in a leakage of the transport containment. Both for solid and liquid substances more than 90% (by mass) of a specimen prepared for inhalation toxicity shall be in the respirable range as defined above. The result is expressed in milligrams per litre of air for dusts and mists or in millilitres per cubic metre of air (parts per million) for vapours.

Classification and assignment of packing groups

2.2.61.1.4   Substances of Class 6.1 shall be classified in three packing groups according to the degree of danger they present for carriage, as follows:

Packing group I: highly toxic substances
Packing group II: toxic substances
Packing group III: slightly toxic substances.

2.2.61.1.5   Substances, mixtures, solutions and articles classified in Class 6.1 are listed in Table A of Chapter 3.2. The assignment of substances, mixtures and solutions not mentioned by name in Table A of Chapter 3.2 to the relevant entry of sub-section 2.2.61.3 and to the relevant packing group in accordance with the provisions of Chapter 2.1, shall be made according to the following criteria in 2.2.61.1.6 to 2.2.61.1.11.

2.2.61.1.6   To assess the degree of toxicity, account shall be taken of human experience of instances of accidental poisoning, as well as special properties possessed by any individual substances: liquid state, high volatility, any special likelihood of cutaneous absorption, and special biological effects.

2.2.61.1.7   In the absence of observations on humans, the degree of toxicity shall be assessed using the available data from animal experiments in accordance with the table below:

	Packing group	Oral toxicity LD$_{50}$ (mg/kg)	Dermal toxicity LD$_{50}$ (mg/kg)	Inhalation toxicity by dusts and mists LC$_{50}$ (mg/l)
Highly toxic	I	≤ 5	≤ 0.2	≤ 0.2
Toxic	II	> 5 and ≤ 50	> 50 and ≤ 200	> 0.2 and ≤ 2
Slightly toxic	III [a]	> 50 and ≤ 300	> 200 and ≤ 1 000	> 2 and ≤ 4

[a]   *Tear gas substances shall be included in packing group II even if data concerning their toxicity correspond to packing group III criteria.*

2.2.61.1.7.1 Where a substance exhibits different degrees of toxicity for two or more kinds of exposure, it shall be classified under the highest such degree of toxicity.

2.2.61.1.7.2 Substances meeting the criteria of Class 8 and with an inhalation toxicity of dusts and mists (LC_{50}) leading to packing group I shall only be accepted for an allocation to Class 6.1 if the toxicity through oral ingestion or dermal contact is at least in the range of packing groups I or II. Otherwise an assignment to Class 8 shall be made if appropriate (see footnote [6] in 2.2.8.1.4).

2.2.61.1.7.3 The criteria for inhalation toxicity of dusts and mists are based on LC_{50} data relating to 1-hour exposure, and where such information is available it shall be used. However, where only LC_{50} data relating to 4-hour exposure are available, such figures can be multiplied by four and the product substituted in the above criteria, i.e. LC_{50} value multiplied by four (4 hour) is considered the equivalent of LC_{50} (1 hour).

Inhalation toxicity of vapours

2.2.61.1.8 Liquids giving off toxic vapours shall be classified into the following groups where "V" is the saturated vapour concentration (in ml/m^3 of air) (volatility) at 20 °C and standard atmospheric pressure:

	Packing group	
Highly toxic	I	Where $V \geq 10\ LC_{50}$ and $LC_{50} \leq 1\ 000\ ml/m^3$
Toxic	II	Where $V \geq LC_{50}$ and $LC_{50} \leq 3\ 000\ ml/m^3$ and the criteria for packing group I are not met
Slightly toxic	III[a]	Where $V \geq 1/5\ LC_{50}$ and $LC_{50} \leq 5\ 000\ ml/m^3$ and the criteria for packing groups I and II are not met

[a] *Tear gas substances shall be included in packing group II even if data concerning their toxicity correspond to packing group III criteria.*

These criteria for inhalation toxicity of vapours are based on LC_{50} data relating to 1-hour exposure, and where such information is available, it shall be used.

However, where only LC_{50} data relating to 4-hour exposure to the vapours are available, such figures can be multiplied by two and the product substituted in the above criteria, i.e. LC_{50} (4 hour) $\times$ 2 is considered the equivalent of LC_{50} (1 hour).

Group borderlines inhalation toxicity of vapours

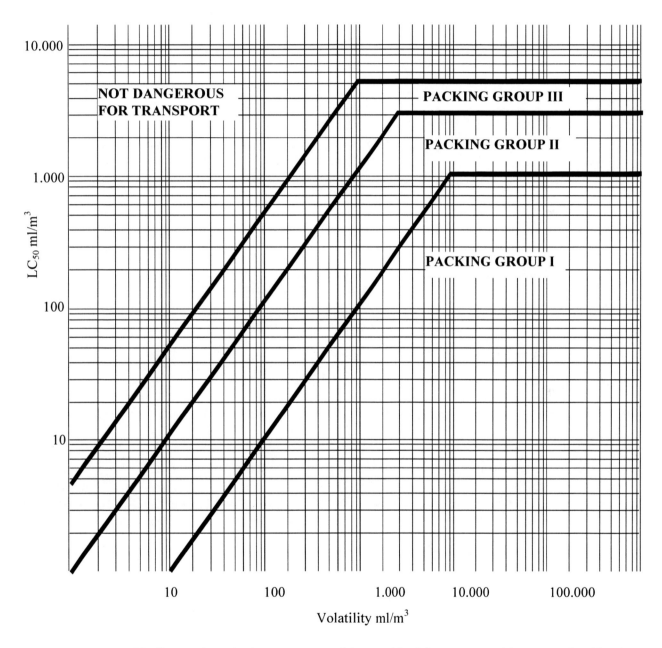

In this figure, the criteria are expressed in graphical form, as an aid to easy classification. However, due to approximations inherent in the use of graphs, substances falling on or near group borderlines shall be checked using numerical criteria.

Mixtures of liquids

2.2.61.1.9   Mixtures of liquids which are toxic on inhalation shall be assigned to packing groups according to the following criteria:

2.2.61.1.9.1   If LC_{50} is known for each of the toxic substances constituting the mixture, the packing group may be determined as follows:

(a)   calculation of the LC_{50} of the mixture:

$$LC_{50} \text{ (mixture)} = \frac{1}{\sum_{i=1}^{1} \frac{f_i}{LC_{50i}}}$$

where f_i =   molar fraction of constituent i of the mixture;

LC_{50i} =   average lethal concentration of constituent i in ml/m^3.

(b)   calculation of volatility of each mixture constituent:

$$V_i = P_i \times \frac{10^6}{101.3} \, (ml/m^3)$$

where P_i =   partial pressure of constituent i in kPa at 20 °C and at standard atmospheric pressure.

(c)   calculation of the ratio of volatility to LC_{50}:

$$R = \sum_{i=1}^{n} \frac{V_i}{LC_{50i}}$$

(d)   the values calculated for LC_{50} (mixture) and R are then used to determine the packing group of the mixture:

Packing group I $R \geq 10$ and LC_{50} (mixture) $\leq 1\,000 \, ml/m^3$;

Packing group II $R \geq 1$ and LC_{50} (mixture) $\leq 3\,000 \, ml/m^3$, if the mixture does not meet the criteria for packing group I;

Packing group III $R \geq 1/5$ and LC_{50} (mixture) $\leq 5\,000 \, ml/m^3$, if the mixture does not meet the criteria of packing groups I or II.

2.2.61.1.9.2   In the absence of LC_{50} data on the toxic constituent substances, the mixture may be assigned to a group based on the following simplified threshold toxicity tests. When these threshold tests are used, the most restrictive group shall be determined and used for carrying the mixture.

2.2.61.1.9.3　　A mixture is assigned to packing group I only if it meets both of the following criteria:

(a)　　A sample of the liquid mixture is vaporized and diluted with air to create a test atmosphere of 1000 ml/m^3 vaporized mixture in air. Ten albino rats (5 male and 5 female) are exposed to the test atmosphere for 1 hour and observed for 14 days. If five or more of the animals die within the 14-day observation period, the mixture is presumed to have an LC$_{50}$ equal to or less than 1000 ml/m^3;

(b)　　A sample of vapour in equilibrium with the liquid mixture is diluted with 9 equal volumes of air to form a test atmosphere. Ten albino rats (5 male and 5 female) are exposed to the test atmosphere for 1 hour and observed for 14 days. If five or more of the animals die within the 14-day observation period, the mixture is presumed to have a volatility equal to or greater than 10 times the mixture LC$_{50}$.

2.2.61.1.9.4　　A mixture is assigned to packing group II only if it meets both of the following criteria, and does not meet the criteria for packing group I:

(a)　　A sample of the liquid mixture is vaporized and diluted with air to create a test atmosphere of 3000 ml/m^3 vaporized mixture in air. Ten albino rats (5 male and 5 female) are exposed to the test atmosphere for 1 hour and observed for 14 days. If five or more of the animals die within the 14-day observation period, the mixture is presumed to have an LC$_{50}$ equal to or less than 3000 ml/m^3;

(b)　　A sample of the vapour in equilibrium with the liquid mixture is used to form a test atmosphere. Ten albino rats (5 male and 5 female) are exposed to the test atmosphere for 1 hour and observed for 14 days. If five or more of the animals die within the 14-day observation period, the mixture is presumed to have a volatility equal to or greater than the mixture LC$_{50}$.

2.2.61.1.9.5　　A mixture is assigned to packing group III only if it meets both of the following criteria, and does not meet the criteria for packing groups I or II:

(a)　　A sample of the liquid mixture is vaporized and diluted with air to create a test atmosphere of 5000 ml/m^3 vaporized mixture in air. Ten albino rats (5 male and 5 female) are exposed to the test atmosphere for 1 hour and observed for 14 days. If five or more of the animals die within the 14-day observation period, the mixture is presumed to have an LC$_{50}$ equal to or less than 5000 ml/m^3;

(b)　　The vapour concentration (volatility) of the liquid mixture is measured and if the vapour concentration is equal to or greater than 1000 ml/m^3, the mixture is presumed to have a volatility equal to or greater than 1/5 the mixture LC$_{50}$.

Methods for determining oral and dermal toxicity of mixtures

2.2.61.1.10　　When classifying and assigning the appropriate packing group to mixtures in Class 6.1 in accordance with the oral and dermal toxicity criteria (see 2.2.61.1.3), it is necessary to determine the acute LD$_{50}$ of the mixture.

2.2.61.1.10.1　　If a mixture contains only one active substance, and the LD$_{50}$ of that constituent is known, in the absence of reliable acute oral and dermal toxicity data on the actual mixture to be carried, the oral or dermal LD$_{50}$ may be obtained by the following method:

$$\text{LD}_{50} \text{ value of preparation} = \frac{\text{LD}_{50} \text{ value of active substance} \times 100}{\text{percentage of active substance by mass}}$$

2.2.61.1.10.2   If a mixture contains more than one active constituent, there are three possible approaches that may be used to determine the oral or dermal LD$_{50}$ of the mixture. The preferred method is to obtain reliable acute oral and dermal toxicity data on the actual mixture to be carried. If reliable, accurate data are not available, then either of the following methods may be performed:

(a)   Classify the formulation according to the most hazardous constituent of the mixture as if that constituent were present in the same concentration as the total concentration of all active constituents; or

(b)   Apply the formula:

$$\frac{C_A}{T_A} + \frac{C_B}{T_B} + ... + \frac{C_Z}{T_Z} = \frac{100}{T_M}$$

where:

C   =   the percentage concentration of constituent A, B, ... Z in the mixture;

T   =   the oral LD$_{50}$ values of constituent A, B, ... Z;

T$_M$ =   the oral LD$_{50}$ value of the mixture.

NOTE: This formula can also be used for dermal toxicities provided that this information is available on the same species for all constituents. The use of this formula does not take into account any potentiation or protective phenomena.

Classification of pesticides

2.2.61.1.11   All active pesticide substances and their preparations for which the LC$_{50}$ and/or LD$_{50}$ values are known and which are classified in Class 6.1 shall be classified under appropriate packing groups in accordance with the criteria given in 2.2.61.1.6 to 2.2.61.1.9. Substances and preparations which are characterized by subsidiary risks shall be classified according to the precedence of hazard Table in 2.1.3.10 with the assignment of appropriate packing groups.

2.2.61.1.11.1   If the oral or dermal LD$_{50}$ value for a pesticide preparation is not known, but the LD$_{50}$ value of its active substance(s) is known, the LD$_{50}$ value for the preparation may be obtained by applying the procedures in 2.2.61.1.10.

NOTE: LD$_{50}$ toxicity data for a number of common pesticides may be obtained from the most current edition of the document "The WHO Recommended Classification of Pesticides by Hazard and Guidelines to Classification" available from the International Programme on Chemical Safety, World Health Organisation (WHO), 1211 Geneva 27, Switzerland. While that document may be used as a source of LD$_{50}$ data for pesticides, its classification system shall not be used for purposes of transport classification of, or assignment of packing groups to, pesticides, which shall be in accordance with the requirements of ADN.

2.2.61.1.11.2   The proper shipping name used in the carriage of the pesticide shall be selected on the basis of the active ingredient, of the physical state of the pesticide and any subsidiary risks it may exhibit (see 3.1.2).

2.2.61.1.12   If substances of Class 6.1, as a result of admixtures, come into categories of risk different from those to which the substances mentioned by name in Table A of Chapter 3.2 belong, these mixtures or solutions shall be assigned to the entries to which they belong on the basis of their actual degree of danger.

NOTE: *For the classification of solutions and mixtures (such as preparations and wastes), see also 2.1.3.*

2.2.61.1.13   On the basis of the criteria of 2.2.61.1.6 to 2.2.61.1.11, it may also be determined whether the nature of a solution or mixture mentioned by name or containing a substance mentioned by name is such that the solution or mixture is not subject to the requirements for this Class.

2.2.61.1.14   Substances, solutions and mixtures, with the exception of substances and preparations used as pesticides, which do not meet the criteria of Directives 67/548/EEC [3] or 88/379/EEC [4] as amended and which are not therefore classified as highly toxic, toxic or harmful according to these directives, as amended, may be considered as substances not belonging to Class 6.1.

2.2.61.2   *Substances not accepted for carriage*

2.2.61.2.1   Chemically unstable substances of Class 6.1 shall not be accepted for carriage unless the necessary steps have been taken to prevent their dangerous decomposition or polymerization during carriage. To this end, it shall in particular be ensured that receptacles and tanks do not contain any substance(s) likely to cause such a reaction.

2.2.61.2.2   The following substances and mixtures shall not be accepted for carriage:

–   Hydrogen cyanide, anhydrous or in solution, which do not meet the descriptions of UN Nos. 1051, 1613, 1614 and 3294;

–   Metal carbonyls, having a flash-point below 23 °C, other than UN Nos. 1259 NICKEL CARBONYL and 1994 IRON PENTACARBONYL;

–   2,3,7,8-TETRACHLORODIBENZO-p-DIOXINE (TCDD) in concentrations considered highly toxic in accordance with the criteria in 2.2.61.1.7;

–   UN No. 2249 DICHLORODIMETHYL ETHER, SYMMETRICAL;

–   Preparations of phosphides without additives inhibiting the emission of toxic flammable gases.

---

[3]   *Council Directive 67/548/EEC of 27 June 1967 on the approximation of laws, regulations and administrative provisions relating to the classification, packaging and labelling of dangerous substances (Official Journal of the European Communities No. L 196 of 16.08.1967, page 1).*
[4]   *Council Directive 88/379/EEC on the approximation of laws, regulations and administrative provisions relating to the classification, packaging and labelling of dangerous preparations (Official Journal of the European Communities No. L 187 of 16.07.1988, page 14).*

2.2.61.3   *List of collective entries*

Toxic substances <u>without</u> subsidiary risk(s)

	liquid [a] **T1**	1583 CHLOROPICRIN MIXTURE, N.O.S. 1602 DYE, LIQUID, TOXIC, N.O.S., or 1602 DYE INTERMEDIATE, LIQUID, TOXIC, N.O.S. 1693 TEAR GAS SUBSTANCE, LIQUID, N.O.S. 1851 MEDICINE, LIQUID, TOXIC, N.O.S. 2206 ISOCYANATES, TOXIC, N.O.S. or 2206 ISOCYANATE SOLUTION, TOXIC, N.O.S. 3140 ALKALOIDS, LIQUID, N.O.S. or 3140 ALKALOID SALTS, LIQUID, N.O.S. 3142 DISINFECTANT, LIQUID, TOXIC, N.O.S. 3144 NICOTINE COMPOUND, LIQUID, N.O.S. or 3144 NICOTINE PREPARATION, LIQUID, N.O.S. 3172 TOXINS, EXTRACTED FROM LIVING SOURCES, LIQUID, N.O.S. 3276 NITRILES, TOXIC, LIQUID, N.O.S 3278 ORGANOPHOSPHORUS COMPOUND, TOXIC, LIQUID, N.O.S. 3381 TOXIC BY INHALATION LIQUID, N.O.S. with an inhalation toxicity lower than or equal to 200 ml/m³ and saturated vapour concentration greater than or equal to 500 LC_{50} 3382 TOXIC BY INHALATION LIQUID, N.O.S. with an inhalation toxicity lower than or equal to 1000 ml/m³ and saturated vapour concentration greater than or equal to 10 LC_{50} 2810 TOXIC LIQUID, ORGANIC, N.O.S.

Organic

	solid [a, b] **T2**	1544 ALKALOIDS, SOLID, N.O.S. or 1544 ALKALOID SALTS, SOLID, N.O.S. 1601 DISINFECTANT, SOLID, TOXIC, N.O.S. 1655 NICOTINE COMPOUND, SOLID, N.O.S., or 1655 NICOTINE PREPARATION, SOLID, N.O.S. 3448 TEAR GAS SUBSTANCE, SOLID, N.O.S. 3143 DYE, SOLID, TOXIC, N.O.S. or 3143 DYE INTERMEDIATE, SOLID, TOXIC, N.O.S. 3462 TOXINS, EXTRACTED FROM LIVING SOURCES, SOLID, N.O.S. 3249 MEDICINE, SOLID, TOXIC, N.O.S. 3464 ORGANOPHOSPHORUS COMPOUND, TOXIC, SOLID, N.O.S. 3439 NITRILES, TOXIC, SOLID, N.O.S. 2811 TOXIC SOLID, ORGANIC, N.O.S.

Organometallic [c, d] **T3**	2026 PHENYLMERCURIC COMPOUND, N.O.S. 2788 ORGANOTIN COMPOUND, LIQUID, N.O.S. 3146 ORGANOTIN COMPOUND, SOLID, N.O.S. 3280 ORGANOARSENIC COMPOUND, LIQUID, N.O.S. 3465 ORGANOARSENIC COMPOUND, SOLID, N.O.S. 3281 METAL CARBONYLS, LIQUID, N.O.S. 3466 METAL CARBONYLS, SOLID, N.O.S. 3282 ORGANOMETALLIC COMPOUND, TOXIC, LIQUID, N.O.S. 3467 ORGANOMETALLIC COMPOUND, TOXIC, SOLID, N.O.S.

(cont'd on next page)

---

[a]   *Substances and preparations containing alkaloids or nicotine used as pesticides shall be classified under UN No. 2588 PESTICIDES, SOLID, TOXIC, N.O.S., UN No. 2902 PESTICIDES, LIQUID, TOXIC, N.O.S. or UN No. 2903 PESTICIDES, LIQUID, TOXIC, FLAMMABLE, N.O.S.*

[b]   *Active substances and triturations or mixtures of substances intended for laboratories and experiments and for the manufacture of pharmaceutical products with other substances shall be classified according to their toxicity (see 2.2.61.1.7 to 2.2.61.1.11).*

[c]   *Self-heating substances, slightly toxic and spontaneously combustible organometallic compounds, are substances of Class 4.2.*

[d]   *Water-reactive substances, slightly toxic, and water-reactive organometallic compounds, are substances of Class 4.3.*

Toxic substances <u>without</u> subsidiary risk(s) *(cont'd)*

Inorganic	**liquid ^e T4**	1556	ARSENIC COMPOUND, LIQUID, N.O.S., inorganic including: Arsenates, n.o.s., Arsenites, n.o.s.; and Arsenic sulphides, n.o.s.
		1935	CYANIDE SOLUTION, N.O.S.
		2024	MERCURY COMPOUND, LIQUID, N.O.S.
		3141	ANTIMONY COMPOUND, INORGANIC, LIQUID, N.O.S.
		3440	SELENIUM COMPOUND, LIQUID, N.O.S.
		3381	TOXIC BY INHALATION LIQUID, N.O.S. with an inhalation toxicity lower than or equal to 200 ml/m^3 and saturated vapour concentration greater than or equal to 500 LC$_{50}$
		3382	TOXIC BY INHALATION LIQUID, N.O.S. with an inhalation toxicity lower than or equal to 1000 ml/m^3 and saturated vapour concentration greater than or equal to 10 LC$_{50}$
		3287	TOXIC LIQUID, INORGANIC, N.O.S.

	solids ^{f, g} T5	1549	ANTIMONY COMPOUND, INORGANIC, SOLID, N.O.S
		1557	ARSENIC COMPOUND, SOLID, N.O.S., including: Arsenates, n.o.s.; Arsenites, n.o.s.; and Arsenic sulphides, n.o.s.
		1564	BARIUM COMPOUND, N.O.S.
		1566	BERYLLIUM COMPOUND, N.O.S.
		1588	CYANIDES, INORGANIC, SOLID, N.O.S.
		1707	THALLIUM COMPOUND, N.O.S.
		2025	MERCURY COMPOUND, SOLID, N.O.S.
		2291	LEAD COMPOUND, SOLUBLE, N.O.S.
		2570	CADMIUM COMPOUND
		2630	SELENATES or
		2630	SELENITES
		2856	FLUOROSILICATES, N.O.S.
		3283	SELENIUM COMPOUND, SOLID, N.O.S.
		3284	TELLURIUM COMPOUND, N.O.S.
		3285	VANADIUM COMPOUND, N.O.S.
		3288	TOXIC SOLID, INORGANIC, N.O.S.

Pesticides	**liquid ^h T6**	2992	CARBAMATE PESTICIDE, LIQUID, TOXIC
		2994	ARSENICAL PESTICIDE, LIQUID, TOXIC
		2996	ORGANOCHLORINE PESTICIDE, LIQUID, TOXIC
		2998	TRIAZINE PESTICIDE, LIQUID, TOXIC
		3006	THIOCARBAMATE PESTICIDE, LIQUID, TOXIC
		3010	COPPER BASED PESTICIDE, LIQUID, TOXIC
		3012	MERCURY BASED PESTICIDE, LIQUID, TOXIC
		3014	SUBSTITUTED NITROPHENOL PESTICIDE, LIQUID, TOXIC
		3016	BIPYRIDILIUM PESTICIDE, LIQUID, TOXIC
		3018	ORGANOPHOSPHORUS PESTICIDE, LIQUID, TOXIC
		3020	ORGANOTIN PESTICIDE, LIQUID, TOXIC
		3026	COUMARIN DERIVATIVE PESTICIDE, LIQUID, TOXIC
		3348	PHENOXYACETIC ACID DERIVATIVE PESTICIDE, LIQUID, TOXIC
		3352	PYRETHROID PESTICIDE, LIQUID, TOXIC
		2902	PESTICIDE, LIQUID, TOXIC, N.O.S.

(cont'd on next page)

---

^e *Mercury fulminate, wetted with not less than 20% water, or mixture of alcohol and water by mass is a substance of Class 1, UN No. 0135.*

^f *Ferricyanides, ferrocyanides, alkaline thiocyanates and ammonium thiocyanates are not subject to the provisions of ADN.*

^g *Lead salts and lead pigments which, when mixed in a ratio of 1:1,000 with 0.07M hydrochloric acid and stirred for one hour at a temperature of 23 °C ± 2 °C, exhibit a solubility of 5% or less, are not subject to the provisions of ADN.*

^h *Articles impregnated with this pesticide, such as fibreboard plates, paper strips, cotton-wool balls, sheets of plastics material, in hermetically closed wrappings, are not subject to the provisions of ADN.*

Toxic substances <u>without</u> subsidiary risk(s) *(cont'd)*

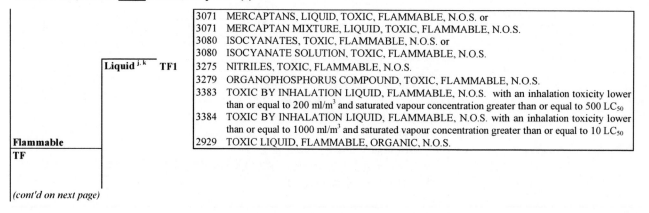

Pesticides *(cont'd)*			
	Solid[h]	T7	2757 CARBAMATE PESTICIDE, SOLID, TOXIC
			2759 ARSENICAL PESTICIDE, SOLID, TOXIC
			2761 ORGANOCHLORINE PESTICIDE, SOLID, TOXIC
			2763 TRIAZINE PESTICIDE, SOLID, TOXIC
			2771 THIOCARBAMATE PESTICIDE, SOLID, TOXIC
			2775 COPPER BASED PESTICIDE, SOLID, TOXIC
			2777 MERCURY BASED PESTICIDE, SOLID, TOXIC
			2779 SUBSTITUTED NITROPHENOL PESTICIDE, SOLID, TOXIC
			2781 BIPYRIDILIUM PESTICIDE, SOLID, TOXIC
			2783 ORGANOPHOSPHORUS PESTICIDE, SOLID, TOXIC
			2786 ORGANOTIN PESTICIDE, SOLID, TOXIC
			3027 COUMARIN DERIVATIVE PESTICIDE, SOLID, TOXIC
			3048 ALUMINIUM PHOSPHIDE PESTICIDE
			3345 PHENOXYACETIC ACID DERIVATIVE PESTICIDE, SOLID, TOXIC
			3349 PYRETHROID PESTICIDE, SOLID, TOXIC
			2588 PESTICIDE, SOLID, TOXIC, N.O.S.
Samples		T8	3315 CHEMICAL SAMPLE, TOXIC
Other toxic substances [i]		T9	3243 SOLIDS CONTAINING TOXIC LIQUID, N.O.S.

Toxic substances <u>with</u> subsidiary risk(s)

Flammable TF	Liquid [j, k]	TF1	3071 MERCAPTANS, LIQUID, TOXIC, FLAMMABLE, N.O.S. or
			3071 MERCAPTAN MIXTURE, LIQUID, TOXIC, FLAMMABLE, N.O.S.
			3080 ISOCYANATES, TOXIC, FLAMMABLE, N.O.S. or
			3080 ISOCYANATE SOLUTION, TOXIC, FLAMMABLE, N.O.S.
			3275 NITRILES, TOXIC, FLAMMABLE, N.O.S.
			3279 ORGANOPHOSPHORUS COMPOUND, TOXIC, FLAMMABLE, N.O.S.
			3383 TOXIC BY INHALATION LIQUID, FLAMMABLE, N.O.S. with an inhalation toxicity lower than or equal to 200 ml/m³ and saturated vapour concentration greater than or equal to 500 LC$_{50}$
			3384 TOXIC BY INHALATION LIQUID, FLAMMABLE, N.O.S. with an inhalation toxicity lower than or equal to 1000 ml/m³ and saturated vapour concentration greater than or equal to 10 LC$_{50}$
			2929 TOXIC LIQUID, FLAMMABLE, ORGANIC, N.O.S.

(cont'd on next page)

---

[h]   *Articles impregnated with this pesticide, such as fibreboard plates, paper strips, cotton-wool balls, sheets of plastics material, in hermetically closed wrappings, are not subject to the provisions of ADN.*

[i]   *Mixtures of solids which are not subject to the provisions of ADN and of toxic liquids may be carried under UN No. 3243 without first applying the classification criteria of Class 6.1, provided there is no free liquid visible at the time the substance is loaded or at the time the packaging, container or transport unit is closed. Each packaging shall correspond to a design type that has passed a leakproofness test at the packing group II level. This entry shall not be used for solids containing a packing group I liquid.*

[j]   *Highly toxic or toxic, flammable liquids having a flash-point below 23 °C excluding substances which are highly toxic on inhalation, i.e. UN Nos. 1051, 1092, 1098, 1143, 1163, 1182, 1185, 1238, 1239, 1244, 1251, 1259, 1613, 1614, 1695, 1994, 2334, 2382, 2407, 2438, 2480, 2482, 2484, 2485, 2606, 2929, 3279 and 3294 are substances of Class 3.*

[k]   *Flammable liquids, slightly toxic, with the exception of substances and preparations used as pesticides, having a flash-point between 23 °C and 60 °C inclusive, are substances of Class 3.*

2.2.61.3 *List of collective entries (cont'd)*

Toxic substances <u>with</u> subsidiary risk(s) (cont'd)

Flammable TF (cont'd)	pesticides, liquid (flashpoint not less than 23 °C)	**TF2**	2991	CARBAMATE PESTICIDE, LIQUID, TOXIC, FLAMMABLE
			2993	ARSENICAL PESTICIDE, LIQUID, TOXIC, FLAMMABLE
			2995	ORGANOCHLORINE PESTICIDE, LIQUID, TOXIC, FLAMMABLE
			2997	TRIAZINE PESTICIDE, LIQUID, TOXIC, FLAMMABLE
			3005	THIOCARBAMATE PESTICIDE, LIQUID, TOXIC, FLAMMABLE
			3009	COPPER BASED PESTICIDE, LIQUID, TOXIC, FLAMMABLE
			3011	MERCURY BASED PESTICIDE, LIQUID, TOXIC, FLAMMABLE
			3013	SUBSTITUTED NITROPHENOL PESTICIDE, LIQUID, TOXIC, FLAMMABLE
			3015	BIPYRIDILIUM PESTICIDE, LIQUID, TOXIC, FLAMMABLE
			3017	ORGANOPHOSPHORUS PESTICIDE, LIQUID, TOXIC, FLAMMABLE
			3019	ORGANOTIN PESTICIDE, LIQUID, TOXIC, FLAMMABLE
			3025	COUMARIN DERIVATIVE PESTICIDE, LIQUID, TOXIC, FLAMMABLE
			3347	PHENOXYACETIC ACID DERIVATIVE PESTICIDE, LIQUID, TOXIC, FLAMMABLE
			3351	PYRETHROID PESTICIDE, LIQUID, TOXIC, FLAMMABLE
			2903	PESTICIDE, LIQUID, TOXIC, FLAMMABLE, N.O.S.
	solid	**TF3**	1700	TEAR GAS CANDLES
			2930	TOXIC SOLID, FLAMMABLE, ORGANIC, N.O.S.
Solid, self-heating [c] **TS**			3124	TOXIC SOLID, SELF-HEATING, N.O.S.
Water-reactive [d] **TW**	liquid	**TW1**	3385	TOXIC BY INHALATION LIQUID, WATER-REACTIVE, N.O.S. with an inhalation toxicity lower than or equal to 200 ml/m^3 and saturated vapour concentration greater than or equal to 500 LC$_{50}$
			3386	TOXIC BY INHALATION LIQUID, WATER-REACTIVE, N.O.S. with an inhalation toxicity lower than or equal to 1000 ml/m^3 and saturated vapour concentration greater than or equal to 10 LC$_{50}$
			3123	TOXIC LIQUID, WATER-REACTIVE, N.O.S.
	solid [n]	**TW2**	3125	TOXIC SOLID, WATER-REACTIVE, N.O.S.
Oxidizing [l] **TO**	liquid	**TO1**	3387	TOXIC BY INHALATION LIQUID, OXIDIZING, N.O.S. with an inhalation toxicity lower than or equal to 200 ml/m^3 and saturated vapour concentration greater than or equal to 500 LC$_{50}$
			3388	TOXIC BY INHALATION LIQUID, OXIDIZING, N.O.S. with an inhalation toxicity lower than or equal to 1000 ml/m^3 and saturated vapour concentration greater than or equal to 10 LC$_{50}$
			3122	TOXIC LIQUID, OXIDIZING, N.O.S.
	solid	**TO2**	3086	TOXIC SOLID, OXIDIZING, N.O.S.
Corrosive [m] **TC** organic	liquid	**TC1**	3277	CHLOROFORMATES, TOXIC, CORROSIVE, N.O.S.
			3361	CHLOROSILANES, TOXIC, CORROSIVE, N.O.S.
			3389	TOXIC BY INHALATION LIQUID, CORROSIVE, N.O.S. with an inhalation toxicity lower than or equal to 200 ml/m^3 and saturated vapour concentration greater than or equal to 500 LC$_{50}$
			3390	TOXIC BY INHALATION LIQUID, CORROSIVE, N.O.S. with an inhalation toxicity lower than or equal to 1000 ml/m^3 and saturated vapour concentration greater than or equal to 10 LC$_{50}$
			2927	TOXIC LIQUID, CORROSIVE, ORGANIC, N.O.S.
	solid	**TC2**	2928	TOXIC SOLID, CORROSIVE, ORGANIC, N.O.S.

(cont'd on next page)

---

[c]   *Self-heating substances, slightly toxic and spontaneously combustible organometallic compounds, are substances of Class 4.2.*

[d]   *Water-reactive substances, slightly toxic, and water-reactive organometallic compounds, are substances of Class 4.3.*

[l]   *Oxidizing substances, slightly toxic, are substances of Class 5.1.*

[m]   *Substances slightly toxic and slightly corrosive, are substances of Class 8.*

[n]   *Metal phosphides assigned to UN Nos. 1360, 1397, 1432, 1714, 2011 and 2013 are substances of Class 4.3.*

2.2.61.3   *List of collective entries (cont'd)*

Toxic substances <u>with</u> subsidiary risk(s) (cont'd)

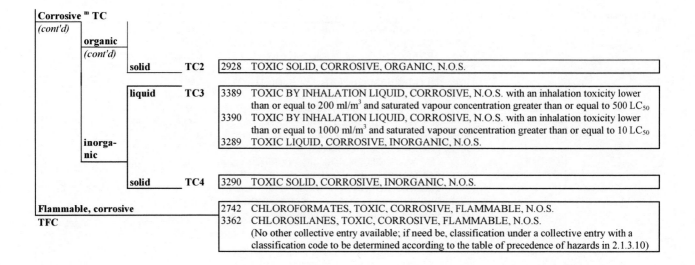

Corrosive [m] TC (cont'd)					
	organic *(cont'd)*	**solid**	**TC2**	2928	TOXIC SOLID, CORROSIVE, ORGANIC, N.O.S.
	inorga-nic	**liquid**	**TC3**	3389	TOXIC BY INHALATION LIQUID, CORROSIVE, N.O.S. with an inhalation toxicity lower than or equal to 200 ml/m^3 and saturated vapour concentration greater than or equal to 500 LC$_{50}$
				3390	TOXIC BY INHALATION LIQUID, CORROSIVE, N.O.S. with an inhalation toxicity lower than or equal to 1000 ml/m^3 and saturated vapour concentration greater than or equal to 10 LC$_{50}$
				3289	TOXIC LIQUID, CORROSIVE, INORGANIC, N.O.S.
		solid	**TC4**	3290	TOXIC SOLID, CORROSIVE, INORGANIC, N.O.S.
Flammable, corrosive TFC				2742	CHLOROFORMATES, TOXIC, CORROSIVE, FLAMMABLE, N.O.S.
				3362	CHLOROSILANES, TOXIC, CORROSIVE, FLAMMABLE, N.O.S.
					(No other collective entry available; if need be, classification under a collective entry with a classification code to be determined according to the table of precedence of hazards in 2.1.3.10)

---

[m]   *Substances slightly toxic and slightly corrosive, are substances of Class 8.*

2.2.62 **Class 6.2** **Infectious substances**

2.2.62.1 *Criteria*

2.2.62.1.1 The heading of Class 6.2 covers infectious substances. For the purposes of ADN, infectious substances are substances which are known or are reasonably expected to contain pathogens. Pathogens are defined as micro-organisms (including bacteria, viruses, rickettsiae, parasites, fungi) and other agents such as prions, which can cause disease in humans or animals.

NOTE 1: Genetically modified micro-organisms and organisms, biological products, diagnostic specimens and infected live animals shall be assigned to this Class if they meet the conditions for this Class.

NOTE 2: Toxins from plant, animal or bacterial sources which do not contain any infectious substances or organisms or which are not contained in them are substances of Class 6.1, UN No. 3172 or 3462.

2.2.62.1.2 Substances of Class 6.2 are subdivided as follows:

I1 Infectious substances affecting humans;

I2 Infectious substances affecting animals only;

I3 Clinical waste;

I4 Biological substances.

Definitions

2.2.62.1.3 For the purposes of ADN,

"Biological products" are those products derived from living organisms which are manufactured and distributed in accordance with the requirements of appropriate national authorities, which may have special licensing requirements, and are used either for prevention, treatment, or diagnosis of disease in humans or animals, or for development, experimental or investigational purposes related thereto. They include, but are not limited to, finished or unfinished products such as vaccines;

"Cultures" are the result of a process by which pathogens are intentionally propagated. This definition does not include human or animal patient specimens as defined in this paragraph;

"Genetically modified micro-organisms and organisms" are micro-organisms and organisms in which genetic material has been purposely altered through genetic engineering in a way that does not occur naturally;

"Medical or clinical wastes" are wastes derived from the medical treatment of animals or humans or from bio-research;

"Patient specimens" are human or animal materials, collected directly from humans or animals, including, but not limited to, excreta, secreta, blood and its components, tissue and tissue fluid swabs, and body parts being carried for purposes such as research, diagnosis, investigational activities, disease treatment and prevention.

Classification

2.2.62.1.4   Infectious substances shall be classified in Class 6.2 and assigned to UN Nos 2814, 2900, 3291 or 3373, as appropriate.

Infectious substances are divided into the following categories:

2.2.62.1.4.1   Category A: An infectious substance which is carried in a form that, when exposure to it occurs, is capable of causing permanent disability, life-threatening or fatal disease in otherwise healthy humans or animals. Indicative examples of substances that meet these criteria are given in the table in this paragraph.

NOTE : *An exposure occurs when an infectious substance is released outside of the protective packaging, resulting in physical contact with humans or animals.*

(a)   Infectious substances meeting these criteria which cause disease in humans or both in humans and animals shall be assigned to UN No. 2814. Infectious substances which cause disease only in animals shall be assigned to UN No. 2900;

(b)   Assignment to UN No. 2814 or UN No. 2900 shall be based on the known medical history and symptoms of the source human or animal, endemic local conditions, or professional judgement concerning individual circumstances of the source human or animal.

NOTE 1: *The proper shipping name for UN No. 2814 is "INFECTIOUS SUBSTANCE, AFFECTING HUMANS". The proper shipping name for UN No. 2900 is "INFECTIOUS SUBSTANCE, AFFECTING ANIMALS only".*

NOTE 2: *The following table is not exhaustive. Infectious substances, including new or emerging pathogens, which do not appear in the table but which meet the same criteria shall be assigned to Category A. In addition, if there is doubt as to whether or not a substance meets the criteria it shall be included in Category A.*

NOTE 3: *In the following table, the micro-organisms written in italics are bacteria, mycoplasmas, rickettsia or fungi.*

INDICATIVE EXAMPLES OF INFECTIOUS SUBSTANCES INCLUDED IN CATEGORY A IN ANY FORM UNLESS OTHERWISE INDICATED (2.2.62.1.4.1)	
UN Number and name	**Micro-organism**
UN No. 2814 Infectious substances affecting humans	*Bacillus anthracis* (cultures only)
	Brucella abortus (cultures only)
	Brucella melitensis (cultures only)
	Brucella suis (cultures only)
	Burkholderia mallei - Pseudomonas mallei – Glanders (cultures only)
	Burkholderia pseudomallei – Pseudomonas pseudomallei (cultures only)
	Chlamydia psittaci - avian strains (cultures only)
	Clostridium botulinum (cultures only)
	Coccidioides immitis (cultures only)
	Coxiella burnetii (cultures only)
	Crimean-Congo haemorrhagic fever virus
	Dengue virus (cultures only)
	Eastern equine encephalitis virus (cultures only)
	Escherichia coli, verotoxigenic (cultures only)*
	Ebola virus
	Flexal virus
	Francisella tularensis (cultures only)
	Guanarito virus
	Hantaan virus
	Hantaviruses causing haemorrhagicfever with renal syndrome
	Hendra virus
	Hepatitis B virus (cultures only)
	Herpes B virus (cultures only)
	Human immunodeficiency virus (cultures only)
	Highly pathogenic avian influenza virus (cultures only)
	Japanese Encephalitis virus (cultures only)
	Junin virus
	Kyasanur Forest disease virus
	Lassa virus
	Machupo virus
	Marburg virus
	Monkeypox virus
	Mycobacterium tuberculosis (cultures only)*
	Nipah virus
	Omsk haemorrhagic fever virus
	Poliovirus (cultures only)
	Rabies virus (cultures only)
	Rickettsia prowazekii (cultures only)
	Rickettsia rickettsii (cultures only)
	Rift Valley fever virus (cultures only)
	Russian spring-summer encephalitis virus (cultures only)
	Sabia virus
	Shigella dysenteriae type 1 (cultures only)*
	Tick-borne encephalitis virus (cultures only)
	Variola virus
	Venezuelan equine encephalitis virus (cultures only)
	West Nile virus (cultures only)
	Yellow fever virus (cultures only)
	Yersinia pestis (cultures only)

INDICATIVE EXAMPLES OF INFECTIOUS SUBSTANCES INCLUDED IN CATEGORY A IN ANY FORM UNLESS OTHERWISE INDICATED (2.2.62.1.4.1)	
UN Number and name	**Micro-organism**
UN No. 2900 Infectious substances affecting animals only	African swine fever virus (cultures only) Avian paramyxovirus Type 1 - Velogenic Newcastle disease virus (cultures only) Classical swine fever virus (cultures only) Foot and mouth disease virus (cultures only) Lumpy skin disease virus (cultures only) *Mycoplasma mycoides* - Contagious bovine pleuropneumonia (cultures only) Peste des petits ruminants virus (cultures only) Rinderpest virus (cultures only) Sheep-pox virus (cultures only) Goatpox virus (cultures only) Swine vesicular disease virus (cultures only) Vesicular stomatitis virus (cultures only)

*/ *Nevertheless, when the cultures are intended for diagnostic or clinical purposes, they may be classified as infectious substances of Category B.*

2.2.62.1.4.2   Category B: An infectious substance which does not meet the criteria for inclusion in Category A. Infectious substances in Category B shall be assigned to UN No. 3373.

NOTE: *The proper shipping name of UN No. 3373 is "BIOLOGICAL SUBSTANCE, CATEGORY B".*

2.2.62.1.5   *Exemptions*

2.2.62.1.5.1   Substances which do not contain infectious substances or substances which are unlikely to cause disease in humans or animals are not subject to the provisions of ADN unless they meet the criteria for inclusion in another class.

2.2.62.1.5.2   Substances containing microorganisms which are non-pathogenic to humans or animals are not subject to ADN unless they meet the criteria for inclusion in another class.

2.2.62.1.5.3   Substances in a form that any present pathogens have been neutralized or inactivated such that they no longer pose a health risk are not subject to ADN unless they meet the criteria for inclusion in another class.

2.2.62.1.5.4   Substances where the concentration of pathogens is at a level naturally encountered (including foodstuff and water samples) and which are not considered to pose a significant risk of infection are not subject to ADN unless they meet the criteria for inclusion in another class.

2.2.62.1.5.5   Dried blood spots, collected by applying a drop of blood onto absorbent material, or faecal occult blood screening tests and blood or blood components which have been collected for the purposes of transfusion or for the preparation of blood products to be used for transfusion or transplantation and any tissues or organs intended for use in transplantation are not subject to the provisions of ADN.

2.2.62.1.5.6   Human or animal specimens for which there is minimal likelihood that pathogens are present are not subject to ADN if the specimen is carried in a packaging which will prevent any leakage and which is marked with the words "Exempt human specimen" or "Exempt animal specimen", as appropriate.

The packaging is deemed to comply with the above requirements if it meets the following conditions:

(a)   The packaging consists of three components:

    (i)   a leak-proof primary receptacle(s);

    (ii)   a leak-proof secondary packaging; and

    (iii)   an outer packaging of adequate strength for its capacity, mass and intended use, and with at least one surface having minimum dimensions of 100 mm × 100 mm;

(b)   For liquids, absorbent material in sufficient quantity to absorb the entire contents is be placed between the primary receptacle(s) and the secondary packaging so that, during carriage, any release or leak of a liquid substance will not reach the outer packaging and will not compromise the integrity of the cushioning material;

(c)   When multiple fragile primary receptacles are placed in a single secondary packaging, they are either individually wrapped or separated to prevent contact between them.

NOTE: An element of professional judgment is required to determine if a substance is exempt under this paragraph. That judgment should be based on the known medical history, symptoms and individual circumstances of the source, human or animal, and endemic local conditions. Examples of specimens which may be carried under this paragraph include the blood or urine tests to monitor cholesterol levels, blood glucose levels, hormone levels, or prostate specific antibodies (PSA); those required to monitor organ function such as heart, liver or kidney function for humans or animals with non-infectious diseases, or for therapeutic drug monitoring; those conducted for insurance or employment purposes and are intended to determine the presence of drugs or alcohol; pregnancy test; biopsies to detect cancer; and antibody detection in humans or animals.

2.2.62.1.6-
2.2.62.1.8   *(Reserved).*

2.2.62.1.9   *Biological products*

For the purposes of ADN, biological products are divided into the following groups:

(a)   those which are manufactured and packaged in accordance with the requirements of appropriate national authorities and carried for the purposes of final packaging or distribution, and use for personal health care by medical professionals or individuals. Substances in this group are not subject to the provisions of ADN;

(b)   those which do not fall under paragraph (a) and are known or reasonably believed to contain infectious substances and which meet the criteria for inclusion in Category A or Category B. Substances in this group shall be assigned to UN No. 2814, UN No. 2900 or UN No. 3373, as appropriate.

NOTE: Some licensed biological products may present a biohazard only in certain parts of the world. In that case, competent authorities may require these biological products to be in compliance with local requirements for infectious substances or may impose other restrictions.

2.2.62.1.10   *Genetically modified micro-organisms and organisms*

Genetically modified micro-organisms not meeting the definition of infectious substance shall be classified according to section 2.2.9.

2.2.62.1.11   *Medical or clinical wastes*

2.2.62.1.11.1   Medical or clinical wastes containing Category A infectious substances shall be assigned to UN No. 2814 or UN No. 2900 as appropriate.   Medical or clinical wastes containing infectious substances in Category B shall be assigned to UN No. 3291.

NOTE: Medical or clinical wastes assigned to number 18 01 03 (Wastes from human or animal health care and/or related research – wastes from natal care, diagnosis, treatment or prevention of disease in humans – wastes whose collection and disposal is subject to special requirement in order to prevent infection) or 18 02 02 (Wastes from human or animal health care and/or related research – wastes from research, diagnosis, treatment or prevention of disease involving animals – wastes whose collection and disposal is subject to special requirements in order to prevent infection) according to the list of wastes annexed to the Commission Decision 2000/532/EC[5] as amended, shall be classified according to the provisions set out in this paragraph, based on the medical or veterinary diagnosis concerning the patient or the animal.

2.2.62.1.11.2   Medical or clinical wastes which are reasonably believed to have a low probability of containing infectious substances shall be assigned to UN No. 3291.

NOTE 1:   The proper shipping name for UN No. 3291 is "CLINICAL WASTE, UNSPECIFIED, N.O.S." or "(BIO) MEDICAL WASTE, N.O.S". or "REGULATED MEDICAL WASTE, N.O.S.".

NOTE 2: Notwithstanding the classification criteria set out above, medical or clinical wastes assigned to number 18 01 04 (Wastes from human or animal health care and/or related research – wastes from natal care, diagnosis, treatment or prevention of disease in humans – wastes whose collection and disposal is not subject to special requirements in order to prevent infection) or 18 02 03 (Wastes from human or animal health care and/or related research – wastes from research, diagnosis, treatment or prevention of disease involving animals – wastes whose collection and disposal is not subject to special requirements in order to prevent infection) according to the list of wastes annexed to the Commission Decision 2000/532/EC[5] as amended, are not subject to the provisions of ADN.

2.2.62.1.11.3   Decontaminated medical or clinical wastes which previously contained infectious substances are not subject to the provisions of ADN unless they meet the criteria for inclusion in another class.

2.2.62.1.11.4   Medical or clinical wastes assigned to UN No. 3291 are assigned to packing group II.

2.2.62.1.12   *Infected animals*

2.2.62.1.12.1   Unless an infectious substance cannot be consigned by any other means, live animals shall not be used to consign such a substance. A live animal which has been intentionally infected

---

[5]   *Commission Decision 2000/532/EC of 3 May 2000 replacing Decision 94/3/EC establishing a list of wastes pursuant to Article 1(a) of Council Directive 75/442/EEC on waste and Council Decision 94/904/EC establishing a list of hazardous waste pursuant to Article 1(4) of Council Directive 91/689/EEC on hazardous waste (Official Journal of the European Communities No. L 226 of 6 September 2000, page 3).*

and is known or suspected to contain an infectious substance shall only be carried under terms and conditions approved by the competent authority[6].

2.2.62.1.12.2   Animal carcasses affected by pathogens of Category A or which would be assigned to Category A in cultures only, shall be assigned to UN 2814 or UN 2900 as appropriate.

Other animal carcasses affected by pathogens included in Category B shall be carried in accordance with provisions determined by the competent authority[7].

2.2.62.2   *Substances not accepted for carriage*

Live vertebrate or invertebrate animals shall not be used to carry an infectious agent unless the agent cannot be carried by other means or unless this carriage has been approved by the competent authority (see 2.2.62.1.12.1).

2.2.62.3   *List of collective entries*

Effects on humans	I1	2814	INFECTIOUS SUBSTANCE, AFFECTING HUMANS
Effects on animals only	I2	2900	INFECTIOUS SUBSTANCE, AFFECTING ANIMALS only
Clinical waste	I3	3291 3291 3291	CLINICAL WASTE, UNSPECIFIED, N.O.S. or (BIO)MEDICAL WASTE, N.O.S. or REGULATED MEDICAL WASTE, N.O.S.
Biological substances	I4	3373	BIOLOGICAL SUBSTANCE, CATEGORY

---

[6]   *Such regulations are contained in, e.g. Directive 91/628/EEC (Official Journal of the European Communities No. L 340 of 11 December 1991, p. 17) and in the Recommendations of the Council of Europe (Ministerial Committee) on the carriage of certain animal species.*

[7]   *Regulations for dead infected animals are contained e.g. in Regulation (EC) No. 1774/2002 of the European Parliament and of the Council of 3 October 2002 laying down health rules concerning animal by-products not intended for human consumption (Official Journal of the European Communities, No. L 273 of 10.10.2002, p. 1).*

2.2.7   **Class 7   Radioactive material**

2.2.7.1   *Definition of Class 7*

2.2.7.1.1   *Radioactive material* means any material containing radionuclides where both the activity concentration and the total activity in the consignment exceed the values specified in 2.2.7.7.2.1 to 2.2.7.7.2.6.

2.2.7.1.2   The following radioactive materials are not included in Class 7 for the purposes of ADN:

(a)   Radioactive material that is an integral part of the means of transport;

(b)   (Reserved);

(c)   Radioactive material implanted or incorporated into a person or live animal for diagnosis or treatment;

(d)   Radioactive material in consumer products which have received regulatory approval, following their sale to the end user;

(e)   Natural material and ores containing naturally occurring radionuclides which are either in their natural state, or have only been processed for purposes other than for extraction of the radionuclides, and which are not intended to be processed for use of these radionuclides provided the activity concentration of the material does not exceed 10 times the values specified in 2.2.7.7.2.1 (b), or calculated in accordance with 2.2.7.7.2.2 to 2.2.7.7.2.6;

(f)   Non-radioactive solid objects with radioactive substances present on any surfaces in quantities not in excess of the limit set out in the definition for "contamination" in 2.2.7.2.

2.2.7.2   *Definitions*

A_1 and A_2

A_1 means the activity value of special form radioactive material which is listed in Table 2.2.7.7.2.1 or derived in 2.2.7.7.2 and is used to determine the activity limits for the requirements of ADN.

A_2 means the activity value of radioactive material, other than special form radioactive material, which is listed in Table 2.2.7.7.2.1 or derived in 2.2.7.7.2 and is used to determine the activity limits for the requirements of ADN.

Approval

Multilateral approval means approval by the relevant competent authority of the country of origin of the design or shipment, as applicable and also, where the consignment is to be carried through or into any other country, approval by the competent authority of that country.

Unilateral approval means an approval of a design which is required to be given by the competent authority of the country of origin of the design only. If the country of origin is not a Contracting Party to ADN, the approval shall require validation by the competent authority of the first country Contracting Party to ADN reached by the consignment (see 6.4.22.6).

Confinement system means the assembly of fissile material and packaging components specified by the designer and agreed to by the competent authority as intended to preserve criticality safety.

Containment system means the assembly of components of the packaging specified by the designer as intended to retain the radioactive material during carriage.
Contamination:

Contamination means the presence of a radioactive substance on a surface in quantities in excess of 0.4 Bq/cm^2 for beta and gamma emitters and low toxicity alpha emitters, or 0.04 Bq/cm^2 for all other alpha emitters.

Non-fixed contamination means contamination that can be removed from a surface during routine conditions of carriage.

Fixed contamination means contamination other than non-fixed contamination.

Criticality safety index (CSI) assigned to a package, overpack or container containing fissile material means a number which is used to provide control over the accumulation of packages, overpacks or containers containing fissile material.

Design means the description of special form radioactive material, low dispersible radioactive material, package or packaging which enables such an item to be fully identified. The description may include specifications, engineering drawings, reports demonstrating compliance with regulatory requirements, and other relevant documentation.

Exclusive use means the sole use, by a single consignor, of a vehicle or of a large container, in respect of which all initial, intermediate and final loading and unloading is carried out in accordance with the directions of the consignor or consignee.

Fissile material means uranium-233, uranium-235, plutonium-239, plutonium-241, or any combination of these radionuclides. Excepted from this definition is:

(a)   natural uranium or depleted uranium which is unirradiated, and

(b)   natural uranium or depleted uranium which has been irradiated in thermal reactors only.

Large container means a container which is not a small container according to the definitions of this Sub-section.

Low dispersible radioactive material means either a solid radioactive material or a solid radioactive material in a sealed capsule, that has limited dispersibility and is not in powder form.

NOTE: *Low dispersible radioactive material may be carried by air in Type B(U) or B(M) packages in quantities as authorised for the package design as specified in the certificate of approval. This definition is included here since such packages carrying low dispersible radioactive material may also be carried by inland waterways.*

Low specific activity (LSA) material, see 2.2.7.3.

Low toxicity alpha emitters are: natural uranium; depleted uranium; natural thorium; uranium-235 or uranium-238; thorium-232; thorium-228 and thorium-230 when contained in ores or physical and chemical concentrates; or alpha emitters with a half-life of less than 10 days.

Maximum normal operating pressure means the maximum pressure above atmospheric pressure at mean sea-level that would develop in the containment system in a period of one year under the conditions of temperature and solar radiation corresponding to environmental conditions in the absence of venting, external cooling by an ancillary system, or operational controls during carriage.

Package in the case of radioactive material means the packaging with its radioactive contents as presented for carriage. The types of packages covered by ADN, which are subject to the activity limits and material restrictions of 2.2.7.7 and meet the corresponding requirements, are:

(a)   Excepted package;

(b)   Industrial package Type 1 (Type IP-1 package);

(c)   Industrial package Type 2 (Type IP-2 package);

(d)   Industrial package Type 3 (Type IP-3 package);

(e)   Type A package;

(f)   Type B(U) package;

(g)   Type B(M) package;

(h)   Type C package.

Packages containing fissile material or uranium hexafluoride are subject to additional requirements (see 2.2.7.7.1.7 and 2.2.7.7.1.8).

NOTE: *For "packages" for other dangerous goods see definitions under 1.2.1.*

Packaging in the case of radioactive material means the assembly of components necessary to enclose the radioactive contents completely. It may, in particular, consist of one or more receptacles, absorbent materials, spacing structures, radiation shielding and service equipment for filling, emptying, venting and pressure relief; devices for cooling, absorbing mechanical shocks, handling and tie-down, thermal insulation; and service devices integral to the package. The packaging may be a box, drum or similar receptacle, or may also be a container, tank or intermediate bulk container (IBC).

NOTE: *For "packagings" for other dangerous goods see definitions under 1.2.1*

Radiation level means the corresponding dose rate expressed in millisieverts per hour.

Radioactive contents mean the radioactive material together with any contaminated or activated solids, liquids, and gases within the packaging.

Shipment means the specific movement of a consignment from origin to destination.

Small container means a container which has either any overall outer dimension less than 1.5 m, or an internal volume of not more than 3 m^3.

Special form radioactive material, see 2.2.7.4.1.

Specific activity of a radionuclide means the activity per unit mass of that nuclide. The specific activity of a material shall mean the activity per unit mass of the material in which the radionuclides are essentially uniformly distributed.

Surface contaminated object (SCO), see 2.2.7.5.

Transport index (TI) assigned to a package, overpack or container, or to unpackaged LSA-1 or SCO-1, means a number which is used to provide control over radiation exposure.
Unirradiated thorium means thorium containing not more than 10^{-7} g of uranium-233 per gram of thorium-232.

Unirradiated uranium means uranium containing not more than 2×10^3 Bq of plutonium per gram of uranium-235, not more than 9×10^6 Bq of fission products per gram of uranium-235 and not more than 5×10^{-3} g of uranium-236 per gram of uranium-235.

Uranium - natural, depleted, enriched means the following:

Natural uranium means uranium (which may be chemically separated) containing the naturally occurring distribution of uranium isotopes (approximately 99.28% uranium-238, and 0.72% uranium-235 by mass).

Depleted uranium means uranium containing a lesser mass percentage of uranium-235 than in natural uranium.

Enriched uranium means uranium containing a greater mass percentage of uranium-235 than 0.72%.

In all cases, a very small mass percentage of uranium-234 is present.

2.2.7.3   *Low specific activity (LSA) material, determination of groups*

2.2.7.3.1   Radioactive material which by its nature has a limited specific activity, or radioactive material for which limits of estimated average specific activity apply, is termed low specific activity or LSA material. External shielding materials surrounding the LSA material shall not be considered in determining the estimated average specific activity.

2.2.7.3.2   LSA material shall be in one of three groups:

(a)   LSA-I

(i)   uranium and thorium ores and concentrates of such ores, and other ores containing naturally occurring radionuclides which are intended to be processed for the use of these radionuclides;

(ii)   natural uranium, depleted uranium, natural thorium or their compounds or mixtures, providing they are unirradiated and in solid or liquid form;

(iii)   radioactive material for which the A_2 value is unlimited, excluding fissile material in quantities not excepted under 6.4.11.2 of ADR; or

(iv)   other radioactive material in which the activity is distributed throughout and the estimated average specific activity does not exceed 30 times the values for activity concentration specified in 2.2.7.7.2.1 to 2.2.7.7.2.6, excluding fissile material in quantities not excepted under 6.4.11.2 of ADR.

(b)   LSA-II

(i)   water with tritium concentration up to 0.8 TBq/l; or

(ii)   other material in which the activity is distributed throughout and the estimated average specific activity does not exceed 10^{-4} A_2/g for solids and gases, and 10^{-5} A_2/g for liquids;

(c)   LSA-III - Solids (e.g. consolidated wastes, activated materials), excluding powders, in which:

(i)   the radioactive material is distributed throughout a solid or a collection of solid objects, or is essentially uniformly distributed in a solid compact binding agent (such as concrete, bitumen, ceramic, etc.);

(ii)   the radioactive material is relatively insoluble, or it is intrinsically contained in a relatively insoluble matrix, so that, even under loss of packaging, the loss of radioactive material per package by leaching when placed in water for seven days would not exceed 0.1 A_2; and

(iii)   the estimated average specific activity of the solid, excluding any shielding material, does not exceed 2×10^{-3} A_2/g.

2.2.7.3.3   LSA-III material shall be a solid of such a nature that if the entire contents of a package were subjected to the test specified in 2.2.7.3.4 the activity in the water would not exceed 0.1 A_2.

2.2.7.3.4   LSA-III material shall be tested as follows:

A solid material sample representing the entire contents of the package shall be immersed for 7 days in water at ambient temperature. The volume of water to be used in the test shall be sufficient to ensure that at the end of the 7 day test period the free volume of the unabsorbed and unreacted water remaining shall be at least 10% of the volume of the solid test sample itself. The water shall have an initial pH of 6-8 and a maximum conductivity of 1 mS/m at 20 °C. The total activity of the free volume of water shall be measured following the 7 day immersion of the test sample.

2.2.7.3.5   Demonstration of compliance with the performance standards in 2.2.7.3.4 shall be in accordance with 6.4.12.1 and 6.4.12.2 of ADR.

2.2.7.4   ***Requirements for special form radioactive material***

2.2.7.4.1   *Special form radioactive material* means either:

(a)   an indispersible solid radioactive material; or

(b)   a sealed capsule containing radioactive material that shall be so manufactured that it can be opened only by destroying the capsule.

Special form radioactive material shall have at least one dimension not less than 5 mm.

2.2.7.4.2   Special form radioactive material shall be of such a nature or shall be so designed that if it is subjected to the tests specified in 2.2.7.4.4 to 2.2.7.4.8, it shall meet the following requirements:

(a)   It would not break or shatter under the impact, percussion and bending tests 2.2.7.4.5 (a)(b)(c), 2.2.7.4.6 (a) as applicable;

(b) It would not melt or disperse in the applicable heat test 2.2.7.4.5 (d) or 2.2.7.4.6 (b) as applicable; and

(c) The activity in the water from the leaching tests specified in 2.2.7.4.7 and 2.2.7.4.8 would not exceed 2 kBq; or alternatively for sealed sources, the leakage rate for the volumetric leakage assessment test specified in ISO 9978:1992 "Radiation Protection - Sealed Radioactive Sources - Leakage Test Methods", would not exceed the applicable acceptance threshold acceptable to the competent authority.

2.2.7.4.3 Demonstration of compliance with the performance standards in 2.2.7.4.2 shall be in accordance with 6.4.12.1 and 6.4.12.2 of ADR.

2.2.7.4.4 Specimens that comprise or simulate special form radioactive material shall be subjected to the impact test, the percussion test, the bending test, and the heat test specified in 2.2.7.4.5 or alternative tests as authorized in 2.2.7.4.6. A different specimen may be used for each of the tests. Following each test, a leaching assessment or volumetric leakage test shall be performed on the specimen by a method no less sensitive than the methods given in 2.2.7.4.7 for indispersible solid material or 2.2.7.4.8 for encapsulated material.

2.2.7.4.5 The relevant test methods are:

(a) Impact test: The specimen shall drop onto the target from a height of 9 m. The target shall be as defined in 6.4.14 of ADR;

(b) Percussion test: The specimen shall be placed on a sheet of lead which is supported by a smooth solid surface and struck by the flat face of a mild steel bar so as to cause an impact equivalent to that resulting from a free drop of 1.4 kg through 1 m. The lower part of the bar shall be 25 mm in diameter with the edges rounded off to a radius of (3.0 ± 0.3) mm. The lead, of hardness number 3.5 to 4.5 on the Vickers scale and not more than 25 mm thick, shall cover an area greater than that covered by the specimen. A fresh surface of lead shall be used for each impact. The bar shall strike the specimen so as to cause maximum damage;

(c) Bending test: The test shall apply only to long, slender sources with both a minimum length of 10 cm and a length to minimum width ratio of not less than 10. The specimen shall be rigidly clamped in a horizontal position so that one half of its length protrudes from the face of the clamp. The orientation of the specimen shall be such that the specimen will suffer maximum damage when its free end is struck by the flat face of a steel bar. The bar shall strike the specimen so as to cause an impact equivalent to that resulting from a free vertical drop of 1.4 kg through 1 m. The lower part of the bar shall be 25 mm in diameter with the edges rounded off to a radius of (3.0 ± 0.3) mm;

(d) Heat test: The specimen shall be heated in air to a temperature of 800°C and held at that temperature for a period of 10 minutes and shall then be allowed to cool.

2.2.7.4.6 Specimens that comprise or simulate radioactive material enclosed in a sealed capsule may be excepted from:

(a) The tests prescribed in 2.2.7.4.5 (a) and (b) provided the mass of the special form radioactive material:

(i) is less than 200 g and they are alternatively subjected to the Class 4 impact test prescribed in ISO 2919:1999 "Radiation protection - Sealed radioactive sources - General requirements and classification"; or

(ii)   is less than 500 g and they are alternatively subjected to the Class 5 impact test prescribed in ISO 2919:1999 "Radiation protection - Sealed Radioactive Sources – General requirements and Classification; and

(b)   The test prescribed in 2.2.7.4.5 (d) provided they are alternatively subjected to the Class 6 temperature test specified in ISO 2919:1999 " Radiation protection - Sealed radioactive sources – General requirements and classification ".

2.2.7.4.7   For specimens which comprise or simulate indispersible solid material, a leaching assessment shall be performed as follows:

(a)   The specimen shall be immersed for 7 days in water at ambient temperature. The volume of water to be used in the test shall be sufficient to ensure that at the end of the 7 day test period the free volume of the unabsorbed and unreacted water remaining shall be at least 10 % of the volume of the solid test sample itself. The water shall have an initial pH of 6-8 and a maximum conductivity of 1 mS/m at 20 °C;

(b)   The water with specimen shall then be heated to a temperature of (50 ± 5) °C and maintained at this temperature for 4 hours;

(c)   The activity of the water shall then be determined;

(d)   The specimen shall then be kept for at least 7 days in still air at not less than 30 °C and relative humidity not less than 90%;

(e)   The specimen shall then be immersed in water of the same specification as in (a) above and the water with the specimen heated to (50 ± 5) °C and maintained at this temperature for 4 hours;

(f)   The activity of the water shall then be determined.

2.2.7.4.8   For specimens which comprise or simulate radioactive material enclosed in a sealed capsule, either a leaching assessment or a volumetric leakage assessment shall be performed as follows:

(a)   The leaching assessment shall consist of the following steps:

(i)   the specimen shall be immersed in water at ambient temperature. The water shall have an initial pH of 6-8 with a maximum conductivity of 1 mS/m at 20 °C;

(ii)   the water and specimen shall be heated to a temperature of (50 ± 5) °C and maintained at this temperature for 4 hours;

(iii)   the activity of the water shall then be determined;

(iv)   the specimen shall then be kept for at least 7 days in still air at not less than 30 °C and relative humidity of not less than 90%;

(v)   the process in (i), (ii) and (iii) shall be repeated;

(b)   The alternative volumetric leakage assessment shall comprise any of the tests prescribed in ISO 9978:1992 "Radiation Protection - Sealed radioactive sources - Leakage test methods", which are acceptable to the competent authority.

2.2.7.5 *Surface contaminated object (SCO), determination of groups*

Surface contaminated object (SCO) means a solid object which is not itself radioactive but which has radioactive material distributed on its surfaces. SCO is classified in one of two groups:

(a) SCO-I: A solid object on which:

  (i) the non-fixed contamination on the accessible surface averaged over 300 cm^2 (or the area of the surface if less than 300 cm^2) does not exceed 4 Bq/cm^2 for beta and gamma emitters and low toxicity alpha emitters, or 0.4 Bq/cm^2 for all other alpha emitters; and

  (ii) the fixed contamination on the accessible surface averaged over 300 cm^2 (or the area of the surface if less than 300 cm^2) does not exceed 4×10^4 Bq/cm^2 for beta and gamma emitters and low toxicity alpha emitters, or 4×10^3 Bq/cm^2 for all other alpha emitters; and

  (iii) the non-fixed contamination plus the fixed contamination on the inaccessible surface averaged over 300 cm^2 (or the area of the surface if less than 300 cm^2) does not exceed 4×10^4 Bq/cm^2 for beta and gamma emitters and low toxicity alpha emitters, or 4×10^3 Bq/cm^2 for all other alpha emitters;

(b) SCO-II: A solid object on which either the fixed or non-fixed contamination on the surface exceeds the applicable limits specified for SCO-I in (a) above and on which:

  (i) the non-fixed contamination on the accessible surface averaged over 300 cm^2 (or the area of the surface if less than 300 cm^2) does not exceed 400 Bq/cm^2 for beta and gamma emitters and low toxicity alpha emitters, or 40 Bq/cm^2 for all other alpha emitters; and

  (ii) the fixed contamination on the accessible surface, averaged over 300 cm^2 (or the area of the surface if less than 300 cm^2) does not exceed 8×10^5 Bq/cm^2 for beta and gamma emitters and low toxicity alpha emitters, or 8×10^4 Bq/cm^2 for all other alpha emitters; and

  (iii) the non-fixed contamination plus the fixed contamination on the inaccessible surface averaged over 300 cm^2 (or the area of the surface if less than 300 cm^2) does not exceed 8×10^5 Bq/cm^2 for beta and gamma emitters and low toxicity alpha emitters, or 8×10^4 Bq/cm^2 for all other alpha emitters.

2.2.7.6 *Determination of transport index (TI) and criticality safety index (CSI)*

2.2.7.6.1 *Determination of transport index*

2.2.7.6.1.1 The transport index (TI) for a package, overpack or container, or for unpackaged LSA-I or SCO-I, shall be the number derived in accordance with the following procedure:

(a) Determine the maximum radiation level in units of millisieverts per hour (mSv/h) at a distance of 1 m from the external surfaces of the package, overpack, container, or unpackaged LSA-I and SCO-I. The value determined shall be multiplied by 100 and the resulting number is the transport index. For uranium and thorium ores and their concentrates, the maximum radiation level at any point 1 m from the external surface of the load may be taken as:

0.4 mSv/h   for ores and physical concentrates of uranium and thorium;

0.3 mSv/h for chemical concentrates of thorium;

0.02 mSv/h for chemical concentrates of uranium, other than uranium hexafluoride;

(b) For tanks, containers and unpackaged LSA-I and SCO-I, the value determined in step (a) above shall be multiplied by the appropriate factor from Table 2.2.7.6.1.1;

(c) The value obtained in steps (a) and (b) above shall be rounded up to the first decimal place (e.g. 1.13 becomes 1.2), except that a value of 0.05 or less may be considered as zero.

Table 2.2.7.6.1.1

Multiplication factor for tanks, containers and unpackaged LSA-I and SCO-I

Size of load [a]	Multiplication factor
size of load ≤ 1 m^2	1
1 m^2 < size of load ≤ 5 m^2	2
5 m^2 < size of load ≤ 20 m^2	3
20 m^2 < size of load	10

[a] *Largest cross-sectional area of the load being measured.*

2.2.7.6.1.2   The transport index for each overpack, container, vehicle shall be determined as either the sum of the TIs of all the packages contained, or by direct measurement of radiation level, except in the case of non-rigid overpacks for which the transport index shall be determined only as the sum of the TIs of all the packages.

2.2.7.6.2   *Determination of criticality safety index (CSI)*

2.2.7.6.2.1   The criticality safety index (CSI) for packages containing fissile material shall be obtained by dividing the number 50 by the smaller of the two values of N derived in 6.4.11.11 and 6.4.11.12 of ADR (i.e. CSI = 50/N). The value of the criticality safety index may be zero, provided that an unlimited number of packages is subcritical (i.e. N is effectively equal to infinity in both cases).

2.2.7.6.2.2   The criticality safety index for each overpack or container shall be determined as the sum of the CSIs of all the packages contained. The same procedure shall be followed for determining the total sum of the CSIs in a consignment or aboard a conveyance.

2.2.7.7   **Activity limits and material restrictions**

2.2.7.7.1   *Contents limits for packages*

2.2.7.7.1.1   General

The quantity of radioactive material in a package shall not exceed the relevant limits for the package type as specified below.

2.2.7.7.1.2   Excepted packages

2.2.7.7.1.2.1   For radioactive material other than articles manufactured of natural uranium, depleted uranium or natural thorium, an excepted package shall not contain activities greater than the following:

(a)   Where the radioactive material is enclosed in or is included as a component part of an instrument or other manufactured article, such as a clock or electronic apparatus, the limits specified in columns 2 and 3 of Table 2.2.7.7.1.2.1 for each individual item and each package, respectively; and

(b)   Where the radioactive material is not so enclosed in or is not included as a component of an instrument or other manufactured article, the package limits specified in column 4 of Table 2.2.7.7.1.2.1.

Table 2.2.7.7.1.2.1

Activity limits for excepted packages

Physical state of contents	Instruments or article		Materials
	Item limits [a]	Package limits [a]	Package limits
Solids			
special form	$10^{-2} A_1$	A_1	$10^{-3} A_1$
other form	$10^{-2} A_2$	A_2	$10^{-3} A_2$
Liquids	$10^{-3} A_2$	$10^{-1} A_2$	$10^{-4} A_2$
Gases			
tritium	$2 \times 10^{-2} A_2$	$2 \times 10^{-1} A_2$	$2 \times 10^{-2} A_2$
special form	$10^{-3} A_1$	$10^{-2} A_1$	$10^{-3} A_1$
other forms	$10^{-3} A_2$	$10^{-2} A_2$	$10^{-3} A_2$

[a]   *For mixtures of radionuclides, see 2.2.7.7.2.4 to 2.2.7.7.2.6.*

2.2.7.7.1.2.2   For articles manufactured of natural uranium, depleted uranium or natural thorium, an excepted package may contain any quantity of such material provided that the outer surface of the uranium or thorium is enclosed in an inactive sheath made of metal or some other substantial material.

2.2.7.7.1.3   Industrial packages

The radioactive contents in a single package of LSA material or in a single package of SCO shall be so restricted that the radiation level specified in 4.1.9.2.1 of ADR shall not be exceeded, and the activity in a single package shall also be so restricted that the activity limits for a vehicle specified in 7.5.11, CV33 (2) of ADR shall not be exceeded.

2.2.7.7.1.4   Type A packages

2.2.7.7.1.4.1   Type A packages shall not contain activities greater than the following:

(a)   For special form radioactive material - A_1; or

(b)   For all other radioactive material - A$_2$.

2.2.7.7.1.4.2   For mixtures of radionuclides whose identities and respective activities are known, the following condition shall apply to the radioactive contents of a Type A package:

$$\sum_i \frac{B(i)}{A_1(i)} + \sum_j \frac{C(j)}{A_2(j)} \leq 1$$

where

B(i)   is the activity of radionuclide i as special form radioactive material and A$_1$(i) is the A$_1$ value for radionuclide i; and

C(j)   is the activity of radionuclide j as other than special form radioactive material and A$_2$(j) is the A$_2$ value for radionuclide j.

2.2.7.7.1.5   Type B(U) and Type B(M) packages

2.2.7.7.1.5.1   Type B(U) and Type B(M) packages shall not contain:

(a)   activities greater than those authorized for the package design;

(b)   radionuclides different from those authorized for the package design; or

(c)   contents in a form, or a physical or chemical state different from those authorized for the package design;

as specified in their certificates of approval.

2.2.7.7.1.6   Type C packages

NOTE: Type C packages may be transported by air carrying radioactive material in quantities exceeding either 3 000A$_1$ or 100 000A$_2$,whichever is the lower for special form radioactive material, or 3 000A$_2$ for all other radioactive material. Whilst Type C packages are not required for carriage of radioactive material by inland waterways in such quantities (Type B(U) or Type B(M) packages suffice), the following requirements are presented since such packages may also be carried by inland waterways.

Type C packages shall not contain:

(a)   activities greater than those authorized for the package design;

(b)   radionuclides different from those authorized for the package design; or

(c)   contents in a form, or physical or chemical state different from those authorized for the package design;

as specified in their certificates of approval.

2.2.7.7.1.7   Packages containing fissile material

Unless excepted by 6.4.11.2 of ADR, packages containing fissile material shall not contain:

(a)   A mass of fissile material different from that authorized for the package design;

(b)   Any radionuclide or fissile material different from those authorized for the package design; or

(c)   Contents in a form or physical or chemical state, or in a spatial arrangement, different from those authorized for the package design;

as specified in their certificates of approval where appropriate.

2.2.7.7.1.8   Packages containing uranium hexafluoride

Packages containing uranium hexafluoride shall not contain:

(a)   a mass of uranium hexafluoride different from that authorized for the package design;

(b)   a mass of uranium hexafluoride greater than a value that would lead to an ullage smaller than 5 % at the maximum temperature of the package as specified for the plant systems where the package shall be used; or

(c)   uranium hexafluoride other than in solid form or at an internal pressure above atmospheric pressure when presented for carriage.

2.2.7.7.2   *Activity levels*

2.2.7.7.2.1   The following basic values for individual radionuclides are given in Table 2.2.7.7.2.1:

(a) A_1 and A_2 in TBq;

(b)   Activity concentration for exempt material in Bq/g; and

(c)   Activity limits for exempt consignments in Bq.

Table 2.2.7.7.2.1

Radionuclide (atomic number)	A_1 (TBq)	A_2 (TBq)	Activity concentration for exempt material (Bq/g)	Activity limit for an exempt consignment (Bq)
Actinium (89)				
Ac-225 (a)	$8 \text{ H } 10^{-1}$	$6 \text{ H } 10^{-3}$	$1 \text{ H } 10^{1}$	$1 \text{ H } 10^{4}$
Ac-227 (a)	$9 \text{ H } 10^{-1}$	$9 \text{ H } 10^{-5}$	$1 \text{ H } 10^{-1}$	$1 \text{ H } 10^{3}$
Ac-228	$6 \text{ H } 10^{-1}$	$5 \text{ H } 10^{-1}$	$1 \text{ H } 10^{1}$	$1 \text{ H } 10^{6}$
Silver (47)				
Ag-105	$2 \text{ H } 10^{0}$	$2 \text{ H } 10^{0}$	$1 \text{ H } 10^{2}$	$1 \text{ H } 10^{6}$
Ag-108m (a)	$7 \text{ H } 10^{-1}$	$7 \text{ H} 10^{-1}$	$1 \text{ H} 10^{1}$ (b)	$1 \text{ H } 10^{6}$ (b)
Ag-110m (a)	$4 \text{ H } 10^{-1}$	$4 \text{ H } 10^{-1}$	$1 \text{ H } 10^{1}$	$1 \text{ H } 10^{6}$
Ag-111	$2 \text{ H } 10^{0}$	$6 \text{ H } 10^{-1}$	$1 \text{ H } 10^{3}$	$1 \text{ H } 10^{6}$
Aluminium (13)				
Al-26	$1 \text{ H } 10^{-1}$	1×10^{-1}	1×10^{1}	1×10^{5}
Americium (95)				
Am-241	1×10^{1}	1×10^{-3}	1×10^{0}	1×10^{4}
Am-242m (a)	1×10^{1}	1×10^{-3}	1×10^{0} (b)	1×10^{4} (b)
Am-243 (a)	5×10^{0}	1×10^{-3}	1×10^{0} (b)	1×10^{3} (b)
Argon (18)				
Ar-37	4×10^{1}	4×10^{1}	1×10^{6}	1×10^{8}
Ar-39	4×10^{1}	2×10^{1}	1×10^{7}	1×10^{4}
Ar-41	3×10^{-1}	3×10^{-1}	1×10^{2}	1×10^{9}
Arsenic (33)				
As-72	3×10^{-1}	3×10^{-1}	1×10^{1}	1×10^{5}
As-73	4×10^{1}	4×10^{1}	1×10^{3}	1×10^{7}
As-74	1×10^{0}	9×10^{-1}	1×10^{1}	1×10^{6}
As-76	3×10^{-1}	3×10^{-1}	1×10^{2}	1×10^{5}
As-77	2×10^{1}	7×10^{-1}	1×10^{3}	1×10^{6}
Astatine (85)				
At-211 (a)	2×10^{1}	5×10^{-1}	1×10^{3}	1×10^{7}

Radionuclide (atomic number)	A_1 (TBq)	A_2 (TBq)	Activity concentration for exempt material (Bq/g)	Activity limit for an exempt consignment (Bq)
Gold (79)				
Au-193	7×10^0	2×10^0	1×10^2	1×10^7
Au-194	1×10^0	1×10^0	1×10^1	1×10^6
Au-195	1×10^1	6×10^0	1×10^2	1×10^7
Au-198	1×10^0	6×10^{-1}	1×10^2	1×10^6
Au-199	1×10^1	6×10^{-1}	1×10^2	1×10^6
Barium (56)				
Ba-131 (a)	2×10^0	2×10^0	1×10^2	1×10^6
Ba-133	3×10^0	3×10^0	1×10^2	1×10^6
Ba-133m	2×10^1	6×10^{-1}	1×10^2	1×10^6
Ba-140 (a)	5×10^{-1}	3×10^{-1}	1×10^1 (b)	1×10^5 (b)
Beryllium (4)				
Be-7	2×10^1	2×10^1	1×10^3	1×10^7
Be-10	4×10^1	6×10^{-1}	1×10^4	1×10^6
Bismuth (83)				
Bi-205	7×10^{-1}	7×10^{-1}	1×10^1	1×10^6
Bi-206	3×10^{-1}	3×10^{-1}	1×10^1	1×10^5
Bi-207	7×10^{-1}	7×10^{-1}	1×10^1	1×10^6
Bi-210	1×10^0	6×10^{-1}	1×10^3	1×10^6
Bi-210m (a)	6×10^{-1}	2×10^{-2}	1×10^1	1×10^5
Bi-212 (a)	7×10^{-1}	6×10^{-1}	1×10^1 (b)	1×10^5 (b)
Berkelium (97)				
Bk-247	8×10^0	8×10^{-4}	1×10^0	1×10^4
Bk-249 (a)	4×10^1	3×10^{-1}	1×10^3	1×10^6
Bromine (35)				
Br-76	4×10^{-1}	4×10^{-1}	1×10^1	1×10^5
Br-77	3×10^0	3×10^0	1×10^2	1×10^6
Br-82	4×10^{-1}	4×10^{-1}	1×10^1	1×10^6
Carbon (6)				
C-11	1×10^0	6×10^{-1}	1×10^1	1×10^6
C-14	4×10^1	3×10^0	1×10^4	1×10^7

Radionuclide (atomic number)	A_1 (TBq)	A_2 (TBq)	Activity concentration for exempt material (Bq/g)	Activity limit for an exempt consignment (Bq)
Calcium (20)				
Ca-41	Unlimited	Unlimited	1×10^5	1×10^7
Ca-45	4×10^1	1×10^0	1×10^4	1×10^7
Ca-47 (a)	3×10^0	3×10^{-1}	1×10^1	1×10^6
Cadmium (48)				
Cd-109	3×10^1	2×10^0	1×10^4	1×10^6
Cd-113m	4×10^1	5×10^{-1}	1×10^3	1×10^6
Cd-115 (a)	3×10^0	4×10^{-1}	1×10^2	1×10^6
Cd-115m	5×10^{-1}	5×10^{-1}	1×10^3	1×10^6
Cerium (58)				
Ce-139	7×10^0	2×10^0	1×10^2	1×10^6
Ce-141	2×10^1	6×10^{-1}	1×10^2	1×10^7
Ce-143	9×10^{-1}	6×10^{-1}	1×10^2	1×10^6
Ce-144 (a)	2×10^{-1}	2×10^{-1}	1×10^2 (b)	1×10^5 (b)
Californium (98)				
Cf-248	4×10^1	6×10^{-3}	1×10^1	1×10^4
Cf-249	3×10^0	8×10^{-4}	1×10^0	1×10^3
Cf-250	2×10^1	2×10^{-3}	1×10^1	1×10^4
Cf-251	7×10^0	7×10^{-4}	1×10^0	1×10^3
Cf-252	1×10^{-1}	3×10^{-3}	1×10^1	1×10^4
Cf-253 (a)	4×10^1	4×10^{-2}	1×10^2	1×10^5
Cf-254	1×10^{-3}	1×10^{-3}	1×10^0	1×10^3
Chlorine (17)				
Cl-36	1×10^1	6×10^{-1}	1×10^4	1×10^6
Cl-38	2×10^{-1}	2×10^{-1}	1×10^1	1×10^5

Radionuclide (atomic number)	A_1 (TBq)	A_2 (TBq)	Activity concentration for exempt material (Bq/g)	Activity limit for an exempt consignment (Bq)
Curium (96)				
Cm-240	4×10^1	2×10^{-2}	1×10^2	1×10^5
Cm-241	2×10^0	1×10^0	1×10^2	1×10^6
Cm-242	4×10^1	1×10^{-2}	1×10^2	1×10^5
Cm-243	9×10^0	1×10^{-3}	1×10^0	1×10^4
Cm-244	2×10^1	2×10^{-3}	1×10^1	1×10^4
Cm-245	9×10^0	9×10^{-4}	1×10^0	1×10^3
Cm-246	9×10^0	9×10^{-4}	1×10^0	1×10^3
Cm-247 (a)	3×10^0	1×10^{-3}	1×10^0	1×10^4
Cm-248	2×10^{-2}	3×10^{-4}	1×10^0	1×10^3
Cobalt (27)				
Co-55	5×10^{-1}	5×10^{-1}	1×10^1	1×10^6
Co-56	3×10^{-1}	3×10^{-1}	1×10^1	1×10^5
Co-57	1×10^1	1×10^1	1×10^2	1×10^6
Co-58	1×10^0	1×10^0	1×10^1	1×10^6
Co-58m	4×10^1	4×10^1	1×10^4	1×10^7
Co-60	4×10^{-1}	4×10^{-1}	1×10^1	1×10^5
Chromium (24)				
Cr-51	3×10^1	3×10^1	1×10^3	1×10^7
Caesium (55)				
Cs-129	4×10^0	4×10^0	1×10^2	1×10^5
Cs-131	3×10^1	3×10^1	1×10^3	1×10^6
Cs-132	1×10^0	1×10^0	1×10^1	1×10^5
Cs-134	7×10^{-1}	7×10^{-1}	1×10^1	1×10^4
Cs-134m	4×10^1	6×10^{-1}	1×10^3	1×10^5
Cs-135	4×10^1	1×10^0	1×10^4	1×10^7
Cs-136	5×10^{-1}	5×10^{-1}	1×10^1	1×10^5
Cs-137 (a)	2×10^0	6×10^{-1}	1×10^1 (b)	1×10^4 (b)

Radionuclide (atomic number)	A_1 (TBq)	A_2 (TBq)	Activity concentration for exempt material (Bq/g)	Activity limit for an exempt consignment (Bq)
Copper (29)				
Cu-64	6×10^0	1×10^0	1×10^2	1×10^6
Cu-67	1×10^1	7×10^{-1}	1×10^2	1×10^6
Dysprosium (66)				
Dy-159	2×10^1	2×10^1	1×10^3	1×10^7
Dy-165	9×10^{-1}	6×10^{-1}	1×10^3	1×10^6
Dy-166 (a)	9×10^{-1}	3×10^{-1}	1×10^3	1×10^6
Erbium (68)				
Er-169	4×10^1	1×10^0	1×10^4	1×10^7
Er-171	8×10^{-1}	5×10^{-1}	1×10^2	1×10^6
Europium (63)				
Eu-147	2×10^0	2×10^0	1×10^2	1×10^6
Eu-148	5×10^{-1}	5×10^{-1}	1×10^1	1×10^6
Eu-149	2×10^1	2×10^1	1×10^2	1×10^7
Eu-150(short lived)	2×10^0	7×10^{-1}	1×10^3	1×10^6
Eu-150(long lived)	7×10^{-1}	7×10^{-1}	1×10^1	1×10^6
Eu-152	1×10^0	1×10^0	1×10^1	1×10^6
Eu-152m	8×10^{-1}	8×10^{-1}	1×10^2	1×10^6
Eu-154	9×10^{-1}	6×10^{-1}	1×10^1	1×10^6
Eu-155	2×10^1	3×10^0	1×10^2	1×10^7
Eu-156	7×10^{-1}	7×10^{-1}	1×10^1	1×10^6
Fluorine (9)				
F-18	1×10^0	6×10^{-1}	1×10^1	1×10^6
Iron (26)				
Fe-52 (a)	3×10^{-1}	3×10^{-1}	1×10^1	1×10^6
Fe-55	4×10^1	4×10^1	1×10^4	1×10^6
Fe-59	9×10^{-1}	9×10^{-1}	1×10^1	1×10^6
Fe-60 (a)	4×10^1	2×10^{-1}	1×10^2	1×10^5

Radionuclide (atomic number)	A_1 (TBq)	A_2 (TBq)	Activity concentration for exempt material (Bq/g)	Activity limit for an exempt consignment (Bq)
Gallium (31)				
Ga-67	7×10^0	3×10^0	1×10^2	1×10^6
Ga-68	5×10^{-1}	5×10^{-1}	1×10^1	1×10^5
Ga-72	4×10^{-1}	4×10^{-1}	1×10^1	1×10^5
Gadolinium (64)				
Gd-146 (a)	5×10^{-1}	5×10^{-1}	1×10^1	1×10^6
Gd-148	2×10^1	2×10^{-3}	1×10^1	1×10^4
Gd-153	1×10^1	9×10^0	1×10^2	1×10^7
Gd-159	3×10^0	6×10^{-1}	1×10^3	1×10^6
Germanium (32)				
Ge-68 (a)	5×10^{-1}	5×10^{-1}	1×10^1	1×10^5
Ge-71	4×10^1	4×10^1	1×10^4	1×10^8
Ge-77	3×10^{-1}	3×10^{-1}	1×10^1	1×10^5
Hafnium (72)				
Hf-172 (a)	6×10^{-1}	6×10^{-1}	1×10^1	1×10^6
Hf-175	3×10^0	3×10^0	1×10^2	1×10^6
Hf-181	2×10^0	5×10^{-1}	1×10^1	1×10^6
Hf-182	Unlimited	Unlimited	1×10^2	1×10^6
Mercury (80)				
Hg-194 (a)	1×10^0	1×10^0	1×10^1	1×10^6
Hg-195m (a)	3×10^0	7×10^{-1}	1×10^2	1×10^6
Hg-197	2×10^1	1×10^1	1×10^2	1×10^7
Hg-197m	1×10^1	4×10^{-1}	1×10^2	1×10^6
Hg-203	5×10^0	1×10^0	1×10^2	1×10^5
Holmium (67)				
Ho-166	4×10^{-1}	4×10^{-1}	1×10^3	1×10^5
Ho-166m	6×10^{-1}	5×10^{-1}	1×10^1	1×10^6

Radionuclide (atomic number)	A_1 (TBq)	A_2 (TBq)	Activity concentration for exempt material (Bq/g)	Activity limit for an exempt consignment (Bq)
Iodine (53)				
I-123	6×10^0	3×10^0	1×10^2	1×10^7
I-124	1×10^0	1×10^0	1×10^1	1×10^6
I-125	2×10^1	3×10^0	1×10^3	1×10^6
I-126	2×10^0	1×10^0	1×10^2	1×10^6
I-129	Unlimited	Unlimited	1×10^2	1×10^5
I-131	3×10^0	7×10^{-1}	1×10^2	1×10^6
I-132	4×10^{-1}	4×10^{-1}	1×10^1	1×10^5
I-133	7×10^{-1}	6×10^{-1}	1×10^1	1×10^6
I-134	3×10^{-1}	3×10^{-1}	1×10^1	1×10^5
I-135 (a)	6×10^{-1}	6×10^{-1}	1×10^1	1×10^6
Indium (49)				
In-111	3×10^0	3×10^0	1×10^2	1×10^6
In-113m	4×10^0	2×10^0	1×10^2	1×10^6
In-114m (a)	1×10^1	5×10^{-1}	1×10^2	1×10^6
In-115m	7×10^0	1×10^0	1×10^2	1×10^6
Iridium (77)				
Ir-189 (a)	1×10^1	1×10^1	1×10^2	1×10^7
Ir-190	7×10^{-1}	7×10^{-1}	1×10^1	1×10^6
Ir-192	1×10^0(c)	6×10^{-1}	1×10^1	1×10^4
Ir-194	3×10^{-1}	3×10^{-1}	1×10^2	1×10^5
Potassium (19)				
K-40	9×10^{-1}	9×10^{-1}	1×10^2	1×10^6
K-42	2×10^{-1}	2×10^{-1}	1×10^2	1×10^6
K-43	7×10^{-1}	6×10^{-1}	1×10^1	1×10^6
Krypton (36)				
Kr-79	4×10^0	1×10^0	1×10^3	1×10^5
Kr-81	4×10^1	4×10^1	1×10^4	1×10^7
Kr-85	1×10^1	1×10^1	1×10^5	1×10^4
Kr-85m	8×10^0	3×10^0	1×10^3	1×10^{10}
Kr-87	2×10^{-1}	2×10^{-1}	1×10^2	1×10^9

Radionuclide (atomic number)	A_1 (TBq)	A_2 (TBq)	Activity concentration for exempt material (Bq/g)	Activity limit for an exempt consignment (Bq)
Lanthanum (57)				
La-137	3×10^1	6×10^0	1×10^3	1×10^7
La-140	4×10^{-1}	4×10^{-1}	1×10^1	1×10^5
Lutetium (71)				
Lu-172	6×10^{-1}	6×10^{-1}	1×10^1	1×10^6
Lu-173	8×10^0	8×10^0	1×10^2	1×10^7
Lu-174	9×10^0	9×10^0	1×10^2	1×10^7
Lu-174m	2×10^1	1×10^1	1×10^2	1×10^7
Lu-177	3×10^1	7×10^{-1}	1×10^3	1×10^7
Magnesium (12)				
Mg-28 (a)	3×10^{-1}	3×10^{-1}	1×10^1	1×10^5
Manganese (25)				
Mn-52	3×10^{-1}	3×10^{-1}	1×10^1	1×10^5
Mn-53	Unlimited	Unlimited	1×10^4	1×10^9
Mn-54	1×10^0	1×10^0	1×10^1	1×10^6
Mn-56	3×10^{-1}	3×10^{-1}	1×10^1	1×10^5
Molybdenum (42)				
Mo-93	4×10^1	2×10^1	1×10^3	1×10^8
Mo-99 (a)	1×10^0	6×10^{-1}	1×10^2	1×10^6
Nitrogen (7)				
N-13	9×10^{-1}	6×10^{-1}	1×10^2	1×10^9
Sodium (11)				
Na-22	5×10^{-1}	5×10^{-1}	1×10^1	1×10^6
Na-24	2×10^{-1}	2×10^{-1}	1×10^1	1×10^5
Niobium (41)				
Nb-93m	4×10^1	3×10^1	1×10^4	1×10^7
Nb-94	7×10^{-1}	7×10^{-1}	1×10^1	1×10^6
Nb-95	1×10^0	1×10^0	1×10^1	1×10^6
Nb-97	9×10^{-1}	6×10^{-1}	1×10^1	1×10^6

Radionuclide (atomic number)	A_1 (TBq)	A_2 (TBq)	Activity concentration for exempt material (Bq/g)	Activity limit for an exempt consignment (Bq)
Neodymium (60)				
Nd-147	6×10^0	6×10^{-1}	1×10^2	1×10^6
Nd-149	6×10^{-1}	5×10^{-1}	1×10^2	1×10^6
Nickel (28)				
Ni-59	Unlimited	Unlimited	1×10^4	1×10^8
Ni-63	4×10^1	3×10^1	1×10^5	1×10^8
Ni-65	4×10^{-1}	4×10^{-1}	1×10^1	1×10^6
Neptunium (93)				
Np-235	4×10^1	4×10^1	1×10^3	1×10^7
Np-236(short-lived)	2×10^1	2×10^0	1×10^3	1×10^7
Np-236(long-lived)	9×10^0	2×10^{-2}	1×10^2	1×10^5
Np-237	2×10^1	2×10^{-3}	1×10^0 (b)	1×10^3 (b)
Np-239	7×10^0	4×10^{-1}	1×10^2	1×10^7
Osmium (76)				
Os-185	1×10^0	1×10^0	1×10^1	1×10^6
Os-191	1×10^1	2×10^0	1×10^2	1×10^7
Os-191m	4×10^1	3×10^1	1×10^3	1×10^7
Os-193	2×10^0	6×10^{-1}	1×10^2	1×10^6
Os-194 (a)	3×10^{-1}	3×10^{-1}	1×10^2	1×10^5
Phosphorus (15)				
P-32	5×10^{-1}	5×10^{-1}	1×10^3	1×10^5
P-33	4×10^1	1×10^0	1×10^5	1×10^8
Protactinium (91)				
Pa-230 (a)	2×10^0	7×10^{-2}	1×10^1	1×10^6
Pa-231	4×10^0	4×10^{-4}	1×10^0	1×10^3
Pa-233	5×10^0	7×10^{-1}	1×10^2	1×10^7

Radionuclide (atomic number)	A_1 (TBq)	A_2 (TBq)	Activity concentration for exempt material (Bq/g)	Activity limit for an exempt consignment (Bq)
Lead (82)				
Pb-201	1×10^0	1×10^0	1×10^1	1×10^6
Pb-202	4×10^1	2×10^1	1×10^3	1×10^6
Pb-203	4×10^0	3×10^0	1×10^2	1×10^6
Pb-205	Unlimited	Unlimited	1×10^4	1×10^7
Pb-210 (a)	1×10^0	5×10^{-2}	1×10^1 (b)	1×10^4 (b)
Pb-212 (a)	7×10^{-1}	2×10^{-1}	1×10^1 (b)	1×10^5 (b)
Palladium (46)				
Pd-103 (a)	4×10^1	4×10^1	1×10^3	1×10^8
Pd-107	Unlimited	Unlimited	1×10^5	1×10^8
Pd-109	2×10^0	5×10^{-1}	1×10^3	1×10^6
Promethium (61)				
Pm-143	3×10^0	3×10^0	1×10^2	1×10^6
Pm-144	7×10^{-1}	7×10^{-1}	1×10^1	1×10^6
Pm-145	3×10^1	1×10^1	1×10^3	1×10^7
Pm-147	4×10^1	2×10^0	1×10^4	1×10^7
Pm-148m (a)	8×10^{-1}	7×10^{-1}	1×10^1	1×10^6
Pm-149	2×10^0	6×10^{-1}	1×10^3	1×10^6
Pm-151	2×10^0	6×10^{-1}	1×10^2	1×10^6
Polonium (84)				
Po-210	4×10^1	2×10^{-2}	1×10^1	1×10^4
Praseodymium (59)				
Pr-142	4×10^{-1}	4×10^{-1}	1×10^2	1×10^5
Pr-143	3×10^0	6×10^{-1}	1×10^4	1×10^6

Radionuclide (atomic number)	A_1 (TBq)	A_2 (TBq)	Activity concentration for exempt material (Bq/g)	Activity limit for an exempt consignment (Bq)
Platinum (78)				
Pt-188 (a)	1×10^0	8×10^{-1}	1×10^1	1×10^6
Pt-191	4×10^0	3×10^0	1×10^2	1×10^6
Pt-193	4×10^1	4×10^1	1×10^4	1×10^7
Pt-193m	4×10^1	5×10^{-1}	1×10^3	1×10^7
Pt-195m	1×10^1	5×10^{-1}	1×10^2	1×10^6
Pt-197	2×10^1	6×10^{-1}	1×10^3	1×10^6
Pt-197m	1×10^1	6×10^{-1}	1×10^2	1×10^6
Plutonium (94)				
Pu-236	3×10^1	3×10^{-3}	1×10^1	1×10^4
Pu-237	2×10^1	2×10^1	1×10^3	1×10^7
Pu-238	1×10^1	1×10^{-3}	1×10^0	1×10^4
Pu-239	1×10^1	1×10^{-3}	1×10^0	1×10^4
Pu-240	1×10^1	1×10^{-3}	1×10^0	1×10^3
Pu-241 (a)	4×10^1	6×10^{-2}	1×10^2	1×10^5
Pu-242	1×10^1	1×10^{-3}	1×10^0	1×10^4
Pu-244 (a)	4×10^{-1}	1×10^{-3}	1×10^0	1×10^4
Radium (88)				
Ra-223 (a)	4×10^{-1}	7×10^{-3}	1×10^2 (b)	1×10^5 (b)
Ra-224 (a)	4×10^{-1}	2×10^{-2}	1×10^1 (b)	1×10^5 (b)
Ra-225 (a)	2×10^{-1}	4×10^{-3}	1×10^2	1×10^5
Ra-226 (a)	2×10^{-1}	3×10^{-3}	1×10^1 (b)	1×10^4 (b)
Ra-228 (a)	6×10^{-1}	2×10^{-2}	1×10^1 (b)	1×10^5 (b)
Rubidium (37)				
Rb-81	2×10^0	8×10^{-1}	1×10^1	1×10^6
Rb-83 (a)	2×10^0	2×10^0	1×10^2	1×10^6
Rb-84	1×10^0	1×10^0	1×10^1	1×10^6
Rb-86	5×10^{-1}	5×10^{-1}	1×10^2	1×10^5
Rb-87	Unlimited	Unlimited	1×10^4	1×10^7
Rb(nat)	Unlimited	Unlimited	1×10^4	1×10^7

Radionuclide (atomic number)	A_1 (TBq)	A_2 (TBq)	Activity concentration for exempt material (Bq/g)	Activity limit for an exempt consignment (Bq)
Rhenium (75)				
Re-184	1×10^0	1×10^0	1×10^1	1×10^6
Re-184m	3×10^0	1×10^0	1×10^2	1×10^6
Re-186	2×10^0	6×10^{-1}	1×10^3	1×10^6
Re-187	Unlimited	Unlimited	1×10^6	1×10^9
Re-188	4×10^{-1}	4×10^{-1}	1×10^2	1×10^5
Re-189 (a)	3×10^0	6×10^{-1}	1×10^2	1×10^6
Re(nat)	Unlimited	Unlimited	1×10^6	1×10^9
Rhodium (45)				
Rh-99	2×10^0	2×10^0	1×10^1	1×10^6
Rh-101	4×10^0	3×10^0	1×10^2	1×10^7
Rh-102	5×10^{-1}	5×10^{-1}	1×10^1	1×10^6
Rh-102m	2×10^0	2×10^0	1×10^2	1×10^6
Rh-103m	4×10^1	4×10^1	1×10^4	1×10^8
Rh-105	1×10^1	8×10^{-1}	1×10^2	1×10^7
Radon (86)				
Rn-222 (a)	3×10^{-1}	4×10^{-3}	1×10^1 (b)	1×10^8 (b)
Ruthenium (44)				
Ru-97	5×10^0	5×10^0	1×10^2	1×10^7
Ru-103 (a)	2×10^0	2×10^0	1×10^2	1×10^6
Ru-105	1×10^0	6×10^{-1}	1×10^1	1×10^6
Ru-106 (a)	2×10^{-1}	2×10^{-1}	1×10^2 (b)	1×10^5 (b)
Sulphur (16)				
S-35	4×10^1	3×10^0	1×10^5	1×10^8
Antimony (51)				
Sb-122	4×10^{-1}	4×10^{-1}	1×10^2	1×10^4
Sb-124	6×10^{-1}	6×10^{-1}	1×10^1	1×10^6
Sb-125	2×10^0	1×10^0	1×10^2	1×10^6
Sb-126	4×10^{-1}	4×10^{-1}	1×10^1	1×10^5

Radionuclide (atomic number)	A_1 (TBq)	A_2 (TBq)	Activity concentration for exempt material (Bq/g)	Activity limit for an exempt consignment (Bq)
Scandium (21)				
Sc-44	5×10^{-1}	5×10^{-1}	1×10^{1}	1×10^{5}
Sc-46	5×10^{-1}	5×10^{-1}	1×10^{1}	1×10^{6}
Sc-47	1×10^{1}	7×10^{-1}	1×10^{2}	1×10^{6}
Sc-48	3×10^{-1}	3×10^{-1}	1×10^{1}	1×10^{5}
Selenium (34)				
Se-75	3×10^{0}	3×10^{0}	1×10^{2}	1×10^{6}
Se-79	4×10^{1}	2×10^{0}	1×10^{4}	1×10^{7}
Silicon (14)				
Si-31	6×10^{-1}	6×10^{-1}	1×10^{3}	1×10^{6}
Si-32	4×10^{1}	5×10^{-1}	1×10^{3}	1×10^{6}
Samarium (62)				
Sm-145	1×10^{1}	1×10^{1}	1×10^{2}	1×10^{7}
Sm-147	Unlimited	Unlimited	1×10^{1}	1×10^{4}
Sm-151	4×10^{1}	1×10^{1}	1×10^{4}	1×10^{8}
Sm-153	9×10^{0}	6×10^{-1}	1×10^{2}	1×10^{6}
Tin (50)				
Sn-113 (a)	4×10^{0}	2×10^{0}	1×10^{3}	1×10^{7}
Sn-117m	7×10^{0}	4×10^{-1}	1×10^{2}	1×10^{6}
Sn-119m	4×10^{1}	3×10^{1}	1×10^{3}	1×10^{7}
Sn-121m (a)	4×10^{1}	9×10^{-1}	1×10^{3}	1×10^{7}
Sn-123	8×10^{-1}	6×10^{-1}	1×10^{3}	1×10^{6}
Sn-125	4×10^{-1}	4×10^{-1}	1×10^{2}	1×10^{5}
Sn-126 (a)	6×10^{-1}	4×10^{-1}	1×10^{1}	1×10^{5}

Radionuclide (atomic number)	A_1 (TBq)	A_2 (TBq)	Activity concentration for exempt material (Bq/g)	Activity limit for an exempt consignment (Bq)
Strontium (38)				
Sr-82 (a)	2×10^{-1}	2×10^{-1}	1×10^1	1×10^5
Sr-85	2×10^0	2×10^0	1×10^2	1×10^6
Sr-85m	5×10^0	5×10^0	1×10^2	1×10^7
Sr-87m	3×10^0	3×10^0	1×10^2	1×10^6
Sr-89	6×10^{-1}	6×10^{-1}	1×10^3	1×10^6
Sr-90 (a)	3×10^{-1}	3×10^{-1}	1×10^2 (b)	1×10^4 (b)
Sr-91 (a)	3×10^{-1}	3×10^{-1}	1×10^1	1×10^5
Sr-92 (a)	1×10^0	3×10^{-1}	1×10^1	1×10^6
Tritium (1)				
T(H-3)	4×10^1	4×10^1	1×10^6	1×10^9
Tantalum (73)				
Ta-178(long-lived)	1×10^0	8×10^{-1}	1×10^1	1×10^6
Ta-179	3×10^1	3×10^1	1×10^3	1×10^7
Ta-182	9×10^{-1}	5×10^{-1}	1×10^1	1×10^4
Terbium (65)				
Tb-157	4×10^1	4×10^1	1×10^4	1×10^7
Tb-158	1×10^0	1×10^0	1×10^1	1×10^6
Tb-160	1×10^0	6×10^{-1}	1×10^1	1×10^6
Technetium (43)				
Tc-95m (a)	2×10^0	2×10^0	1×10^1	1×10^6
Tc-96	4×10^{-1}	4×10^{-1}	1×10^1	1×10^6
Tc-96m (a)	4×10^{-1}	4×10^{-1}	1×10^3	1×10^7
Tc-97	Unlimited	Unlimited	1×10^3	1×10^8
Tc-97m	4×10^1	1×10^0	1×10^3	1×10^7
Tc-98	8×10^{-1}	7×10^{-1}	1×10^1	1×10^6
Tc-99	4×10^1	9×10^{-1}	1×10^4	1×10^7
Tc-99m	1×10^1	4×10^0	1×10^2	1×10^7

Radionuclide (atomic number)	A_1 (TBq)	A_2 (TBq)	Activity concentration for exempt material (Bq/g)	Activity limit for an exempt consignment (Bq)
Tellurium (52)				
Te-121	2×10^0	2×10^0	1×10^1	1×10^6
Te-121m	5×10^0	3×10^0	1×10^2	1×10^6
Te-123m	8×10^0	1×10^0	1×10^2	1×10^7
Te-125m	2×10^1	9×10^{-1}	1×10^3	1×10^7
Te-127	2×10^1	7×10^{-1}	1×10^3	1×10^6
Te-127m (a)	2×10^1	5×10^{-1}	1×10^3	1×10^7
Te-129	7×10^{-1}	6×10^{-1}	1×10^2	1×10^6
Te-129m (a)	8×10^{-1}	4×10^{-1}	1×10^3	1×10^6
Te-131m (a)	7×10^{-1}	5×10^{-1}	1×10^1	1×10^6
Te-132 (a)	5×10^{-1}	4×10^{-1}	1×10^2	1×10^7
Thorium (90)				
Th-227	1×10^1	5×10^{-3}	1×10^1	1×10^4
Th-228 (a)	5×10^{-1}	1×10^{-3}	1×10^0 (b)	1×10^4 (b)
Th-229	5×10^0	5×10^{-4}	1×10^0 (b)	1×10^3 (b)
Th-230	1×10^1	1×10^{-3}	1×10^0	1×10^4
Th-231	4×10^1	2×10^{-2}	1×10^3	1×10^7
Th-232	Unlimited	Unlimited	1×10^1	1×10^4
Th-234 (a)	3×10^{-1}	3×10^{-1}	1×10^3 (b)	1×10^5 (b)
Th(nat)	Unlimited	Unlimited	1×10^0 (b)	1×10^3 (b)
Titanium (22)				
Ti-44 (a)	5×10^{-1}	4×10^{-1}	1×10^1	1×10^5
Thallium (81)				
Tl-200	9×10^{-1}	9×10^{-1}	1×10^1	1×10^6
Tl-201	1×10^1	4×10^0	1×10^2	1×10^6
Tl-202	2×10^0	2×10^0	1×10^2	1×10^6
Tl-204	1×10^1	7×10^{-1}	1×10^4	1×10^4
Thulium (69)				
Tm-167	7×10^0	8×10^{-1}	1×10^2	1×10^6
Tm-170	3×10^0	6×10^{-1}	1×10^3	1×10^6
Tm-171	4×10^1	4×10^1	1×10^4	1×10^8

Radionuclide (atomic number)	A_1 (TBq)	A_2 (TBq)	Activity concentration for exempt material (Bq/g)	Activity limit for an exempt consignment (Bq)
Uranium (92)				
U-230 (fast lung absorption) (a)(d)	4×10^1	1×10^{-1}	1×10^1 (b)	1×10^5 (b)
U-230 (medium lung absorption) (a)(e)	4×10^1	4×10^{-3}	1×10^1	1×10^4
U-230 (slow lung absorption) (a)(f)	3×10^1	3×10^{-3}	1×10^1	1×10^4
U-232 (fast lung absorption) (d)	4×10^1	1×10^{-2}	1×10^0 (b)	1×10^3 (b)
U-232 (medium lung absorption) (e)	4×10^1	7×10^{-3}	1×10^1	1×10^4
U-232 (slow lung absorption) (f)	1×10^1	1×10^{-3}	1×10^1	1×10^4
U-233 (fast lung absorption) (d)	4×10^1	9×10^{-2}	1×10^1	1×10^4
U-233 (medium lung absorption) (e)	4×10^1	2×10^{-2}	1×10^2	1×10^5
U-233 (slow lung absorption) (f)	4×10^1	6×10^{-3}	1×10^1	1×10^5
U-234 (fast lung absorption) (d)	4×10^1	9×10^{-2}	1×10^1	1×10^4
U-234 (medium lung absorption) (e)	4×10^1	2×10^{-2}	1×10^2	1×10^5
U-234 (slow lung absorption) (f)	4×10^1	6×10^{-3}	1×10^1	1×10^5
U-235 (all lung absorption types) (a)(d)(e)(f)	Unlimited	Unlimited	1×10^1 (b)	1×10^4 (b)
U-236 (fast lung absorption) (d)	Unlimited	Unlimited	1×10^1	1×10^4
U-236 (medium lung absorption) (e)	4×10^1	2×10^{-2}	1×10^2	1×10^5
U-236 (slow lung absorption) (f)	4×10^1	6×10^{-3}	1×10^1	1×10^4
U-238 (all lung absorption types) (d)(e)(f)	Unlimited	Unlimited	1×10^1 (b)	1×10^4 (b)
U (nat)	Unlimited	Unlimited	1×10^0 (b)	1×10^3 (b)
U (enriched to 20% or less) (g)	Unlimited	Unlimited	1×10^0	1×10^3
U (dep)	Unlimited	Unlimited	1×10^0	1×10^3
Vanadium (23)				
V-48	4×10^{-1}	4×10^{-1}	1×10^1	1×10^5
V-49	4×10^1	4×10^1	1×10^4	1×10^7
Tungsten (74)				
W-178 (a)	9×10^0	5×10^0	1×10^1	1×10^6
W-181	3×10^1	3×10^1	1×10^3	1×10^7
W-185	4×10^1	8×10^{-1}	1×10^4	1×10^7
W-187	2×10^0	6×10^{-1}	1×10^2	1×10^6

Radionuclide (atomic number)	A_1 (TBq)	A_2 (TBq)	Activity concentration for exempt material (Bq/g)	Activity limit for an exempt consignment (Bq)
W-188 (a)	4×10^{-1}	3×10^{-1}	1×10^2	1×10^5
Xenon (54)				
Xe-122 (a)	4×10^{-1}	4×10^{-1}	1×10^2	1×10^9
Xe-123	2×10^0	7×10^{-1}	1×10^2	1×10^9
Xe-127	4×10^0	2×10^0	1×10^3	1×10^5
Xe-131m	4×10^1	4×10^1	1×10^4	1×10^4
Xe-133	2×10^1	1×10^1	1×10^3	1×10^4
Xe-135	3×10^0	2×10^0	1×10^3	1×10^{10}
Yttrium (39)				
Y-87 (a)	1×10^0	1×10^0	1×10^1	1×10^6
Y-88	4×10^{-1}	4×10^{-1}	1×10^1	1×10^6
Y-90	3×10^{-1}	3×10^{-1}	1×10^3	1×10^5
Y-91	6×10^{-1}	6×10^{-1}	1×10^3	1×10^6
Y-91m	2×10^0	2×10^0	1×10^2	1×10^6
Y-92	2×10^{-1}	2×10^{-1}	1×10^2	1×10^5
Y-93	3×10^{-1}	3×10^{-1}	1×10^2	1×10^5
Ytterbium (70)				
Yb-169	4×10^0	1×10^0	1×10^2	1×10^7
Yb-175	3×10^1	9×10^{-1}	1×10^3	1×10^7
Zinc (30)				
Zn-65	2×10^0	2×10^0	1×10^1	1×10^6
Zn-69	3×10^0	6×10^{-1}	1×10^4	1×10^6
Zn-69m (a)	3×10^0	6×10^{-1}	1×10^2	1×10^6
Zirconium (40)				
Zr-88	3×10^0	3×10^0	1×10^2	1×10^6
Zr-93	Unlimited	Unlimited	1×10^3 (b)	1×10^7 (b)
Zr-95 (a)	2×10^0	8×10^{-1}	1×10^1	1×10^6
Zr-97 (a)	4×10^{-1}	4×10^{-1}	1×10^1 (b)	1×10^5 (b)

(a) A_1 and/or A_2 values for these parent radionuclides include contributions from daughter radionuclides with half-lives less than 10 days, as listed in the following:

Mg-28	Al-28
Ar-42	K-42
Ca-47	Sc-47
Ti-44	Sc-44
Fe-52	Mn-52m
Fe-60	Co-60m
Zn-69m	Zn-69
Ge-68	Ga-68
Rb-83	Kr-83m
Sr-82	Rb-82
Sr-90	Y-90
Sr-91	Y-91m
Sr-92	Y-92
Y-87	Sr-87m
Zr-95	Nb-95m
Zr-97	Nb-97m, Nb-97
Mo-99	Tc-99m
Tc-95m	Tc-95
Tc-96m	Tc-96
Ru-103	Rh-103m
Ru-106	Rh-106
Pd-103	Rh-103m
Ag-108m	Ag-108
Ag-110m	Ag-110
Cd-115	In-115m
In-114m	In-114
Sn-113	In-113m
Sn-121m	Sn-121
Sn-126	Sb-126m
Te-118	Sb-118
Te-127m	Te-127
Te-129m	Te-129
Te-131m	Te-131
Te-132	I-132
I-135	Xe-135m
Xe-122	I-122
Cs-137	Ba-137m
Ba-131	Cs-131
Ba-140	La-140
Ce-144	Pr-144m, Pr-144
Pm-148m	Pm-148
Gd-146	Eu-146
Dy-166	Ho-166
Hf-172	Lu-172
W-178	Ta-178
W-188	Re-188
Re-189	Os-189m
Os-194	Ir-194
Ir-189	Os-189m
Pt-188	Ir-188
Hg-194	Au-194
Hg-195m	Hg-195
Pb-210	Bi-210
Pb-212	Bi-212, Tl-208, Po-212
Bi-210m	Tl-206
Bi-212	Tl-208, Po-212

At-211	Po-211
Rn-222	Po-218, Pb-214, At-218, Bi-214, Po-214
Ra-223	Rn-219, Po-215, Pb-211, Bi-211, Po-211, Tl-207
Ra-224	Rn-220, Po-216, Pb-212, Bi-212, Tl-208, Po-212
Ra-225	Ac-225, Fr-221, At-217, Bi-213, Tl-209, Po-213, Pb-209
Ra-226	Rn-222, Po-218, Pb-214, At-218, Bi-214, Po-214
Ra-228	Ac-228
Ac-225	Fr-221, At-217, Bi-213, Tl-209, Po-213, Pb-209
Ac-227	Fr-223
Th-228	Ra-224, Rn-220, Po-216, Pb-212, Bi-212, Tl-208, Po-212
Th-234	Pa-234m, Pa-234
Pa-230	Ac-226, Th-226, Fr-222, Ra-222, Rn-218, Po-214
U-230	Th-226, Ra-222, Rn-218, Po-214
U-235	Th-231
Pu-241	U-237
Pu-244	U-240, Np-240m
Am-242m	Am-242, Np-238
Am-243	Np-239
Cm-247	Pu-243
Bk-249	Am-245
Cf-253	Cm-249"

(b) Parent nuclides and their progeny included in secular equilibrium are listed in the following:

Sr-90	Y-90
Zr-93	Nb-93m
Zr-97	Nb-97
Ru-106	Rh-106
Ag-108m	Ag-108
Cs-137	Ba-137m
Ce-144	Pr-144
Ba-140	La-140
Bi-212	Tl-208 (0.36), Po-212 (0.64)
Pb-210	Bi-210, Po-210
Pb-212	Bi-212, Tl-208 (0.36), Po-212 (0.64)
Rn-222	Po-218, Pb-214, Bi-214, Po-214
Ra-223	Rn-219, Po-215, Pb-211, Bi-211, Tl-207
Ra-224	Rn-220, Po-216, Pb-212, Bi-212, Tl-208 (0.36), Po-212 (0.64)
Ra-226	Rn-222, Po-218, Pb-214, Bi-214, Po-214, Pb-210, Bi-210, Po-210
Ra-228	Ac-228
Th-228	Ra-224, Rn-220, Po-216, Pb212, Bi-212, Tl208 (0.36), Po-212 (0.64)
Th-229	Ra-225, Ac-225, Fr-221, At-217, Bi-213, Po-213, Pb-209
Th-nat	Ra-228, Ac-228, Th-228, Ra-224, Rn-220, Po-216, Pb-212, Bi-212, l208(0.36), Po-212 (0.64)
Th-234	Pa-234m
U-230	Th-226, Ra-222, Rn-218, Po-214
U-232	Th-228, Ra-224, Rn-220, Po-216, Pb-212, Bi-212, Tl-208 (0.36), Po-212 (0.64)
U-235	Th-231
U-238	Th-234, Pa-234m
U-nat	Th-234, Pa-234m, U-234, Th-230, Ra-226, Rn-222, Po-218, Pb-214, Bi-214, Po-214, Pb-210, Bi-210, Po-210
	Np-237 Pa-233
Am-242m	Am-242
Am-243	Np-239

(c)   The quantity may be determined from a measurement of the rate of decay or a measurement of the radiation level at a prescribed distance from the source.

(d)   These values apply only to compounds of uranium that take the chemical form of UF_6, UO_2F_2 and $UO_2(NO_3)_2$ in both normal and accident conditions of carriage.

(e)   These values apply only to compounds of uranium that take the chemical form of UO_3, UF_4, UCl_4 and hexavalent compounds in both normal and accident conditions of carriage.

(f)   These values apply to all compounds of uranium other than those specified in (d) and (e) above.

(g)   These values apply to unirradiated uranium only.

2.2.7.7.2.2   For individual radionuclides which are not listed in Table 2.2.7.7.2.1 the determination of the basic radionuclide values referred to in 2.2.7.7.2.1 shall require, multilateral approval. It is permissible to use an A_2 value calculated using a dose coefficient for the appropriate lung absorption type as recommended by the International Commission on Radiological Protection, if the chemical forms of each radionuclide under both normal and accident conditions of carriage are taken into consideration. Alternatively, the radionuclide values in Table 2.2.7.7.2.2 may be used without obtaining competent authority approval.

Table 2.2.7.7.2.2

Basic radionuclide values for unknown radionuclides or mixtures

Radioactive contents	A_1 TBq	A_2 TBq	Activity concentration for exempt material Bq/g	Activity limit for an exempt consignment Bq
Only beta or gamma emitting nuclides are known to be present	0.1	0.02	1×10^1	1×10^4
Alpha emitting nuclides but no neutron emitters are known to be present	0.2	9×10^{-5}	1×10^{-1}	1×10^3
Neutron emitting nuclides are known to be present or no relevant data are available	0.001	9×10^{-5}	1×10^{-1}	1×10^3

2.2.7.7.2.3   In the calculations of A_1 and A_2 for a radionuclide not in Table 2.2.7.7.2.1, a single radioactive decay chain in which the radionuclides are present in their naturally occurring proportions, and in which no daughter nuclide has a half-life either longer than 10 days or longer than that of the parent nuclide, shall be considered as a single radionuclide; and the activity to be taken into account and the A_1 or A_2 value to be applied shall be those corresponding to the parent nuclide of that chain. In the case of radioactive decay chains in which any daughter nuclide has a half-life either longer than 10 days or greater than that of the parent nuclide, the parent and such daughter nuclides shall be considered as mixtures of different nuclides.

2.2.7.7.2.4   For mixtures of radionuclides, the determination of the basic radionuclide values referred to in 2.2.7.7.2.1 may be determined as follows:

$$X_m = \frac{1}{\sum_i \frac{f(i)}{X(i)}}$$

where,

f(i) is the fraction of activity or activity concentration of radionuclide i in the mixture;

X(i) is the appropriate value of A_1 or A_2, or the activity concentration for exempt material or the activity limit for an exempt consignment as appropriate for the radionuclide i; and

X_m is the derived value of A_1 or A_2, or the activity concentration for exempt material or the activity limit for an exempt consignment in the case of a mixture.

2.2.7.7.2.5 When the identity of each radionuclide is known but the individual activities of some of the radionuclides are not known, the radionuclides may be grouped and the lowest radionuclide value, as appropriate, for the radionuclides in each group may be used in applying the formulas in 2.2.7.7.2.4 and 2.2.7.7.1.4.2. Groups may be based on the total alpha activity and the total beta/gamma activity when these are known, using the lowest radionuclide values for the alpha emitters or beta/gamma emitters, respectively.

2.2.7.7.2.6 For individual radionuclides or for mixtures of radionuclides for which relevant data are not available, the values shown in Table 2.2.7.7.2.2 shall be used.

2.2.7.8 *Limits on transport index (TI), criticality safety index (CSI), radiation levels for packages and overpacks*

2.2.7.8.1 Except for consignments under exclusive use, the transport index of any package or overpack shall not exceed 10, nor shall the criticality safety index of any package or overpack exceed 50.

2.2.7.8.2 Except for packages or overpacks carried under exclusive use by rail or by road under the conditions specified in 7.1.4.14.7.3.5 a), the maximum radiation level at any point on any external surface of a package or overpack shall not exceed 2 mSv/h.

2.2.7.8.3 The maximum radiation level at any point on any external surface of a package or overpack under exclusive use shall not exceed 10 mSv/h.

2.2.7.8.4 Packages and overpacks shall be assigned to either category I-WHITE, II-YELLOW or III-YELLOW in accordance with the conditions specified in Table 2.2.7.8.4 and with the following requirements:

(a) For a package or overpack, both the transport index and the surface radiation level conditions shall be taken into account in determining which is the appropriate category. Where the transport index satisfies the condition for one category but the surface radiation level satisfies the condition for a different category, the package or overpack shall be assigned to the higher category. For this purpose, category I-WHITE shall be regarded as the lowest category;

(b) The transport index shall be determined following the procedures specified in 2.2.7.6.1.1 and 2.2.7.6.1.2;

(c) If the surface radiation level is greater than 2 mSv/h, the package or overpack shall be carried under exclusive use and under the provisions of 7.5.11, CV33 (3.5) (a);

(d)   A package carried under a special arrangement shall be assigned to category III-YELLOW except under the provisions of 2.2.7.8.5;

(e)   An overpack which contains packages carried under special arrangement shall be assigned to category III-YELLOW except under the provisions of 2.2.7.8.5.

2.2.7.8.5   In case of international carriage of packages requiring competent authority design or shipment approval, for which different approval types apply in the different countries concerned by the shipment, assignment to the category as required in 2.2.7.8.4 shall be in accordance with the certificate of the country of origin of design.

Table 2.2.7.8.4

CATEGORIES OF PACKAGES AND OVERPACKS

Conditions		Category
Transport index (TI)	**Maximum radiation level at any point on external surface**	
0 [a]	Not more than 0.005 mSv/h	I-WHITE
More than 0 but not more than 1 [a]	More than 0.005 mSv/h but not more than 0.5 mSv/h	II-YELLOW
More than 1 but not more than 10	More than 0.5 mSv/h but not more than 2 mSv/h	III-YELLOW
More than 10	More than 2 mSv/h but not more than 10 mSv/h	III-YELLOW [b]

[a]   *If the measured TI is not greater than 0.05, the value quoted may be zero in accordance with 2.2.7.6.1.1(c).*

[b]   *Shall also be carried under exclusive use.*

2.2.7.9   Requirements and controls for carriage of excepted packages

2.2.7.9.1   Excepted packages which may contain radioactive material in limited quantities, instruments, manufactured articles as specified in 2.2.7.7.1.2 and empty packagings as specified in 2.2.7.9.6 may be carried under the following conditions:

(a)   The applicable requirements specified in 2.2.7.9.2, 3.3.1 (special provision 290, if relevant), 4.1.9.1.2 of ADR, 5.2.1.2, 5.2.1.7.1, 5.2.1.7.2, 5.2.1.7.3, 5.4.1.1.1 (a), 7.1.4.14.7.5.2, and, as applicable 2.2.7.9.3 to 2.2.7.9.6;

(b)   The requirements for excepted packages specified in 6.4.4 of ADR;

(c)   If the excepted package contains fissile material, one of the fissile exceptions provided by 6.4.11.2 of ADR shall apply and the requirement of 6.4.7.2 of ADR shall be met.

2.2.7.9.2   The radiation level at any point on the external surface of an excepted package shall not exceed 5 μSv/h.

2.2.7.9.3   Radioactive material which is enclosed in or is included as a component part of an instrument or other manufactured article, with activity not exceeding the item and package limits specified in columns 2 and 3 respectively of Table 2.2.7.7.1.2.1, may be carried in an excepted package provided that:

(a) the radiation level at 10 cm from any point on the external surface of any unpackaged instrument or article is not greater than 0.1 mSv/h; and

(b) each instrument or manufactured article bears the marking "RADIOACTIVE" except:

    (i) radioluminescent time-pieces or devices;

    (ii) consumer products that either have received regulatory approval according to 2.2.7.1.2 (d) or do not individually exceed the activity limit for an exempt consignment in column (5) of Table 2.2.7.7.2.1, provided such products are carried in a package that bears the marking "RADIOACTIVE" on an internal surface in such a manner that warning of the presence of radioactive material is visible on opening the package; and

(c) the active material is completely enclosed by non-active components (a device performing the sole function of containing radioactive material shall not be considered to be an instrument or manufactured article).

2.2.7.9.4   Radioactive material in forms other than as specified in 2.2.7.9.3, with an activity not exceeding the limit specified in column 4 of Table 2.2.7.7.1.2.1, may be carried in an excepted package provided that:

(a) the package retains its radioactive contents under routine conditions of carriage; and

(b) the package bears the marking "RADIOACTIVE" on an internal surface in such a manner that a warning of the presence of radioactive material is visible on opening the package.

2.2.7.9.5   A manufactured article in which the sole radioactive material is unirradiated natural uranium, unirradiated depleted uranium or unirradiated natural thorium may be carried as an excepted package provided that the outer surface of the uranium or thorium is enclosed in an inactive sheath made of metal or some other substantial material.

2.2.7.9.6   An empty packaging which had previously contained radioactive material may be carried as an excepted package provided that:

(a) It is in a well maintained condition and securely closed;

(b) The outer surface of any uranium or thorium in its structure is covered with an inactive sheath made of metal or some other substantial material; and

(c) The level of internal non-fixed contamination does not exceed one hundred times the levels specified in 4.1.9.1.2 of ADR.

2.2.7.9.7   The following provisions do not apply to excepted packages and the controls for carriage of excepted packages:

Chapter 1.10, 2.2.7.4.1, 2.2.7.4.2, 4.1.9.1.3 of ADR, 4.1.9.1.4 of ADR, 5.1.3.2, 5.1.5.1.1, 5.1.5.1.2, 5.2.2.1.11.1, 5.4.1.1.1 except (a), 5.4.1.2.5.1, 5.4.1.2.5.2, 5.4.3, 6.4.6.1 of ADR, 7.1.4.14.7 with the exception of 7.1.4.14.7.5.2.

2.2.7.10   *(Reserved).*

2.2.8   **Class 8**   **Corrosive substances**

2.2.8.1   *Criteria*

2.2.8.1.1   The heading of Class 8 covers substances and articles containing substances of this class which by chemical action attack epithelial tissue - of skin or mucous membranes - with which they are in contact, or which in the event of leakage are capable of damaging or destroying other goods, or means of transport. The heading of this class also covers other substances which form a corrosive liquid only in the presence of water, or which produce corrosive vapour or mist in the presence of natural moisture of the air.

2.2.8.1.2   Substances and articles of Class 8 are subdivided as follows:

C1-C10   Corrosive substances without subsidiary risk:

          C1-C4   Acid substances:
                  C1   Inorganic, liquid;
                  C2   Inorganic, solid;
                  C3   Organic, liquid;
                  C4   Organic, solid;

          C5-C8   Basic substances:
                  C5   Inorganic, liquid;
                  C6   Inorganic, solid;
                  C7   Organic, liquid;
                  C8   Organic, solid;

          C9-C10   Other corrosive substances:
                  C9   Liquid;
                  C10   Solid;

C11   Articles;

CF   Corrosive substances, flammable:
          CF1   Liquid;
          CF2   Solid;

CS   Corrosive substances, self-heating:
          CS1   Liquid;
          CS2   Solid;

CW   Corrosive substances which, in contact with water, emit flammable gases:
          CW1   Liquid;
          CW2   Solid;

CO   Corrosive substances, oxidizing:
          CO1   Liquid;
          CO2   Solid;

CT   Corrosive substances, toxic:

          CT1   Liquid;
          CT2   Solid;

CFT   Corrosive substances, flammable, liquid, toxic;
COT   Corrosive substances, oxidizing, toxic.

Classification and assignment of packing groups

2.2.8.1.3   Substances of Class 8 shall be classified in three packing groups according to the degree of danger they present for carriage, as follows:

   Packing group I: highly corrosive substances
   Packing group II: corrosive substances
   Packing group III: slightly corrosive substances.

2.2.8.1.4   Substances and articles classified in Class 8 are listed in Table A of Chapter 3.2. Allocation of substances to packing groups I, II and III has been made on the basis of experience taking into account such additional factors as inhalation risk (see 2.2.8.1.5) and reactivity with water (including the formation of dangerous decomposition products).

2.2.8.1.5   A substance or preparation meeting the criteria of Class 8 having an inhalation toxicity of dusts and mists (LC_{50}) in the range of packing group I, but toxicity through oral ingestion or dermal contact only in the range of packing group III or less, shall be allocated to Class 8.

2.2.8.1.6   Substances, including mixtures, not mentioned by name in Table A of Chapter 3.2 can be assigned to the relevant entry of sub-section 2.2.8.3, and to the relevant packing group on the basis of the length of time of contact necessary to produce full thickness destruction of human skin in accordance with the criteria of (a) to (c) below.

   Liquids, and solids which may become liquid during carriage, which are judged not to cause full thickness destruction of human skin shall still be considered for their potential to cause corrosion to certain metal surfaces. In assigning the packing group, account shall be taken of human experience in instances of accidental exposure. In the absence of human experience, the grouping shall be based on data obtained from experiments in accordance with OECD Guideline 404 [8].

   (a)   Packing group I is assigned to substances that cause full thickness destruction of intact skin tissue within an observation period up to 60 minutes starting after the exposure time of 3 minutes or less.

   (b)   Packing group II is assigned to substances that cause full thickness destruction of intact skin tissue within an observation period up to 14 days starting after the exposure time of more than 3 minutes but not more than 60 minutes.

   (c)   Packing group III is assigned to substances that:

   –   cause full thickness destruction of intact skin tissue within an observation period up to 14 days starting after the exposure time of more than 60 minutes but not more than 4 hours; or

   –   are judged not to cause full thickness destruction of intact skin tissue, but which exhibit a corrosion rate on steel or aluminium surfaces exceeding 6.25 mm a year at a test temperature of 55 °C. For the purposes of testing steel, type S235JR+CR (1.0037 resp. St 37-2), S275J2G3+CR (1.0144 resp. St 44-3), ISO 3574, Unified Numbering System (UNS) G10200 or SAE 1020, and for testing aluminium, non-clad, types 7075-T6 or AZ5GU-T6 shall be used. An acceptable test is prescribed in the Manual of Tests and Criteria, Part III, Section 37.

---

[8]   *OECD guidelines for Testing of Chemicals, No. 404 " Acute Dermal Irritation/Corrosion" (1992).*

2.2.8.1.7   If substances of Class 8, as a result of admixtures, come into categories of risk different from those to which the substances mentioned by name in Table A of Chapter 3.2 belong, these mixtures or solutions shall be assigned to the entries to which they belong, on the basis of their actual degree of danger.

NOTE: For the classification of solutions and mixtures (such as preparations and wastes), see also 2.1.3.

2.2.8.1.8   On the basis of the criteria set out in paragraph 2.2.8.1.6, it may also be determined whether the nature of a solution or mixture mentioned by name or containing a substance mentioned by name is such that the solution or mixture is not subject to the provisions for this class.

2.2.8.1.9   Substances, solutions and mixtures, which

–   do not meet the criteria of Directives 67/548/EEC[3] or 88/379/EEC[4] as amended and therefore are not classified as corrosive according to these directives, as amended; and

–   do not exhibit a corrosive effect on steel or aluminium,

may be considered as substances not belonging to Class 8.

NOTE: UN No. 1910 calcium oxide and UN No. 2812 sodium aluminate, listed in the UN Model Regulations, are not subject to the provisions of ADN.

2.2.8.2   *Substances not accepted for carriage*

2.2.8.2.1   The chemically unstable substances of Class 8 shall not be accepted for carriage unless the necessary steps have been taken to prevent their dangerous decomposition or polymerization during carriage. To this end it shall in particular be ensured that receptacles and tanks do not contain any substance liable to promote these reactions.

2.2.8.2.2   The following substances shall not be accepted for carriage:

–   UN No. 1798 NITROHYDROCHLORIC ACID;

–   chemically unstable mixtures of spent sulphuric acid;

–   chemically unstable mixtures of nitrating acid or mixtures of residual sulphuric and nitric acids, not denitrated;

–   perchloric acid aqueous solution with more than 72 % pure acid, by mass, or mixtures of perchloric acid with any liquid other than water.

---

[3]   *Council Directive 67/548/EEC of 27 June 1967 on the approximation of laws, regulations and administrative provisions relating to the classification, packaging and labelling of dangerous substances (Official Journal of the European Communities No. L 196 of 16.08.1967, page 1).*

[4]   *Council Directive 88/379/EEC on the approximation of laws, regulations and administrative provisions relating to the classification, packaging and labelling of dangerous preparations (Official Journal of the European Communities No. L.187 of 16.07.1988, page 14).*

2.2.8.3   *List of collective entries*

Corrosive substances <u>without</u> subsidiary risk

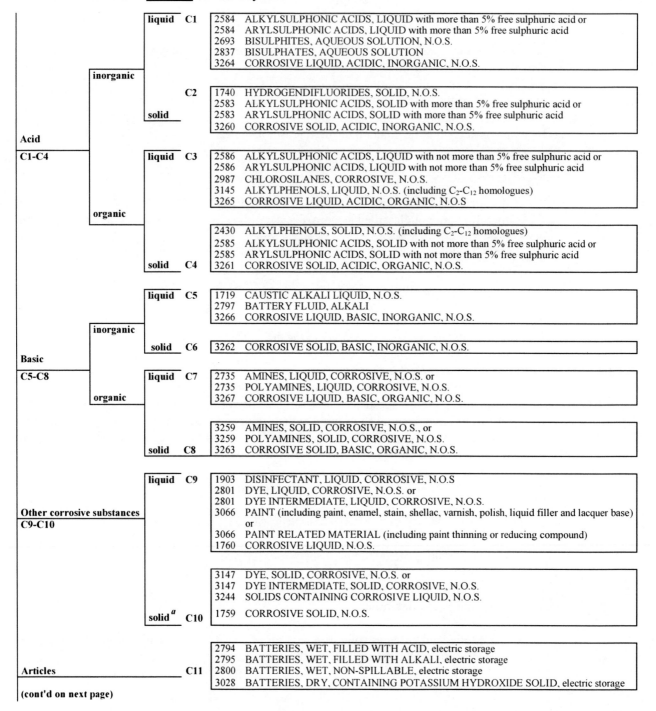

Acid **C1-C4**	**inorganic**	**liquid**	**C1**	2584 ALKYLSULPHONIC ACIDS, LIQUID with more than 5% free sulphuric acid or 2584 ARYLSULPHONIC ACIDS, LIQUID with more than 5% free sulphuric acid 2693 BISULPHITES, AQUEOUS SOLUTION, N.O.S. 2837 BISULPHATES, AQUEOUS SOLUTION 3264 CORROSIVE LIQUID, ACIDIC, INORGANIC, N.O.S.
		solid	**C2**	1740 HYDROGENDIFLUORIDES, SOLID, N.O.S. 2583 ALKYLSULPHONIC ACIDS, SOLID with more than 5% free sulphuric acid or 2583 ARYLSULPHONIC ACIDS, SOLID with more than 5% free sulphuric acid 3260 CORROSIVE SOLID, ACIDIC, INORGANIC, N.O.S.
	organic	**liquid**	**C3**	2586 ALKYLSULPHONIC ACIDS, LIQUID with not more than 5% free sulphuric acid or 2586 ARYLSULPHONIC ACIDS, LIQUID with not more than 5% free sulphuric acid 2987 CHLOROSILANES, CORROSIVE, N.O.S. 3145 ALKYLPHENOLS, LIQUID, N.O.S. (including C_2-C_{12} homologues) 3265 CORROSIVE LIQUID, ACIDIC, ORGANIC, N.O.S
		solid	**C4**	2430 ALKYLPHENOLS, SOLID, N.O.S. (including C_2-C_{12} homologues) 2585 ALKYLSULPHONIC ACIDS, SOLID with not more than 5% free sulphuric acid or 2585 ARYLSULPHONIC ACIDS, SOLID with not more than 5% free sulphuric acid 3261 CORROSIVE SOLID, ACIDIC, ORGANIC, N.O.S.
Basic **C5-C8**	**inorganic**	**liquid**	**C5**	1719 CAUSTIC ALKALI LIQUID, N.O.S. 2797 BATTERY FLUID, ALKALI 3266 CORROSIVE LIQUID, BASIC, INORGANIC, N.O.S.
		solid	**C6**	3262 CORROSIVE SOLID, BASIC, INORGANIC, N.O.S.
	organic	**liquid**	**C7**	2735 AMINES, LIQUID, CORROSIVE, N.O.S. or 2735 POLYAMINES, LIQUID, CORROSIVE, N.O.S. 3267 CORROSIVE LIQUID, BASIC, ORGANIC, N.O.S.
		solid	**C8**	3259 AMINES, SOLID, CORROSIVE, N.O.S., or 3259 POLYAMINES, SOLID, CORROSIVE, N.O.S. 3263 CORROSIVE SOLID, BASIC, ORGANIC, N.O.S.
Other corrosive substances **C9-C10**		**liquid**	**C9**	1903 DISINFECTANT, LIQUID, CORROSIVE, N.O.S 2801 DYE, LIQUID, CORROSIVE, N.O.S. or 2801 DYE INTERMEDIATE, LIQUID, CORROSIVE, N.O.S. 3066 PAINT (including paint, enamel, stain, shellac, varnish, polish, liquid filler and lacquer base) or 3066 PAINT RELATED MATERIAL (including paint thinning or reducing compound) 1760 CORROSIVE LIQUID, N.O.S.
		solid [a]	**C10**	3147 DYE, SOLID, CORROSIVE, N.O.S. or 3147 DYE INTERMEDIATE, SOLID, CORROSIVE, N.O.S. 3244 SOLIDS CONTAINING CORROSIVE LIQUID, N.O.S. 1759 CORROSIVE SOLID, N.O.S.
Articles			**C11**	2794 BATTERIES, WET, FILLED WITH ACID, electric storage 2795 BATTERIES, WET, FILLED WITH ALKALI, electric storage 2800 BATTERIES, WET, NON-SPILLABLE, electric storage 3028 BATTERIES, DRY, CONTAINING POTASSIUM HYDROXIDE SOLID, electric storage

(cont'd on next page)

---

[a]   *Mixtures of solids which are not subject to the provisions of ADR and of corrosive liquids may be carried under UN No. 3244 without being subject to the classification criteria of Class 8, provided there is no free liquid visible at the time the substance is loaded or at the time the packaging, container or transport unit is closed. Each packaging shall correspond to a design type which has passed the leakproofness test for Packing group II level.*

Corrosive substances <u>with</u> subsidiary risk(s)

(cont'd)	liquid	CF1	3470	PAINT, CORROSIVE, FLAMMABLE (including paint, enamel, stain, shellac, varnish, polish, liquid filler and lacquer base) or
			3470	PAINT RELATED MATERIAL, CORROSIVE, FLAMMABLE (including paint thinning or reducing compound)
			2734	AMINES, LIQUID, CORROSIVE, FLAMMABLE, N.O.S. or
			2734	POLYAMINES, LIQUID, CORROSIVE, FLAMMABLE, N.O.S.
			2986	CHLOROSILANES, CORROSIVE, FLAMMABLE, N.O.S.
Flammable [b]			2920	CORROSIVE LIQUID, FLAMMABLE, N.O.S.
CF	solid	CF2	2921	CORROSIVE SOLID, FLAMMABLE, N.O.S.
Self-heating	liquid		3301	CORROSIVE LIQUID, SELF-HEATING, N.O.S.
CS	solid		3095	CORROSIVE SOLID, SELF-HEATING, N.O.S.
Water-reactive	liquid [b]		3094	CORROSIVE LIQUID, WATER-REACTIVE, N.O.S.
CW	solid		3096	CORROSIVE SOLID, WATER-REACTIVE, N.O.S.
Oxidizing	liquid		3093	CORROSIVE LIQUID, OXIDIZING, N.O.S.
CO	solid		3084	CORROSIVE SOLID, OXIDIZING, N.O.S.
Toxic [d]	liquid [c]	CT1	3471	HYDROGENDIFLUORIDES SOLUTION, N.O.S.
			2922	CORROSIVE LIQUID, TOXIC, N.O.S.
CT	solid [e]		2923	CORROSIVE SOLID, TOXIC, N.O.S.
Flammable, liquid, toxic [d]		CFT	No collective entry with this classification code available; if need be, classification under a collective entry with a classification code to be determined according to table of precedence of hazard in 2.1.3.10.	
Oxidizing, toxic [d, e]		COT	No collective entry with this classification code available; if need be, classification under a collective entry with a classification code to be determined according to table of precedence of hazard in 2.1.3.10.	

---

[b]   *Chlorosilanes which, in contact with water or moist air, emit flammable gases, are substances of Class 4.3.*

[c]   *Chloroformates having predominantly toxic properties are substances of Class 6.1.*

[d]   *Corrosive substances which are highly toxic by inhalation, as defined in 2.2.61.1.4 to 2.2.61.1.9 are substances of Class 6.1.*

[e]   *UN No. 2505 AMMONIUM FLUORIDE, UN No. 1812 POTASSIUM FLUORIDE, UN No. 1690 SODIUM FLUORIDE, SOLD, UN No. 2674 SODIUM FLUOROSILICATE, UN No. 2856 FLUOROSILICATES, N.O.S. , UN No. 3415 SODIUM FLUORIDE SOLUTION and UN No. 3422 POTASSIUM FLUORIDE SOLUTION are substances of Class 6.1.*

2.2.9   **Class 9 Miscellaneous dangerous substances and articles**

2.2.9.1   *Criteria*

2.2.9.1.1   The heading of Class 9 covers substances and articles which, during carriage, present a danger not covered by the heading of other classes.

2.2.9.1.2   The substances and articles of Class 9 are subdivided as follows:

   M1   Substances which, on inhalation as fine dust, may endanger health;

   M2   Substances and apparatus which, in the event of fire, may form dioxins;

   M3   Substances evolving flammable vapour;

   M4   Lithium batteries;

   M5   Life-saving appliances;

   M6-M8   Environmentally hazardous substances:

         M6   Pollutant to the aquatic environment, liquid;
         M7   Pollutant to the aquatic environment, solid;
         M8   Genetically modified micro-organisms and organisms;

   M9-M10   Elevated temperature substances:

         M9   Liquid;
         M10   Solid;

   M11   Other substances presenting a danger during carriage, but not meeting the definitions of another class.

Definitions and classification

2.2.9.1.3   Substances and articles classified in Class 9 are listed in Table A of Chapter 3.2. The assignment of substances and articles not mentioned by name in Table A of Chapter 3.2 to the relevant entry of that Table or of sub-section 2.2.9.3 shall be done in accordance with 2.2.9.1.4 to 2.2.9.1.14 below.

Substances which, on inhalation as fine dust, may endanger health

2.2.9.1.4   Substances which, on inhalation as fine dust, may endanger health include asbestos and mixtures containing asbestos.

Substances and apparatus which, in the event of fire, may form dioxins

2.2.9.1.5   Substances and apparatus which, in the event of fire, may form dioxins include polychlorinated biphenyls (PCBs) and terphenyls (PCTs) and polyhalogenated biphenyls and terphenyls and mixtures containing these substances, as well as apparatus such as transformers, condensers and apparatus containing those substances or mixtures.

   NOTE: *Mixtures with a PCB or PCT content of not more than 50 mg/kg are not subject to the provisions of ADN.*

Substances evolving flammable vapour

2.2.9.1.6   Substances evolving flammable vapour include polymers containing flammable liquids with a flash-point not exceeding 55 °C.

Lithium batteries

2.2.9.1.7   Lithium cells and batteries may be assigned to Class 9 if they meet the requirements of special provision 230 of Chapter 3.3. They are not subject to the provisions of ADN if they meet the requirements of special provision 188 of Chapter 3.3. They shall be classified in accordance with the procedures of Section 38.3 of the Manual of Tests and Criteria.

Life-saving appliances

2.2.9.1.8   Life-saving appliances include life-saving appliances and motor vehicle components which meet the descriptions of special provisions 235 or 296 of Chapter 3.3.

Environmentally hazardous substances

2.2.9.1.9   Environmentally hazardous substances include liquid or solid substances pollutant to the aquatic environment and solutions and mixtures of such substances (such as preparations and wastes), which cannot be classified in the other classes or under any other entry of Class 9 listed in Table A of Chapter 3.2. It also includes genetically modified micro-organisms and organisms.

Pollutants to the aquatic environment

2.2.9.1.10   Assignment of a substance to the entries UN No. 3082 ENVIRONMENTALLY HAZARDOUS SUBSTANCE, LIQUID, N.O.S and UN No. 3077 ENVIRONMEN-TALLY HAZARDOUS SUBSTANCE, SOLID, N.O.S. as pollutant to the aquatic environment shall be as indicated in 2.3.5.

Notwithstanding the provisions of 2.3.5, substances which cannot be assigned to other classes of ADN or to other entries of Class 9, and which are not identified in Council Directive 67/548/EEC of 27 June 1967 on the approximation of laws, regulations and administrative provisions relating to the classification, packaging and labelling of dangerous substances[9], as amended, as substances to which letter N "Environmentally hazardous" (R50; R50/53; R51/53) has been allocated, are not subject to ADN.

Notwithstanding the provisions of 2.1.3.8, solutions and mixtures (such as preparations and wastes) of substances to which letter N "Environmentally hazardous" (R50; R50/53; R51/53) has been allocated in Directive 67/548/EEC, as amended, need only be assigned to UN Nos. 3077 or 3082 if, according to Directive 1999/45/EC of the European Parliament and of the Council of 31 May 1999 concerning the approximation of the laws, regulations and administrative provisions of the Member States relating to the classification, packaging and labelling of dangerous preparations[10], as amended, they are also allocated letter N "Environmentally hazardous" (R50; R50/53; R51/53) and they cannot be assigned to one of classes 1 to 8 or to any other entry of Class 9.

---

[9]   *Official Journal of the European Communities No.196, of 16 August 1967, pp. 1 – 5.*
[10]   *Official Journal of the European Communities No. L 200, of 30 July 1999, pp. 1 – 68.*

Genetically modified micro-organisms or organisms

2.2.9.1.11   Genetically modified micro-organisms (GMMOs) and genetically modified organisms (GMOs) are micro-organisms and organisms in which genetic material has been purposely altered through genetic engineering in a way that does not occur naturally. They are assigned to Class 9 (UN No. 3245) if they do not meet the definition of infectious substances, but are capable of altering animals, plants or microbiological substances in a way not normally the result of natural reproduction.

> ***NOTE 1***: *GMMOs which are infectious are substances of Class 6.2 (UN Nos. 2814 and 2900).*

> ***NOTE 2***: *GMMOs or GMOs are not subject to the provisions of ADN when authorized for use by the competent authorities of the countries of origin, transit and destination[11].*

> **NOTE 3**: Live animals shall not be used to carry genetically modified micro-organisms classified in Class 9 unless the substance can be carried no other way.

2.2.9.1.12   Genetically modified organisms, which are known or suspected to be dangerous to the environment shall be carried in accordance with conditions specified by the competent authority of the country of origin.

Elevated temperature substances

2.2.9.1.13   Elevated temperature substances include substances which are carried or handed over for carriage in the liquid state at or above 100 °C and, in the case of those with a flash-point, below their flash-point. They also include solids which are carried or handed over for carriage at or above 240 °C.

> **NOTE 1:** *Elevated temperature substances may be assigned to Class 9 only if they do not meet the criteria of any other class.*

> **NOTE 2:** *Substances having a flash-point above 60 °C which are carried or handed over for carriage within a range of 15 K below the flash-point are substances of Class 3, identification number 9001.*

> *Other substances presenting a danger during carriage but not meeting the definitions of another class.*

2.2.9.1.14   The following other miscellaneous substances not meeting the definitions of another class are assigned to Class 9:

Solid ammonia compounds having a flash-point below 60 °C
Low hazard dithionites
Highly volatile liquids
Substances emitting noxious fumes
Substances containing allergens
Chemical kits and first aid kits

---

[11]   *See in particular Part C of Directive 2001/18/EC of the European Parliament and of the Council on the deliberate release into the environment of genetically modified organisms and repealing Council Directive 90/220/EEC (Official Journal of the European Communities, No. L 106, of 17 April 2001, pp. 8-14), which sets out the authorization procedures for the European Community.*

The following miscellaneous substances not meeting the definition of another class are assigned to Class 9 when they are carried in bulk or in tank vessels:

– UN 2071 AMMONIUM NITRATE FERTILIZERS: uniform non-segregating mixtures of nitrogen phosphate or nitrogen potash types or complete fertilizers of nitrogen phosphate potash type, with not more than 70% ammonium nitrate and not more than 0.4% total combustible material, or with not more than 45% ammonium nitrate irrespective of their content of combustible material;

NOTE 1: In determining the ammonium nitrate content, all nitrate ions for which a molecular equivalent of ammonium ions is present in the mixture shall be calculated as ammonium nitrate mass.

NOTE 2: Ammonium nitrate fertilizers of Class 9 are not subject to ADN if:

– *the results of the trough test (see Manual of Tests and Criteria, Part III, sub-section 38.2) show that they are not capable of undergoing self-sustaining decomposition; and*

– *the calculation referred to in NOTE 1 does not give an excess of nitrate greater than 10% by mass, calculated in KNO_3.*

– UN 2216 FISH MEAL, STABILIZED (humidity between 5% by mass and 12% by mass with not more than 15% fat by mass); or

– UN 2216 FISH SCRAP, STABILIZED (humidity between 5% by mass and 12% by mass with not more than 15% fat by mass);

– Identification No. 9003 SUBSTANCES HAVING A FLASH-POINT ABOVE 60° C AND NOT MORE THAN 100° C which cannot be assigned to another class or another entry of Class 9;

– Identification No. 9004, 4,4' DIPHENYLMETHANE DIISOCYANATE.

NOTE: UN No. 1845 carbon dioxide, solid (dry ice), UN No. 2807 magnetized material, UN No. 3166 engine, internal combustion or vehicle, flammable gas powered or vehicle, flammable liquid powered, UN No. 3171 battery-powered vehicle or 3171 battery-powered equipment (wet battery), UN No. 3334 aviation regulated liquid, n.o.s., UN No. 3335 aviation regulated solid, n.o.s. and UN No. 3363 dangerous goods in machinery or dangerous goods in apparatus, listed in the UN Model Regulations, are not subject to the provisions of ADN.

Assignment of the packing groups

2.2.9.1.15   The substances and articles of Class 9 listed as such in Table A of Chapter 3.2 shall be assigned to one of the following packing groups according to their degree of danger:

Packing group II: substances presenting medium danger

Packing group III: substances presenting low danger.

2.2.9.2   *Substances and articles not accepted for carriage*

The following substances and articles shall not be accepted for carriage:

–   Lithium batteries which do not meet the relevant conditions of special provisions 188, 230 or 636 of Chapter 3.3;

–   Uncleaned empty containment vessels for apparatus such as transformers, condensers and hydraulic apparatus containing substances assigned to UN Nos. 2315, 3151, 3152 or 3432..

2.2.9.3   *List of collective entries*

Substances which, on inhalation as fine dust, may endanger health	M1	2212 BLUE ASBESTOS (crocidolite) or 2212 BROWN ASBESTOS (amosite, mysorite) 2590 WHITE ASBESTOS (chrysotile, actinolite, anthophyllite, tremolite)
Substances and apparatus which, in the event of fire, may form dioxins	M2	2315 POLYCHLORINATED BIPHENYLS, LIQUID 3432 POLYCHLORINATED BIPHENYLS, SOLID 3151 POLYHALOGENATED BIPHENYLS, LIQUID or 3151 POLYHALOGENATED TERPHENYLS, LIQUID 3152 POLYHALOGENATED BIPHENYLS, SOLID or 3152 POLYHALOGENATED TERPHENYLS, SOLID
Substances evolving flammable vapour	M3	2211 POLYMERIC BEADS, EXPANDABLE, evolving flammable vapour 3314 PLASTICS MOULDING COMPOUND in dough, sheet or extruded rope form evolving flammable vapour
Lithium batteries	M4	3090 LITHIUM BATTERIES 3091 LITHIUM BATTERIES CONTAINED IN EQUIPMENT or 3091 LITHIUM BATTERIES PACKED WITH EQUIPMENT
Live-saving appliances	M5	2990 LIFE-SAVING APPLIANCES, SELF-INFLATING 3072 LIFE-SAVING APPLIANCES NOT SELF-INFLATING containing dangerous goods as equipment 3268 AIR BAG INFLATORS or 3268 AIR BAG MODULES or 3268 SEAT-BELT PRETENSIONERS

Environmentally hazardous substances	pollutant to the aquatic environment, liquid	M6	3082 ENVIRONMENTALLY HAZARDOUS SUBSTANCE, LIQUID, N.O.S.
	pollutant to the aquatic environment, solid	M7	3077 ENVIRONMENTALLY HAZARDOUS SUBSTANCE, SOLID, N.O.S.
	genetically modified micro-organisms and organisms	M8	3245 GENETICALLY MODIFIED MICROORGANISMS or 3245 GENETICALLY MODIFIED ORGANISMS
Elevated temperature substances	liquid	M9	3257 ELEVATED TEMPERATURE LIQUID, N.O.S., at or above 100 °C and below its flash-point (including molten metal, molten salts, etc.)
	solid	M10	3258 ELEVATED TEMPERATURE SOLID, N.O.S., at or above 240 °C

Other substances or articles presenting a danger during carriage, but not meeting the definitions of another class	M11	No collective entry available. Only substances listed in Table A of Chapter 3.2 are subject to the provisions for Class 9 under this classification code, as follows: 1841 ACETALDEHYDE AMMONIA 1931 ZINC DITHIONITE (ZINC HYDROSULPHITE) 1941 DIBROMODIFLUOROMETHANE 1990 BENZALDEHYDE 2969 CASTOR BEANS, or 2969 CASTOR MEAL, or 2969 CASTOR POMACE, or 2969 CASTOR FLAKE 3316 CHEMICAL KIT, or 3316 FIRST AID KIT 3359 FUMIGATED UNIT

CHAPTER 2.3

TEST METHODS

2.3.0   **General**

Unless otherwise provided for in Chapter 2.2 or in this Chapter, the test methods to be used for the classification of dangerous goods are those described in the Manual of Tests and Criteria.

2.3.1   **Exudation test for blasting explosives of Type A**

2.3.1.1   Blasting explosives of type A (UN No. 0081) shall, if they contain more than 40 % liquid nitric ester, in addition to the testing specified in the Manual of Tests and Criteria, satisfy the following exudation test.

2.3.1.2   The apparatus for testing blasting explosive for exudation (figs. 1 to 3) consists of a hollow bronze cylinder. This cylinder, which is closed at one end by a plate of the same metal, has an internal diameter of 15.7 mm and a depth of 40 mm.

It is pierced by 20 holes 0.5 mm in diameter (four sets of five holes) on the circumference. A bronze piston, cylindrically fashioned over a length of 48 mm and having a total length of 52 mm, slides into the vertically placed cylinder.

The piston, whose diameter is 15.6 mm, is loaded with a mass of 2 220 g so that a pressure of 120 kPa (1.20 bar) is exerted on the base of the cylinder.

2.3.1.3   A small plug of blasting explosive weighing 5 to 8 g, 30 mm long and 15 mm in diameter, is wrapped in very fine gauze and placed in the cylinder; the piston and its loading mass are then placed on it so that the blasting explosive is subjected to a pressure of 120 kPa (1.20 bar). The time taken for the appearance of the first signs of oily droplets (nitroglycerine) at the outer orifices of the cylinder holes is noted.

2.3.1.4   The blasting explosive is considered satisfactory if the time elapsing before the appearance of the liquid exudations is more than five minutes, the test having been carried out at a temperature of 15 °C to 25 °C.

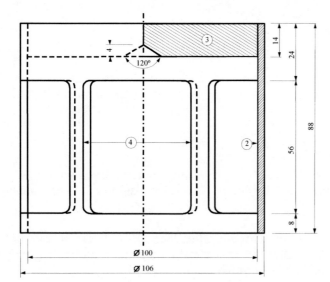

Fig.1: Bell-form charge, mass 2220 g, capable of being suspended from a bronze piston

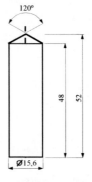

Fig.2: Cylindrical bronze piston, dimensions in mm

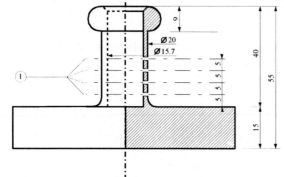

Fig.3: Hollow bronze cylinder, closed at one end; Plan and cut dimensions in mm

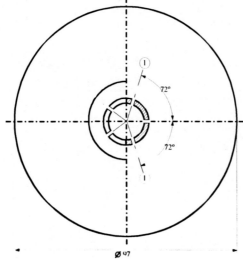

Fig. 1 to 3
----
(1) 4 series of 5 holes at 0.5 N
(2) copper
(3) iron plate with centre cone at the inferior face
(4) 4 openings, approximately 46x56, set at even intervals on the periphery

2.3.2   **Tests relating to nitrated cellulose mixtures of Class 4.1**

2.3.2.1   Nitrocellulose heated for half an hour at 132 °C shall not give off visible yellowish-brown nitrous fumes (nitrous gases). The ignition temperature shall be above 180 °C. See 2.3.2.3 to 2.3.2.8, 2.3.2.9 (a) and 2.3.2.10 below.

2.3.2.2   3 g of plasticized nitrocellulose, heated for one hour at 132 °C, shall not give off visible yellowish-brown nitrous fumes (nitrous gases). The ignition temperature shall be above 170 °C. See 2.3.2.3 to 2.3.2.8, 2.3.2.9 (b) and 2.3.2.10 below.

2.3.2.3   The test procedures set out below are to be applied when differences of opinion arise as to the acceptability of substances for carriage by road.

2.3.2.4   If other methods or test procedures are used to verify the conditions of stability prescribed above in this section, those methods shall lead to the same findings as could be reached by the methods specified below.

2.3.2.5   In carrying out the stability tests by heating described below, the temperature of the oven containing the sample under test shall not deviate by more than 2 °C from the prescribed temperature; the prescribed duration of a 30-minute or 60-minute test shall be observed to within two minutes. The oven shall be such that the required temperature is restored not more than five minutes after insertion of the sample.

2.3.2.6   Before undergoing the tests in 2.3.2.9 and 2.3.2.10, the samples shall be dried for not less than 15 hours at the ambient temperature in a vacuum desiccator containing fused and granulated calcium chloride, the sample substance being spread in a thin layer; for this purpose, substances which are neither in powder form nor fibrous shall be ground, or grated, or cut into small pieces. The pressure in the desiccator shall be brought below 6.5 kPa (0.065 bar).

2.3.2.7   Before being dried as prescribed in 2.3.2.6 above, substances conforming to 2.3.2.2 shall undergo preliminary drying in a well-ventilated oven, with its temperature set at 70 °C, until the loss of mass per quarter-hour is less than 0.3 % of the original mass.

2.3.2.8   Weakly nitrated nitrocellulose conforming to 2.3.2.1 shall first undergo preliminary drying as prescribed in 2.3.2.7 above; drying shall then be completed by keeping the nitrocellulose for at least 15 hours over concentrated sulphuric acid in a desiccator.

2.3.2.9   *Test of chemical stability under heat*

(a)   *Test of the substance listed in paragraph 2.3.2.1 above.*

(i)   In each of two glass test tubes having the following dimensions:

length	350	mm
internal diameter	16	mm
thickness of wall	1.5	mm

is placed 1 g of substance dried over calcium chloride (if necessary the drying shall be carried out after reducing the substance to pieces weighing not more than 0.05 g each).

Both test tubes, completely covered with loose-fitting closures, are then so placed in an oven that at least four-fifths of their length is visible, and are kept at a constant temperature of 132 °C for 30 minutes. It is observed whether nitrous gases in the form of yellowish-brown fumes clearly visible against a white background are given off during this time.

(ii)   In the absence of such fumes the substance is deemed to be stable.

(b)   *Test of plasticized nitrocellulose (see 2.3.2.2)*

(i)   3 g of plasticized nitrocellulose are placed in glass test tubes, similar to those referred to in (a), which are then placed in an oven kept at a constant temperature of 132 °C.

(ii)   The test tubes containing the plasticized nitrocellulose are kept in the oven for one hour. During this time no yellowish-brown nitrous fumes (nitrous gases) shall be visible. Observation and appraisal as in (a).

2.3.2.10   *Ignition temperature (see 2.3.2.1 and 2.3.2.2)*

(a)   The ignition temperature is determined by heating 0.2 g of substance enclosed in a glass test tube immersed in a Wood's alloy bath. The test tube is placed in the bath when the latter has reached 100 °C. The temperature of the bath is then progressively increased by 5 °C per minute;

(b)   The test tubes must have the following dimensions:

length   125 mm
internal diameter   15 mm
thickness of wall   0.5 mm

and shall be immersed to a depth of 20 mm;

(c)   The test shall be repeated three times, the temperature at which ignition of the substance occurs, i.e., slow or rapid combustion, deflagration or detonation, being noted each time;

(d)   The lowest temperature recorded in the three tests is the ignition temperature.

2.3.3   **Tests relating to flammable liquids of Classes 3, 6.1 and 8**

2.3.3.1   *Test for determining flash-point*

2.3.3.1.1   The flash-point shall be determined by means of one of the following types of apparatus:

(a)   Abel;

(b)   Abel-Pensky;

(c)   Tag;

(d)   Pensky-Martens;

(e)   Apparatus in accordance with ISO 3679:1983 or ISO 3680:1983.

2.3.3.1.2   To determine the flash-point of paints, gums and similar viscous products containing solvents, only apparatus and test methods suitable for determining the flash-point for viscous liquids shall be used, in accordance with the following standards:

(a)   International Standard ISO 3679:1983;

(b)   International Standard ISO 3680:1983;

(c)   International Standard ISO 1523:1983;

(d)   German Standard DIN 53213:1978, Part 1.

2.3.3.1.3   The test procedure shall be either according to an equilibrium method or according to a non-equilibrium method.

2.3.3.1.4   For the procedure according to an equilibrium method, see:

(a)   International Standard ISO 1516:1981;

(b)   International Standard ISO 3680:1983;

(c)   International Standard ISO 1523:1983;

(d)   International Standard ISO 3679:1983.

2.3.3.1.5   The procedure according to a non-equilibrium method shall be:

(a)   for the Abel apparatus, see:

(i)   British Standard BS 2000 Part 170:1995;

(ii)   French Standard NF MO7-011:1988;

(iii)   French Standard NF T66-009:1969;

(b)   for the Abel-Pensky apparatus, see:

(i)   German Standard DIN 51755, Part 1:1974 (for temperatures from 5 °C to 65 °C);

(ii)   German Standard DIN 51755, Part 2:1978 (for temperatures below 5 °C);

(iii)   French Standard NF MO7-036:1984;

(c)   for the Tag apparatus, see American Standard ASTM D 56:1993;

(d)   for the Pensky-Martens apparatus, see:

(i)   International Standard ISO 2719:1988;

(ii)   European Standard EN 22719 in each of its national versions (e.g. BS 2000, part 404/EN 22719):1994;

(iii)   American Standard ASTM D 93:1994;

(iv)   Institute of Petroleum Standard IP 34:1988.

2.3.3.1.6   The test methods listed in 2.3.3.1.4 and 2.3.3.1.5 shall only be used for flash-point ranges which are specified in the individual methods. The possibility of chemical reactions between the substance and the sample holder shall be considered when selecting the method to be used. The apparatus shall, as far as is consistent with safety, be placed in a draught-free position. For safety, a method utilizing a small sample size, around 2 ml, shall be used for organic peroxides and self-reactive substances (also known as "energetic" substances), or for toxic substances.

2.3.3.1.7   When the flash-point, determined by a non-equilibrium method in accordance with 2.3.3.1.5 is found to be 23 ± 2 °C or 60 ± 2 °C, it shall be confirmed for each temperature range by an equilibrium method in accordance with 2.3.3.1.4.

2.3.3.1.8   In the event of a dispute as to the classification of a flammable liquid, the classification proposed by the consignor shall be accepted if a check-test of the flash-point, yields a result not differing by more than 2 °C from the limits (23 °C and 60 °C respectively) stated in 2.2.3.1. If the difference is more than 2 °C, a second check-test shall be carried out, and the lowest figure of the flash-points obtained in either check-test shall be adopted.

2.3.3.2   *Test for determining peroxide content*

To determine the peroxide content of a liquid, the procedure is as follows:

A quantity p (about 5 g, weighed to the nearest 0.01 g) of the liquid to be titrated is placed in an Erlenmeyer flask; 20 cm^3 of acetic anhydride and about 1 g of powdered solid potassium iodide are added; the flask is shaken and, after 10 minutes, heated for 3 minutes to about 60 °C. When it has been left to cool for 5 minutes, 25 cm^3 of water are added. After this, it is left standing for half an hour, then the liberated iodine is titrated with a decinormal solution of sodium thiosulphate, no indicator being added; complete discoloration indicates the end of the reaction. If n is the number of cm^3 of thiosulphate solution required, the percentage of peroxide (calculated as H_2O_2) present in the sample is obtained by the formula:

$$\frac{17n}{100p}$$

2.3.4   **Test for determining fluidity**

To determine the fluidity of liquid, viscous or pasty substances and mixtures, the following test method shall be used.

2.3.4.1   *Test apparatus*

Commercial penetrometer conforming to ISO 2137:1985, with a guide rod of 47.5 g ± 0.05 g; sieve disc of duralumin with conical bores and a mass of 102.5 g ± 0.05 g (see Figure 1); penetration vessel with an inside diameter of 72 mm to 80 mm for reception of the sample.

2.3.4.2   *Test procedure*

The sample is poured into the penetration vessel not less than half an hour before the measurement. The vessel is then hermetically closed and left standing until the measurement. The sample in the hermetically closed penetration vessel is heated to 35 °C ± 0.5 °C and is placed on the penetrometer table immediately prior to measurement (not more than two minutes). The point S of the sieve disc is then brought into contact with the surface of the liquid and the rate of penetration is measured.

2.3.4.3   *Evaluation of test results*

A substance is pasty if, after the centre S has been brought into contact with the surface of the sample, the penetration indicated by the dial gauge:

(a)   after a loading time of 5 s ± 0.1 s, is less than 15.0 mm ± 0.3 mm; or

(b)   after a loading time of 5 s ± 0.1 s, is greater than 15.0 mm ± 0.3 mm, but the additional penetration after another 55 s ± 0.5 s is less than 5.0 mm ± 0.5 mm.

NOTE: In the case of samples having a flow point, it is often impossible to produce a steady level surface in the penetration vessel and, hence, to establish satisfactory initial measuring conditions for the contact of the point S. Furthermore, with some samples, the impact of the sieve disc can cause an elastic deformation of the surface and, in the first few seconds, simulate a deeper penetration. In all these cases, it may be appropriate to make the evaluation in paragraph (b) above.

Figure 1 – Penetrometer

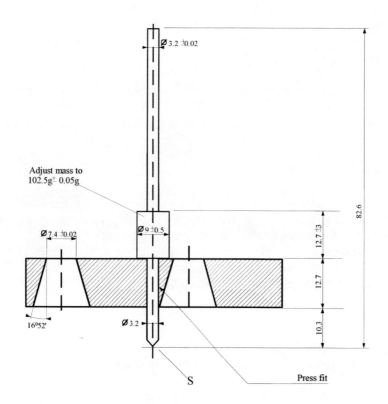

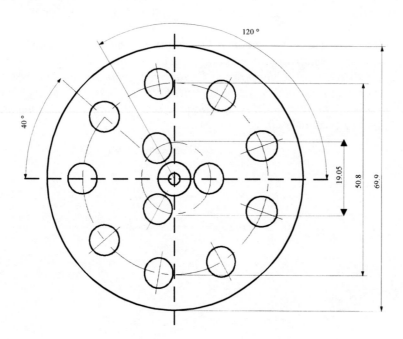

Tolerances not specified are ± 0.1 mm.

2.3.5   **Test for determining the ecotoxicity, persistence and bioaccumulation of substances in the aquatic environment for assignment to Class 9**

NOTE: The test methods used shall be those adopted by the Organization for Economic Cooperation and Development (OECD) and the European Commission (EC). If other methods are used, they shall be internationally recognized, be equivalent to the OECD/EC tests and be referenced in test reports.

2.3.5.1   *Acute toxicity for fish*

The object is to determine the concentration which causes 50% mortality in the test species; this is the (LC_{50}) value, namely, the concentration of the substance in water which will cause the death of 50% of a test group of fish during a continuous period of testing of at least 96 hours. Appropriate types of fish include: striped brill (<u>Brachydanio rerio</u>), fathead minnow (<u>Pimephales promelas</u>) and rainbow trout (<u>Oncorhynchus mykiss</u>).

The fish are exposed to the test substance added to the water in varying concentrations (+1 control). Observations are recorded at least every 24 hours. At the end of the 96-hour activity and, if possible, at each observation, the concentration causing the death of 50% of the fish is calculated. The no observed effect concentration (NOEC) at 96 hours is also determined.

2.3.5.2   *Acute toxicity for daphnia*

The object is to determine the effective concentration of the substance in water which renders 50% of the daphnia unable to swim (EC_{50}). The appropriate test organisms are <u>daphnia magna</u> and <u>daphnia pulex</u>. The daphnia are exposed for 48 hours to the test substance added to the water in varying concentrations. The no observed effect concentration (NOEC) at 48 hours is also determined.

2.3.5.3   *Algal growth inhibition*

The object is to determine the effect of a chemical on the growth of algae under standard conditions. The change in biomass and the rate of growth with algae under the same conditions, but without the presence of the test chemical, are compared over 72 hours. The results are expressed as the effective concentration which reduces the rate of algal growth by 50%, IC_{50r}, and also the formation of the biomass, IC_{50b}.

2.3.5.4   *Tests for ready biodegradability*

The object is to determine the degree of biodegradation under standard aerobic conditions. The test substance is added in low concentrations to a nutrient solution containing aerobic bacteria. The progress of degradation is followed for 28 days by determining the parameter specified in the test method used. Several equivalent test methods are available. The parameters include reduction of dissolved organic carbon (DOC), carbon dioxide (CO_2) generation of oxygen (O_2) depletion.

A substance is considered to be readily biodegradable if within not more than 28 days the following criteria are satisfied - within 10 days from when degradation first reaches 10%:

Reduction of DOC: 70%
Generation of CO_2:   60% of theoretical CO_2 production
Depletion of O_2:   60% of theoretical O_2 requirement.

The test may be continued beyond 28 days if the above criteria are not satisfied, but the result will represent the inherent biodegradability of the test substance. For assignment purposes, the "ready" result is normally required.

Where only COD and BOD_5 data are available, a substance is considered to be readily biodegradable if:

$$\frac{BOD_5}{COD} \geq 0.5$$

BOD (Biochemical Oxygen Demand) is defined as the mass of dissolved oxygen required by a specific volume of solution of the substance for the process of biochemical oxidation under prescribed conditions. The result is expressed as grams of BOD per gram of test substance. The normal test period is five days (BOD_5) using a national standard test procedure.

COD (Chemical Oxygen Demand) is a measure of the oxidizability of a substance, expressed as the equivalent amount in oxygen of an oxidizing reagent consumed by the substance under fixed laboratory conditions. The results are expressed in grams of COD per gram of substance. A national standard procedure may be used.

2.3.5.5   ***Tests for bioaccumulation potential***

2.3.5.5.1   The object is to determine the potential for bioaccumulation either by the ratio at equilibrium of the concentration (c) of a substance in a solvent to that in water or by the bioconcentration factor (BCF).

2.3.5.5.2   The ratio at equilibrium of the concentration (c) of a substance in a solvent to that in water is normally expressed as a $\log_{10}$. The solvent and water shall have negligible miscibility and the substance shall not ionize in water. The solvent normally used is n-octanol.

In the case of n-octanol and water, the result is:

$$\log P_{ow} = \log_{10} [c_o/c_w]$$

where P_{ow} is the partition coefficient obtained by dividing the concentration of the substance in n-octanol (c_o) by the concentration of the substance in water (C_w).
If $\log P_{ow} \geq 3.0$ then the substance has a potential to bioaccumulate.

2.3.5.5.3   The bioconcentration factor (BCF) is defined as the ratio of the concentration of the test substance in the test fish (c_f) to the concentration in the test water (c_w) at steady state:

$$BCF = (c_f) / (c_w).$$

The principle of the test involves exposing fish to a solution or dispersion at known concentrations of the test substance in water. Continuous flow, static or semi-static procedures may be used according to the test procedure selected, based on the properties of the test substances. Fish are exposed to the test substances over a given period of time, followed by a period of no further exposure. During the second period, measurements are made of the rate of increase in the water of the test substance (i.e. the rate of excretion or depuration).

(Full details of the various test procedures and the calculation method for the BCF are given in the OECD Guidelines for Testing of Chemicals, methods 305A to 305E, 12 May 1981).

2.3.5.5.4   A substance may have a $\log P_{ow}$ greater than 3 and a BCF less than 100 which would indicate little or no potential to bioaccumulate. In cases of doubt, the BCF value takes precedence over $\log P_{ow}$, as indicated in the flow chart of the procedure in 2.3.5.7.

2.3.5.6   *Criteria*

A substance may be regarded as a pollutant to the aquatic environment if it satisfies one of the following criteria:

The lowest of the values of the 96-hour LC_{50} for fish, the 48-hour EC_{50} for daphnia or the 72-hour IC_{50} for algae

–   is less than or equal to 1 mg/l;

–   is greater than 1 mg/l but less than or equal to 10 mg/l, and the substance is not biodegradable;

–   is greater than 1 mg/l but less than or equal to 10 mg/l, and the log P_{ow} is greater than or equal to 3.0 (unless the experimentally determined BCF is less than or equal to 100).

2.3.5.7   *Procedure to be followed*

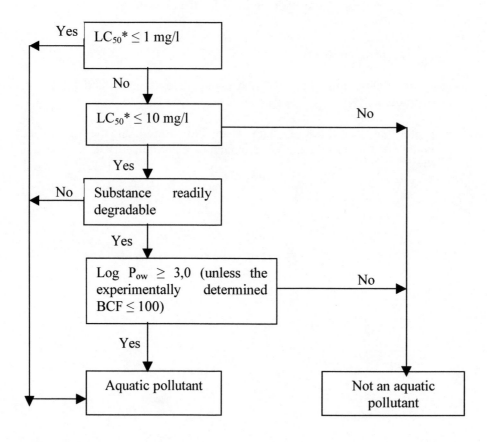

* Lowest value of 96-hour LC_{50}, 48-hour EC_{50} or 72-hour IC_{50} as appropriate.

BCF = bioconcentration factor

2.3.6   **Classification of organometallic substances in Classes 4.2 and 4.3**

Depending on their properties as determined in accordance with tests N.1 to N.5 of the Manual of Tests and Criteria, Part III, section 33, organometallic substances may be classified in Classes 4.2 or 4.3, as appropriate, in accordance with the flowchart scheme given in Figure 2.3.6.

NOTE 1:   *Depending on their other properties and on the precedence of hazard table (see 2.1.3.10), organometallic substances may have to be classified in other classes as appropriate.*

NOTE 2:   *Flammable solutions with organometallic compounds in concentrations which are not liable to spontaneous combustion or, in contact with water, do not emit flammable gases in dangerous quantities, are substances of Class 3.*

Figure 2.3.6   Flowchart scheme for the classification of organometallic substances in Classes 4.2 and 4.3 [b]

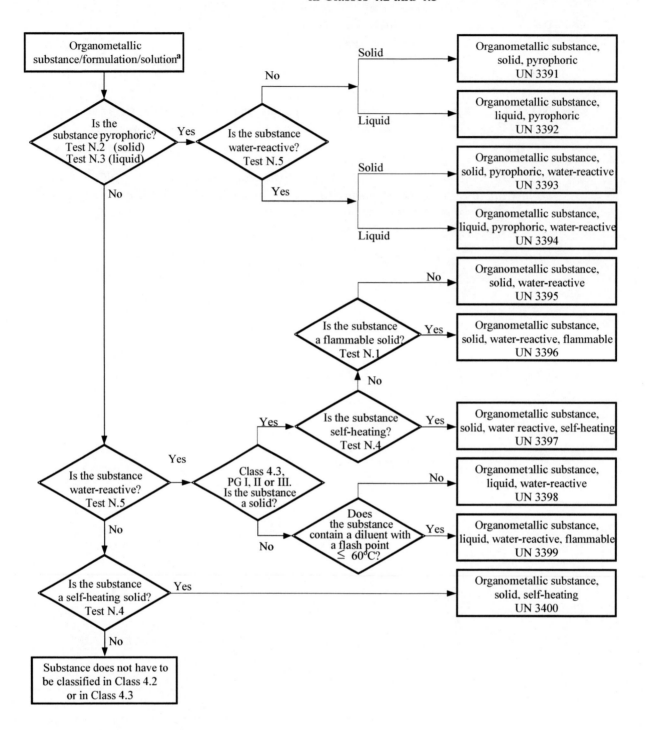

---

[a]   *If applicable and testing is relevant, taking into account reactivity properties, class 6.1 and 8 properties should be considered according to the precedence of hazard table of 2.1.3.10.*
[b]   *Test methods N.1 to N.5 can be found in the Manual of Tests and Criteria, Part III, Section 33.*

PART 3

Dangerous goods list, special provisions and exemptions related to dangerous goods packed in limited quantities

CHAPTER 3.1

GENERAL

3.1.1   Introduction

In addition to the provisions referred to or given in the tables of this Part, the general requirements of each Part, Chapter and/or Section are to be observed. These general requirements are not given in the tables. When a general requirement is contradictory to a special provision, the special provision prevails.

3.1.2   Proper shipping name

NOTE: For proper shipping names used for the carriage of samples, see 2.1.4.1.

3.1.2.1   The proper shipping name is that portion of the entry most accurately describing the goods in Table A or Table C in Chapter 3.2, which is shown in upper case characters (plus any numbers, Greek letters, "sec", "tert", and the letters "m", "n", "o", "p", which form an integral part of the name). Particulars concerning the vapour pressure (vp) and the boiling point (bp) in column (2) of Table C in chapter 3.2 are part of the proper shipping name. An alternative proper shipping name may be shown in brackets following the main proper shipping name. In Table A, it is shown in upper case characters (e.g., ETHANOL (ETHYL ALCOHOL)). In Table C, it is shown in lower case characters (e.g. ACETONITRILE (methyl cyanide)). Portions of an entry appearing in lower case need not be considered as part of the proper shipping name unless otherwise stated above.

3.1.2.2   When conjunctions such as "and" or "or" are in lower case or when segments of the name are punctuated by commas, the entire name of the entry need not necessarily be shown in the transport document or package markings. This is the case particularly when a combination of several distinct entries are listed under a single UN Number. Examples illustrating the selection of the proper shipping name for such entries are:

     (a)   UN 1057 LIGHTERS or LIGHTER REFILLS - The proper shipping name is the most appropriate of the following possible combinations:

         LIGHTERS
         LIGHTER REFILLS;

     (b)   UN 2793 FERROUS METAL BORINGS, SHAVINGS, TURNINGS or CUTTINGS in a form liable to self-heating. The proper shipping name is the most appropriate of the following combinations:

         FERROUS METAL BORINGS

         FERROUS METAL SHAVINGS

         FERROUS METAL TURNINGS

         FERROUS METAL CUTTINGS.

3.1.2.3   Proper shipping names may be used in the singular or plural as appropriate. In addition, when qualifying words are used as part of the proper shipping name, their sequence on documentation or package markings is optional. For instance, "DIMETHYLAMINE AQUEOUS SOLUTION" may alternatively be shown "AQUEOUS SOLUTION OF DIMETHYLAMINE". Commercial or military names for goods of Class 1 which contain the proper shipping name supplemented by additional descriptive text may be used.

3.1.2.4 Many substances have an entry for both the liquid and solid state (see definitions for liquid and solid in 1.2.1), or for the solid and solution. These are allocated separate UN numbers which are not necessarily adjacent to each other[1].

3.1.2.5 Unless it is already included in capital letters in the name indicated in Table A or Table C in Chapter 3.2, the qualifying word "MOLTEN" shall be added as part of the proper shipping name when a substance, which is a solid in accordance with the definition in 1.2.1, is offered for carriage in the molten state (e.g. ALKYLPHENOL, SOLID, N.O.S., MOLTEN).

3.1.2.6 Except for self-reactive substances and organic peroxides and unless it is already included in capital letters in the name indicated in Column (2) of Table A of Chapter 3.2, the word "STABILIZED" shall be added as part of the proper shipping name of a substance which without stabilization would be forbidden from carriage in accordance with paragraphs 2.2.X.2 due to it being liable to dangerously react under conditions normally encountered in carriage (e.g.: "TOXIC LIQUID, ORGANIC, N.O.S., STABILIZED").

When temperature control is used to stabilize such substances to prevent the development of any dangerous excess pressure, then:

(a) For liquids: where the SADT is less than or equal to 50 °C, the provisions of 2.2.41.1.17, the special provision V8 of Chapter 7.2, the special provision S4 of Chapter 8.5 and the requirements of Chapter 9.6 shall apply; for carriage in IBCs and tanks, all the provisions applicable to UN No. 3239 apply (see in particular 4.1.7.2 of ADR, packing instruction IBC520 and 4.2.1.13 of ADR);

(b) For gases: the conditions of carriage shall be approved by the competent authority.

3.1.2.7 Hydrates may be carried under the proper shipping name for the anhydrous substance.

3.1.2.8 *Generic or "not otherwise specified" (N.O.S.) names*

3.1.2.8.1 Generic and "not otherwise specified" proper shipping names that are assigned to special provision 274 in Column (6) of Table A in Chapter 3.2 or remark 27 in column (20) of Table C in Chapter 3.2 shall be supplemented with the technical name of the goods unless a national law or international convention prohibits its disclosure if it is a controlled substance. For explosives of Class 1, the dangerous goods description may be supplemented by additional descriptive text to indicate commercial or military names. Technical names shall be entered in brackets immediately following the proper shipping name. An appropriate modifier, such as "contains" or "containing" or other qualifying words such as "mixture", "solution", etc. and the percentage of the technical constituent may also be used. For example: "UN 1993 FLAMMABLE LIQUID, N.O.S. (CONTAINS XYLENE AND BENZENE), 3, II".

3.1.2.8.1.1 The technical name shall be a recognized chemical name, if relevant a biological name, or other name currently used in scientific and technical handbooks, journals and texts. Trade names shall not be used for this purpose. In the case of pesticides, only ISO common name(s), other name(s) in the World Health Organization (WHO) Recommended Classification of Pesticides by Hazard and Guidelines to Classification, or the name(s) of the active substance(s) may be used.

---

[1] *Details are provided in the alphabetical index (Table B of Chapter 3.2), e.g.:*
NITROXYLENES, LIQUID   6.1   1665
NITROXYLENES, SOLID   6.1   3447

3.1.2.8.1.2   When a mixture of dangerous goods is described by one of the "N.O.S." or "generic" entries to which special provision 274 has been allocated in Column (6) of Table A in Chapter 3.2, or remark 27 has been allocated in column (20) of Table C in Chapter 3.2, not more than the two constituents which most predominantly contribute to the hazard or hazards of a mixture need to be shown, excluding controlled substances when their disclosure is prohibited by national law or international convention. If a package containing a mixture is labelled with any subsidiary risk label, one of the two technical names shown in parentheses shall be the name of the constituent which compels the use of the subsidiary risk label.

NOTE: see 5.4.1.2.2.

3.1.2.8.1.3   Examples illustrating the selection of the proper shipping name supplemented with the technical name of goods for such N.O.S. entries are:

UN 2902 PESTICIDE, LIQUID, TOXIC, N.O.S. (drazoxolon);

UN 3394 ORGANOMETALLIC SUBSTANCE, LIQUID, PYROPHORIC, WATER-REACTIVE (trimethylgallium).

3.1.2.8.1.4   Examples illustrating how the proper shipping name is supplemented in the indication of the vapour pressure or the boiling-point for N.O.S entries for carriage in tank vessels are:

UN No. 1268 PETROLEUM DISTILLATES, N.O.S. or PETROLEUM PRODUCTS, N.O.S., $110 \text{ kPa} < pv50 \leq 150 \text{ kpa}$;

UN No. 1993 FLAMMABLE LIQUID, N.O.S. (ACETONE with more than 10% BENZENE), $pv \ 50 \leq 110 \text{ kPa}, 85° \text{ C} < bp \leq 115° \text{ C}$.

3.1.2.9   *Mixtures and solutions containing one dangerous substance*

When mixtures and solutions have to be regarded as the dangerous substance mentioned by name in accordance with the classification requirements of 2.1.3.3, the qualifying word "SOLUTION" or "MIXTURE", as appropriate, shall be added as part of the proper shipping name, e.g. "ACETONE SOLUTION". In addition, the concentration of the solution or mixture may also be indicated, e.g. "ACETONE 75% SOLUTION".

CHAPTER 3.2

DANGEROUS GOODS LIST

3.2.1 **List of dangerous goods in numerical order**

Explanations concerning Table A:

As a rule, each row of Table A of this Chapter deals with the substance(s) or article(s) covered by a specific UN number or an identification number. However, when substances or articles belonging to the same UN number have different chemical properties, physical properties and/or carriage conditions, several consecutive rows may be used for that UN number or identification number.

Each column of Table A is dedicated to a specific subject as indicated in the explanatory notes below. The intersection of columns and rows (cell) contains information concerning the subject treated in that column, for the substance(s) or article(s) of that row:

– The first four cells identify the substance(s) or article(s) belonging to that row (additional information in that respect may be given by the special provisions referred to in Column (6));

– The following cells give the applicable special provisions, either in the form of complete information or in coded form. The codes cross-refer to detailed information that is to be found in the numbers indicated in the explanatory notes below. An empty cell means either that there is no special provision and that only the general requirements apply, or that the carriage restriction indicated in the explanatory notes is in force.

The applicable general requirements are not referred to in the corresponding cells.

Explanatory notes for each column:

Column (1)   "UN number/identification number".

Contains the UN number or the identification number:

– of the dangerous substance or article if the substance or article has been assigned its own specific UN number or identification number, or

– of the generic or n.o.s. entry to which the dangerous substances or articles not mentioned by name shall be assigned in accordance with the criteria ("decision trees") of Part 2.

Column (2)   "Name and description"

Contains, in upper case characters, the name of the substance or article, if the substance or article has been assigned its own specific UN number or identification number, or of the generic or n.o.s. entry to which it has been assigned in accordance with the criteria ("decision trees") of Part 2. This name shall be used as the proper shipping name or, when applicable, as part of the proper shipping name (see 3.1.2 for further details on the proper shipping name).

A descriptive text in lower case characters is added after the proper shipping name to clarify the scope of the entry if the classification and/or carriage conditions of the substance or article may be different under certain conditions.

Column (3a)   "Class"

Contains the number of the Class, whose heading covers the dangerous substance or article. This Class number is assigned in accordance with the procedures and criteria of Part 2.

Column (3b)   "Classification code"

Contains the classification code of the dangerous substance or article.

–   For dangerous substances or articles of Class 1, the code consists of a division number and compatibility group letter, which are assigned in accordance with the procedures and criteria of 2.2.1.1.4.

–   For dangerous substances or articles of Class 2, the code consists of a number and one or more letters representing the hazardous property group, which are explained in 2.2.2.1.2 and 2.2.2.1.3.

–   For dangerous substances or articles of Classes 3, 4.1, 4.2, 4.3, 5.1, 5.2, 6.1, 6.2, 8 and 9, the codes are explained in 2.2.x.1.2. [1]

–   Dangerous substances or articles of Class 7 do not have a classification code.

Column (4)   "Packing group"

Contains the packing group number(s) (I, II or III) assigned to the dangerous substance. These packing group numbers are assigned on the basis of the procedures and criteria of Part 2. Certain articles and substances are not assigned to packing groups.

Column (5)   "Labels"

Contains the model number of the labels/placards (see 5.2.2.2 and 5.3.1.1.7) that have to be affixed to packages, containers, tank-containers, portable tanks, MEGCs and vehicles. However:

–   For substances or articles of Class 7, 7X means label model No. 7A, 7B or 7C as appropriate according to the category (see 2.2.7.8.4 and 5.2.2.1.11.1) or placard No. 7D (see 5.3.1.1.3 and 5.3.1.1.7.2).

The general provisions on labelling/placarding (e.g. number of labels, their location) are to be found in 5.2.2.1 for packages, and in 5.3.1, for containers, tank-containers, MEGCs, portable tanks and vehicles.

_____

[1]   *x = the Class number of the dangerous substance or article, without dividing point if applicable.*

NOTE: Special provisions, indicated in Column (6), may change the above labelling provisions.

Column (6)　　　"Special provisions"

Contains the numeric codes of special provisions that have to be met. These provisions concern a wide array of subjects, mainly connected with the contents of Columns (1) to (5) (e.g. carriage prohibitions, exemptions from requirements, explanations concerning the classification of certain forms of the dangerous goods concerned and additional labelling or marking provisions), and are listed in Chapter 3.3 in numerical order. If Column (6) is empty, no special provisions apply to the contents of Columns (1) to (5) for the dangerous goods concerned. Special provisions specific to inland navigation begin at 800.

Column (7)　　　"Limited quantities"

Contains an alphanumeric code with the following meaning:

–　　　"LQ0" signifies that no exemption from the provisions of ADN exists for the dangerous goods packed in limited quantities;

–　　　All the other alphanumeric codes starting with the letters "LQ" signify that the provisions of ADN are not applicable if the conditions indicated in Chapter 3.4 are fulfilled (general conditions of 3.4.1 and conditions of 3.4.3, 3.4.4, 3.4.5 and 3.4.6, as appropriate, for the relevant code).

Column (8)　　　"Carriage permitted"

This column contains the alphanumeric codes concerning the permitted form of carriage in inland navigation vessels.

If column (8) is empty, the substance or article may only be carried in packages.

If column (8) contains code "B", carriage is permitted in packages or in bulk (see 7.1.1.11).

If column (8) contains code "T", carriage is permitted in packages and in tank vessels. In the event of carriage in tank vessels, the requirements of Table C are applicable (see 7.2.1.21).

If "carriage prohibited" appears in column (8), carriage is not permitted.

If "free" appears in column (8), the substance is not subject to the requirements of ADN.

Column (9)　　　"Equipment required"

This column contains the alphanumeric codes for the equipment required for the carriage of the dangerous substance or article (see 8.1.5).

Column (10)　　　"Ventilation"

This column contains the alphanumeric codes of the special requirements concerning ventilation applicable to carriage with the following meaning:

– alphanumeric codes starting with the letters "VE" mean that special additional conditions are applicable to carriage. These can be found in 7.1.6.12 and establish special requirements.

Column (11)   "Provisions concerning loading, unloading and carriage"

This column contains the alphanumeric codes of the special requirements applicable to carriage with the following meaning:

– alphanumeric codes starting with the letters "CO", "ST" and "RA" mean that special additional conditions are applicable to carriage in bulk. These can be found in 7.1.6.11 and establish special requirements.

– alphanumeric codes starting with the letters "LO" mean that special additional conditions are applicable prior to loading. These can be found in 7.1.6.13 and establish special requirements.

– alphanumeric codes starting with the letters "HA" mean that special additional conditions are applicable to the handling and stowage of the cargo. These can be found in 7.1.6.14 and establish special requirements.

– alphanumeric codes starting with the letters "IN" mean that special additional conditions are applicable to the inspection of holds during carriage. These can be found in 7.1.6.16 and establish special requirements.

Column (12)   "Number of cones/blue lights"

This column contains the number of cones/lights which should constitute the marking of the vessel during the carriage of this dangerous substance or article (see 7.1.5).

Column (13)   "Additional requirements/Remarks"

This column contains additional requirements or observations concerning the carriage of this dangerous substance or article

UN No. or ID No.	Name and description	Class	Classification Code	Packing group	Labels	Special provisions	Limited quantities	Carriage permitted	Equipment required	Ventilation	Provisions concerning loading, unloading and carriage	Number of cones, blue lights	Remarks
	3.1.2	2.2	2.2	2.1.1.3	5.2.2	3.3	3.4.6	3.2.1	8.1.5	7.1.6	7.1.6	7.1.5	3.2.1
(1)	(2)	(3a)	(3b)	(4)	(5)	(6)	(7)	(8)	(9)	(10)	(11)	(12)	(13)
0004	AMMONIUM PICRATE dry or wetted with less than 10% water, by mass	1	1.1D		1		LQ0		PP		LO01 HA01, HA02, HA03, HA04, HA05, HA06	3	
0005	CARTRIDGES FOR WEAPONS with bursting charge	1	1.1F		1		LQ0		PP		LO01 HA01, HA02, HA03, HA04, HA05, HA06	3	
0006	CARTRIDGES FOR WEAPONS with bursting charge	1	1.1E		1		LQ0		PP		LO01 HA01, HA02, HA03, HA04, HA05, HA06	3	
0007	CARTRIDGES FOR WEAPONS with bursting charge	1	1.2F		1		LQ0		PP		LO01 HA01, HA02, HA03, HA04, HA05, HA06	3	
0009	AMMUNITION, INCENDIARY with or without burster, expelling charge or propelling charge	1	1.2G		1		LQ0		PP		LO01 HA01, HA03, HA04, HA05, HA06	3	
0010	AMMUNITION, INCENDIARY with or without burster, expelling charge or propelling charge	1	1.3G		1		LQ0		PP		LO01 HA01, HA03, HA04, HA05, HA06	3	

UN No. or ID No.	Name and description	Class	Classification Code	Packing group	Labels	Special provisions	Limited quantities	Carriage permitted	Equipment required	Ventilation	Provisions concerning loading, unloading and carriage	Number of cones, blue lights	Remarks
3.1.2	3.1.2	2.2	2.2	2.1.1.3	5.2.2	3.3	3.4.6	3.2.1	8.1.5	7.1.6	7.1.6	7.1.5	3.2.1
(1)	(2)	(3a)	(3b)	(4)	(5)	(6)	(7)	(8)	(9)	(10)	(11)	(12)	(13)
0012	CARTRIDGES FOR WEAPONS, INERT PROJECTILE or CARTRIDGES, SMALL ARMS	1	1.4S		1.4		LQ0		PP		LO01 HA01, HA03, HA04, HA05, HA06	0	
0014	CARTRIDGES FOR WEAPONS, BLANK or CARTRIDGES, SMALL ARMS, BLANK	1	1.4S		1.4		LQ0		PP		LO01 HA01, HA03, HA04, HA05, HA06	0	
0015	AMMUNITION, SMOKE with or without burster, expelling charge or propelling charge	1	1.2G		1		LQ0		PP		LO01 HA01, HA03, HA04, HA05, HA06	3	
0015	AMMUNITION, SMOKE with or without burster, expelling charge or propelling charge, containing corrosive substances	1	1.2G		1 +8		LQ0		PP		LO01 HA01, HA03, HA04, HA05, HA06	3	
0016	AMMUNITION, SMOKE with or without burster, expelling charge or propelling charge	1	1.3G		1		LQ0		PP		LO01 HA01, HA03, HA04, HA05, HA06	3	
0016	AMMUNITION, SMOKE with or without burster, expelling charge or propelling charge, containing corrosive substances	1	1.3G		1 +8		LQ0		PP		LO01 HA01, HA03, HA04, HA05, HA06	3	
0018	AMMUNITION, TEAR-PRODUCING with burster, expelling charge or propelling charge	1	1.2G		1+6.1+8	802	LQ0		PP		LO01 HA01, HA03, HA04, HA05, HA06	3	

UN No. or ID No.	Name and description	Class	Classification Code	Packing group	Labels	Special provisions	Limited quantities	Carriage permitted	Equipment required	Ventilation	Provisions concerning loading, unloading and carriage	Number of cones, blue lights	Remarks
	3.1.2	2.2	2.2	2.1.1.3	5.2.2	3.3	3.4.6	3.2.1	8.1.5	7.1.6	7.1.6	7.1.5	3.2.1
(1)	(2)	(3a)	(3b)	(4)	(5)	(6)	(7)	(8)	(9)	(10)	(11)	(12)	(13)
0019	AMMUNITION, TEAR-PRODUCING with burster, expelling charge or propelling charge	1	1.3G		1+6.1+8	802	LQ0		PP		LO01 HA01, HA03, HA04, HA05, HA06	3	
0020	AMMUNITION, TOXIC with burster, expelling charge or propelling charge	1	1.2K					CARRIAGE PROHIBITED					
0021	AMMUNITION, TOXIC with burster, expelling charge or propelling charge	1	1.3K					CARRIAGE PROHIBITED					
0027	BLACK POWDER (GUNPOWDER), granular or as a meal	1	1.1D		1		LQ0		PP		LO01 HA01, HA02, HA03, HA04, HA05, HA06	3	
0028	BLACK POWDER (GUNPOWDER), COMPRESSED or BLACK POWDER (GUNPOWDER), IN PELLETS	1	1.1D		1		LQ0		PP		LO01 HA01, HA02, HA03, HA04, HA05, HA06	3	
0029	DETONATORS, NON-ELECTRIC for blasting	1	1.1B		1		LQ0		PP		LO01 HA01, HA02, HA03, HA04, HA05, HA06	3	
0030	DETONATORS, ELECTRIC for blasting	1	1.1B		1		LQ0		PP		LO01 HA01, HA02, HA03, HA04, HA05, HA06	3	

UN No. or ID No.	Name and description	Class	Classification Code	Packing group	Labels	Special provisions	Limited quantities	Carriage permitted	Equipment required	Ventilation	Provisions concerning loading, unloading and carriage	Number of cones, blue lights	Remarks
3.1.2	3.1.2	2.2	2.2	2.1.1.3	5.2.2	3.3	3.4.6	3.2.1	8.1.5	7.1.6	7.1.6	7.1.5	3.2.1
(1)	(2)	(3a)	(3b)	(4)	(5)	(6)	(7)	(8)	(9)	(10)	(11)	(12)	(13)
0033	BOMBS with bursting charge	1	1.1F		1		LQ0		PP		LO01 HA01, HA02, HA03, HA04, HA05, HA06	3	
0034	BOMBS with bursting charge	1	1.1D		1		LQ0		PP		LO01 HA01, HA02, HA03, HA04, HA05, HA06	3	
0035	BOMBS with bursting charge	1	1.2D		1		LQ0		PP		LO01 HA01, HA03, HA04, HA05, HA06	3	
0037	BOMBS, PHOTO-FLASH	1	1.1F		1		LQ0		PP		LO01 HA01, HA02, HA03, HA04, HA05, HA06	3	
0038	BOMBS, PHOTO-FLASH	1	1.1D		1		LQ0		PP		LO01 HA01, HA02, HA03, HA04, HA05, HA06	3	
0039	BOMBS, PHOTO-FLASH	1	1.2G		1		LQ0		PP		LO01 HA01, HA03, HA04, HA05, HA06	3	

UN No. or ID No.	Name and description	Class	Classification Code	Packing group	Labels	Special provisions	Limited quantities	Carriage permitted	Equipment required	Ventilation	Provisions concerning loading, unloading and carriage	Number of cones, blue lights	Remarks
3.1.2	3.1.2	2.2	2.2	2.1.1.3	5.2.2	3.3	3.4.6	3.2.1	8.1.5	7.1.6	7.1.6	7.1.5	3.2.1
(1)	(2)	(3a)	(3b)	(4)	(5)	(6)	(7)	(8)	(9)	(10)	(11)	(12)	(13)
0042	BOOSTERS without detonator	1	1.1D		1		LQ0		PP		LO01 HA01, HA02, HA03, HA04, HA05, HA06	3	
0043	BURSTERS, explosive	1	1.1D		1		LQ0		PP		LO01 HA01, HA02, HA03, HA04, HA05, HA06	3	
0044	PRIMERS, CAP TYPE	1	1.4S		1.4		LQ0		PP		LO01 HA01, HA03, HA04, HA05, HA06	0	
0048	CHARGES, DEMOLITION	1	1.1D		1		LQ0		PP		LO01 HA01, HA02, HA03, HA04, HA05, HA06	3	
0049	CARTRIDGES, FLASH	1	1.1G		1		LQ0		PP		LO01 HA01, HA02, HA03, HA04, HA05, HA06	3	
0050	CARTRIDGES, FLASH	1	1.3G		1		LQ0		PP		LO01 HA01, HA03, HA04, HA05, HA06	3	

UN No. or ID No.	Name and description	Class	Classification Code	Packing group	Labels	Special provisions	Limited quantities	Carriage permitted	Equipment required	Ventilation	Provisions concerning loading, unloading and carriage	Number of cones, blue lights	Remarks
3.1.2	3.1.2	2.2	2.2	2.1.1.3	5.2.2	3.3	3.4.6	3.2.1	8.1.5	7.1.6	7.1.6	7.1.5	3.2.1
(1)	(2)	(3a)	(3b)	(4)	(5)	(6)	(7)	(8)	(9)	(10)	(11)	(12)	(13)
0054	CARTRIDGES, SIGNAL	1	1.3G		1		LQ0		PP		LO01 HA01, HA03, HA04, HA05, HA06	3	
0055	CASES, CARTRIDGE, EMPTY, WITH PRIMER	1	1.4S		1.4		LQ0		PP		LO01 HA01, HA03, HA04, HA05, HA06	0	
0056	CHARGES, DEPTH	1	1.1D		1		LQ0		PP		LO01 HA01, HA02, HA03, HA04, HA05, HA06	3	
0059	CHARGES, SHAPED without detonator	1	1.1D		1		LQ0		PP		LO01 HA01, HA02, HA03, HA04, HA05, HA06	3	
0060	CHARGES, SUPPLEMENTARY, EXPLOSIVE	1	1.1D		1		LQ0		PP		LO01 HA01, HA02, HA03, HA04, HA05, HA06	3	
0065	CORD, DETONATING, flexible	1	1.1D		1		LQ0		PP		LO01 HA01, HA02, HA03, HA04, HA05, HA06	3	

UN No. or ID No.	Name and description	Class	Classification Code	Packing group	Labels	Special provisions	Limited quantities	Carriage permitted	Equipment required	Ventilation	Provisions concerning loading, unloading and carriage	Number of cones, blue lights	Remarks
	3.1.2	2.2	2.2	2.1.1.3	5.2.2	3.3	3.4.6	3.2.1	8.1.5	7.1.6	7.1.6	7.1.5	3.2.1
(1)	(2)	(3a)	(3b)	(4)	(5)	(6)	(7)	(8)	(9)	(10)	(11)	(12)	(13)
0066	CORD. IGNITER	1	1.4G		1.4		LQ0		PP		LO01 HA01, HA03, HA04, HA05, HA06	1	
0070	CUTTERS. CABLE, EXPLOSIVE	1	1.4S		1.4		LQ0		PP		LO01 HA01, HA03, HA04, HA05, HA06	0	
0072	CYCLOTRIMETHYLENE-TRINITRAMINE (CYCLONITE: HEXOGEN; RDX), WETTED with not less than 15% water, by mass	1	1.1D		1	266	LQ0		PP		LO01 HA01, HA02, HA03, HA04, HA05, HA06	3	
0073	DETONATORS FOR AMMUNITION	1	1.1B		1		LQ0		PP		LO01 HA01, HA02, HA03, HA04, HA05, HA06	3	
0074	DIAZODINITROPHENOL, WETTED with not less than 40% water, or mixture of alcohol and water, by mass	1	1.1A		1	266	LQ0		PP		LO01 HA01, HA02, HA03, HA04, HA05, HA06	3	
0075	DIETHYLENEGLYCOL DINITRATE, DESENSITIZED with not less than 25% non-volatile, water-insoluble phlegmatizer, by mass	1	1.1D		1	266	LQ0		PP		LO01 HA01, HA02, HA03, HA04, HA05, HA06	3	

UN No. or ID No.	Name and description	Class	Classification Code	Packing group	Labels	Special provisions	Limited quantities	Carriage permitted	Equipment required	Ventilation	Provisions concerning loading, unloading and carriage	Number of cones, blue lights	Remarks
3.1.2	3.1.2	2.2	2.2	2.1.1.3	5.2.2	3.3	3.4.6	3.2.1	8.1.5	7.1.6	7.1.6	7.1.5	3.2.1
(1)	(2)	(3a)	(3b)	(4)	(5)	(6)	(7)	(8)	(9)	(10)	(11)	(12)	(13)
0076	DINITROPHENOL, dry or wetted with less than 15% water, by mass	1	1.1D		1+6.1	802	LQ0		PP		LO01 HA01, HA02, HA03, HA04, HA05, HA06	3	
0077	DINITROPHENOLATES, alkali metals, dry or wetted with less than 15% water, by mass	1	1.3C		1+6.1	802	LQ0		PP		LO01 HA01, HA03, HA04, HA05, HA06	3	
0078	DINITRORESORCINOL, dry or wetted with less than 15% water, by mass	1	1.1D		1		LQ0		PP		LO01 HA01, HA02, HA03, HA04, HA05, HA06	3	
0079	HEXANITRODIPHENYLAMINE (DIPICRYLAMINE; HEXYL)	1	1.1D		1		LQ0		PP		LO01 HA01, HA02, HA03, HA04, HA05, HA06	3	
0081	EXPLOSIVE, BLASTING, TYPE A	1	1.1D		1	616 617	LQ0		PP		LO01 HA01, HA02, HA03, HA04, HA05, HA06	3	
0082	EXPLOSIVE, BLASTING, TYPE B	1	1.1D		1	617	LQ0		PP		LO01 HA01, HA02, HA03, HA04, HA05, HA06	3	

UN No. or ID No.	Name and description	Class	Classification Code	Packing group	Labels	Special provisions	Limited quantities	Carriage permitted	Equipment required	Ventilation	Provisions concerning loading, unloading and carriage	Number of cones, blue lights	Remarks
	3.1.2	2.2	2.2	2.1.1.3	5.2.2	3.3	3.4.6	3.2.1	8.1.5	7.1.6	7.1.6	7.1.5	3.2.1
(1)	(2)	(3a)	(3b)	(4)	(5)	(6)	(7)	(8)	(9)	(10)	(11)	(12)	(13)
0083	EXPLOSIVE, BLASTING, TYPE C	1	1.1D		1	267 617	LQ0		PP		LO01 HA01, HA02, HA03, HA04, HA05, HA06	3	
0084	EXPLOSIVE, BLASTING, TYPE D	1	1.1D		1	617	LQ0		PP		LO01 HA01, HA02, HA03, HA04, HA05, HA06	3	
0092	FLARES, SURFACE	1	1.3G		1		LQ0		PP		LO01 HA01, HA03, HA04, HA05, HA06	3	
0093	FLARES, AERIAL	1	1.3G		1		LQ0		PP		LO01 HA01, HA03, HA04, HA05, HA06	3	
0094	FLASH POWDER	1	1.1G		1		LQ0		PP		LO01 HA01, HA02, HA03, HA04, HA05, HA06	3	
0099	FRACTURING DEVICES, EXPLOSIVE without detonator, for oil wells	1	1.1D		1		LQ0		PP		LO01 HA01, HA02, HA03, HA04, HA05, HA06	3	

UN No. or ID No.	Name and description	Class	Classification Code	Packing group	Labels	Special provisions	Limited quantities	Carriage permitted	Equipment required	Ventilation	Provisions concerning loading, unloading and carriage	Number of cones, blue lights	Remarks
3.1.2	3.1.2	2.2	2.2	2.1.1.3	5.2.2	3.3	3.4.6	3.2.1	8.1.5	7.1.6	7.1.6	7.1.5	3.2.1
(1)	(2)	(3a)	(3b)	(4)	(5)	(6)	(7)	(8)	(9)	(10)	(11)	(12)	(13)
0101	FUSE, NON-DETONATING	1	1.3G		1		LQ0		PP		LO01 HA01, HA03, HA04, HA05, HA06	3	
0102	CORD (FUSE), DETONATING, metal clad	1	1.2D		1		LQ0		PP		LO01 HA01, HA03, HA04, HA05, HA06	3	
0103	FUSE, IGNITER, tubular, metal clad	1	1.4G		1.4		LQ0		PP		LO01 HA01, HA03, HA04, HA05, HA06	1	
0104	CORD (FUSE), DETONATING, MILD EFFECT, metal clad	1	1.4D		1.4		LQ0		PP		LO01 HA01, HA03, HA04, HA05, HA06	1	
0105	FUSE, SAFETY	1	1.4S		1.4		LQ0		PP		LO01 HA01, HA03, HA04, HA05, HA06	0	
0106	FUZES, DETONATING	1	1.1B		1		LQ0		PP		LO01 HA01, HA02, HA03, HA04, HA05, HA06	3	
0107	FUZES, DETONATING	1	1.2B		1		LQ0		PP		LO01 HA01, HA02, HA03, HA04, HA05, HA06	3	

UN No. or ID No.	Name and description	Class	Classification Code	Packing group	Labels	Special provisions	Limited quantities	Carriage permitted	Equipment required	Ventilation	Provisions concerning loading, unloading and carriage	Number of cones, blue lights	Remarks
3.1.2	3.1.2	2.2	2.2	2.1.1.3	5.2.2	3.3	3.4.6	3.2.1	8.1.5	7.1.6	7.1.6	7.1.5	3.2.1
(1)	(2)	(3a)	(3b)	(4)	(5)	(6)	(7)	(8)	(9)	(10)	(11)	(12)	(13)
0110	GRENADES, PRACTICE, hand or rifle	1	1.4S		1.4		LQ0		PP			0	
0113	GUANYLNITROSAMINO-GUANYLIDENE HYDRAZINE, WETTED with not less than 30% water, by mass	1	1.1A		1	266	LQ0		PP		LO01 HA01, HA02, HA03, HA04, HA05, HA06	3	
0114	GUANYLNITROSAMINO-GUANYLTETRAZENE (TETRAZENE), WETTED with not less than 30% water, or mixture of alcohol and water, by mass	1	1.1A		1	266	LQ0		PP		LO01 HA01, HA02, HA03, HA04, HA05, HA06	3	
0118	HEXOLITE (HEXOTOL), dry or wetted with less than 15% water, by mass	1	1.1D		1		LQ0		PP		LO01 HA01, HA02, HA03, HA04, HA05, HA06	3	
0121	IGNITERS	1	1.1G		1		LQ0		PP		LO01 HA01, HA02, HA03, HA04, HA05, HA06	3	
0124	JET PERFORATING GUNS, CHARGED, oil well, without detonator	1	1.1D		1		LQ0		PP		LO01 HA01, HA02, HA03, HA04, HA05, HA06	3	

UN No. or ID No. (1) 3.1.2	Name and description (2) 3.1.2	Class (3a) 2.2	Classification Code (3b) 2.2	Packing group (4) 2.1.1.3	Labels (5) 5.2.2	Special provisions (6) 3.3	Limited quantities (7) 3.4.6	Carriage permitted (8) 3.2.1	Equipment required (9) 8.1.5	Ventilation (10) 7.1.6	Provisions concerning loading, unloading and carriage (11) 7.1.6	Number of cones, blue lights (12) 7.1.5	Remarks (13) 3.2.1
0129	LEAD AZIDE, WETTED with not less than 20% water, or mixture of alcohol and water, by mass	1	1.1A		1	266	LQ0		PP		LO01 HA01, HA02, HA03, HA04, HA05, HA06	3	
0130	LEAD TRINITRORESORCINATE (LEAD STYPHNATE), WETTED with not less than 20% water, or mixture of alcohol and water, by mass	1	1.1A		1	266	LQ0		PP		LO01 HA01, HA02, HA03, HA04, HA05, HA06	3	
0131	LIGHTERS, FUSE	1	1.4S		1.4		LQ0		PP		LO01 HA01, HA03, HA04, HA05, HA06	0	
0132	DEFLAGRATING METAL SALTS OF AROMATIC NITRODERIVATIVES, N.O.S.	1	1.3C		1	274	LQ0		PP		LO01 HA01, HA03, HA04, HA05, HA06	3	
0133	MANNITOL HEXANITRATE (NITROMANNITE), WETTED with not less than 40% water, or mixture of alcohol and water, by mass	1	1.1D		1	266	LQ0		PP		LO01 HA01, HA02, HA03, HA04, HA05, HA06	3	
0135	MERCURY FULMINATE, WETTED with not less than 20% water, or mixture of alcohol and water, by mass	1	1.1A		1	266	LQ0		PP		LO01 HA01, HA02, HA03, HA04, HA05, HA06	3	

UN No. or ID No.	Name and description	Class	Classification Code	Packing group	Labels	Special provisions	Limited quantities	Carriage permitted	Equipment required	Ventilation	Provisions concerning loading, unloading and carriage	Number of cones, blue lights	Remarks
	3.1.2	2.2	2.2	2.1.1.3	5.2.2	3.3	3.4.6	3.2.1	8.1.5	7.1.6	7.1.6	7.1.5	3.2.1
(1)	(2)	(3a)	(3b)	(4)	(5)	(6)	(7)	(8)	(9)	(10)	(11)	(12)	(13)
0136	MINES with bursting charge	1	1.1F		1		LQ0		PP		LO01 HA01, HA02, HA03, HA04, HA05, HA06	3	
0137	MINES with bursting charge	1	1.1D		1		LQ0		PP		LO01 HA01, HA02, HA03, HA04, HA05, HA06	3	
0138	MINES with bursting charge	1	1.2D		1		LQ0		PP		LO01 HA01, HA03, HA04, HA05, HA06	3	
0143	NITROGLYCERIN, DESENSITIZED with not less than 40% non-volatile water-insoluble phlegmatizer, by mass	1	1.1D		1+6.1	266 271 802	LQ0		PP		LO01 HA01, HA02, HA03, HA04, HA05, HA06	3	
0144	NITROGLYCERIN SOLUTION IN ALCOHOL with more than 1% but not more than 10% nitroglycerin	1	1.1D		1	500	LQ0		PP		LO01 HA01, HA02, HA03, HA04, HA05, HA06	3	
0146	NITROSTARCH, dry or wetted with less than 20% water, by mass	1	1.1D		1		LQ0		PP		LO01 HA01, HA02, HA03, HA04, HA05, HA06	3	

UN No. or ID No.	Name and description	Class	Classification Code	Packing group	Labels	Special provisions	Limited quantities	Carriage permitted	Equipment required	Ventilation	Provisions concerning loading, unloading and carriage	Number of cones, blue lights	Remarks
	3.1.2	2.2	2.2	2.1.1.3	5.2.2	3.3	3.4.6	3.2.1	8.1.5	7.1.6	7.1.6	7.1.5	3.2.1
(1)	(2)	(3a)	(3b)	(4)	(5)	(6)	(7)	(8)	(9)	(10)	(11)	(12)	(13)
0147	NITRO UREA	1	1.1D		1		LQ0		PP		LO01 HA01, HA02, HA03, HA04, HA05, HA06	3	
0150	PENTAERYTHRITE TETRANITRATE (PENTAERYTHRITOL TETRANITRATE; PETN), WETTED with not less than 25% water, by mass, or DESENSITIZED with not less than 15% phlegmatizer, by mass	1	1.1D		1	266	LQ0		PP		LO01 HA01, HA02, HA03, HA04, HA05, HA06	3	
0151	PENTOLITE, dry or wetted with less than 15% water, by mass	1	1.1D		1		LQ0		PP		LO01 HA01, HA02, HA03, HA04, HA05, HA06	3	
0153	TRINITROANILINE (PICRAMIDE)	1	1.1D		1		LQ0		PP		LO01 HA01, HA02, HA03, HA04, HA05, HA06	3	
0154	TRINITROPHENOL (PICRIC ACID), dry or wetted with less than 30% water, by mass	1	1.1D		1		LQ0		PP		LO01 HA01, HA02, HA03, HA04, HA05, HA06	1	
0155	TRINITROCHLOROBENZENE (PICRYL CHLORIDE)	1	1.1D		1		LQ0		PP		LO01 HA01, HA02, HA03, HA04, HA05, HA06	3	

UN No. or ID No.	Name and description	Class	Classifi-cation Code	Packing group	Labels	Special provisions	Limited quantities	Carriage permitted	Equipment required	Ventilation	Provisions concerning loading, unloading and carriage	Number of cones, blue lights	Remarks
3.1.2		2.2	2.2	2.1.1.3	5.2.2	3.3	3.4.6	3.2.1	8.1.5	7.1.6	7.1.6	7.1.5	3.2.1
(1)	(2)	(3a)	(3b)	(4)	(5)	(6)	(7)	(8)	(9)	(10)	(11)	(12)	(13)
0159	POWDER CAKE (POWDER PASTE), WETTED with not less than 25% water, by mass	1	1.3C		1	266	LQ0		PP		LO01 HA01, HA03, HA04, HA05, HA06	3	
0160	POWDER, SMOKELESS	1	1.1C		1		LQ0		PP		LO01 HA01, HA02, HA03, HA04, HA05, HA06	3	
0161	POWDER, SMOKELESS	1	1.3C		1		LQ0		PP		LO01 HA01, HA03, HA04, HA05, HA06	3	
0167	PROJECTILES with bursting charge	1	1.1F		1		LQ0		PP		LO01 HA01, HA02, HA03, HA04, HA05, HA06	3	
0168	PROJECTILES with bursting charge	1	1.1D		1		LQ0		PP		LO01 HA01, HA02, HA03, HA04, HA05, HA06	3	
0169	PROJECTILES with bursting charge	1	1.2D		1		LQ0		PP		LO01 HA01, HA03, HA04, HA05, HA06	3	
0171	AMMUNITION, ILLUMINATING with or without burster, expelling charge or propelling charge	1	1.2G		1		LQ0		PP		LO01 HA01, HA03, HA04, HA05, HA06	3	

UN No. or ID No. (1)	Name and description 3.1.2 (2)	Class 2.2 (3a)	Classification Code 2.2 (3b)	Packing group 2.1.1.3 (4)	Labels 5.2.2 (5)	Special provisions 3.3 (6)	Limited quantities 3.4.6 (7)	Carriage permitted 3.2.1 (8)	Equipment required 8.1.5 (9)	Ventilation 7.1.6 (10)	Provisions concerning loading, unloading and carriage 7.1.6 (11)	Number of cones, blue lights 7.1.5 (12)	Remarks 3.2.1 (13)
0173	RELEASE DEVICES, EXPLOSIVE	1	1.4S		1.4		LQ0		PP		LO01 HA01, HA03, HA04, HA05, HA06	0	
0174	RIVETS, EXPLOSIVE	1	1.4S		1.4		LQ0		PP		LO01 HA01, HA03, HA04, HA05, HA06	0	
0180	ROCKETS with bursting charge	1	1.1F		1		LQ0		PP		LO01 HA01, HA02, HA03, HA04, HA05, HA06	3	
0181	ROCKETS with bursting charge	1	1.1E		1		LQ0		PP		LO01 HA01, HA02, HA03, HA04, HA05, HA06	3	
0182	ROCKETS with bursting charge	1	1.2E		1		LQ0		PP		LO01 HA01, HA02, HA03, HA04, HA05, HA06	3	
0183	ROCKETS with inert head	1	1.3C		1		LQ0		PP		LO01 HA01, HA03, HA04, HA05, HA06	3	
0186	ROCKET MOTORS	1	1.3C		1		LQ0		PP		LO01 HA01, HA03, HA04, HA05, HA06	3	

UN No. or ID No.	Name and description	Class	Classification Code	Packing group	Labels	Special provisions	Limited quantities	Carriage permitted	Equipment required	Ventilation	Provisions concerning loading, unloading and carriage	Number of cones, blue lights	Remarks
3.1.2	3.1.2	2.2	2.2	2.1.1.3	5.2.2	3.3	3.4.6	3.2.1	8.1.5	7.1.6	7.1.6	7.1.5	3.2.1
(1)	(2)	(3a)	(3b)	(4)	(5)	(6)	(7)	(8)	(9)	(10)	(11)	(12)	(13)
0190	SAMPLES, EXPLOSIVE, other than initiating explosive	1				16 274	LQ0		PP		LO01 HA01, HA02, HA03, HA04, HA05, HA06	3	
0191	SIGNAL DEVICES, HAND	1	1.4G		1.4		LQ0		PP		LO01 HA01, HA03, HA04, HA05, HA06	1	
0192	SIGNALS, RAILWAY TRACK, EXPLOSIVE	1	1.1G		1		LQ0		PP		LO01 HA01, HA02, HA03, HA04, HA05, HA06	3	
0193	SIGNALS, RAILWAY TRACK, EXPLOSIVE	1	1.4S		1.4		LQ0		PP		LO01 HA01, HA03, HA04, HA05, HA06	0	
0194	SIGNALS, DISTRESS, ship	1	1.1G		1		LQ0		PP		LO01 HA01, HA02, HA03, HA04, HA05, HA06	3	
0195	SIGNALS, DISTRESS, ship	1	1.3G		1		LQ0		PP		LO01 HA01, HA03, HA04, HA05, HA06	3	

UN No. or ID No.	Name and description	Class	Classification Code	Packing group	Labels	Special provisions	Limited quantities	Carriage permitted	Equipment required	Ventilation	Provisions concerning loading, unloading and carriage	Number of cones, blue lights	Remarks
3.1.2	3.1.2	2.2	2.2	2.1.1.3	5.2.2	3.3	3.4.6	3.2.1	8.1.5	7.1.6	7.1.6	7.1.5	3.2.1
(1)	(2)	(3a)	(3b)	(4)	(5)	(6)	(7)	(8)	(9)	(10)	(11)	(12)	(13)
0196	SIGNALS, SMOKE	1	1.1G		1		LQ0		PP		LO01 HA01, HA02, HA03, HA04, HA05, HA06	3	
0197	SIGNALS, SMOKE	1	1.4G		1.4				PP		LO01 HA01, HA03, HA04, HA05, HA06	1	
0204	SOUNDING DEVICES, EXPLOSIVE	1	1.2F		1		LQ0		PP		LO01 HA01, HA02, HA03, HA04, HA05, HA06	3	
0207	TETRANITROANILINE	1	1.1D		1		LQ0		PP		LO01 HA01, HA02, HA03, HA04, HA05, HA06	3	
0208	TRINITROPHENYLMETHYL-NITRAMINE (TETRYL)	1	1.1D		1		LQ0		PP		LO01 HA01, HA02, HA03, HA04, HA05, HA06	3	
0209	TRINITROTOLUENE (TNT), dry or wetted with less than 30% water, by mass	1	1.1D		1		LQ0		PP		LO01 HA01, HA02, HA03, HA04, HA05, HA06	3	

UN No. or ID No.	Name and description	Class	Classification Code	Packing group	Labels	Special provisions	Limited quantities	Carriage permitted	Equipment required	Ventilation	Provisions concerning loading, unloading and carriage	Number of cones, blue lights	Remarks
3.1.2	3.1.2	2.2	2.2	2.1.1.3	5.2.2	3.3	3.4.6	3.2.1	8.1.5	7.1.6	7.1.6	7.1.5	3.2.1
(1)	(2)	(3a)	(3b)	(4)	(5)	(6)	(7)	(8)	(9)	(10)	(11)	(12)	(13)
0212	TRACERS FOR AMMUNITION	1	1.3G		1		LQ0		PP		LO01 HA01, HA03, HA04, HA05, HA06	3	
0213	TRINITROANISOLE	1	1.1D		1		LQ0		PP		LO01 HA01, HA02, HA03, HA04, HA05, HA06	3	
0214	TRINITROBENZENE, dry or wetted with less than 30% water, by mass	1	1.1D		1		LQ0		PP		LO01 HA01, HA02, HA03, HA04, HA05, HA06	3	
0215	TRINITROBENZOIC ACID, dry or wetted with less than 30% water, by mass	1	1.1D		1		LQ0		PP		LO01 HA01, HA02, HA03, HA04, HA05, HA06	3	
0216	TRINITRO-m-CRESOL	1	1.1D		1		LQ0		PP		LO01 HA01, HA02, HA03, HA04, HA05, HA06	3	
0217	TRINITRONAPHTHALENE	1	1.1D		1		LQ0		PP		LO01 HA01, HA02, HA03, HA04, HA05, HA06	3	

UN No. or ID No.	Name and description	Class	Classification Code	Packing group	Labels	Special provisions	Limited quantities	Carriage permitted	Equipment required	Ventilation	Provisions concerning loading, unloading and carriage	Number of cones, blue lights	Remarks
3.1.2	3.1.2	2.2	2.2	2.1.1.3	5.2.2	3.3	3.4.6	3.2.1	8.1.5	7.1.6	7.1.6	7.1.5	3.2.1
(1)	(2)	(3a)	(3b)	(4)	(5)	(6)	(7)	(8)	(9)	(10)	(11)	(12)	(13)
0218	TRINITROPHENETOLE	1	1.1D		1		LQ0		PP		LO01 HA01, HA02, HA03, HA04, HA05, HA06	3	
0219	TRINITRORESORCINOL (STYPHNIC ACID), dry or wetted with less than 20% water, or mixture of alcohol and water, by mass	1	1.1D		1		LQ0		PP		LO01 HA01, HA02, HA03, HA04, HA05, HA06	3	
0220	UREA NITRATE, dry or wetted with less than 20% water, by mass	1	1.1D		1		LQ0		PP		LO01 HA01, HA02, HA03, HA04, HA05, HA06	3	
0221	WARHEADS, TORPEDO with bursting charge	1	1.1D		1		LQ0		PP		LO01 HA01, HA02, HA03, HA04, HA05, HA06	3	
0222	AMMONIUM NITRATE with more than 0.2% combustible substances, including any organic substance calculated as carbon, to the exclusion of any other added substance	1	1.1D		1		LQ0		PP		LO01 HA01, HA02, HA03, HA04, HA05, HA06	3	
0224	BARIUM AZIDE, dry or wetted with less than 50% water, by mass	1	1.1A		1+6.1	802	LQ0		PP		LO01 HA01, HA02, HA03, HA04, HA05, HA06	3	

UN No. or ID No.	Name and description	Class	Classification Code	Packing group	Labels	Special provisions	Limited quantities	Carriage permitted	Equipment required	Ventilation	Provisions concerning loading, unloading and carriage	Number of cones, blue lights	Remarks
3.1.2	3.1.2	2.2	2.2	2.1.1.3	5.2.2	3.3	3.4.6	3.2.1	8.1.5	7.1.6	7.1.6	7.1.5	3.2.1
(1)	(2)	(3a)	(3b)	(4)	(5)	(6)	(7)	(8)	(9)	(10)	(11)	(12)	(13)
0225	BOOSTERS WITH DETONATOR	1	1.1B		1		LQ0		PP		LO01 HA01, HA02, HA03, HA04, HA05, HA06	3	
0226	CYCLOTETRAMETHYLENE-TETRANITRAMINE (HMX; OCTOGEN), WETTED with not less than 15% water, by mass	1	1.1D		1	266	LQ0		PP		LO01 HA01, HA02, HA03, HA04, HA05, HA06	3	
0234	SODIUM DINITRO-o-CRESOLATE, dry or wetted with less than 15% water, by mass	1	1.3C		1		LQ0		PP		LO01 HA01, HA03, HA04, HA05, HA06	3	
0235	SODIUM PICRAMATE, dry or wetted with less than 20% water, by mass	1	1.3C		1		LQ0		PP		LO01 HA01, HA03, HA04, HA05, HA06	3	
0236	ZIRCONIUM PICRAMATE, dry or wetted with less than 20% water, by mass	1	1.3C		1		LQ0		PP		LO01 HA01, HA03, HA04, HA05, HA06	3	
0237	CHARGES, SHAPED, FLEXIBLE, LINEAR	1	1.4D		1.4		LQ0		PP		LO01 HA01, HA03, HA04, HA05, HA06	1	
0238	ROCKETS, LINE-THROWING	1	1.2G		1		LQ0		PP		LO01 HA01, HA03, HA04, HA05, HA06	3	

UN No. or ID No.	Name and description	Class	Classification Code	Packing group	Labels	Special provisions	Limited quantities	Carriage permitted	Equipment required	Ventilation	Provisions concerning loading, unloading and carriage	Number of cones, blue lights	Remarks
3.1.2	3.1.2	2.2	2.2	2.1.1.3	5.2.2	3.3	3.4.6	3.2.1	8.1.5	7.1.6	7.1.6	7.1.5	3.2.1
(1)	(2)	(3a)	(3b)	(4)	(5)	(6)	(7)	(8)	(9)	(10)	(11)	(12)	(13)
0240	ROCKETS, LINE-THROWING	1	1.3G		1		LQ0		PP		LO01 HA01, HA03, HA04, HA05, HA06	3	
0241	EXPLOSIVE, BLASTING, TYPE E	1	1.1D		1	617	LQ0		PP		LO01 HA01, HA02, HA03, HA04, HA05, HA06	3	
0242	CHARGES, PROPELLING, FOR CANNON	1	1.3C		1		LQ0		PP		LO01 HA01, HA03, HA04, HA05, HA06	3	
0243	AMMUNITION, INCENDIARY, WHITE PHOSPHORUS with burster, expelling charge or propelling charge	1	1.2H		1		LQ0		PP		LO01 HA01, HA03, HA04, HA05, HA06	3	
0244	AMMUNITION, INCENDIARY, WHITE PHOSPHORUS with burster, expelling charge or propelling charge	1	1.3H		1		LQ0		PP		LO01 HA01, HA03, HA04, HA05, HA06	3	
0245	AMMUNITION, SMOKE, WHITE PHOSPHORUS with burster, expelling charge or propelling charge	1	1.2H		1		LQ0		PP		LO01 HA01, HA03, HA04, HA05, HA06	3	
0246	AMMUNITION, SMOKE, WHITE PHOSPHORUS with burster, expelling charge or propelling charge	1	1.3H		1		LQ0		PP		LO01 HA01, HA03, HA04, HA05, HA06	3	

UN No. or ID No.	Name and description	Class	Classification Code	Packing group	Labels	Special provisions	Limited quantities	Carriage permitted	Equipment required	Ventilation	Provisions concerning loading, unloading and carriage	Number of cones, blue lights	Remarks
3.1.2	3.1.2	2.2	2.2	2.1.1.3	5.2.2	3.3	3.4.6	3.2.1	8.1.5	7.1.6	7.1.6	7.1.5	3.2.1
(1)	(2)	(3a)	(3b)	(4)	(5)	(6)	(7)	(8)	(9)	(10)	(11)	(12)	(13)
0247	AMMUNITION, INCENDIARY, liquid or gel, with burster, expelling charge or propelling charge	1	1.3J		1		LQ0		PP		LO01 HA01, HA03, HA04, HA05, HA06	3	
0248	CONTRIVANCES, WATER-ACTIVATED with burster, expelling charge or propelling charge	1	1.2L		1	274	LQ0		PP		LO01 HA01, HA03, HA04, HA05, HA06	3	
0249	CONTRIVANCES, WATER-ACTIVATED with burster, expelling charge or propelling charge	1	1.3L		1	274	LQ0		PP		LO01 HA01, HA03, HA04, HA05, HA06	3	
0250	ROCKET MOTORS WITH HYPERGOLIC LIQUIDS with or without expelling charge	1	1.3L		1		LQ0		PP		LO01 HA01, HA03, HA04, HA05, HA06	3	
0254	AMMUNITION, ILLUMINATING with or without burster, expelling charge or propelling charge	1	1.3G		1		LQ0		PP		LO01 HA01, HA03, HA04, HA05, HA06	3	
0255	DETONATORS, ELECTRIC for blasting	1	1.4B		1.4		LQ0		PP		LO01 HA01, HA02, HA03, HA04, HA05, HA06	1	
0257	FUZES, DETONATING	1	1.4B		1.4		LQ0		PP		LO01 HA01, HA02, HA03, HA04, HA05, HA06	1	

UN No. or ID No.	Name and description	Class	Classification Code	Packing group	Labels	Special provisions	Limited quantities	Carriage permitted	Equipment required	Ventilation	Provisions concerning loading, unloading and carriage	Number of cones, blue lights	Remarks
3.1.2	3.1.2	2.2	2.2	2.1.1.3	5.2.2	3.3	3.4.6	3.2.1	8.1.5	7.1.6	7.1.6	7.1.5	3.2.1
(1)	(2)	(3a)	(3b)	(4)	(5)	(6)	(7)	(8)	(9)	(10)	(11)	(12)	(13)
0266	OCTOLITE (OCTOL), dry or wetted with less than 15% water, by mass	1	1.1D		1		LQ0		PP		LO01 HA01, HA02, HA03, HA04, HA05, HA06	3	
0267	DETONATORS, NON-ELECTRIC for blasting	1	1.4B		1.4		LQ0		PP		LO01 HA01, HA02, HA03, HA04, HA05, HA06	1	
0268	BOOSTERS WITH DETONATOR	1	1.2B		1		LQ0		PP		LO01 HA01, HA02, HA03, HA04, HA05, HA06	3	
0271	CHARGES, PROPELLING	1	1.1C		1		LQ0		PP		LO01 HA01, HA02, HA03, HA04, HA05, HA06	3	
0272	CHARGES, PROPELLING	1	1.3C		1		LQ0		PP		LO01 HA01, HA03, HA04, HA05, HA06	3	
0275	CARTRIDGES, POWER DEVICE	1	1.3C		1		LQ0		PP		LO01 HA01, HA03, HA04, HA05, HA06	3	

UN No. or ID No.	Name and description	Class	Classifi-cation Code	Packing group	Labels	Special provisions	Limited quantities	Carriage permitted	Equipment required	Ventilation	Provisions concerning loading, unloading and carriage		Number of cones, blue lights	Remarks
	3.1.2	2.2	2.2	2.1.1.3	5.2.2	3.3	3.4.6	3.2.1	8.1.5	7.1.6	7.1.6		7.1.5	3.2.1
(1)	(2)	(3a)	(3b)	(4)	(5)	(6)	(7)	(8)	(9)	(10)	(11)		(12)	(13)
0276	CARTRIDGES, POWER DEVICE	1	1.4C		1.4		LQ0		PP		LO01	HA01, HA03, HA04, HA05, HA06	1	
0277	CARTRIDGES, OIL WELL	1	1.3C		1		LQ0		PP		LO01	HA01, HA03, HA04, HA05, HA06	3	
0278	CARTRIDGES, OIL WELL	1	1.4C		1.4		LQ0		PP		LO01	HA01, HA03, HA04, HA05, HA06	1	
0279	CHARGES, PROPELLING, FOR CANNON	1	1.1C		1		LQ0		PP		LO01	HA01, HA02, HA03, HA04, HA05, HA06	3	
0280	ROCKET MOTORS	1	1.1C		1		LQ0		PP		LO01	HA01, HA02, HA03, HA04, HA05, HA06	3	
0281	ROCKET MOTORS	1	1.2C		1		LQ0		PP		LO01	HA01, HA03, HA04, HA05, HA06	3	
0282	NITROGUANIDINE (PICRITE), dry or wetted with less than 20% water, by mass	1	1.1D		1		LQ0		PP		LO01	HA01, HA02, HA03, HA04, HA05, HA06	3	

UN No. or ID No. (1)	Name and description 3.1.2 (2)	Class 2.2 (3a)	Classification Code 2.2 (3b)	Packing group 2.1.1.3 (4)	Labels 5.2.2 (5)	Special provisions 3.3 (6)	Limited quantities 3.4.6 (7)	Carriage permitted 3.2.1 (8)	Equipment required 8.1.5 (9)	Ventilation 7.1.6 (10)	Provisions concerning loading, unloading and carriage 7.1.6 (11)	Number of cones, blue lights 7.1.5 (12)	Remarks 3.2.1 (13)
0283	BOOSTERS without detonator	1	1.2D		1		LQ0		PP		LO01 HA01, HA03, HA04, HA05, HA06	3	
0284	GRENADES, hand or rifle, with bursting charge	1	1.1D		1		LQ0		PP		LO01 HA01, HA02, HA03, HA04, HA05, HA06	3	
0285	GRENADES, hand or rifle, with bursting charge	1	1.2D		1		LQ0		PP		LO01 HA01, HA03, HA04, HA05, HA06	3	
0286	WARHEADS, ROCKET with bursting charge	1	1.1D		1		LQ0		PP		LO01 HA01, HA02, HA03, HA04, HA05, HA06	3	
0287	WARHEADS, ROCKET with bursting charge	1	1.2D		1		LQ0		PP		LO01 HA01, HA03, HA04, HA05, HA06	3	
0288	CHARGES, SHAPED, FLEXIBLE, LINEAR	1	1.1D		1		LQ0		PP		LO01 HA01, HA02, HA03, HA04, HA05, HA06	3	
0289	CORD, DETONATING, flexible	1	1.4D		1.4		LQ0		PP		LO01 HA01, HA03, HA04, HA05, HA06	1	

UN No. or ID No.	Name and description	Class	Classification Code	Packing group	Labels	Special provisions	Limited quantities	Carriage permitted	Equipment required	Ventilation	Provisions concerning loading, unloading and carriage	Number of cones, blue lights	Remarks
3.1.2	3.1.2	2.2	2.2	2.1.1.3	5.2.2	3.3	3.4.6	3.2.1	8.1.5	7.1.6	7.1.6	7.1.5	3.2.1
(1)	(2)	(3a)	(3b)	(4)	(5)	(6)	(7)	(8)	(9)	(10)	(11)	(12)	(13)
0290	CORD (FUSE), DETONATING, metal clad	1	1.1D		1		LQ0		PP		LO01 HA01, HA02, HA03, HA04, HA05, HA06	3	
0291	BOMBS with bursting charge	1	1.2F		1		LQ0		PP		LO01 HA01, HA02, HA03, HA04, HA05, HA06	3	
0292	GRENADES, hand or rifle, with bursting charge	1	1.1F		1		LQ0		PP		LO01 HA01, HA02, HA03, HA04, HA05, HA06	3	
0293	GRENADES, hand or rifle, with bursting charge	1	1.2F		1		LQ0		PP		LO01 HA01, HA02, HA03, HA04, HA05, HA06	3	
0294	MINES with bursting charge	1	1.2F		1		LQ0		PP		LO01 HA01, HA02, HA03, HA04, HA05, HA06	3	
0295	ROCKETS with bursting charge	1	1.2F		1		LQ0		PP		LO01 HA01, HA02, HA03, HA04, HA05, HA06	3	

UN No. or ID No.	Name and description	Class	Classification Code	Packing group	Labels	Special provisions	Limited quantities	Carriage permitted	Equipment required	Ventilation	Provisions concerning loading, unloading and carriage	Number of cones, blue lights	Remarks
3.1.2	3.1.2	2.2	2.2	2.1.1.3	5.2.2	3.3	3.4.6	3.2.1	8.1.5	7.1.6	7.1.6	7.1.5	3.2.1
(1)	(2)	(3a)	(3b)	(4)	(5)	(6)	(7)	(8)	(9)	(10)	(11)	(12)	(13)
0296	SOUNDING DEVICES, EXPLOSIVE	1	1.1F		1		LQ0		PP		LO01 HA01, HA02, HA03, HA04, HA05, HA06	3	
0297	AMMUNITION, ILLUMINATING with or without burster, expelling charge or propelling charge	1	1.4G		1.4		LQ0		PP		LO01 HA01, HA03, HA04, HA05, HA06	1	
0299	BOMBS, PHOTO-FLASH	1	1.3G		1		LQ0		PP		LO01 HA01, HA03, HA04, HA05, HA06	3	
0300	AMMUNITION, INCENDIARY with or without burster, expelling charge or propelling charge	1	1.4G		1.4		LQ0		PP		LO01 HA01, HA03, HA04, HA05, HA06	1	
0301	AMMUNITION, TEAR-PRODUCING with burster, expelling charge or propelling charge	1	1.4G		1.4+6.1+8	802	LQ0		PP		LO01 HA01, HA03, HA04, HA05, HA06	1	
0303	AMMUNITION, SMOKE with or without burster, expelling charge or propelling charge	1	1.4G		1.4		LQ0		PP		LO01 HA01, HA03, HA04, HA05, HA06	1	
0303	AMMUNITION, SMOKE with or without burster, expelling charge or propelling charge, containing corrosive substances	1	1.4G		1.4 +8		LQ0		PP		LO01 HA01, HA03, HA04, HA05, HA06	1	

UN No. or ID No.	Name and description	Class	Classification Code	Packing group	Labels	Special provisions	Limited quantities	Carriage permitted	Equipment required	Ventilation	Provisions concerning loading, unloading and carriage	Number of cones, blue lights	Remarks
	3.1.2	2.2	2.2	2.1.1.3	5.2.2	3.3	3.4.6	3.2.1	8.1.5	7.1.6	7.1.6	7.1.5	3.2.1
(1)	(2)	(3a)	(3b)	(4)	(5)	(6)	(7)	(8)	(9)	(10)	(11)	(12)	(13)
0305	FLASH POWDER	1	1.3G		1		LQ0		PP		LO01 HA01, HA03, HA04, HA05, HA06	3	
0306	TRACERS FOR AMMUNITION	1	1.4G		1.4		LQ0		PP		LO01 HA01, HA03, HA04, HA05, HA06	1	
0312	CARTRIDGES, SIGNAL	1	1.4G		1.4		LQ0		PP		LO01 HA01, HA03, HA04, HA05, HA06	1	
0313	SIGNALS, SMOKE	1	1.2G		1		LQ0		PP		LO01 HA01, HA03, HA04, HA05, HA06	3	
0314	IGNITERS	1	1.2G		1		LQ0		PP		LO01 HA01, HA03, HA04, HA05, HA06	3	
0315	IGNITERS	1	1.3G		1		LQ0		PP		LO01 HA01, HA03, HA04, HA05, HA06	3	
0316	FUZES, IGNITING	1	1.3G		1		LQ0		PP		LO01 HA01, HA03, HA04, HA05, HA06	3	

UN No. or ID No.	Name and description	Class	Classification Code	Packing group	Labels	Special provisions	Limited quantities	Carriage permitted	Equipment required	Ventilation	Provisions concerning loading, unloading and carriage	Number of cones, blue lights	Remarks
3.1.2	3.1.2	2.2	2.2	2.1.1.3	5.2.2	3.3	3.4.6	3.2.1	8.1.5	7.1.6	7.1.6	7.1.5	3.2.1
(1)	(2)	(3a)	(3b)	(4)	(5)	(6)	(7)	(8)	(9)	(10)	(11)	(12)	(13)
0317	FUZES, IGNITING	1	1.4G		1.4		LQ0		PP		LO01 HA01, HA03, HA04, HA05, HA06	1	
0318	GRENADES, PRACTICE, hand or rifle	1	1.3G		1		LQ0		PP		LO01 HA01, HA03, HA04, HA05, HA06	3	
0319	PRIMERS, TUBULAR	1	1.3G		1		LQ0		PP		LO01 HA01, HA03, HA04, HA05, HA06	3	
0320	PRIMERS, TUBULAR	1	1.4G		1.4		LQ0		PP		LO01 HA01, HA03, HA04, HA05, HA06	1	
0321	CARTRIDGES FOR WEAPONS with bursting charge	1	1.2E		1		LQ0		PP		LO01 HA01, HA03, HA04, HA05, HA06	3	
0322	ROCKET MOTORS WITH HYPERGOLIC LIQUIDS with or without expelling charge	1	1.2L		1		LQ0		PP		LO01 HA01, HA03, HA04, HA05, HA06	3	
0323	CARTRIDGES, POWER DEVICE	1	1.4S		1.4		LQ0		PP		LO01 HA01, HA03, HA04, HA05, HA06	0	

UN No. or ID No.	Name and description	Class	Classification Code	Packing group	Labels	Special provisions	Limited quantities	Carriage permitted	Equipment required	Ventilation	Provisions concerning loading, unloading and carriage	Number of cones, blue lights	Remarks
3.1.2	3.1.2	2.2	2.2	2.1.1.3	5.2.2	3.3	3.4.6	3.2.1	8.1.5	7.1.6	7.1.6	7.1.5	3.2.1
(1)	(2)	(3a)	(3b)	(4)	(5)	(6)	(7)	(8)	(9)	(10)	(11)	(12)	(13)
0324	PROJECTILES with bursting charge	1	1.2F		1		LQ0		PP		LO01 HA01, HA02, HA03, HA04, HA05, HA06	3	
0325	IGNITERS	1	1.4G		1.4		LQ0		PP		LO01 HA01, HA03, HA04, HA05, HA06	1	
0326	CARTRIDGES FOR WEAPONS, BLANK	1	1.1C		1		LQ0		PP		LO01 HA01, HA02, HA03, HA04, HA05, HA06	3	
0327	CARTRIDGES FOR WEAPONS, BLANK or CARTRIDGES, SMALL ARMS, BLANK	1	1.3C		1		LQ0		PP		LO01 HA01, HA03, HA04, HA05, HA06	3	
0328	CARTRIDGES FOR WEAPONS, INERT PROJECTILE	1	1.2C		1		LQ0		PP		LO01 HA01, HA03, HA04, HA05, HA06	3	
0329	TORPEDOES with bursting charge	1	1.1E		1		LQ0		PP		LO01 HA01, HA02, HA03, HA04, HA05, HA06	3	

UN No. or ID No.	Name and description	Class	Classification Code	Packing group	Labels	Special provisions	Limited quantities	Carriage permitted	Equipment required	Ventilation	Provisions concerning loading, unloading and carriage	Number of cones, blue lights	Remarks
3.1.2	3.1.2	2.2	2.2	2.1.1.3	5.2.2	3.3	3.4.6	3.2.1	8.1.5	7.1.6	7.1.6	7.1.5	3.2.1
(1)	(2)	(3a)	(3b)	(4)	(5)	(6)	(7)	(8)	(9)	(10)	(11)	(12)	(13)
0330	TORPEDOES with bursting charge	1	1.1F		1		LQ0		PP		LO01 HA01, HA02, HA03, HA04, HA05, HA06	3	
0331	EXPLOSIVE, BLASTING, TYPE B (AGENT, BLASTING, TYPE B)	1	1.5D		1.5	617	LQ0		PP		LO01 HA01, HA03, HA04, HA05, HA06	3	
0332	EXPLOSIVE, BLASTING, TYPE E (AGENT, BLASTING, TYPE B)	1	1.5D		1.5	617	LQ0		PP		LO01 HA01, HA03, HA04, HA05, HA06	3	
0333	FIREWORKS	1	1.1G		1	645	LQ0		PP		LO01 HA01, HA02, HA03, HA04, HA05, HA06	3	
0334	FIREWORKS	1	1.2G		1	645	LQ0		PP		LO01 HA01, HA03, HA04, HA05, HA06	3	
0335	FIREWORKS	1	1.3G		1	645	LQ0		PP		LO01 HA01, HA03, HA04, HA05, HA06	3	
0336	FIREWORKS	1	1.4G		1.4	645 651	LQ0		PP		LO01 HA01, HA03, HA04, HA05, HA06	1	

UN No. or ID No.	Name and description	Class	Classification Code	Packing group	Labels	Special provisions	Limited quantities	Carriage permitted	Equipment required	Ventilation	Provisions concerning loading, unloading and carriage	Number of cones, blue lights	Remarks
3.1.2	3.1.2	2.2	2.2	2.1.1.3	5.2.2	3.3	3.4.6	3.2.1	8.1.5	7.1.6	7.1.6	7.1.5	3.2.1
(1)	(2)	(3a)	(3b)	(4)	(5)	(6)	(7)	(8)	(9)	(10)	(11)	(12)	(13)
0337	FIREWORKS	1	1.4S		1.4	645	LQ0		PP		LO01 HA01, HA03, HA04, HA05, HA06	0	
0338	CARTRIDGES FOR WEAPONS, BLANK or CARTRIDGES, SMALL ARMS, BLANK	1	1.4C		1.4		LQ0		PP		LO01 HA01, HA03, HA04, HA05, HA06	1	
0339	CARTRIDGES FOR WEAPONS, INERT PROJECTILE or CARTRIDGES, SMALL ARMS	1	1.4C		1.4		LQ0		PP		LO01 HA01, HA03, HA04, HA05, HA06	1	
0340	NITROCELLULOSE, dry or wetted with less than 25% water (or alcohol), by mass	1	1.1D		1		LQ0		PP		LO01 HA01, HA02, HA03, HA04, HA05, HA06	3	
0341	NITROCELLULOSE, unmodified or plasticized with less than 18% plasticizing substance, by mass	1	1.1D		1		LQ0		PP		LO01 HA01, HA02, HA03, HA04, HA05, HA06	3	
0342	NITROCELLULOSE, WETTED with not less than 25% alcohol, by mass	1	1.3C		1	105	LQ0		PP		LO01 HA01, HA03, HA04, HA05, HA06	3	
0343	NITROCELLULOSE, PLASTICIZED with not less than 18% plasticizing substance, by mass	1	1.3C		1	105	LQ0		PP		LO01 HA01, HA03, HA04, HA05, HA06	3	

UN No. or ID No.	Name and description	Class	Classification Code	Packing group	Labels	Special provisions	Limited quantities	Carriage permitted	Equipment required	Ventilation	Provisions concerning loading, unloading and carriage	Number of cones, blue lights	Remarks
	3.1.2	2.2	2.2	2.1.1.3	5.2.2	3.3	3.4.6	3.2.1	8.1.5	7.1.6	7.1.6	7.1.5	3.2.1
(1)	(2)	(3a)	(3b)	(4)	(5)	(6)	(7)	(8)	(9)	(10)	(11)	(12)	(13)
0344	PROJECTILES with bursting charge	1	1.4D		1.4		LQ0		PP		LO01 HA01, HA03, HA04, HA05, HA06	1	
0345	PROJECTILES, inert with tracer	1	1.4S		1.4		LQ0		PP		LO01 HA01, HA03, HA04, HA05, HA06	0	
0346	PROJECTILES with burster or expelling charge	1	1.2D		1		LQ0		PP		LO01 HA01, HA03, HA04, HA05, HA06	3	
0347	PROJECTILES with burster or expelling charge	1	1.4D		1.4		LQ0		PP		LO01 HA01, HA03, HA04, HA05, HA06	1	
0348	CARTRIDGES FOR WEAPONS with bursting charge	1	1.4F		1.4		LQ0		PP		LO01 HA01, HA02, HA03, HA04, HA05, HA06	1	
0349	ARTICLES, EXPLOSIVE, N.O.S.	1	1.4S		1.4	178 274	LQ0		PP		LO01 HA01, HA03, HA04, HA05, HA06	0	
0350	ARTICLES, EXPLOSIVE, N.O.S.	1	1.4B		1.4	178 274	LQ0		PP		LO01 HA01, HA02, HA03, HA04, HA05, HA06	1	

UN No. or ID No. (1)	Name and description 3.1.2 (2)	Class 2.2 (3a)	Classification Code 2.2 (3b)	Packing group 2.1.1.3 (4)	Labels 5.2.2 (5)	Special provisions 3.3 (6)	Limited quantities 3.4.6 (7)	Carriage permitted 3.2.1 (8)	Equipment required 8.1.5 (9)	Ventilation 7.1.6 (10)	Provisions concerning loading, unloading and carriage 7.1.6 (11)	Number of cones, blue lights 7.1.5 (12)	Remarks 3.2.1 (13)
0351	ARTICLES, EXPLOSIVE, N.O.S.	1	1.4C		1.4	178 274	LQ0		PP		LO01 HA01, HA03, HA04, HA05, HA06	1	
0352	ARTICLES, EXPLOSIVE, N.O.S.	1	1.4D		1.4	178 274	LQ0		PP		LO01 HA01, HA03, HA04, HA05, HA06	1	
0353	ARTICLES, EXPLOSIVE, N.O.S.	1	1.4G		1.4	178 274	LQ0		PP		LO01 HA01, HA03, HA04, HA05, HA06	1	
0354	ARTICLES, EXPLOSIVE, N.O.S.	1	1.1L		1	178 274	LQ0		PP		LO01 HA01, HA02, HA03, HA04, HA05, HA06	3	
0355	ARTICLES, EXPLOSIVE, N.O.S.	1	1.2L		1	178 274	LQ0		PP		LO01 HA01, HA03, HA04, HA05, HA06	3	
0356	ARTICLES, EXPLOSIVE, N.O.S.	1	1.3L		1	178 274	LQ0		PP		LO01 HA01, HA03, HA04, HA05, HA06	3	
0357	SUBSTANCES, EXPLOSIVE, N.O.S.	1	1.1L		1	178 274	LQ0		PP		LO01 HA01, HA02, HA03, HA04, HA05, HA06	3	

UN No. or ID No.	Name and description	Class	Classification Code	Packing group	Labels	Special provisions	Limited quantities	Carriage permitted	Equipment required	Ventilation	Provisions concerning loading, unloading and carriage	Number of cones, blue lights	Remarks
3.1.2	3.1.2	2.2	2.2	2.1.1.3	5.2.2	3.3	3.4.6	3.2.1	8.1.5	7.1.6	7.1.6	7.1.5	3.2.1
(1)	(2)	(3a)	(3b)	(4)	(5)	(6)	(7)	(8)	(9)	(10)	(11)	(12)	(13)
0358	SUBSTANCES, EXPLOSIVE, N.O.S.	1	1.2L		1	178 274	LQ0		PP		LO01 HA01, HA03, HA04, HA05, HA06	3	
0359	SUBSTANCES, EXPLOSIVE, N.O.S.	1	1.3L		1	178 274	LQ0		PP		LO01 HA01, HA03, HA04, HA05, HA06	3	
0360	DETONATOR ASSEMBLIES, NON-ELECTRIC for blasting	1	1.1B		1		LQ0		PP		LO01 HA01, HA02, HA03, HA04, HA05, HA06	3	
0361	DETONATOR ASSEMBLIES, NON-ELECTRIC for blasting	1	1.4B		1.4		LQ0		PP		LO01 HA01, HA02, HA03, HA04, HA05, HA06	1	
0362	AMMUNITION, PRACTICE	1	1.4G		1.4		LQ0		PP		LO01 HA01, HA03, HA04, HA05, HA06	1	
0363	AMMUNITION, PROOF	1	1.4G		1.4		LQ0		PP		LO01 HA01, HA03, HA04, HA05, HA06	1	
0364	DETONATORS FOR AMMUNITION	1	1.2B		1		LQ0		PP		LO01 HA01, HA02, HA03, HA04, HA05, HA06	3	

UN No. or ID No.	Name and description	Class	Classification Code	Packing group	Labels	Special provisions	Limited quantities	Carriage permitted	Equipment required	Ventilation	Provisions concerning loading, unloading and carriage	Number of cones, blue lights	Remarks
3.1.2	3.1.2	2.2	2.2	2.1.1.3	5.2.2	3.3	3.4.6	3.2.1	8.1.5	7.1.6	7.1.6	7.1.5	3.2.1
(1)	(2)	(3a)	(3b)	(4)	(5)	(6)	(7)	(8)	(9)	(10)	(11)	(12)	(13)
0365	DETONATORS FOR AMMUNITION	1	1.4B		1.4		LQ0		PP		LO01 HA01, HA02, HA03, HA04, HA05, HA06	1	
0366	DETONATORS FOR AMMUNITION	1	1.4S		1.4		LQ0		PP		LO01 HA01, HA03, HA04, HA05, HA06	0	
0367	FUZES, DETONATING	1	1.4S		1.4		LQ0		PP		LO01 HA01, HA03, HA04, HA05, HA06	0	
0368	FUZES, IGNITING	1	1.4S		1.4		LQ0		PP		LO01 HA01, HA03, HA04, HA05, HA06	0	
0369	WARHEADS, ROCKET with bursting charge	1	1.1F		1		LQ0		PP		LO01 HA01, HA02, HA03, HA04, HA05, HA06	3	
0370	WARHEADS, ROCKET with burster or expelling charge	1	1.4D		1.4		LQ0		PP		LO01 HA01, HA03, HA04, HA05, HA06	1	
0371	WARHEADS, ROCKET with burster or expelling charge	1	1.4F		1.4		LQ0		PP		LO01 HA01, HA02, HA03, HA04, HA05, HA06	1	

UN No. or ID No.	Name and description	Class	Classification Code	Packing group	Labels	Special provisions	Limited quantities	Carriage permitted	Equipment required	Ventilation	Provisions concerning loading, unloading and carriage	Number of cones, blue lights	Remarks
3.1.2	3.1.2	2.2	2.2	2.1.1.3	5.2.2	3.3	3.4.6	3.2.1	8.1.5	7.1.6	7.1.6	7.1.5	3.2.1
(1)	(2)	(3a)	(3b)	(4)	(5)	(6)	(7)	(8)	(9)	(10)	(11)	(12)	(13)
0372	GRENADES, PRACTICE, hand or rifle	1	1.2G		1		LQ0		PP		LO01 HA01, HA03, HA04, HA05, HA06	3	
0373	SIGNAL DEVICES, HAND	1	1.4S		1.4		LQ0		PP		LO01 HA01, HA03, HA04, HA05, HA06	0	
0374	SOUNDING DEVICES, EXPLOSIVE	1	1.1D		1		LQ0		PP		LO01 HA01, HA02, HA03, HA04, HA05, HA06	3	
0375	SOUNDING DEVICES, EXPLOSIVE	1	1.2D		1		LQ0		PP		LO01 HA01, HA03, HA04, HA05, HA06	3	
0376	PRIMERS, TUBULAR	1	1.4S		1.4		LQ0		PP		LO01 HA01, HA03, HA04, HA05, HA06	0	
0377	PRIMERS, CAP TYPE	1	1.1B		1		LQ0		PP		LO01 HA01, HA02, HA03, HA04, HA05, HA06	3	
0378	PRIMERS, CAP TYPE	1	1.4B		1.4		LQ0		PP		LO01 HA01, HA02, HA03, HA04, HA05, HA06	1	

UN No. or ID No. (1)	Name and description 3.1.2 (2)	Class 2.2 (3a)	Classification Code 2.2 (3b)	Packing group 2.1.1.3 (4)	Labels 5.2.2 (5)	Special provisions 3.3 (6)	Limited quantities 3.4.6 (7)	Carriage permitted 3.2.1 (8)	Equipment required 8.1.5 (9)	Ventilation 7.1.6 (10)	Provisions concerning loading, unloading and carriage 7.1.6 (11)	Number of cones, blue lights 7.1.5 (12)	Remarks 3.2.1 (13)
0379	CASES, CARTRIDGE, EMPTY, WITH PRIMER	1	1.4C		1.4		LQ0		PP		LO01 HA01, HA03, HA04, HA05, HA06	1	
0380	ARTICLES, PYROPHORIC	1	1.2L		1		LQ0		PP		LO01 HA01, HA03, HA04, HA05, HA06	3	
0381	CARTRIDGES, POWER DEVICE	1	1.2C		1		LQ0		PP		LO01 HA01, HA03, HA04, HA05, HA06	3	
0382	COMPONENTS, EXPLOSIVE TRAIN, N.O.S.	1	1.2B		1	178 274	LQ0		PP		LO01 HA01, HA02, HA03, HA04, HA05, HA06	3	
0383	COMPONENTS, EXPLOSIVE TRAIN, N.O.S.	1	1.4B		1.4	178 274	LQ0		PP		LO01 HA01, HA02, HA03, HA04, HA05, HA06	1	
0384	COMPONENTS, EXPLOSIVE TRAIN, N.O.S.	1	1.4S		1.4	178 274	LQ0		PP		LO01 HA01, HA03, HA04, HA05, HA06	0	
0385	5-NITROBENZOTRIAZOL	1	1.1D		1		LQ0		PP		LO01 HA01, HA02, HA03, HA04, HA05, HA06	3	

UN No. or ID No.	Name and description	Class	Classification Code	Packing group	Labels	Special provisions	Limited quantities	Carriage permitted	Equipment required	Ventilation	Provisions concerning loading, unloading and carriage	Number of cones, blue lights	Remarks
3.1.2	3.1.2	2.2	2.2	2.1.1.3	5.2.2	3.3	3.4.6	3.2.1	8.1.5	7.1.6	7.1.6	7.1.5	3.2.1
(1)	(2)	(3a)	(3b)	(4)	(5)	(6)	(7)	(8)	(9)	(10)	(11)	(12)	(13)
0386	TRINITROBENZENE-SULPHONIC ACID	1	1.1D		1		LQ0		PP		LO01 HA01, HA02, HA03, HA04, HA05, HA06	3	
0387	TRINITROFLUORENONE	1	1.1D		1		LQ0		PP		LO01 HA01, HA02, HA03, HA04, HA05, HA06	3	
0388	TRINITROTOLUENE (TNT) AND TRINITROBENZENE MIXTURE or TRINITROTOLUENE (TNT) AND HEXANITROSTILBENE MIXTURE	1	1.1D		1		LQ0		PP		LO01 HA01, HA02, HA03, HA04, HA05, HA06	3	
0389	TRINITROTOLUENE (TNT) MIXTURE CONTAINING TRINITROBENZENE AND HEXANITROSTILBENE	1	1.1D		1		LQ0		PP		LO01 HA01, HA02, HA03, HA04, HA05, HA06	3	
0390	TRITONAL	1	1.1D		1		LQ0		PP		LO01 HA01, HA02, HA03, HA04, HA05, HA06	3	

UN No. or ID No.	Name and description	Class	Classification Code	Packing group	Labels	Special provisions	Limited quantities	Carriage permitted	Equipment required	Ventilation	Provisions concerning loading, unloading and carriage	Number of cones, blue lights	Remarks
3.1.2		2.2	2.2	2.1.1.3	5.2.2	3.3	3.4.6	3.2.1	8.1.5	7.1.6	7.1.6	7.1.5	3.2.1
(1)	(2)	(3a)	(3b)	(4)	(5)	(6)	(7)	(8)	(9)	(10)	(11)	(12)	(13)
0391	CYCLOTRIMETHYLENE-TRINITRAMINE (CYCLONITE; HEXOGEN; RDX) AND CYCLOTETRAMETHYLENE-TETRANITRAMINE (HMX; OCTOGEN) MIXTURE, WETTED with not less than 15% water, by mass or DESENSITIZED with not less than 10% phlegmatiser by mass	1	1.1D		1	266	LQ0		PP		LO01 HA01, HA02, HA03, HA04, HA05, HA06	3	
0392	HEXANITROSTILBENE	1	1.1D		1		LQ0		PP		LO01 HA01, HA02, HA03, HA04, HA05, HA06	3	
0393	HEXOTONAL	1	1.1D		1		LQ0		PP		LO01 HA01, HA02, HA03, HA04, HA05, HA06	3	
0394	TRINITRORESORCINOL (STYPHNIC ACID), WETTED with not less than 20% water, or mixture of alcohol and water, by mass	1	1.1D		1		LQ0		PP		LO01 HA01, HA02, HA03, HA04, HA05, HA06	3	
0395	ROCKET MOTORS, LIQUID FUELLED	1	1.2J		1		LQ0		PP		LO01 HA01, HA03, HA04, HA05, HA06	3	
0396	ROCKET MOTORS, LIQUID FUELLED	1	1.3J		1		LQ0		PP		LO01 HA01, HA03, HA04, HA05, HA06	3	

UN No. or ID No.	Name and description	Class	Classification Code	Packing group	Labels	Special provisions	Limited quantities	Carriage permitted	Equipment required	Ventilation	Provisions concerning loading, unloading and carriage	Number of cones, blue lights	Remarks
3.1.2	3.1.2	2.2	2.2	2.1.1.3	5.2.2	3.3	3.4.6	3.2.1	8.1.5	7.1.6	7.1.6	7.1.5	3.2.1
(1)	(2)	(3a)	(3b)	(4)	(5)	(6)	(7)	(8)	(9)	(10)	(11)	(12)	(13)
0397	ROCKETS, LIQUID FUELLED with bursting charge	1	1.1J		1		LQ0		PP		LO01 HA01, HA02, HA03, HA04, HA05, HA06	3	
0398	ROCKETS, LIQUID FUELLED with bursting charge	1	1.2J		1		LQ0		PP		LO01 HA01, HA03, HA04, HA05, HA06	3	
0399	BOMBS WITH FLAMMABLE LIQUID with bursting charge	1	1.1J		1		LQ0		PP		LO01 HA01, HA02, HA03, HA04, HA05, HA06	3	
0400	BOMBS WITH FLAMMABLE LIQUID with bursting charge	1	1.2J		1		LQ0		PP		LO01 HA01, HA03, HA04, HA05, HA06	3	
0401	DIPICRYL SULPHIDE, dry or wetted with less than 10% water, by mass	1	1.1D		1		LQ0		PP		LO01 HA01, HA02, HA03, HA04, HA05, HA06	3	
0402	AMMONIUM PERCHLORATE	1	1.1D		1	152	LQ0		PP		LO01 HA01, HA02, HA03, HA04, HA05, HA06	3	

UN No. or ID No.	Name and description	Class	Classification Code	Packing group	Labels	Special provisions	Limited quantities	Carriage permitted	Equipment required	Ventilation	Provisions concerning loading, unloading and carriage	Number of cones, blue lights	Remarks
3.1.2	3.1.2	2.2	2.2	2.1.1.3	5.2.2	3.3	3.4.6	3.2.1	8.1.5	7.1.6	7.1.6	7.1.5	3.2.1
(1)	(2)	(3a)	(3b)	(4)	(5)	(6)	(7)	(8)	(9)	(10)	(11)	(12)	(13)
0403	FLARES, AERIAL	1	1.4G		1.4		LQ0		PP		LO01 HA01, HA03, HA04, HA05, HA06	1	
0404	FLARES, AERIAL	1	1.4S		1.4		LQ0		PP		LO01 HA01, HA03, HA04, HA05, HA06	0	
0405	CARTRIDGES, SIGNAL	1	1.4S		1.4		LQ0		PP		LO01 HA01, HA03, HA04, HA05, HA06	0	
0406	DINITROSOBENZENE	1	1.3C		1		LQ0		PP		LO01 HA01, HA03, HA04, HA05, HA06	3	
0407	TETRAZOL-1-ACETIC ACID	1	1.4C		1.4		LQ0		PP		LO01 HA01, HA03, HA04, HA05, HA06	1	
0408	FUZES, DETONATING with protective features	1	1.1D		1		LQ0		PP		LO01 HA01, HA02, HA03, HA04, HA05, HA06	3	
0409	FUZES, DETONATING with protective features	1	1.2D		1		LQ0		PP		LO01 HA01, HA03, HA04, HA05, HA06	3	

UN No. or ID No.	Name and description	Class	Classification Code	Packing group	Labels	Special provisions	Limited quantities	Carriage permitted	Equipment required	Ventilation	Provisions concerning loading, unloading and carriage	Number of cones, blue lights	Remarks
	3.1.2	2.2	2.2	2.1.1.3	5.2.2	3.3	3.4.6	3.2.1	8.1.5	7.1.6	7.1.6	7.1.5	3.2.1
(1)	(2)	(3a)	(3b)	(4)	(5)	(6)	(7)	(8)	(9)	(10)	(11)	(12)	(13)
0410	FUZES, DETONATING with protective features	1	1.4D		1.4		LQ0		PP		LO01 HA01, HA03, HA04, HA05, HA06	1	
0411	PENTAERYTHRITE TETRANITRATE (PENTAERYTHRITOL TETRANITRATE; PETN) with not less than 7% wax, by mass	1	1.1D		1	131	LQ0		PP		LO01 HA01, HA02, HA03, HA04, HA05, HA06	3	
0412	CARTRIDGES FOR WEAPONS with bursting charge	1	1.4E		1.4		LQ0		PP		LO01 HA01, HA03, HA04, HA05, HA06	1	
0413	CARTRIDGES FOR WEAPONS, BLANK	1	1.2C		1		LQ0		PP		LO01 HA01, HA03, HA04, HA05, HA06	3	
0414	CHARGES, PROPELLING, FOR CANNON	1	1.2C		1		LQ0		PP		LO01 HA01, HA03, HA04, HA05, HA06	3	
0415	CHARGES, PROPELLING	1	1.2C		1		LQ0		PP		LO01 HA01, HA03, HA04, HA05, HA06	3	
0417	CARTRIDGES FOR WEAPONS, INERT PROJECTILE or CARTRIDGES, SMALL ARMS	1	1.3C		1		LQ0		PP		LO01 HA01, HA03, HA04, HA05, HA06	3	

UN No. or ID No.	Name and description	Class	Classification Code	Packing group	Labels	Special provisions	Limited quantities	Carriage permitted	Equipment required	Ventilation	Provisions concerning loading, unloading and carriage	Number of cones, blue lights	Remarks
3.1.2	3.1.2	2.2	2.2	2.1.1.3	5.2.2	3.3	3.4.6	3.2.1	8.1.5	7.1.6	7.1.6	7.1.5	3.2.1
(1)	(2)	(3a)	(3b)	(4)	(5)	(6)	(7)	(8)	(9)	(10)	(11)	(12)	(13)
0418	FLARES, SURFACE	1	1.1G		1		LQ0		PP		LO01 HA01, HA02, HA03, HA04, HA05, HA06	3	
0419	FLARES, SURFACE	1	1.2G		1		LQ0		PP		LO01 HA01, HA03, HA04, HA05, HA06	3	
0420	FLARES, AERIAL	1	1.1G		1		LQ0		PP		LO01 HA01, HA02, HA03, HA04, HA05, HA06	3	
0421	FLARES, AERIAL	1	1.2G		1		LQ0		PP		LO01 HA01, HA03, HA04, HA05, HA06	3	
0424	PROJECTILES, inert with tracer	1	1.3G		1		LQ0		PP		LO01 HA01, HA03, HA04, HA05, HA06	3	
0425	PROJECTILES, inert with tracer	1	1.4G		1.4		LQ0		PP		LO01 HA01, HA03, HA04, HA05, HA06	1	
0426	PROJECTILES with burster or expelling charge	1	1.2F		1		LQ0		PP		LO01 HA01, HA02, HA03, HA04, HA05, HA06	3	

UN No. or ID No.	Name and description	Class	Classification Code	Packing group	Labels	Special provisions	Limited quantities	Carriage permitted	Equipment required	Ventilation	Provisions concerning loading, unloading and carriage	Number of cones, blue lights	Remarks
3.1.2	3.1.2	2.2	2.2	2.1.1.3	5.2.2	3.3	3.4.6	3.2.1	8.1.5	7.1.6	7.1.6	7.1.5	3.2.1
(1)	(2)	(3a)	(3b)	(4)	(5)	(6)	(7)	(8)	(9)	(10)	(11)	(12)	(13)
0427	PROJECTILES with burster or expelling charge	1	1.4F		1.4		LQ0		PP		LO01 HA01, HA02, HA03, HA04, HA05, HA06	1	
0428	ARTICLES, PYROTECHNIC for technical purposes	1	1.1G		1		LQ0		PP		LO01 HA01, HA02, HA03, HA04, HA05, HA06	3	
0429	ARTICLES, PYROTECHNIC for technical purposes	1	1.2G		1		LQ0		PP		LO01 HA01, HA03, HA04, HA05, HA06	3	
0430	ARTICLES, PYROTECHNIC for technical purposes	1	1.3G		1		LQ0		PP		LO01 HA01, HA03, HA04, HA05, HA06	3	
0431	ARTICLES, PYROTECHNIC for technical purposes	1	1.4G		1.4		LQ0		PP		LO01 HA01, HA03, HA04, HA05, HA06	1	
0432	ARTICLES, PYROTECHNIC for technical purposes	1	1.4S		1.4		LQ0		PP		LO01 HA01, HA03, HA04, HA05, HA06	0	
0433	POWDER CAKE (POWDER PASTE), WETTED with not less than 17% alcohol, by mass	1	1.1C		1	266	LQ0		PP		LO01 HA01, HA02, HA03, HA04, HA05, HA06	3	

UN No. or ID No.	Name and description	Class	Classification Code	Packing group	Labels	Special provisions	Limited quantities	Carriage permitted	Equipment required	Ventilation	Provisions concerning loading, unloading and carriage	Number of cones, blue lights	Remarks
	3.1.2	2.2	2.2	2.1.1.3	5.2.2	3.3	3.4.6	3.2.1	8.1.5	7.1.6	7.1.6	7.1.5	3.2.1
(1)	(2)	(3a)	(3b)	(4)	(5)	(6)	(7)	(8)	(9)	(10)	(11)	(12)	(13)
0434	PROJECTILES with burster or expelling charge	1	1.2G		1		LQ0		PP		LO01 HA01, HA03, HA04, HA05, HA06	3	
0435	PROJECTILES with burster or expelling charge	1	1.4G		1.4		LQ0		PP		LO01 HA01, HA03, HA04, HA05, HA06	1	
0436	ROCKETS with expelling charge	1	1.2C		1		LQ0		PP		LO01 HA01, HA03, HA04, HA05, HA06	3	
0437	ROCKETS with expelling charge	1	1.3C		1		LQ0		PP		LO01 HA01, HA03, HA04, HA05, HA06	3	
0438	ROCKETS with expelling charge	1	1.4C		1.4		LQ0		PP		LO01 HA01, HA03, HA04, HA05, HA06	1	
0439	CHARGES, SHAPED, without detonator	1	1.2D		1		LQ0		PP		LO01 HA01, HA03, HA04, HA05, HA06	3	
0440	CHARGES, SHAPED, without detonator	1	1.4D		1.4		LQ0		PP		LO01 HA01, HA03, HA04, HA05, HA06	1	

UN No. or ID No.	Name and description	Class	Classification Code	Packing group	Labels	Special provisions	Limited quantities	Carriage permitted	Equipment required	Ventilation	Provisions concerning loading, unloading and carriage	Number of cones, blue lights	Remarks
3.1.2	3.1.2	2.2	2.2	2.1.1.3	5.2.2	3.3	3.4.6	3.2.1	8.1.5	7.1.6	7.1.6	7.1.5	3.2.1
(1)	(2)	(3a)	(3b)	(4)	(5)	(6)	(7)	(8)	(9)	(10)	(11)	(12)	(13)
0441	CHARGES, SHAPED, without detonator	1	1.4S		1.4		LQ0		PP		LO01 HA01, HA03, HA04, HA05, HA06	0	
0442	CHARGES, EXPLOSIVE, COMMERCIAL without detonator	1	1.1D		1		LQ0		PP		LO01 HA01, HA02, HA03, HA04, HA05, HA06	3	
0443	CHARGES, EXPLOSIVE, COMMERCIAL without detonator	1	1.2D		1		LQ0		PP		LO01 HA01, HA03, HA04, HA05, HA06	3	
0444	CHARGES, EXPLOSIVE, COMMERCIAL without detonator	1	1.4D		1.4		LQ0		PP		LO01 HA01, HA03, HA04, HA05, HA06	1	
0445	CHARGES, EXPLOSIVE, COMMERCIAL without detonator	1	1.4S		1.4		LQ0		PP		LO01 HA01, HA03, HA04, HA05, HA06	0	
0446	CASES, COMBUSTIBLE, EMPTY, WITHOUT PRIMER	1	1.4C		1.4		LQ0		PP		LO01 HA01, HA03, HA04, HA05, HA06	1	
0447	CASES, COMBUSTIBLE, EMPTY, WITHOUT PRIMER	1	1.3C		1		LQ0		PP		LO01 HA01, HA03, HA04, HA05, HA06	3	

UN No. or ID No.	Name and description	Class	Classification Code	Packing group	Labels	Special provisions	Limited quantities	Carriage permitted	Equipment required	Ventilation	Provisions concerning loading, unloading and carriage	Number of cones, blue lights	Remarks
3.1.2	3.1.2	2.2	2.2	2.1.1.3	5.2.2	3.3	3.4.6	3.2.1	8.1.5	7.1.6	7.1.6	7.1.5	3.2.1
(1)	(2)	(3a)	(3b)	(4)	(5)	(6)	(7)	(8)	(9)	(10)	(11)	(12)	(13)
0448	5-MERCAPTOTETRAZOL-1-ACETIC ACID	1	1.4C		1.4		LQ0		PP		LO01 HA01, HA03, HA04, HA05, HA06	1	
0449	TORPEDOES, LIQUID FUELLED with or without bursting charge	1	1.1J		1		LQ0		PP		LO01 HA01, HA02, HA03, HA04, HA05, HA06	3	
0450	TORPEDOES, LIQUID FUELLED with inert head	1	1.3J		1		LQ0		PP		LO01 HA01, HA03, HA04, HA05, HA06	3	
0451	TORPEDOES with bursting charge	1	1.1D		1		LQ0		PP		LO01 HA01, HA02, HA03, HA04, HA05, HA06	3	
0452	GRENADES, PRACTICE, hand or rifle	1	1.4G		1.4		LQ0		PP		LO01 HA01, HA03, HA04, HA05, HA06	1	
0453	ROCKETS, LINE-THROWING	1	1.4G		1.4		LQ0		PP		LO01 HA01, HA03, HA04, HA05, HA06	1	
0454	IGNITERS	1	1.4S		1.4		LQ0		PP		LO01 HA01, HA03, HA04, HA05, HA06	0	

UN No. or ID No.	Name and description	Class	Classification Code	Packing group	Labels	Special provisions	Limited quantities	Carriage permitted	Equipment required	Ventilation	Provisions concerning loading, unloading and carriage	Number of cones, blue lights	Remarks
3.1.2	3.1.2	2.2	2.2	2.1.1.3	5.2.2	3.3	3.4.6	3.2.1	8.1.5	7.1.6	7.1.6	7.1.5	3.2.1
(1)	(2)	(3a)	(3b)	(4)	(5)	(6)	(7)	(8)	(9)	(10)	(11)	(12)	(13)
0455	DETONATORS, NON-ELECTRIC for blasting	1	1.4S		1.4		LQ0		PP		LO01	HA01, HA03, HA04, HA05, HA06	0	
0456	DETONATORS, ELECTRIC for blasting	1	1.4S		1.4		LQ0		PP		LO01	HA01, HA03, HA04, HA05, HA06	0	
0457	CHARGES, BURSTING, PLASTICS BONDED	1	1.1D		1		LQ0		PP		LO01	HA01, HA02, HA03, HA04, HA05, HA06	3	
0458	CHARGES, BURSTING, PLASTICS BONDED	1	1.2D		1		LQ0		PP		LO01	HA01, HA03, HA04, HA05, HA06	3	
0459	CHARGES, BURSTING, PLASTICS BONDED	1	1.4D		1.4		LQ0		PP		LO01	HA01, HA03, HA04, HA05, HA06	1	
0460	CHARGES, BURSTING, PLASTICS BONDED	1	1.4S		1.4		LQ0		PP		LO01	HA01, HA03, HA04, HA05, HA06	0	
0461	COMPONENTS, EXPLOSIVE TRAIN, N.O.S.	1	1.1B		1	178 274	LQ0		PP		LO01	HA01, HA02, HA03, HA04, HA05, HA06	3	

UN No. or ID No.	Name and description	Class	Classification Code	Packing group	Labels	Special provisions	Limited quantities	Carriage permitted	Equipment required	Ventilation	Provisions concerning loading, unloading and carriage	Number of cones, blue lights	Remarks
	3.1.2	2.2	2.2	2.1.1.3	5.2.2	3.3	3.4.6	3.2.1	8.1.5	7.1.6	7.1.6	7.1.5	3.2.1
(1)	(2)	(3a)	(3b)	(4)	(5)	(6)	(7)	(8)	(9)	(10)	(11)	(12)	(13)
0462	ARTICLES, EXPLOSIVE, N.O.S.	1	1.1C		1	178 274	LQ0		PP		LO01 HA01, HA02, HA03, HA04, HA05, HA06	3	
0463	ARTICLES, EXPLOSIVE, N.O.S.	1	1.1D		1	178 274	LQ0		PP		LO01 HA01, HA02, HA03, HA04, HA05, HA06	3	
0464	ARTICLES, EXPLOSIVE, N.O.S.	1	1.1E		1	178 274	LQ0		PP		LO01 HA01, HA02, HA03, HA04, HA05, HA06	3	
0465	ARTICLES, EXPLOSIVE, N.O.S.	1	1.1F		1	178 274	LQ0		PP		LO01 HA01, HA02, HA03, HA04, HA05, HA06	3	
0466	ARTICLES, EXPLOSIVE, N.O.S.	1	1.2C		1	178 274	LQ0		PP		LO01 HA01, HA03, HA04, HA05, HA06	3	
0467	ARTICLES, EXPLOSIVE, N.O.S.	1	1.2D		1	178 274	LQ0		PP		LO01 HA01, HA03, HA04, HA05, HA06	3	

UN No. or ID No.	Name and description	Class	Classification Code	Packing group	Labels	Special provisions	Limited quantities	Carriage permitted	Equipment required	Ventilation	Provisions concerning loading, unloading and carriage	Number of cones, blue lights	Remarks
3.1.2	3.1.2	2.2	2.2	2.1.1.3	5.2.2	3.3	3.4.6	3.2.1	8.1.5	7.1.6	7.1.6	7.1.5	3.2.1
(1)	(2)	(3a)	(3b)	(4)	(5)	(6)	(7)	(8)	(9)	(10)	(11)	(12)	(13)
0468	ARTICLES, EXPLOSIVE, N.O.S.	1	1.2E		1	178 274	LQ0		PP		LO01 HA01, HA03, HA04, HA05, HA06	3	
0469	ARTICLES, EXPLOSIVE, N.O.S.	1	1.2F		1	178 274	LQ0		PP		LO01 HA01, HA02, HA03, HA04, HA05, HA06	3	
0470	ARTICLES, EXPLOSIVE, N.O.S.	1	1.3C		1	178 274	LQ0		PP		LO01 HA01, HA03, HA04, HA05, HA06	3	
0471	ARTICLES, EXPLOSIVE, N.O.S.	1	1.4E		1.4	178 274	LQ0		PP		LO01 HA01, HA03, HA04, HA05, HA06	1	
0472	ARTICLES, EXPLOSIVE, N.O.S.	1	1.4F		1.4	178 274	LQ0		PP		LO01 HA01, HA03, HA04, HA05, HA06	1	
0473	SUBSTANCES, EXPLOSIVE, N.O.S.	1	1.1A		1	178 274	LQ0		PP		LO01 HA01, HA02, HA03, HA04, HA05, HA06	3	

UN No. or ID No. (1)	Name and description 3.1.2 (2)	Class 2.2 (3a)	Classification Code 2.2 (3b)	Packing group 2.1.1.3 (4)	Labels 5.2.2 (5)	Special provisions 3.3 (6)	Limited quantities 3.4.6 (7)	Carriage permitted 3.2.1 (8)	Equipment required 8.1.5 (9)	Ventilation 7.1.6 (10)	Provisions concerning loading, unloading and carriage 7.1.6 (11)	Number of cones, blue lights 7.1.5 (12)	Remarks 3.2.1 (13)
0474	SUBSTANCES, EXPLOSIVE, N.O.S.	1	1.1C		1	178 274	LQ0		PP		LO01 HA01, HA02, HA03, HA04, HA05, HA06	3	
0475	SUBSTANCES, EXPLOSIVE, N.O.S.	1	1.1D		1	178 274	LQ0		PP		LO01 HA01, HA02, HA03, HA04, HA05, HA06	3	
0476	SUBSTANCES, EXPLOSIVE, N.O.S.	1	1.1G		1	178 274	LQ0		PP		LO01 HA01, HA02, HA03, HA04, HA05, HA06	3	
0477	SUBSTANCES, EXPLOSIVE, N.O.S.	1	1.3C		1	178 274	LQ0		PP		LO01 HA01, HA03, HA04, HA05, HA06	3	
0478	SUBSTANCES, EXPLOSIVE, N.O.S.	1	1.3G		1	178 274	LQ0		PP		LO01 HA01, HA03, HA04, HA05, HA06	3	
0479	SUBSTANCES, EXPLOSIVE, N.O.S.	1	1.4C		1.4	178 274	LQ0		PP		LO01 HA01, HA03, HA04, HA05, HA06	1	
0480	SUBSTANCES, EXPLOSIVE, N.O.S.	1	1.4D		1.4	178 274	LQ0		PP		LO01 HA01, HA03, HA04, HA05, HA06	1	

UN No. or ID No.	Name and description	Class	Classification Code	Packing group	Labels	Special provisions	Limited quantities	Carriage permitted	Equipment required	Ventilation	Provisions concerning loading, unloading and carriage	Number of cones, blue lights	Remarks
3.1.1	3.1.2	2.2	2.2	2.1.1.3	5.2.2	3.3	3.4.6	3.2.1	8.1.5	7.1.6	7.1.6	7.1.5	3.2.1
(1)	(2)	(3a)	(3b)	(4)	(5)	(6)	(7)	(8)	(9)	(10)	(11)	(12)	(13)
0481	SUBSTANCES, EXPLOSIVE, N.O.S.	1	1.4S		1.4	178 274	LQ0		PP		LO01 HA01, HA03, HA04, HA05, HA06	0	
0482	SUBSTANCES, EXPLOSIVE, VERY INSENSITIVE (SUBSTANCES, EVI), N.O.S.	1	1.5D		1.5	178 274	LQ0		PP		LO01 HA01, HA03, HA04, HA05, HA06	3	
0483	CYCLOTRIMETHYLENE-TRINITRAMINE (CYCLONITE; HEXOGEN; RDX), DESENSITIZED	1	1.1D		1		LQ0		PP		LO01 HA01, HA02, HA03, HA04, HA05, HA06	3	
0484	CYCLOTETRAMETHYLENE-TETRANITRAMINE (HMX; OCTOGEN), DESENSITIZED	1	1.1D		1		LQ0		PP		LO01 HA01, HA02, HA03, HA04, HA05, HA06	3	
0485	SUBSTANCES, EXPLOSIVE, N.O.S.	1	1.4G		1.4	178 274	LQ0		PP		LO01 HA01, HA03, HA04, HA05, HA06	1	
0486	ARTICLES, EXPLOSIVE, EXTREMELY INSENSITIVE (ARTICLES, EEI)	1	1.6N		1.6		LQ0		PP		LO01 HA01, HA03, HA04, HA05, HA06	3	
0487	SIGNALS, SMOKE	1	1.3G		1		LQ0		PP		LO01 HA01, HA03, HA04, HA05, HA06	3	

UN No. or ID No.	Name and description	Class	Classification Code	Packing group	Labels	Special provisions	Limited quantities	Carriage permitted	Equipment required	Ventilation	Provisions concerning loading, unloading and carriage	Number of cones, blue lights	Remarks
3.1.2	3.1.2	2.2	2.2	2.1.1.3	5.2.2	3.3	3.4.6	3.2.1	8.1.5	7.1.6	7.1.6	7.1.5	3.2.1
(1)	(2)	(3a)	(3b)	(4)	(5)	(6)	(7)	(8)	(9)	(10)	(11)	(12)	(13)
0488	AMMUNITION, PRACTICE	1	1.3G		1		LQ0		PP		LO01 HA01, HA03, HA04, HA05, HA06	3	
0489	DINITROGLYCOLURIL (DINGU)	1	1.1D		1		LQ0		PP		LO01 HA01, HA02, HA03, HA04, HA05, HA06	3	
0490	NITROTRIAZOLONE (NTO)	1	1.1D		1		LQ0		PP		LO01 HA01, HA02, HA03, HA04, HA05, HA06	3	
0491	CHARGES, PROPELLING	1	1.4C		1.4		LQ0		PP		LO01 HA01, HA03, HA04, HA05, HA06	1	
0492	SIGNALS, RAILWAY TRACK, EXPLOSIVE	1	1.3G		1		LQ0		PP		LO01 HA01, HA03, HA04, HA05, HA06	3	
0493	SIGNALS, RAILWAY TRACK, EXPLOSIVE	1	1.4G		1.4		LQ0		PP		LO01 HA01, HA03, HA04, HA05, HA06	1	
0494	JET PERFORATING GUNS, CHARGED, oil well, without detonator	1	1.4D		1.4		LQ0		PP		LO01 HA01, HA03, HA04, HA05, HA06	1	

UN No. or ID No.	Name and description	Class	Classification Code	Packing group	Labels	Special provisions	Limited quantities	Carriage permitted	Equipment required	Ventilation	Provisions concerning loading, unloading and carriage	Number of cones, blue lights	Remarks
3.1.2	3.1.2	2.2	2.2	2.1.1.3	5.2.2	3.3	3.4.6	3.2.1	8.1.5	7.1.6	7.1.6	7.1.5	3.2.1
(1)	(2)	(3a)	(3b)	(4)	(5)	(6)	(7)	(8)	(9)	(10)	(11)	(12)	(13)
0495	PROPELLANT, LIQUID	1	1.3C		1	224	LQ0		PP		LO01 HA01, HA03, HA04, HA05, HA06	3	
0496	OCTONAL	1	1.1D		1		LQ0		PP		LO01 HA01, HA02, HA03, HA04, HA05, HA06	3	
0497	PROPELLANT, LIQUID	1	1.1C		1	224	LQ0		PP		LO01 HA01, HA02, HA03, HA04, HA05, HA06	3	
0498	PROPELLANT, SOLID	1	1.1C		1		LQ0		PP		LO01 HA01, HA02, HA03, HA04, HA05, HA06	3	
0499	PROPELLANT, SOLID	1	1.3C		1		LQ0		PP		LO01 HA01, HA03, HA04, HA05, HA06	3	
0500	DETONATOR ASSEMBLIES, NON-ELECTRIC for blasting	1	1.4S		1.4		LQ0		PP		LO01 HA01, HA03, HA04, HA05, HA06	0	
0501	PROPELLANT, SOLID	1	1.4C		1.4		LQ0		PP		LO01 HA01, HA03, HA04, HA05, HA06	1	

UN No. or ID No.	Name and description	Class	Classification Code	Packing group	Labels	Special provisions	Limited quantities	Carriage permitted	Equipment required	Ventilation	Provisions concerning loading, unloading and carriage	Number of cones, blue lights	Remarks
3.1.1	3.1.2	2.2	2.2	2.1.1.3	5.2.2	3.3	3.4.6	3.2.1	8.1.5	7.1.6	7.1.6	7.1.5	3.2.1
(1)	(2)	(3a)	(3b)	(4)	(5)	(6)	(7)	(8)	(9)	(10)	(11)	(12)	(13)
0502	ROCKETS with inert head	1	1.2C		1		LQ0		PP		LO01 HA01, HA03, HA04, HA05, HA06	3	
0503	AIR BAG INFLATORS or AIR BAG MODULES or SEAT-BELT PRETENSIONERS	1	1.4G		1.4	235 289	LQ0		PP		LO01 HA01, HA03, HA04, HA05, HA06	1	
0504	1H-TETRAZOLE	1	1.1D		1		LQ0		PP		LO01 HA01, HA02, HA03, HA04, HA05, HA06	3	
1001	ACETYLENE, DISSOLVED	2	4F		2.1		LQ0		PP, EX, A	VE01		1	
1002	AIR, COMPRESSED	2	1A		2.2	292	LQ1		PP			0	
1003	AIR, REFRIGERATED LIQUID	2	3O		2.2+5.1		LQ0		PP			0	
1005	AMMONIA, ANHYDROUS	2	2TC		2.3+8	23	LQ0	T	PP, EP, TOX, A	VE02		2	
1006	ARGON, COMPRESSED	2	1A		2.2		LQ1		PP			0	
1008	BORON TRIFLUORIDE	2	2TC		2.3+8		LQ0		PP, EP, TOX, A	VE02		2	
1009	BROMOTRIFLUOROMETHANE (REFRIGERANT GAS R 13B1)	2	2A		2.2		LQ1		PP			0	
1010	BUTADIENES, STABILIZED or BUTADIENES AND HYDROCARBON MIXTURE, STABILIZED, having a vapour pressure at 70 °C not exceeding 1.1 MPa (11 bar) and a density at 50 °C not lower than 0.525 kg/l	2	2F		2.1	618	LQ0	T	PP, EX, A	VE01		1	
1011	BUTANE	2	2F		2.1		LQ0	T	PP, EX, A	VE01		1	

UN No. or ID No.	Name and description	Class	Classification Code	Packing group	Labels	Special provisions	Limited quantities	Carriage permitted	Equipment required	Ventilation	Provisions concerning loading, unloading and carriage	Number of cones, blue lights	Remarks
(1)	3.1.2	2.2	2.2	2.1.1.3	5.2.2	3.3	3.4.6	3.2.1	8.1.5	7.1.6	7.1.6	7.1.5	3.2.1
(1)	(2)	(3a)	(3b)	(4)	(5)	(6)	(7)	(8)	(9)	(10)	(11)	(12)	(13)
1012	BUTYLENES MIXTURE or 1-BUTYLENE or CIS-2-BUTYLENE or TRANS-2-BUTYLENE	2	2F		2.1		LQ0	T	PP, EX, A	VE01		1	
1013	CARBON DIOXIDE	2	2A		2.2	584 653	LQ1		PP			0	
1016	CARBON MONOXIDE, COMPRESSED	2	1TF		2.3+2.1		LQ0		PP, EP, EX, TOX, A	VE01, VE02		2	
1017	CHLORINE	2	2TC		2.3+8		LQ0		PP, EP, TOX, A	VE02		2	
1018	CHLORODIFLUOROMETHANE (REFRIGERANT GAS R 22)	2	2A		2.2		LQ1		PP			0	
1020	CHLOROPENTAFLUORO-ETHANE (REFRIGERANT GAS R 115)	2	2A		2.2		LQ1	T	PP			0	
1021	1-CHLORO-1,2,2,2-TETRAFLUOROETHANE (REFRIGERANT GAS R 124)	2	2A		2.2		LQ1		PP			0	
1022	CHLOROTRIFLUORO-METHANE (REFRIGERANT GAS R 13)	2	2A		2.2		LQ1		PP			0	
1023	COAL GAS, COMPRESSED	2	1TF		2.3+2.1		LQ0		PP, EP, EX, TOX, A	VE01, VE02		2	
1026	CYANOGEN	2	2TF		2.3+2.1		LQ0		PP, EP, EX, TOX, A	VE01, VE02		2	
1027	CYCLOPROPANE	2	2F		2.1		LQ0		PP, EX, A	VE01		1	
1028	DICHLORODIFLUORO-METHANE (REFRIGERANT GAS R 12)	2	2A		2.2		LQ1		PP			0	
1029	DICHLOROFLUORO-METHANE (REFRIGERANT GAS R 21)	2	2A		2.2		LQ1		PP			0	
1030	1,1-DIFLUOROETHANE (REFRIGERANT GAS R 152a)	2	2F		2.1		LQ0	T	PP, EX, A	VE01		1	
1032	DIMETHYLAMINE, ANHYDROUS	2	2F		2.1		LQ0		PP, EX, A	VE01		1	
1033	DIMETHYL ETHER	2	2F		2.1		LQ0	T	PP, EX, A	VE01		1	
1035	ETHANE	2	2F		2.1		LQ0		PP, EX, A	VE01		1	

(1) UN No. or ID No. 3.1.2	(2) Name and description 3.1.2	(3a) Class 2.2	(3b) Classification Code 2.2	(4) Packing group 2.1.1.3	(5) Labels 5.2.2	(6) Special provisions 3.3	(7) Limited quantities 3.4.6	(8) Carriage permitted 3.2.1	(9) Equipment required 8.1.5	(10) Ventilation 7.1.6	(11) Provisions concerning loading, unloading and carriage 7.1.6	(12) Number of cones, blue lights 7.1.5	(13) Remarks 3.2.1
1036	ETHYLAMINE	2	2F		2.1		LQ0		PP, EX, A	VE01		1	
1037	ETHYL CHLORIDE	2	2F		2.1		LQ0		PP, EX, A	VE01		1	
1038	ETHYLENE, REFRIGERATED LIQUID	2	3F		2.1		LQ0		PP, EX, A	VE01		1	
1039	ETHYL METHYL ETHER	2	2F		2.1		LQ0		PP, EX, A	VE01		1	
1040	ETHYLENE OXIDE	2	2TF		2.3+2.1		LQ0		PP, EP, EX, TOX, A	VE01, VE02		2	
1040	ETHYLENE OXIDE WITH NITROGEN up to a total pressure of 1 MPa (10 bar) at 50 °C	2	2TF		2.3+2.1		LQ0	T	PP, EP, EX, TOX, A	VE01, VE02		2	
1041	ETHYLENE OXIDE AND CARBON DIOXIDE MIXTURE with more than 9% but not more than 87% ethylene oxide	2	2F		2.1		LQ0		PP, EX, A	VE01		1	
1043	FERTILIZER AMMONIATING SOLUTION with free ammonia	2	4A		2.2		LQ1		PP			0	
1044	FIRE EXTINGUISHERS with compressed or liquefied gas	2	6A		2.2	225 594	LQ0		PP			0	
1045	FLUORINE, COMPRESSED	2	1TOC		2.3+5.1+8		LQ0		PP, EP, TOX, A	VE02		2	
1046	HELIUM, COMPRESSED	2	1A		2.2		LQ1		PP			0	
1048	HYDROGEN BROMIDE, ANHYDROUS	2	2TC		2.3+8		LQ0		PP, EP, TOX, A	VE02		2	
1049	HYDROGEN, COMPRESSED	2	1F		2.1		LQ0		PP, EX, A	VE01		1	
1050	HYDROGEN CHLORIDE, ANHYDROUS	2	2TC		2.3+8		LQ0		PP, EP, TOX, A	VE02		2	
1051	HYDROGEN CYANIDE, STABILIZED containing less than 3% water	6.1	TF1	I	6.1+3	603 802	LQ0		PP, EP, EX, TOX, A	VE01, VE02		2	
1052	HYDROGEN FLUORIDE, ANHYDROUS	8	CT1	I	8+6.1	802	LQ0		PP, EP, TOX, A	VE02		2	
1053	HYDROGEN SULPHIDE	2	2TF		2.3+2.1		LQ0		PP, EP, EX, TOX, A	VE01, VE02		2	
1055	ISOBUTYLENE	2	2F		2.1		LQ0	T	PP, EX, A	VE01		1	
1056	KRYPTON, COMPRESSED	2	1A		2.2		LQ1		PP			0	
1057	LIGHTERS or LIGHTER REFILLS containing flammable gas	2	6F		2.1	201	LQ0		PP, EX, A	VE01		1	

UN No. or ID No. (1) 3.1.2	Name and description (2) 3.1.2	Class (3a) 2.2	Classification Code (3b) 2.2	Packing group (4) 2.1.1.3	Labels (5) 5.2.2	Special provisions (6) 3.3	Limited quantities (7) 3.4.6	Carriage permitted (8) 3.2.1	Equipment required (9) 8.1.5	Ventilation (10) 7.1.6	Provisions concerning loading, unloading and carriage (11) 7.1.6	Number of cones, blue lights (12) 7.1.5	Remarks (13) 3.2.1
1058	LIQUEFIED GASES, non-flammable, charged with nitrogen, carbon dioxide or air	2	2A		2.2		LQ1		PP			0	
1060	METHYLACETYLENE AND PROPADIENE MIXTURE, STABILIZED such as mixture P1 or mixture P2	2	2F		2.1	581	LQ0		PP, EX, A	VE01		1	
1061	METHYLAMINE, ANHYDROUS	2	2F		2.1		LQ0		PP, EX, A	VE01		1	
1062	METHYL BROMIDE with not more than 2% chloropicrin	2	2T		2.3	23	LQ0		PP, EP, TOX, A	VE02		2	
1063	METHYL CHLORIDE (REFRIGERANT GAS R 40)	2	2F		2.1		LQ0	T	PP, EX, A	VE01		1	
1064	METHYL MERCAPTAN	2	2TF		2.3+2.1		LQ0		PP, EP, EX, TOX, A	VE01, VE02		2	
1065	NEON, COMPRESSED	2	1A		2.2		LQ1		PP			0	
1066	NITROGEN, COMPRESSED	2	1A		2.2		LQ1		PP			0	
1067	DINITROGEN TETROXIDE (NITROGEN DIOXIDE)	2	2TOC		2.3+5.1+8		LQ0		PP, EP, TOX, A	VE02		2	
1069	NITROSYL CHLORIDE	2	2TC		2.3+8		LQ0		PP, EP, TOX, A	VE02		2	
1070	NITROUS OXIDE	2	2O		2.2+5.1	584	LQ0		PP			0	
1071	OIL GAS, COMPRESSED	2	1TF		2.3+2.1		LQ0		PP, EP, EX, TOX, A	VE01, VE02		2	
1072	OXYGEN, COMPRESSED	2	1O		2.2+5.1		LQ0		PP			0	
1073	OXYGEN, REFRIGERATED LIQUID	2	3O		2.2+5.1		LQ0		PP			0	
1075	PETROLEUM GASES, LIQUEFIED	2	2F		2.1	274 583 639	LQ0		PP, EX, A	VE01		1	
1076	PHOSGENE	2	2TC		2.3+8		LQ0		PP, EP, TOX, A	VE02		2	
1077	PROPYLENE	2	2F		2.1		LQ0	T	PP, EX, A	VE01		1	
1078	REFRIGERANT GAS, N.O.S., such as mixture F1, mixture F2 or mixture F3	2	2A		2.2	274 582	LQ1		PP			0	
1079	SULPHUR DIOXIDE	2	2TC		2.3+8		LQ0		PP, EP, TOX, A	VE02		2	

UN No. or ID No.	Name and description	Class	Classification Code	Packing group	Labels	Special provisions	Limited quantities	Carriage permitted	Equipment required	Ventilation	Provisions concerning loading, unloading and carriage	Number of cones, blue lights	Remarks
	3.1.2	2.2	2.2	2.1.1.3	5.2.2	3.3	3.4.6	3.2.1	8.1.5	7.1.6	7.1.6	7.1.5	3.2.1
(1)	(2)	(3a)	(3b)	(4)	(5)	(6)	(7)	(8)	(9)	(10)	(11)	(12)	(13)
1080	SULPHUR HEXAFLUORIDE	2	2A		2.2		LQ1		PP			0	
1081	TETRAFLUOROETHYLENE, STABILIZED	2	2F		2.1		LQ0		PP, EX, A	VE01		1	
1082	TRIFLUOROCHLORO-ETHYLENE, STABILIZED	2	2TF		2.3+2.1		LQ0		PP, EP, EX, TOX, A	VE01, VE02		2	
1083	TRIMETHYLAMINE, ANHYDROUS	2	2F		2.1		LQ0	T	PP, EX, A	VE01		1	
1085	VINYL BROMIDE, STABILIZED	2	2F		2.1		LQ0		PP, EX, A	VE01		1	
1086	VINYL CHLORIDE, STABILIZED	2	2F		2.1		LQ0	T	PP, EX, A	VE01		1	
1087	VINYL METHYL ETHER, STABILIZED	2	2F		2.1		LQ0		PP, EX, A	VE01		1	
1088	ACETAL	3	F1	II	3		LQ4	T	PP, EX, A	VE01		1	
1089	ACETALDEHYDE	3	F1	I	3		LQ3	T	PP, EX, A	VE01		1	
1090	ACETONE	3	F1	II	3		LQ4	T	PP, EX, A	VE01		1	
1091	ACETONE OILS	3	F1	II	3		LQ4		PP, EX, A	VE01		1	
1092	ACROLEIN, STABILIZED	6.1	TF1	I	6.1+3	802	LQ0	T	PP, EP, EX, TOX, A	VE01, VE02		2	
1093	ACRYLONITRILE, STABILIZED	3	FT1	I	3+6.1	802	LQ0	T	PP, EP, EX, TOX, A	VE01, VE02		2	
1098	ALLYL ALCOHOL	6.1	TF1	I	6.1+3	802	LQ0	T	PP, EP, EX, TOX, A	VE01, VE02		2	
1099	ALLYL BROMIDE	3	FT1	I	3+6.1	802	LQ0		PP, EP, EX, TOX, A	VE01, VE02		2	
1100	ALLYL CHLORIDE	3	FT1	I	3+6.1	802	LQ0	T	PP, EP, EX, TOX, A	VE01, VE02		2	
1104	AMYL ACETATES	3	F1	III	3		LQ7		PP, EX, A	VE01		0	
1105	PENTANOLS	3	F1	II	3		LQ4		PP, EX, A	VE01		1	
1105	PENTANOLS	3	F1	III	3		LQ7	T	PP, EX, A	VE01		0	
1106	AMYLAMINE	3	FC	II	3+8		LQ4	T	PP, EP, EX, A	VE01		1	
1106	AMYLAMINE	3	FC	III	3+8		LQ7		PP, EP, EX, A	VE01		0	
1107	AMYL CHLORIDE	3	F1	II	3		LQ4	T	PP, EX, A	VE01		1	
1108	1-PENTENE (n-AMYLENE)	3	F1	I	3		LQ3	T	PP, EX, A	VE01		1	
1109	AMYL FORMATES	3	F1	III	3		LQ7		PP, EX, A	VE01		0	
1110	n-AMYL METHYL KETONE	3	F1	III	3		LQ7		PP, EX, A	VE01		0	

UN No. or ID No.	Name and description	Class	Classification Code	Packing group	Labels	Special provisions	Limited quantities	Carriage permitted	Equipment required	Ventilation	Provisions concerning loading, unloading and carriage	Number of cones, blue lights	Remarks
(1)	3.1.2 (2)	2.2 (3a)	2.2 (3b)	2.1.1.3 (4)	5.2.2 (5)	3.3 (6)	3.4.6 (7)	3.2.1 (8)	8.1.5 (9)	7.1.6 (10)	7.1.6 (11)	7.1.5 (12)	3.2.1 (13)
1111	AMYL MERCAPTAN	3	F1	II	3		LQ4		PP, EX, A	VE01		1	
1112	AMYL NITRATE	3	F1	III	3		LQ7		PP, EX, A	VE01		0	
1113	AMYL NITRITE	3	F1	II	3		LQ4		PP, EX, A	VE01		1	
1114	BENZENE	3	F1	II	3		LQ4	T	PP, EX, A	VE01		1	
1120	BUTANOLS	3	F1	II	3		LQ4	T	PP, EX, A	VE01		1	
1120	BUTANOLS	3	F1	III	3		LQ7	T	PP, EX, A	VE01		0	
1123	BUTYL ACETATES	3	F1	II	3		LQ4	T	PP, EX, A	VE01		1	
1123	BUTYL ACETATES	3	F1	III	3		LQ7	T	PP, EX, A	VE01		0	
1125	n-BUTYLAMINE	3	FC	II	3+8		LQ4	T	PP, EP, EX, A	VE01		1	
1126	1-BROMOBUTANE	3	F1	II	3		LQ4		PP, EX, A	VE01		1	
1127	CHLOROBUTANES	3	F1	II	3		LQ4	T	PP, EX, A	VE01		1	
1128	n-BUTYL FORMATE	3	F1	II	3		LQ4		PP, EX, A	VE01		1	
1129	BUTYRALDEHYDE	3	F1	II	3		LQ4	T	PP, EX, A	VE01		1	
1130	CAMPHOR OIL	3	F1	III	3		LQ7		PP, EX, A	VE01		0	
1131	CARBON DISULPHIDE	3	FT1	I	3+6.1	802	LQ0	T	PP, EP, EX, TOX, A	VE01, VE02		2	
1133	ADHESIVES containing flammable liquid	3	F1	I	3		LQ3		PP, EX, A	VE01		1	
1133	ADHESIVES containing flammable liquid (vapour pressure at 50 °C more than 110 kPa)	3	F1	II	3	640C	LQ6		PP, EX, A	VE01		1	
1133	ADHESIVES containing flammable liquid (vapour pressure at 50 °C not more than 110 kPa)	3	F1	II	3	640D	LQ6		PP, EX, A	VE01		1	
1133	ADHESIVES containing flammable liquid	3	F1	III	3	640E	LQ7		PP, EX, A	VE01		0	
1133	ADHESIVES containing flammable liquid (having a flash-point below 23 °C and viscous according to 2.2.3.1.4) (boiling point not more than 35° C)	3	F1	III	3	640F	LQ7		PP, EX, A	VE01		0	

- 248 -

UN No. or ID No.	Name and description	Class	Classification Code	Packing group	Labels	Special provisions	Limited quantities	Carriage permitted	Equipment required	Ventilation	Provisions concerning loading, unloading and carriage	Number of cones, blue lights	Remarks
3.1.2	3.1.2	2.2	2.2	2.1.1.3	5.2.2	3.3	3.4.6	3.2.1	8.1.5	7.1.6	7.1.6	7.1.5	3.2.1
(1)	(2)	(3a)	(3b)	(4)	(5)	(6)	(7)	(8)	(9)	(10)	(11)	(12)	(13)
1133	ADHESIVES containing flammable liquid (having a flash-point below 23 °C and viscous according to 2.2.3.1.4) (vapour pressure at 50 °C more than 110 kPa , boiling point of more than 35° C)	3	F1	III	3	640G	LQ7		PP, EX, A	VE01		0	
1133	ADHESIVES containing flammable liquid (having a flash-point below 23 °C and viscous according to 2.2.3.1.4) (vapour pressure at 50 °C not more than 110 kPa)	3	F1	III	3	640H	LQ7		PP, EX, A	VE01		0	
1134	CHLOROBENZENE	3	F1	III	3		LQ7	T	PP, EX, A	VE01		0	
1135	ETHYLENE CHLOROHYDRIN	6.1	TF1	I	6.1+3	802	LQ0	T	PP, EP, EX, TOX, A	VE01, VE02		2	
1136	COAL TAR DISTILLATES, FLAMMABLE	3	F1	II	3		LQ4		PP, EX, A	VE01		1	
1136	COAL TAR DISTILLATES, FLAMMABLE	3	F1	III	3		LQ7		PP, EX, A	VE01		0	
1139	COATING SOLUTION (includes surface treatments or coatings used for industrial or other purposes such as vehicle under coating, drum or barrel lining)	3	F1	I	3		LQ3		PP, EX, A	VE01		1	
1139	COATING SOLUTION (includes surface treatments or coatings used for industrial or other purposes such as vehicle under coating, drum or barrel lining) (vapour pressure at 50 °C more than 110 kPa)	3	F1	II	3	640C	LQ6		PP, EX, A	VE01		1	
1139	COATING SOLUTION (includes surface treatments or coatings used for industrial or other purposes such as vehicle under coating, drum or barrel lining) (vapour pressure at 50 °C not more than 110 kPa)	3	F1	II	3	640D	LQ6		PP, EX, A	VE01		1	

UN No. or ID No. (1)	Name and description (2)	Class (3a)	Classification Code (3b)	Packing group (4)	Labels (5)	Special provisions (6)	Limited quantities (7)	Carriage permitted (8)	Equipment required (9)	Ventilation (10)	Provisions concerning loading, unloading and carriage (11)	Number of cones, blue lights (12)	Remarks (13)
	3.1.2	2.2	2.2	2.1.1.3	5.2.2	3.3	3.4.6	3.2.1	8.1.5	7.1.6	7.1.6	7.1.5	3.2.1
1139	COATING SOLUTION (includes surface treatments or coatings used for industrial or other purposes such as vehicle under coating, drum or barrel lining)	3	F1	III	3	640E	LQ7		PP, EX, A	VE01		0	
1139	COATING SOLUTION (includes surface treatments or coatings used for industrial or other purposes such as vehicle under coating, drum or barrel lining) (having a flash-point below 23 °C and viscous according to 2.2.3.1.4) (boiling point not more than 35° C)	3	F1	III	3	640F	LQ7		PP, EX, A	VE01		0	
1139	COATING SOLUTION (includes surface treatments or coatings used for industrial or other purposes such as vehicle under coating, drum or barrel lining) (having a flash-point below 23 °C and viscous according to 2.2.3.1.4) (vapour pressure at 50 °C, boiling point of more than 35° C)	3	F1	III	3	640G	LQ7		PP, EX, A	VE01		0	
1139	COATING SOLUTION (includes surface treatments or coatings used for industrial or other purposes such as vehicle under coating, drum or barrel lining) (having a flash-point below 23 °C and viscous according to 2.2.3.1.4) (vapour pressure at 50 °C not more than 110 kPa)	3	F1	III	3	640H	LQ7		PP, EX, A	VE01		0	
1143	CROTONALDEHYDE or CROTONALDEHYDE, STABILIZED	6.1	TF1	I	6.1+3	324 802	LQ0	T	PP, EP, EX, TOX, A	VE01, VE02		2	
1144	CROTONYLENE	3	F1	I	3		LQ3		PP, EX, A	VE01		1	
1145	CYCLOHEXANE	3	F1	II	3		LQ4	T	PP, EX, A	VE01		1	
1146	CYCLOPENTANE	3	F1	II	3		LQ4	T	PP, EX, A	VE01		1	
1147	DECAHYDRO-NAPHTHALENE	3	F1	III	3		LQ7		PP, EX, A	VE01		0	

(1) UN No. or ID No.	(2) Name and description	(3a) Class	(3b) Classification Code	(4) Packing group	(5) Labels	(6) Special provisions	(7) Limited quantities	(8) Carriage permitted	(9) Equipment required	(10) Ventilation	(11) Provisions concerning loading, unloading and carriage	(12) Number of cones, blue lights	(13) Remarks
		2.2	2.2	2.1.1.3	5.2.2	3.3	3.4.6	3.2.1	8.1.5	7.1.6	7.1.6	7.1.5	3.2.1
1148	DIACETONE ALCOHOL	3	F1	II	3		LQ4		PP, EX, A	VE01		1	
1148	DIACETONE ALCOHOL	3	F1	III	3		LQ7		PP, EX, A	VE01		0	
1149	DIBUTYL ETHERS	3	F1	III	3		LQ7		PP, EX, A	VE01		0	
1150	1,2-DICHLOROETHYLENE	3	F1	II	3		LQ4	T	PP, EX, A	VE01		1	
1152	DICHLOROPENTANES	3	F1	III	3		LQ7		PP, EX, A	VE01		0	
1153	ETHYLENE GLYCOL DIETHYL ETHER	3	F1	II	3		LQ4	T	PP, EX, A	VE01		1	
1153	ETHYLENE GLYCOL DIETHYL ETHER	3	F1	III	3		LQ7	T	PP, EX, A	VE01		0	
1154	DIETHYLAMINE	3	FC	II	3+8		LQ4	T	PP, EP, EX, A	VE01		1	
1155	DIETHYL ETHER (ETHYL ETHER)	3	F1	I	3		LQ3	T	PP, EX, A	VE01		1	
1156	DIETHYL KETONE	3	F1	II	3		LQ4		PP, EX, A	VE01		1	
1157	DIISOBUTYL KETONE	3	F1	III	3		LQ7	T	PP, EX, A	VE01		0	
1158	DIISOPROPYLAMINE	3	FC	II	3+8		LQ4		PP, EP, EX, A	VE01		1	
1159	DIISOPROPYL ETHER	3	F1	II	3		LQ4	T	PP, EX, A	VE01		1	
1160	DIMETHYLAMINE AQUEOUS SOLUTION	3	FC	II	3+8		LQ4	T	PP, EP, EX, A	VE01		1	
1161	DIMETHYL CARBONATE	3	F1	II	3		LQ4		PP, EP, EX, A	VE01		1	
1162	DIMETHYLDICHLORO-SILANE	3	FC	II	3+8		LQ4		PP, EP, EX, A	VE01		1	
1163	DIMETHYLHYDRAZINE, UNSYMMETRICAL	6.1	TFC	I	6.1+3+8	802	LQ0	T	PP, EP, EX, TOX, A	VE01, VE02		2	
1164	DIMETHYL SULPHIDE	3	F1	II	3		LQ4		PP, EX, A	VE01		1	
1165	DIOXANE	3	F1	II	3		LQ4	T	PP, EX, A	VE01		1	
1166	DIOXOLANE	3	F1	II	3		LQ4		PP, EX, A	VE01		1	
1167	DIVINYL ETHER, STABILIZED	3	F1	I	3		LQ3	T	PP, EX, A	VE01		1	
1169	EXTRACTS, AROMATIC, LIQUID	3	F1	I	3		LQ3		PP, EX, A	VE01		1	
1169	EXTRACTS, AROMATIC, LIQUID (vapour pressure at 50 °C more than 110 kPa)	3	F1	II	3	601 640C	LQ6		PP, EX, A	VE01		1	
1169	EXTRACTS, AROMATIC, LIQUID (vapour pressure at 50 °C not more than 110 kPa)	3	F1	II	3	601 640D	LQ6		PP, EX, A	VE01		1	

UN No. or ID No.	Name and description	Class	Classification Code	Packing group	Labels	Special provisions	Limited quantities	Carriage permitted	Equipment required	Ventilation	Provisions concerning loading, unloading and carriage	Number of cones, blue lights	Remarks
3.1.2	3.1.2	2.2	2.2	2.1.1.3	5.2.2	3.3	3.4.6	3.2.1	8.1.5	7.1.6	7.1.6	7.1.5	3.2.1
(1)	(2)	(3a)	(3b)	(4)	(5)	(6)	(7)	(8)	(9)	(10)	(11)	(12)	(13)
1169	EXTRACTS, AROMATIC, LIQUID	3	F1	III	3	601 640E	LQ7		PP, EX, A	VE01		0	
1169	EXTRACTS, AROMATIC, LIQUID (having a flash-point below 23 °C and viscous according to 2.2.3.1.4) (boiling point not more than 35° C)	3	F1	III	3	601 640F	LQ7		PP, EX, A	VE01		0	
1169	EXTRACTS, AROMATIC, LIQUID (having a flash-point below 23 °C and viscous according to 2.2.3.1.4) (vapour pressure at 50 °C more than 110 kPa , boiling point of more than 35° C)	3	F1	III	3	601 640G	LQ7		PP, EX, A	VE01		0	
1169	EXTRACTS, AROMATIC, LIQUID (having a flash-point below 23 °C and viscous according to 2.2.3.1.4) (vapour pressure at 50 °C not more than 110 kPa)	3	F1	III	3	601 640H	LQ7		PP, EX, A	VE01		0	
1170	ETHANOL (ETHYL ALCOHOL) or ETHANOL SOLUTION (ETHYL ALCOHOL SOLUTION)	3	F1	II	3	144 330 601	LQ4	T	PP, EX, A	VE01		1	
1170	ETHANOL SOLUTION (ETHYL ALCOHOL SOLUTION)	3	F1	III	3	144 330 601	LQ7	T	PP, EX, A	VE01		0	
1171	ETHYLENE GLYCOL MONOETHYL ETHER	3	F1	III	3		LQ7	T	PP, EX, A	VE01		0	
1172	ETHYLENE GLYCOL MONOETHYL ETHER ACETATE	3	F1	III	3		LQ7	T	PP, EX, A	VE01		0	
1173	ETHYL ACETATE	3	F1	II	3		LQ4	T	PP, EX, A	VE01		1	
1175	ETHYLBENZENE	3	F1	II	3		LQ4	T	PP, EX, A	VE01		1	
1176	ETHYL BORATE	3	F1	II	3		LQ4		PP, EX, A	VE01		1	
1177	2-ETHYLBUTYL ACETATE	3	F1	III	3		LQ7	T	PP, EX, A	VE01		0	
1178	2-ETHYLBUTYRALDEHYDE	3	F1	II	3		LQ4		PP, EX, A	VE01		1	
1179	ETHYL BUTYL ETHER	3	F1	II	3		LQ4		PP, EX, A	VE01		1	
1180	ETHYL BUTYRATE	3	F1	III	3		LQ7		PP, EX, A	VE01		0	
1181	ETHYL CHLOROACETATE	6.1	TF1	II	6.1+3	802	LQ17		PP, EP, EX, TOX, A	VE01, VE02		2	

UN No. or ID No.	Name and description	Class	Classification Code	Packing group	Labels	Special provisions	Limited quantities	Carriage permitted	Equipment required	Ventilation	Provisions concerning loading, unloading and carriage	Number of cones, blue lights	Remarks
3.1.2		2.2	2.2	2.1.1.3	5.2.2	3.3	3.4.6	3.2.1	8.1.5	7.1.6	7.1.6	7.1.5	3.2.1
(1)	(2)	(3a)	(3b)	(4)	(5)	(6)	(7)	(8)	(9)	(10)	(11)	(12)	(13)
1182	ETHYL CHLOROFORMATE	6.1	TFC	I	6.1+3+8	802	LQ0		PP, EP, EX, TOX, A	VE01, VE02		2	
1183	ETHYLDICHLOROSILANE	4.3	WFC	I	4.3+3+8		LQ0		PP, EP, EX, A	VE01	HA08	1	
1184	ETHYLENE DICHLORIDE	3	FT1	II	3+6.1	802	LQ0	T	PP, EP, EX, TOX, A	VE01, VE02		2	
1185	ETHYLENEIMINE, STABILIZED	6.1	TF1	I	6.1+3	802	LQ0	T	PP, EP, EX, TOX, A	VE01, VE02		2	
1188	ETHYLENE GLYCOL MONOMETHYL ETHER	3	F1	III	3		LQ7	T	PP, EX, A	VE01		0	
1189	ETHYLENE GLYCOL MONOMETHYL ETHER ACETATE	3	F1	III	3		LQ7		PP, EX, A	VE01		0	
1190	ETHYL FORMATE	3	F1	II	3		LQ4		PP, EX, A	VE01		1	
1191	OCTYL ALDEHYDES	3	F1	III	3		LQ7		PP, EX, A	VE01		0	
1192	ETHYL LACTATE	3	F1	III	3		LQ7		PP, EX, A	VE01		0	
1193	ETHYL METHYL KETONE (METHYL ETHYL KETONE)	3	F1	II	3		LQ4		PP, EX, A	VE01		1	
1194	ETHYL NITRITE SOLUTION	3	FT1	I	3+6.1	802	LQ0	T	PP, EP, EX, TOX, A	VE01, VE02		2	
1195	ETHYL PROPIONATE	3	F1	II	3		LQ4		PP, EX, A	VE01		1	
1196	ETHYLTRICHLOROSILANE	3	FC	II	3+8		LQ4		PP, EP, EX, A	VE01		1	
1197	EXTRACTS, FLAVOURING, LIQUID	3	F1	I	3		LQ3		PP, EX, A	VE01		1	
1197	EXTRACTS, FLAVOURING, LIQUID (vapour pressure at 50 °C more than 110 kPa)	3	F1	II	3	601 640C	LQ6		PP, EX, A	VE01		1	
1197	EXTRACTS, FLAVOURING, LIQUID (vapour pressure at 50 °C not more than 110 kPa)	3	F1	II	3	601 640D	LQ6		PP, EX, A	VE01		1	
1197	EXTRACTS, FLAVOURING, LIQUID	3	F1	III	3	601 640E	LQ7		PP, EX, A	VE01		0	

UN No. or ID No.	Name and description	Class	Classification Code	Packing group	Labels	Special provisions	Limited quantities	Carriage permitted	Equipment required	Ventilation	Provisions concerning loading, unloading and carriage	Number of cones, blue lights	Remarks
3.1.2	3.1.2	2.2	2.2	2.1.1.3	5.2.2	3.3	3.4.6	3.2.1	8.1.5	7.1.6	7.1.6	7.1.5	3.2.1
(1)	(2)	(3a)	(3b)	(4)	(5)	(6)	(7)	(8)	(9)	(10)	(11)	(12)	(13)
1197	EXTRACTS, FLAVOURING, LIQUID (having a flash-point below 23 °C and viscous according to 2.2.3.1.4) (boiling point not more than 35° C)	3	F1	III	3	601 640F	LQ7		PP, EX, A	VE01		0	
1197	EXTRACTS, FLAVOURING, LIQUID (having a flash-point below 23 °C and viscous according to 2.2.3.1.4) (vapour pressure at 50 °C more than 110 kPa , boiling point of more than 35° C)	3	F1	III	3	601 640G	LQ7		PP, EX, A	VE01		0	
1197	EXTRACTS, FLAVOURING, LIQUID (having a flash-point below 23 °C and viscous according to 2.2.3.1.4) (vapour pressure at 50 °C not more than 110 kPa)	3	F1	III	3	601 640H	LQ7		PP, EX, A	VE01		0	
1198	FORMALDEHYDE SOLUTION, FLAMMABLE	3	FC	III	3+8		LQ7	T	PP, EP, EX, A	VE01		0	
1199	FURALDEHYDES	6.1	TF1	II	6.1+3	802	LQ0	T	PP, EP, EX, TOX, A	VE01, VE02		2	
1201	FUSEL OIL	3	F1	II	3		LQ4		PP, EX, A	VE01		1	
1201	FUSEL OIL	3	F1	III	3		LQ7		PP, EX, A	VE01		0	
1202	GAS OIL or DIESEL FUEL or HEATING OIL, LIGHT (flash-point not more than 60 °C)	3	F1	III	3	640K	LQ7	T	PP, EX, A	VE01		0	
1202	DIESEL FUEL complying with standard EN 590:2004 or GAS OIL or HEATING OIL, LIGHT with a flash-point as specified in EN 590:2004	3	F1	III	3	640L	LQ7	T	PP, EX, A	VE01		0	
1202	GAS OIL or DIESEL FUEL or HEATING OIL, LIGHT (flash-point more than 60 °C and not more than 100 °C)	3	F1	III	3	640M	LQ7	T	PP, EX, A	VE01		0	
1203	MOTOR SPIRIT or GASOLINE or PETROL	3	F1	II	3	243 534	LQ4	T	PP, EX, A	VE01		1	

UN No. or ID No.	Name and description	Class	Classification Code	Packing group	Labels	Special provisions	Limited quantities	Carriage permitted	Equipment required	Ventilation	Provisions concerning loading, unloading and carriage	Number of cones, blue lights	Remarks
	3.1.2	2.2	2.2	2.1.1.3	5.2.2	3.3	3.4.6	3.2.1	8.1.5	7.1.6	7.1.6	7.1.5	3.2.1
(1)	(2)	(3a)	(3b)	(4)	(5)	(6)	(7)	(8)	(9)	(10)	(11)	(12)	(13)
1204	NITROGLYCERIN SOLUTION IN ALCOHOL with not more than 1% nitroglycerin	3	D	II	3		LQ0		PP, EX, A	VE01		1	
1206	HEPTANES	3	F1	II	3		LQ4	T	PP, EX, A	VE01		1	
1207	HEXALDEHYDE	3	F1	III	3		LQ7		PP, EX, A	VE01		0	
1208	HEXANES	3	F1	II	3		LQ4	T	PP, EX, A	VE01		1	
1210	PRINTING INK, flammable or PRINTING INK RELATED MATERIAL (including printing ink thinning or reducing compound), flammable	3	F1	I	3	163	LQ3		PP, EX, A	VE01		1	
1210	PRINTING INK, flammable or PRINTING INK RELATED MATERIAL (including printing ink thinning or reducing compound), flammable (vapour pressure at 50 °C more than 110 kPa)	3	F1	II	3	163 640C	LQ6		PP, EX, A	VE01		1	
1210	PRINTING INK, flammable or PRINTING INK RELATED MATERIAL (including printing ink thinning or reducing compound), flammable (vapour pressure at 50 °C not more than 110 kPa)	3	F1	II	3	163 640D	LQ6		PP, EX, A	VE01		1	
1210	PRINTING INK, flammable or PRINTING INK RELATED MATERIAL (including printing ink thinning or reducing compound), flammable	3	F1	III	3	163 640E	LQ7		PP, EX, A	VE01		0	
1210	PRINTING INK, flammable or PRINTING INK RELATED MATERIAL (including printing ink thinning or reducing compound), flammable (having a flash-point below 23 °C and viscous according to 2.2.3.1.4) (boiling point not more than 35° C)	3	F1	III	3	163 640F	LQ7		PP, EX, A	VE01		0	

UN No. or ID No.	Name and description	Class	Classification Code	Packing group	Labels	Special provisions	Limited quantities	Carriage permitted	Equipment required	Ventilation	Provisions concerning loading, unloading and carriage	Number of cones, blue lights	Remarks
3.1.2	3.1.2	2.2	2.2	2.1.1.3	5.2.2	3.3	3.4.6	3.2.1	8.1.5	7.1.6	7.1.6	7.1.5	3.2.1
(1)	(2)	(3a)	(3b)	(4)	(5)	(6)	(7)	(8)	(9)	(10)	(11)	(12)	(13)
1210	PRINTING INK, flammable or PRINTING INK RELATED MATERIAL (including printing ink thinning or reducing compound), flammable (having a flash-point below 23 °C and viscous according to 2.2.3.1.4) (vapour pressure at 50 °C more than 110 kPa , boiling point of more than 35° C)	3	F1	III	3	163 640G	LQ7		PP, EX, A	VE01		0	
1210	PRINTING INK, flammable or PRINTING INK RELATED MATERIAL (including printing ink thinning or reducing compound), flammable (having a flash-point below 23 °C and viscous according to 2.2.3.1.4) (vapour pressure at 50 °C not more than 110 kPa)	3	F1	III	3	163 640H	LQ7		PP, EX, A	VE01		0	
1212	ISOBUTANOL (ISOBUTYL ALCOHOL)	3	F1	III	3		LQ7	T	PP, EX, A	VE01		0	
1213	ISOBUTYL ACETATE	3	F1	II	3		LQ4	T	PP, EX, A	VE01		1	
1214	ISOBUTYLAMINE	3	FC	II	3+8		LQ4	T	PP, EP, EX, A	VE01		1	
1216	ISOOCTENES	3	F1	II	3		LQ4	T	PP, EX, A	VE01		1	
1218	ISOPRENE, STABILIZED	3	F1	I	3		LQ3	T	PP, EX, A	VE01		1	
1219	ISOPROPANOL (ISOPROPYL ALCOHOL)	3	F1	II	3	601	LQ4	T	PP, EX, A	VE01		1	
1220	ISOPROPYL ACETATE	3	F1	II	3		LQ4	T	PP, EX, A	VE01		1	
1221	ISOPROPYLAMINE	3	FC	I	3+8		LQ3	T	PP, EP, EX, A	VE01		1	
1222	ISOPROPYL NITRATE	3	F1	II	3		LQ4	T	PP, EX, A	VE01		1	
1223	KEROSENE	3	F1	III	3		LQ7	T	PP, EX, A	VE01		0	
1224	KETONES, LIQUID, N.O.S. (vapour pressure at 50 °C more than 110 kPa)	3	F1	II	3	274 640C	LQ4	T	PP, EX, A	VE01		1	
1224	KETONES, LIQUID, N.O.S. (vapour pressure at 50 °C not more than 110 kPa)	3	F1	II	3	274 640D	LQ4	T	PP, EX, A	VE01		1	

UN No. or ID No.	Name and description	Class	Classification Code	Packing group	Labels	Special provisions	Limited quantities	Carriage permitted	Equipment required	Ventilation	Provisions concerning loading, unloading and carriage	Number of cones, blue lights	Remarks
	3.1.2	2.2	2.2	2.1.1.3	5.2.2	3.3	3.4.6	3.2.1	8.1.5	7.1.6	7.1.6	7.1.5	3.2.1
(1)	(2)	(3a)	(3b)	(4)	(5)	(6)	(7)	(8)	(9)	(10)	(11)	(12)	(13)
1224	KETONES, LIQUID, N.O.S.	3	F1	III	3	274	LQ7	T	PP, EX, A	VE01		0	
1228	MERCAPTANS, LIQUID, FLAMMABLE, TOXIC, N.O.S. or MERCAPTAN MIXTURE, LIQUID, FLAMMABLE, TOXIC, N.O.S.	3	FT1	II	3+6.1	274 802	LQ0		PP, EP, EX, TOX, A	VE01, VE02		2	
1228	MERCAPTANS, LIQUID, FLAMMABLE, TOXIC, N.O.S. or MERCAPTAN MIXTURE, LIQUID, FLAMMABLE, TOXIC, N.O.S.	3	FT1	III	3+6.1	274 802	LQ7		PP, EP, EX, TOX, A	VE01, VE02		0	
1229	MESITYL OXIDE	3	F1	III	3		LQ7	T	PP, EX, A	VE01		0	
1230	METHANOL	3	FT1	II	3+6.1	279 802	LQ0	T	PP, EP, EX, TOX, A	VE01, VE02		2	
1231	METHYL ACETATE	3	F1	II	3		LQ4	T	PP, EX, A	VE01		1	
1233	METHYLAMYL ACETATE	3	F1	III	3		LQ7		PP, EX, A	VE01		0	
1234	METHYLAL	3	F1	II	3		LQ4		PP, EX, A	VE01		1	
1235	METHYLAMINE, AQUEOUS SOLUTION	3	FC	II	3+8		LQ4	T	PP, EP, EX, A	VE01		1	
1237	METHYL BUTYRATE	3	F1	II	3		LQ4		PP, EX, A	VE01		1	
1238	METHYL CHLOROFORMATE	6.1	TFC	I	6.1+3+8	802	LQ0		PP, EP, EX, TOX, A	VE01, VE02		2	
1239	METHYL CHLORO-METHYL ETHER	6.1	TF1	I	6.1+3	802	LQ0		PP, EP, EX, TOX, A	VE01, VE02		2	
1242	METHYLDICHLOROSILANE	4.3	WFC	I	4.3+3+8		LQ0		PP, EP, EX, TOX, A	VE01	HA08	1	
1243	METHYL FORMATE	3	F1	I	3		LQ3	T	PP, EX, A	VE01		1	
1244	METHYLHYDRAZINE	6.1	TFC	I	6.1+3+8	802	LQ0	T	PP, EP, EX, TOX, A	VE01, VE02		2	
1245	METHYL ISOBUTYL KETONE	3	F1	II	3		LQ4		PP, EX, A	VE01		1	
1246	METHYL ISOPROPENYL KETONE, STABILIZED	3	F1	II	3		LQ4		PP, EX, A	VE01		1	
1247	METHYL METHACRYLATE MONOMER, STABILIZED	3	F1	II	3		LQ4	T	PP, EX, A	VE01		1	
1248	METHYL PROPIONATE	3	F1	II	3		LQ4		PP, EX, A	VE01		1	
1249	METHYL PROPYL KETONE	3	F1	II	3		LQ4		PP, EX, A	VE01		1	

(1) UN No. or ID No.	(2) Name and description	(3a) Class	(3b) Classifi-cation Code	(4) Packing group	(5) Labels	(6) Special provisions	(7) Limited quantities	(8) Carriage permitted	(9) Equipment required	(10) Ventilation	(11) Provisions concerning loading, unloading and carriage	(12) Number of cones, blue lights	(13) Remarks
	3.1.2	2.2	2.2	2.1.1.3	5.2.2	3.3	3.4.6	3.2.1	8.1.5	7.1.6	7.1.6	7.1.5	3.2.1
1250	METHYLTRICHLOROSILANE	3	FC	I	3+8		LQ3		PP, EP, EX, A	VE01		1	
1251	METHYL VINYL KETONE, STABILIZED	6.1	TFC	I	6.1+3+8	802	LQ0		PP, EP, EX, TOX, A	VE01, VE02		2	
1259	NICKEL CARBONYL	6.1	TF1	I	6.1+3	802	LQ0		PP, EP, EX, TOX, A	VE01, VE02		2	
1261	NITROMETHANE	3	F1	II	3		LQ4		PP, EX, A	VE01		1	
1262	OCTANES	3	F1	II	3		LQ4	T	PP, EX, A	VE01		1	
1263	PAINT (including paint, lacquer, enamel. stain, shellac, varnish, polish, liquid filler and liquid lacquer base) or PAINT RELATED MATERIAL (including paint thinning and reducing compound)	3	F1	I	3	163 650	LQ3		PP, EX, A	VE01		1	
1263	PAINT (including paint, lacquer, enamel. stain, shellac, varnish, polish, liquid filler and liquid lacquer base) or PAINT RELATED MATERIAL (including paint thinning and reducing compound) (vapour pressure at 50 °C more than 110 kPa)	3	F1	II	3	163 640C 650	LQ6		PP, EX, A	VE01		1	
1263	PAINT (including paint, lacquer, enamel. stain, shellac, varnish, polish, liquid filler and liquid lacquer base) or PAINT RELATED MATERIAL (including paint thinning and reducing compound) (vapour pressure at 50 °C not more than 110 kPa)	3	F1	II	3	163 640D 650	LQ6		PP, EX, A	VE01		1	
1263	PAINT (including paint, lacquer, enamel. stain, shellac, varnish, polish, liquid filler and liquid lacquer base) or PAINT RELATED MATERIAL (including paint thinning and reducing compound)	3	F1	III	3	163 640E 650	LQ7		PP, EX, A	VE01		0	

UN No. or ID No.	Name and description	Class	Classification Code	Packing group	Labels	Special provisions	Limited quantities	Carriage permitted	Equipment required	Ventilation	Provisions concerning loading, unloading and carriage		Number of cones, blue lights	Remarks
	3.1.2	2.2	2.2	2.1.1.3	5.2.2	3.3	3.4.6	3.2.1	8.1.5	7.1.6	7.1.6		7.1.5	3.2.1
(1)	(2)	(3a)	(3b)	(4)	(5)	(6)	(7)	(8)	(9)	(10)	(11)		(12)	(13)
1263	PAINT (including paint, lacquer, enamel. stain. shellac, varnish, polish, liquid filler and liquid lacquer base) or PAINT RELATED MATERIAL (including paint thinning and reducing compound) (having a flash-point below 23 °C and viscous according to 2.2.3.1 (boiling point not more than 35° C).	3	F1	III	3	163 640F 650	LQ7		PP, EX. A	VE01			0	
1263	PAINT (including paint, lacquer, enamel. stain. shellac, varnish, polish, liquid filler and liquid lacquer base) or PAINT RELATED MATERIAL (including paint thinning and reducing compound) (having a flash-point below 23 °C and viscous according to 2.2.3.1) (vapour pressure at 50 °C more than 110 kPa , boiling point of more than 35° C).	3	F1	III	3	163 640G 650	LQ7		PP, EX. A	VE01			0	
1263	PAINT (including paint, lacquer, enamel. stain. shellac, varnish, polish, liquid filler and liquid lacquer base) or PAINT RELATED MATERIAL (including paint thinning and reducing compound) (having a flash-point below 23 °C and viscous according to 2.2.3.1) (vapour pressure at 50 °C not more than 110 kPa).	3	F1	III	3	163 640H 650	LQ7		PP, EX. A	VE01			0	
1264	PARALDEHYDE	3	F1	III	3		LQ7	T	PP, EX. A	VE01			0	
1265	PENTANES. liquid	3	F1	I	3		LQ3	T	PP, EX. A	VE01			1	
1265	PENTANES. liquid	3	F1	II	3		LQ4	T	PP, EX. A	VE01			1	
1266	PERFUMERY PRODUCTS with flammable solvents	3	F1	I	3		LQ3		PP, EX. A	VE01			1	

UN No. or ID No.	Name and description	Class	Classification Code	Packing group	Labels	Special provisions	Limited quantities	Carriage permitted	Equipment required	Ventilation	Provisions concerning loading, unloading and carriage	Number of cones, blue lights	Remarks
3.1.2	3.1.2	2.2	2.2	2.1.1.3	5.2.2	3.3	3.4.6	3.2.1	8.1.5	7.1.6	7.1.6	7.1.5	3.2.1
(1)	(2)	(3a)	(3b)	(4)	(5)	(6)	(7)	(8)	(9)	(10)	(11)	(12)	(13)
1266	PERFUMERY PRODUCTS with flammable solvents (vapour pressure at 50 °C more than 110 kPa)	3	F1	II	3	640C	LQ6		PP, EX, A	VE01		1	
1266	PERFUMERY PRODUCTS with flammable solvents (vapour pressure at 50 °C not more than 110 kPa)	3	F1	II	3	640D	LQ6		PP, EX, A	VE01		1	
1266	PERFUMERY PRODUCTS with flammable solvents	3	F1	III	3	640E	LQ7		PP, EX, A	VE01		0	
1266	PERFUMERY PRODUCTS with flammable solvents (having a flash-point below 23 °C and viscous according to 2.2.3.1.4) (boiling point not more than 35° C)	3	F1	III	3	640F	LQ7		PP, EX, A	VE01		0	
1266	PERFUMERY PRODUCTS with flammable solvents (having a flash-point below 23 °C and viscous according to 2.2.3.1.4) (vapour pressure at 50 °C more than 110 kPa . boiling point of more than 35° C)	3	F1	III	3	640G	LQ7		PP, EX, A	VE01		0	
1266	PERFUMERY PRODUCTS with flammable solvents (having a flash-point below 23 °C and viscous according to 2.2.3.1.4) (vapour pressure at 50 °C not more than 110 kPa)	3	F1	III	3	640H	LQ7		PP, EX, A	VE01		0	
1267	PETROLEUM CRUDE OIL	3	F1	I	3	649	LQ3	T	PP, EX, A	VE01		1	
1267	PETROLEUM CRUDE OIL (vapour pressure at 50 °C more than 110 kPa)	3	F1	II	3	640C 649	LQ4	T	PP, EX, A	VE01		1	
1267	PETROLEUM CRUDE OIL (vapour pressure at 50 °C not more than 110 kPa)	3	F1	II	3	640D 649	LQ4	T	PP, EX, A	VE01		1	
1267	PETROLEUM CRUDE OIL	3	F1	III	3		LQ7	T	PP, EX, A	VE01		0	

UN No. or ID No.	Name and description	Class	Classification Code	Packing group	Labels	Special provisions	Limited quantities	Carriage permitted	Equipment required	Ventilation	Provisions concerning loading, unloading and carriage	Number of cones, blue lights	Remarks
	3.1.2	2.2	2.2	2.1.1.3	5.2.2	3.3	3.4.6	3.2.1	8.1.5	7.1.6	7.1.6	7.1.5	3.2.1
(1)	(2)	(3a)	(3b)	(4)	(5)	(6)	(7)	(8)	(9)	(10)	(11)	(12)	(13)
1268	PETROLEUM DISTILLATES, N.O.S. or PETROLEUM PRODUCTS, N.O.S.	3	F1	I	3	649	LQ3	T	PP, EX, A	VE01		1	
1268	PETROLEUM DISTILLATES, N.O.S. or PETROLEUM PRODUCTS, N.O.S. (vapour pressure at 50 °C more than 110 kPa)	3	F1	II	3	640C 649	LQ4	T	PP, EX, A	VE01		1	
1268	PETROLEUM DISTILLATES, N.O.S. or PETROLEUM PRODUCTS, N.O.S. (vapour pressure at 50 °C not more than 110 kPa)	3	F1	II	3	640D 649	LQ4	T	PP, EX, A	VE01		1	
1268	PETROLEUM DISTILLATES, N.O.S. or PETROLEUM PRODUCTS, N.O.S.	3	F1	III	3		LQ7	T	PP, EX, A	VE01		0	
1272	PINE OIL	3	F1	III	3		LQ7		PP, EX, A	VE01		0	
1274	n-PROPANOL (PROPYL ALCOHOL, NORMAL)	3	F1	II	3		LQ4	T	PP, EX, A	VE01		1	
1274	n-PROPANOL (PROPYL ALCOHOL, NORMAL)	3	F1	III	3		LQ7	T	PP, EX, A	VE01		0	
1275	PROPIONALDEHYDE	3	F1	II	3		LQ4	T	PP, EX, A	VE01		1	
1276	n-PROPYL ACETATE	3	F1	II	3		LQ4	T	PP, EX, A	VE01		1	
1277	PROPYLAMINE	3	FC	II	3+8		LQ4	T	PP, EP, EX, A	VE01		1	
1278	1-CHLOROPROPANE	3	F1	II	3		LQ4	T	PP, EX, A	VE01		1	
1279	1,2-DICHLOROPROPANE	3	F1	II	3		LQ4	T	PP, EX, A	VE01		1	
1280	PROPYLENE OXIDE	3	F1	I	3		LQ3	T	PP, EX, A	VE01		1	
1281	PROPYL FORMATES	3	F1	II	3		LQ4		PP, EX, A	VE01		1	
1282	PYRIDINE	3	F1	II	3		LQ4	T	PP, EX, A	VE01		1	
1286	ROSIN OIL	3	F1	I	3		LQ3		PP, EX, A	VE01		1	
1286	ROSIN OIL (vapour pressure at 50 °C more than 110 kPa)	3	F1	II	3	640C	LQ6		PP, EX, A	VE01		1	
1286	ROSIN OIL (vapour pressure at 50 °C not more than 110 kPa)	3	F1	II	3	640D	LQ6		PP, EX, A	VE01		1	
1286	ROSIN OIL	3	F1	III	3	640E	LQ7		PP, EX, A	VE01		0	

(1) UN No. or ID No. 3.1.2	(2) Name and description 3.1.2	(3a) Class 2.2	(3b) Classifi-cation Code 2.2	(4) Packing group 2.1.1.3	(5) Labels 5.2.2	(6) Special provisions 3.3	(7) Limited quantities 3.4.6	(8) Carriage permitted 3.2.1	(9) Equipment required 8.1.5	(10) Ventilation 7.1.6	(11) Provisions concerning loading, unloading and carriage 7.1.6	(12) Number of cones, blue lights 7.1.5	(13) Remarks 3.2.1
1286	ROSIN OIL (having a flash-point below 23 °C and viscous according to 2.2.3.1.4) (boiling point not more than 35°C)	3	F1	III	3	640F	LQ7		PP, EX, A	VE01		0	
1286	ROSIN OIL (having a flash-point below 23 °C and viscous according to 2.2.3.1.4) (vapour pressure at 50 °C more than 110 kPa, boiling point of more than 35° C)	3	F1	III	3	640G	LQ7		PP, EX, A	VE01		0	
1286	ROSIN OIL (having a flash-point below 23 °C and viscous according to 2.2.3.1.4) (vapour pressure at 50 °C not more than 110 kPa)	3	F1	III	3	640H	LQ7		PP, EX, A	VE01		0	
1287	RUBBER SOLUTION	3	F1	I	3		LQ3		PP, EX, A	VE01		1	
1287	RUBBER SOLUTION (vapour pressure at 50 °C more than 110 kPa)	3	F1	II	3	640C	LQ6		PP, EX, A	VE01		1	
1287	RUBBER SOLUTION (vapour pressure at 50 °C not more than 110 kPa)	3	F1	II	3	640D	LQ6		PP, EX, A	VE01		1	
1287	RUBBER SOLUTION	3	F1	III	3	640E	LQ7		PP, EX, A	VE01		0	
1287	RUBBER SOLUTION (having a flash-point below 23 °C and viscous according to 2.2.3.1.4) (boiling point not more than 35° C)	3	F1	III	3	640F	LQ7		PP, EX, A	VE01		0	
1287	RUBBER SOLUTION (having a flash-point below 23 °C and viscous according to 2.2.3.1.4) (vapour pressure at 50 °C more than 110 kPa , boiling point of more than 35° C)	3	F1	III	3	640G	LQ7			VE01		0	
1287	RUBBER SOLUTION (having a flash-point below 23 °C and viscous according to 2.2.3.1.4) (vapour pressure at 50 °C not more than 110 kPa)	3	F1	III	3	640H	LQ7		PP, EX, A	VE01		0	
1288	SHALE OIL	3	F1	II	3		LQ4		PP, EX, A	VE01		1	

UN No. or ID No.	Name and description	Class	Classification Code	Packing group	Labels	Special provisions	Limited quantities	Carriage permitted	Equipment required	Ventilation	Provisions concerning loading, unloading and carriage	Number of cones, blue lights	Remarks
3.1.2	3.1.2	2.2	2.2	2.1.1.3	5.2.2	3.3	3.4.6	3.2.1	8.1.5	7.1.6	7.1.6	7.1.5	3.2.1
(1)	(2)	(3a)	(3b)	(4)	(5)	(6)	(7)	(8)	(9)	(10)	(11)	(12)	(13)
1288	SHALE OIL	3	F1	III	3		LQ7		PP, EX, A	VE01		0	
1289	SODIUM METHYLATE SOLUTION in alcohol	3	FC	II	3+8		LQ4		PP, EP, EX, A	VE01		1	
1289	SODIUM METHYLATE SOLUTION in alcohol	3	FC	III	3+8		LQ7	T	PP, EP, EX, A	VE01		0	
1292	TETRAETHYL SILICATE	3	F1	III	3		LQ7		PP, EX, A	VE01		0	
1293	TINCTURES, MEDICINAL	3	F1	II	3	601	LQ4		PP, EX, A	VE01		1	
1293	TINCTURES, MEDICINAL	3	F1	III	3	601	LQ7		PP, EX, A	VE01		0	
1294	TOLUENE	3	F1	II	3		LQ4	T	PP, EX, A	VE01		1	
1295	TRICHLOROSILANE	4.3	WFC	I	4.3+3+8		LQ0		PP, EP, EX, A	VE01	HA08	1	
1296	TRIETHYLAMINE	3	FC	II	3+8		LQ4	T	PP, EP, EX, A	VE01		1	
1297	TRIMETHYLAMINE, AQUEOUS SOLUTION, not more than 50% trimethylamine, by mass	3	FC	I	3+8		LQ3		PP, EP, EX, A	VE01		1	
1297	TRIMETHYLAMINE, AQUEOUS SOLUTION, not more than 50% trimethylamine, by mass	3	FC	II	3+8		LQ4		PP, EP, EX, A	VE01		1	
1297	TRIMETHYLAMINE, AQUEOUS SOLUTION, not more than 50% trimethylamine, by mass	3	FC	III	3+8		LQ7		PP, EP, EX, A	VE01		0	
1298	TRIMETHYLCHLORO-SILANE	3	FC	II	3+8		LQ4		PP, EP, EX, A	VE01		1	
1299	TURPENTINE	3	F1	III	3		LQ7		PP, EX, A	VE01		0	
1300	TURPENTINE SUBSTITUTE	3	F1	II	3		LQ4		PP, EX, A	VE01		1	
1300	TURPENTINE SUBSTITUTE	3	F1	III	3		LQ7	T	PP, EX, A	VE01		0	
1301	VINYL ACETATE, STABILIZED	3	F1	II	3		LQ4	T	PP, EX, A	VE01		1	
1302	VINYL ETHYL ETHER, STABILIZED	3	F1	I	3		LQ3		PP, EX, A	VE01		1	
1303	VINYLIDENE CHLORIDE, STABILIZED	3	F1	I	3		LQ3		PP, EX, A	VE01		1	
1304	VINYL ISOBUTYL ETHER, STABILIZED	3	F1	II	3		LQ4		PP, EX, A	VE01		1	
1305	VINYLTRICHLOROSILANE, STABILIZED	3	FC	I	3+8		LQ3		PP, EP, EX, A	VE01		1	

UN No. or ID No.	Name and description	Class	Classifi-cation Code	Packing group	Labels	Special provisions	Limited quantities	Carriage permitted	Equipment required	Ventilation	Provisions concerning loading, unloading and carriage	Number of cones, blue lights	Remarks
3.1.2	3.1.2	2.2	2.2	2.1.1.3	5.2.2	3.3	3.4.6	3.2.1	8.1.5	7.1.6	7.1.6	7.1.5	3.2.1
(1)	(2)	(3a)	(3b)	(4)	(5)	(6)	(7)	(8)	(9)	(10)	(11)	(12)	(13)
1306	WOOD PRESERVATIVES, LIQUID (vapour pressure at 50 °C more than 110 kPa)	3	F1	II	3	640C	LQ6		PP, EX, A	VE01		1	
1306	WOOD PRESERVATIVES, LIQUID (vapour pressure at 50 °C not more than 110 kPa)	3	F1	II	3	640D	LQ6		PP, EX, A	VE01		1	
1306	WOOD PRESERVATIVES, LIQUID	3	F1	III	3	640E	LQ7		PP, EX, A	VE01		0	
1306	WOOD PRESERVATIVES, LIQUID (having a flash-point below 23 °C and viscous according to 2.2.3.1.4) (boiling point not more than 35° C)	3	F1	III	3	640F	LQ7		PP, EX, A	VE01		0	
1306	WOOD PRESERVATIVES, LIQUID (having a flash-point below 23 °C and viscous according to 2.2.3.1.4) (vapour pressure at 50 °C more than 110 kPa , boiling point of more than 35° C)	3	F1	III	3	640G	LQ7		PP, EX, A	VE01		0	
1306	WOOD PRESERVATIVES, LIQUID (having a flash-point below 23 °C and viscous according to 2.2.3.1.4) (vapour pressure at 50 °C not more than 110 kPa)	3	F1	III	3	640H	LQ7		PP, EX, A	VE01		0	
1307	XYLENES	3	F1	II	3		LQ4	T	PP, EX, A	VE01		1	
1307	XYLENES	3	F1	III	3		LQ7	T	PP, EX, A	VE01		0	
1308	ZIRCONIUM SUSPENDED IN A FLAMMABLE LIQUID	3	F1	I	3		LQ3		PP, EX, A	VE01		1	
1308	ZIRCONIUM SUSPENDED IN A FLAMMABLE LIQUID (vapour pressure at 50 °C more than 110 kPa)	3	F1	II	3	640C	LQ4		PP, EX, A	VE01		1	
1308	ZIRCONIUM SUSPENDED IN A FLAMMABLE LIQUID (vapour pressure at 50 °C not more than 110 kPa)	3	F1	II	3	640D	LQ4		PP, EX, A	VE01		1	

(1) UN No. or ID No.	(2) Name and description	(3a) Class	(3b) Classification Code	(4) Packing group	(5) Labels	(6) Special provisions	(7) Limited quantities	(8) Carriage permitted	(9) Equipment required	(10) Ventilation	(11) Provisions concerning loading, unloading and carriage	(12) Number of cones, blue lights	(13) Remarks
		2.2	2.2	2.1.1.3	5.2.2	3.3	3.4.6	3.2.1	8.1.5	7.1.6	7.1.6	7.1.5	3.2.1
1308	ZIRCONIUM SUSPENDED IN A FLAMMABLE LIQUID	3	F1	III	3		LQ7		PP, EX, A	VE01		0	
1309	ALUMINIUM POWDER, COATED	4.1	F3	II	4.1		LQ8		PP			1	
1309	ALUMINIUM POWDER, COATED	4.1	F3	III	4.1		LQ9		PP			0	
1310	AMMONIUM PICRATE, WETTED with not less than 10% water, by mass	4.1	D	I	4.1		LQ0		PP			1	
1312	BORNEOL	4.1	F1	III	4.1		LQ9		PP			0	
1313	CALCIUM RESINATE	4.1	F3	III	4.1		LQ9		PP			0	
1314	CALCIUM RESINATE, FUSED	4.1	F3	III	4.1		LQ9		PP			0	
1318	COBALT RESINATE, PRECIPITATED	4.1	F3	III	4.1		LQ9		PP			0	
1320	DINITROPHENOL, WETTED with less than 15% water, by mass	4.1	DT	I	4.1+6.1	802	LQ0		PP			2	
1321	DINITROPHENOLATES, WETTED with not less than 15% water, by mass	4.1	DT	I	4.1+6.1	802	LQ0		PP			2	
1322	DINITRORESORCINOL, WETTED with not less than 15% water, by mass	4.1	D	I	4.1		LQ0		PP			1	
1323	FERROCERIUM	4.1	F3	II	4.1	249	LQ8		PP			1	
1324	FILMS, NITROCELLULOSE BASE, gelatin coated, except scrap	4.1	F1	III	4.1		LQ9		PP			0	
1325	FLAMMABLE SOLID, ORGANIC, N.O.S.	4.1	F1	II	4.1	274	LQ8		PP			1	
1325	FLAMMABLE SOLID, ORGANIC, N.O.S.	4.1	F1	III	4.1	274	LQ9		PP			0	
1326	HAFNIUM POWDER, WETTED with not less than 25% water	4.1	F3	II	4.1	586	LQ8		PP			1	
1327	Hay, Straw or Bhusa	4.1	F1			NOT SUBJECT TO ADN							
1328	HEXAMETHYLENE-TETRAMINE	4.1	F1	III	4.1		LQ9		PP			0	
1330	MANGANESE RESINATE	4.1	F3	III	4.1		LQ9		PP			0	

UN No. or ID No. (1)	Name and description 3.1.2 (2)	Class 2.2 (3a)	Classification Code 2.2 (3b)	Packing group 2.1.1.3 (4)	Labels 5.2.2 (5)	Special provisions 3.3 (6)	Limited quantities 3.4.6 (7)	Carriage permitted 3.2.1 (8)	Equipment required 8.1.5 (9)	Ventilation 7.1.6 (10)	Provisions concerning loading, unloading and carriage 7.1.6 (11)	Number of cones, blue lights 7.1.5 (12)	Remarks 3.2.1 (13)
1331	MATCHES, 'STRIKE ANYWHERE'	4.1	F1	III	4.1	293	LQ9		PP			0	
1332	METALDEHYDE	4.1	F1	III	4.1		LQ9		PP			0	
1333	CERIUM, slabs, ingots or rods	4.1	F3	II	4.1		LQ8		PP			1	
1334	NAPHTHALENE, CRUDE or NAPHTHALENE, REFINED	4.1	F1	III	4.1	501	LQ9	B	PP		CO01	0	
1336	NITROGUANIDINE (PICRITE), WETTED with not less than 20% water, by mass	4.1	D	I	4.1		LQ0		PP			1	
1337	NITROSTARCH, WETTED with not less than 20% water, by mass	4.1	D	I	4.1		LQ0		PP			1	
1338	PHOSPHORUS, AMORPHOUS	4.1	F3	III	4.1		LQ9		PP			0	
1339	PHOSPHORUS HEPTASULPHIDE, free from yellow and white phosphorus	4.1	F3	II	4.1	602	LQ8		PP			1	
1340	PHOSPHORUS PENTASULPHIDE, free from yellow and white phosphorus	4.3	WF2	II	4.3+4.1	602	LQ11		PP, EX, A	VE01	HA08	1	
1341	PHOSPHORUS SESQUISULPHIDE, free from yellow and white phosphorus	4.1	F3	II	4.1	602	LQ8		PP			1	
1343	PHOSPHORUS TRISULPHIDE, free from yellow and white phosphorus	4.1	F3	II	4.1	602	LQ8		PP			1	
1344	TRINITROPHENOL, WETTED with not less than 30% water, by mass	4.1	D	I	4.1		LQ0		PP			1	
1345	RUBBER SCRAP or RUBBER SHODDY, powdered or granulated	4.1	F1	II	4.1		LQ8		PP			1	
1346	SILICON POWDER, AMORPHOUS	4.1	F3	III	4.1	32	LQ9		PP			0	
1347	SILVER PICRATE, WETTED with not less than 30% water, by mass	4.1	D	I	4.1		LQ0		PP			1	
1348	SODIUM DINITRO-o-CRESOLATE, WETTED with not less than 15% water, by mass	4.1	DT	I	4.1+6.1	802	LQ0		PP			2	

UN No. or ID No. (1)	Name and description 3.1.2 (2)	Class 2.2 (3a)	Classification Code 2.2 (3b)	Packing group 2.1.1.3 (4)	Labels 5.2.2 (5)	Special provisions 3.3 (6)	Limited quantities 3.4.6 (7)	Carriage permitted 3.2.1 (8)	Equipment required 8.1.5 (9)	Ventilation 7.1.6 (10)	Provisions concerning loading, unloading and carriage 7.1.6 (11)	Number of cones, blue lights 7.1.5 (12)	Remarks 3.2.1 (13)
1349	SODIUM PICRAMATE, WETTED with not less than 20% water, by mass	4.1	D	I	4.1		LQ0		PP			1	
1350	SULPHUR	4.1	F3	III	4.1	242	LQ9	B	PP			0	
1352	TITANIUM POWDER, WETTED with not less than 25% water	4.1	F3	II	4.1	586	LQ8		PP			1	
1353	FIBRES or FABRICS IMPREGNATED WITH WEAKLY NITRATED NITROCELLULOSE, N.O.S.	4.1	F1	III	4.1	274 502	LQ9		PP			0	
1354	TRINITROBENZENE, WETTED with not less than 30% water, by mass	4.1	D	I	4.1		LQ0		PP			1	
1355	TRINITROBENZOIC ACID, WETTED with not less than 30% water, by mass	4.1	D	I	4.1		LQ0		PP			1	
1356	TRINITROTOLUENE (TNT), WETTED with not less than 30% water, by mass	4.1	D	I	4.1		LQ0		PP			1	
1357	UREA NITRATE, WETTED with not less than 20% water, by mass	4.1	D	I	4.1	227	LQ0		PP			1	
1358	ZIRCONIUM POWDER, WETTED with not less than 25% water	4.1	F3	II	4.1	586	LQ8		PP			1	
1360	CALCIUM PHOSPHIDE	4.3	WT2	I	4.3+6.1	802	LQ0		PP, EP, EX, TOX, A	VE01, VE02	HA08	2	
1361	CARBON, animal or vegetable origin	4.2	S2	II	4.2		LQ0		PP			0	
1361	CARBON, animal or vegetable origin	4.2	S2	III	4.2		LQ0		PP			0	
1362	CARBON, ACTIVATED	4.2	S2	III	4.2	646	LQ0		PP			0	
1363	COPRA	4.2	S2	III	4.2		LQ0	B	PP		IN01, IN02	0	IN01 and IN02 apply only when this substance is carried in bulk or without packaging
1364	COTTON WASTE, OILY	4.2	S2	III	4.2		LQ0	B	PP			0	

UN No. or ID No.	Name and description	Class	Classification Code	Packing group	Labels	Special provisions	Limited quantities	Carriage permitted	Equipment required	Ventilation	Provisions concerning loading, unloading and carriage	Number of cones, blue lights	Remarks
	3.1.2	2.2	2.2	2.1.1.3	5.2.2	3.3	3.4.6	3.2.1	8.1.5	7.1.6	7.1.6	7.1.5	3.2.1
(1)	(2)	(3a)	(3b)	(4)	(5)	(6)	(7)	(8)	(9)	(10)	(11)	(12)	(13)
1365	COTTON, WET	4.2	S2	III	4.2		LQ0	B	PP			0	
1369	p-NITROSODIMETHYL-ANILINE	4.2	S2	II	4.2		LQ0		PP			0	
1372	Fibres, animal or fibres, vegetable burnt, wet or damp	4.2	S2	NOT SUBJECT TO ADN									
1373	FIBRES or FABRICS, ANIMAL or VEGETABLE or SYNTHETIC, N.O.S. with oil	4.2	S2	III	4.2	274	LQ0	B	PP			0	
1374	FISH MEAL (FISH SCRAP), UNSTABILIZED	4.2	S2	II	4.2	300	LQ0		PP			0	
1376	IRON OXIDE, SPENT or IRON SPONGE, SPENT obtained from coal gas purification	4.2	S4	III	4.2	592	LQ0	B	PP			0	
1378	METAL CATALYST, WETTED with a visible excess of liquid	4.2	S4	II	4.2	274	LQ0		PP			0	
1379	PAPER, UNSATURATED OIL TREATED, incompletely dried (including carbon paper)	4.2	S2	III	4.2		LQ0	B	PP			0	
1380	PENTABORANE	4.2	ST3	I	4.2+6.1	802	LQ0		PP, EP, TOX, A	VE02		2	
1381	PHOSPHORUS, WHITE or YELLOW, UNDER WATER or IN SOLUTION	4.2	ST3	I	4.2+6.1	503 802	LQ0		PP, EP, TOX, A	VE02		2	
1381	PHOSPHORUS, WHITE or YELLOW, DRY	4.2	ST4	I	4.2+6.1	503 802	LQ0		PP, EP			2	
1382	POTASSIUM SULPHIDE, ANHYDROUS or POTASSIUM SULPHIDE with less than 30% water of crystallization	4.2	S4	II	4.2	504	LQ0		PP			0	
1383	PYROPHORIC METAL, N.O.S. or PYROPHORIC ALLOY, N.O.S.	4.2	S4	I	4.2	274	LQ0		PP			0	
1384	SODIUM DITHIONITE (SODIUM HYDROSULPHITE)	4.2	S4	II	4.2		LQ0		PP			0	
1385	SODIUM SULPHIDE, ANHYDROUS or SODIUM SULPHIDE with less than 30% water of crystallization	4.2	S4	II	4.2	504	LQ0		PP			0	

UN No. or ID No. (1)	Name and description (2)	Class (3a)	Classification Code (3b)	Packing group (4)	Labels (5)	Special provisions (6)	Limited quantities (7)	Carriage permitted (8)	Equipment required (9)	Ventilation (10)	Provisions concerning loading, unloading and carriage (11)	Number of cones, blue lights (12)	Remarks (13)
1386	SEED CAKE with more than 1.5% oil and not more than 11% moisture	4.2	S2	III	4.2	800	LQ0	B	PP		IN01, IN02	0	IN01 and IN02 apply only when this substance is carried in bulk or without packaging
1387	Wool waste, wet	4.2	S2	NOT SUBJECT TO ADN									
1389	ALKALI METAL AMALGAM, LIQUID	4.3	W1	I	4.3	182 274	LQ0		PP, EX, A	VE01	HA08	0	
1390	ALKALI METAL AMIDES	4.3	W2	II	4.3	182 274 505	LQ11		PP, EX, A	VE01	HA08	0	
1391	ALKALI METAL DISPERSION or ALKALINE EARTH METAL DISPERSION having a flash-point above 60 °C	4.3	W1	I	4.3	182 183 274 506	LQ0		PP, EX, A	VE01	HA08	0	
1391	ALKALI METAL DISPERSION OR ALKALINE EARTH METAL DISPERSION having a flash-point of not more than 60 °C	4.3	WF1	I	4.3 +3	182 183 274 506	LQ0		PP, EX, A	VE01	HA08	0	
1392	ALKALINE EARTH METAL AMALGAM, LIQUID	4.3	W1	I	4.3	182 274 506	LQ0		PP, EX, A	VE01	HA08	0	
1393	ALKALINE EARTH METAL ALLOY, N.O.S.	4.3	W2	II	4.3	183 274 506	LQ11		PP, EX, A	VE01	HA08	0	
1394	ALUMINIUM CARBIDE	4.3	W2	II	4.3		LQ11		PP, EX, A	VE01	HA08	0	
1395	ALUMINIUM FERROSILICON POWDER	4.3	WT2	II	4.3+6.1	802	LQ11		PP, EP, EX, TOX, A	VE01, VE02	HA08	2	
1396	ALUMINIUM POWDER, UNCOATED	4.3	W2	II	4.3		LQ12		PP, EX, A	VE01	HA08	0	
1396	ALUMINIUM POWDER, UNCOATED	4.3	W2	III	4.3		LQ12		PP, EX, A	VE01	HA08	0	
1397	ALUMINIUM PHOSPHIDE	4.3	WT2	I	4.3+6.1	507 802	LQ0		PP, EP, EX, TOX, A	VE01, VE02	HA08	2	

UN No. or ID No. (1) 3.1.2	Name and description (2) 3.1.2	Class (3a) 2.2	Classification Code (3b) 2.2	Packing group (4) 2.1.1.3	Labels (5) 5.2.2	Special provisions (6) 3.3	Limited quantities (7) 3.4.6	Carriage permitted (8) 3.2.1	Equipment required (9) 8.1.5	Ventilation (10) 7.1.6	Provisions concerning loading, unloading and carriage (11) 7.1.6	Number of cones, blue lights (12) 7.1.5	Remarks (13) 3.2.1
1398	ALUMINIUM SILICON POWDER, UNCOATED	4.3	W2	III	4.3	37	LQ12	B	PP, EX, A	VE01, VE03	LO03 HA07, HA08 IN01, IN03	0	VE03, LO03, HA07, IN01 and IN03 apply only when this substance is carried in bulk or without packaging
1400	BARIUM	4.3	W2	II	4.3		LQ11		PP, EX, A	VE01	HA08	0	
1401	CALCIUM	4.3	W2	II	4.3		LQ11		PP, EX, A	VE01	HA08	0	
1402	CALCIUM CARBIDE	4.3	W2	I	4.3		LQ0		PP, EX, A	VE01	HA08	0	
1402	CALCIUM CARBIDE	4.3	W2	II	4.3		LQ11		PP, EX, A	VE01	HA08	0	
1403	CALCIUM CYANAMIDE with more than 0.1% calcium carbide	4.3	W2	III	4.3	38	LQ12		PP, EX, A	VE01	HA08	0	
1404	CALCIUM HYDRIDE	4.3	W2	I	4.3		LQ0		PP, EX, A	VE01	HA08	0	
1405	CALCIUM SILICIDE	4.3	W2	II	4.3		LQ11		PP, EX, A	VE01	HA08	0	
1405	CALCIUM SILICIDE	4.3	W2	III	4.3		LQ12		PP, EX, A	VE01	HA08	0	
1407	CAESIUM	4.3	W2	I	4.3		LQ0		PP, EX, A	VE01	HA08	0	
1408	FERROSILICON with 30% or more but less than 90% silicon	4.3	WT2	III	4.3+6.1	39 802	LQ12	B	PP, EP, EX, TOX, A	VE01, VE02, VE03	LO03 HA07, HA08 IN01, IN02, IN03	0	VE03, LO03, HA07, IN01, IN02 and IN03 apply only when this substance is carried in bulk or without packaging
1409	METAL HYDRIDES, WATER-REACTIVE, N.O.S.	4.3	W2	I	4.3	274 508	LQ0		PP, EX, A	VE01	HA08	0	
1409	METAL HYDRIDES, WATER-REACTIVE, N.O.S.	4.3	W2	II	4.3	274 508	LQ11		PP, EX, A	VE01	HA08	0	
1410	LITHIUM ALUMINIUM HYDRIDE	4.3	W2	I	4.3		LQ0		PP, EX, A	VE01	HA08	0	
1411	LITHIUM ALUMINIUM HYDRIDE, ETHEREAL	4.3	WF1	I	4.3+3		LQ0		PP, EX, A	VE01	HA08	1	
1413	LITHIUM BOROHYDRIDE	4.3	W2	I	4.3		LQ0		PP, EX, A	VE01	HA08	0	
1414	LITHIUM HYDRIDE	4.3	W2	I	4.3		LQ0		PP, EX, A	VE01	HA08	0	
1415	LITHIUM	4.3	W2	I	4.3		LQ0		PP, EX, A	VE01	HA08	0	
1417	LITHIUM SILICON	4.3	W2	II	4.3		LQ11		PP, EX, A	VE01	HA08	0	

UN No. or ID No. (1)	Name and description 3.1.2 (2)	Class 2.2 (3a)	Classification Code 2.2 (3b)	Packing group 2.1.1.3 (4)	Labels 5.2.2 (5)	Special provisions 3.3 (6)	Limited quantities 3.4.6 (7)	Carriage permitted 3.2.1 (8)	Equipment required 8.1.5 (9)	Ventilation 7.1.6 (10)	Provisions concerning loading, unloading and carriage 7.1.6 (11)	Number of cones, blue lights 7.1.5 (12)	Remarks 3.2.1 (13)
1418	MAGNESIUM POWDER or MAGNESIUM ALLOYS POWDER	4.3	WS	I	4.3+4.2		LQ0		PP, EX, A	VE01	HA08	0	
1418	MAGNESIUM POWDER or MAGNESIUM ALLOYS POWDER	4.3	WS	II	4.3+4.2		LQ11		PP, EX, A	VE01	HA08	0	
1418	MAGNESIUM POWDER or MAGNESIUM ALLOYS POWDER	4.3	WS	III	4.3+4.2		LQ12		PP, EX, A	VE01	HA08	0	
1419	MAGNESIUM ALUMINIUM PHOSPHIDE	4.3	WT2	I	4.3+6.1	802	LQ0		PP, EP, EX, TOX, A	VE01, VE02	HA08	2	
1420	POTASSIUM METAL ALLOYS, LIQUID	4.3	W1	I	4.3		LQ0		PP, EX, A	VE01	HA08	0	
1421	ALKALI METAL ALLOY, LIQUID, N.O.S.	4.3	W1	I	4.3	182 274	LQ0		PP, EX, A	VE01	HA08	0	
1422	POTASSIUM SODIUM ALLOYS, LIQUID	4.3	W1	I	4.3		LQ0		PP, EX, A	VE01	HA08	0	
1423	RUBIDIUM	4.3	W2	I	4.3		LQ0		PP, EX, A	VE01	HA08	0	
1426	SODIUM BOROHYDRIDE	4.3	W2	I	4.3		LQ0		PP, EX, A	VE01	HA08	0	
1427	SODIUM HYDRIDE	4.3	W2	I	4.3		LQ0		PP, EX, A	VE01	HA08	0	
1428	SODIUM	4.3	W2	I	4.3		LQ0		PP, EX, A	VE01	HA08	0	
1431	SODIUM METHYLATE	4.2	SC4	II	4.2+8		LQ0		PP	VE01	HA08	0	
1432	SODIUM PHOSPHIDE	4.3	WT2	I	4.3+6.1	802	LQ0		PP, EP, EX, TOX, A	VE01, VE02	HA08	2	
1433	STANNIC PHOSPHIDES	4.3	WT2	I	4.3+6.1	802	LQ0		PP, EP, EX, TOX, A	VE01, VE02	HA08	2	
1435	ZINC ASHES	4.3	W2	III	4.3		LQ12	B	PP, EX, A	VE01, VE03	LO03 HA07, HA08 IN01, IN03	0	VE03, LO03, HA07, IN01 and IN03 apply only when this substance is carried in bulk or without packaging
1436	ZINC POWDER or ZINC DUST	4.3	WS	I	4.3+4.2		LQ0		PP, EX, A	VE01	HA08	0	
1436	ZINC POWDER or ZINC DUST	4.3	WS	II	4.3+4.2		LQ11		PP, EX, A	VE01	HA08	0	
1436	ZINC POWDER or ZINC DUST	4.3	WS	III	4.3+4.2		LQ12		PP, EX, A	VE01	HA08	0	
1437	ZIRCONIUM HYDRIDE	4.1	F3	II	4.1		LQ8		PP			1	

UN No. or ID No.	Name and description	Class	Classification Code	Packing group	Labels	Special provisions	Limited quantities	Carriage permitted	Equipment required	Ventilation	Provisions concerning loading, unloading and carriage	Number of cones, blue lights	Remarks
3.1.2	3.1.2	2.2	2.2	2.1.1.3	5.2.2	3.3	3.4.6	3.2.1	8.1.5	7.1.6	7.1.6	7.1.5	3.2.1
(1)	(2)	(3a)	(3b)	(4)	(5)	(6)	(7)	(8)	(9)	(10)	(11)	(12)	(13)
1438	ALUMINIUM NITRATE	5.1	O2	III	5.1		LQ12	B	PP		CO02, LO04	0	CO02 and LO04 apply only when this substance is carried in bulk or without packaging
1439	AMMONIUM DICHROMATE	5.1	O2	II	5.1		LQ11		PP			0	
1442	AMMONIUM PERCHLORATE	5.1	O2	II	5.1	152	LQ11		PP			0	
1444	AMMONIUM PERSULPHATE	5.1	O2	III	5.1		LQ12		PP			0	
1445	BARIUM CHLORATE, SOLID	5.1	OT2	II	5.1+6.1	802	LQ11		PP			2	
1446	BARIUM NITRATE	5.1	OT2	II	5.1+6.1	802	LQ11		PP			2	
1447	BARIUM PERCHLORATE, SOLID	5.1	OT2	II	5.1+6.1	802	LQ11		PP			2	
1448	BARIUM PERMANGANATE	5.1	OT2	II	5.1+6.1	802	LQ11		PP			2	
1449	BARIUM PEROXIDE	5.1	OT2	II	5.1+6.1	802	LQ11		PP			2	
1450	BROMATES, INORGANIC, N.O.S.	5.1	O2	II	5.1	274 604	LQ11		PP			0	
1451	CAESIUM NITRATE	5.1	O2	III	5.1		LQ12	B	PP		CO02, LO04	0	CO02 and LO04 apply only when this substance is carried in bulk or without packaging
1452	CALCIUM CHLORATE	5.1	O2	II	5.1		LQ11		PP			0	
1453	CALCIUM CHLORITE	5.1	O2	II	5.1		LQ11		PP			0	
1454	CALCIUM NITRATE	5.1	O2	III	5.1	208	LQ12	B	PP		CO02, LO04	0	CO02 and LO04 apply only when this substance is carried in bulk or without packaging
1455	CALCIUM PERCHLORATE	5.1	O2	II	5.1		LQ11		PP			0	
1456	CALCIUM PERMANGANATE	5.1	O2	II	5.1		LQ11		PP			0	
1457	CALCIUM PEROXIDE	5.1	O2	II	5.1		LQ11		PP			0	
1458	CHLORATE AND BORATE MIXTURE	5.1	O2	II	5.1		LQ11		PP			0	
1458	CHLORATE AND BORATE MIXTURE	5.1	O2	III	5.1		LQ12		PP			0	
1459	CHLORATE AND MAGNESIUM CHLORIDE MIXTURE, SOLID	5.1	O2	II	5.1		LQ11		PP			0	

UN No. or ID No.	Name and description	Class	Classification Code	Packing group	Labels	Special provisions	Limited quantities	Carriage permitted	Equipment required	Ventilation	Provisions concerning loading, unloading and carriage	Number of cones, blue lights	Remarks
	3.1.2	2.2	2.2	2.1.1.3	5.2.2	3.3	3.4.6	3.2.1	8.1.5	7.1.6	7.1.6	7.1.5	3.2.1
(1)	(2)	(3a)	(3b)	(4)	(5)	(6)	(7)	(8)	(9)	(10)	(11)	(12)	(13)
1459	CHLORATE AND MAGNESIUM CHLORIDE MIXTURE, SOLID	5.1	O2	III	5.1		LQ12		PP			0	
1461	CHLORATES, INORGANIC, N.O.S.	5.1	O2	II	5.1	274 605	LQ11		PP			0	
1462	CHLORITES, INORGANIC, N.O.S.	5.1	O2	II	5.1	274 509 606	LQ11		PP			0	
1463	CHROMIUM TRIOXIDE, ANHYDROUS	5.1	OTC	II	5.1+6.1+8	510	LQ11		PP			0	
1465	DIDYMIUM NITRATE	5.1	O2	III	5.1		LQ12	B	PP		CO02, LO04	0	CO02 and LO04 apply only when this substance is carried in bulk or without packaging
1466	FERRIC NITRATE	5.1	O2	III	5.1		LQ12	B	PP		CO02, LO04	0	CO02 and LO04 apply only when this substance is carried in bulk or without packaging
1467	GUANIDINE NITRATE	5.1	O2	III	5.1		LQ12	B	PP		CO02, LO04	0	CO02 and LO04 apply only when this substance is carried in bulk or without packaging
1469	LEAD NITRATE	5.1	OT2	II	5.1+6.1	802	LQ11		PP			2	
1470	LEAD PERCHLORATE, SOLID	5.1	OT2	II	5.1+6.1	802	LQ11		PP			2	
1471	LITHIUM HYPOCHLORITE, DRY or LITHIUM HYPOCHLORITE MIXTURE	5.1	O2	II	5.1		LQ11		PP			0	
1472	LITHIUM PEROXIDE	5.1	O2	II	5.1		LQ11		PP			0	
1473	MAGNESIUM BROMATE	5.1	O2	II	5.1		LQ11		PP			0	
1474	MAGNESIUM NITRATE	5.1	O2	III	5.1		LQ12	B	PP		CO02, LO04	0	CO02 and LO04 apply only when this substance is carried in bulk or without packaging
1475	MAGNESIUM PERCHLORATE	5.1	O2	II	5.1		LQ11		PP			0	

UN No. or ID No. (1) 3.1.2	Name and description (2) 3.1.2	Class (3a) 2.2	Classification Code (3b) 2.2	Packing group (4) 2.1.1.3	Labels (5) 5.2.2	Special provisions (6) 3.3	Limited quantities (7) 3.4.6	Carriage permitted (8) 3.2.1	Equipment required (9) 8.1.5	Ventilation (10) 7.1.6	Provisions concerning loading, unloading and carriage (11) 7.1.6	Number of cones, blue lights (12) 7.1.5	Remarks (13) 3.2.1
1476	MAGNESIUM PEROXIDE	5.1	O2	II	5.1		LQ11		PP			0	
1477	NITRATES, INORGANIC, N.O.S.	5.1	O2	II	5.1	274 511	LQ11		PP			0	
1477	NITRATES, INORGANIC, N.O.S.	5.1	O2	III	5.1	274 511	LQ12	B	PP		CO02, LO04	0	CO02 and LO04 apply only when this substance is carried in bulk or without packaging
1479	OXIDIZING SOLID, N.O.S.	5.1	O2	I	5.1	274	LQ0		PP			0	
1479	OXIDIZING SOLID, N.O.S.	5.1	O2	II	5.1	274	LQ11		PP			0	
1479	OXIDIZING SOLID, N.O.S.	5.1	O2	III	5.1	274	LQ12		PP			0	
1481	PERCHLORATES, INORGANIC, N.O.S.	5.1	O2	II	5.1	274	LQ11		PP			0	
1481	PERCHLORATES, INORGANIC, N.O.S.	5.1	O2	III	5.1	274	LQ12		PP			0	
1482	PERMANGANATES, INORGANIC, N.O.S.	5.1	O2	II	5.1	274 608	LQ11		PP			0	
1482	PERMANGANATES, INORGANIC, N.O.S.	5.1	O2	III	5.1	274 608	LQ12		PP			0	
1483	PEROXIDES, INORGANIC, N.O.S.	5.1	O2	II	5.1	274	LQ11		PP			0	
1483	PEROXIDES, INORGANIC, N.O.S.	5.1	O2	III	5.1	274	LQ12		PP			0	
1484	POTASSIUM BROMATE	5.1	O2	II	5.1		LQ11		PP			0	
1485	POTASSIUM CHLORATE	5.1	O2	II	5.1		LQ11		PP			0	
1486	POTASSIUM NITRATE	5.1	O2	III	5.1		LQ12	B	PP		CO02, LO04	0	CO02 and LO04 apply only when this substance is carried in bulk or without packaging
1487	POTASSIUM NITRATE AND SODIUM NITRITE MIXTURE	5.1	O2	II	5.1	607	LQ11		PP			0	
1488	POTASSIUM NITRITE	5.1	O2	II	5.1		LQ11		PP			0	
1489	POTASSIUM PERCHLORATE	5.1	O2	II	5.1		LQ11		PP			0	
1490	POTASSIUM PERMANGANATE	5.1	O2	II	5.1		LQ11		PP			0	
1491	POTASSIUM PEROXIDE	5.1	O2	I	5.1		LQ0		PP			0	
1492	POTASSIUM PERSULPHATE	5.1	O2	III	5.1		LQ12		PP			0	

UN No. or ID No.	Name and description	Class	Classification Code	Packing group	Labels	Special provisions	Limited quantities	Carriage permitted	Equipment required	Ventilation	Provisions concerning loading, unloading and carriage	Number of cones, blue lights	Remarks
3.1.2	3.1.2	2.2	2.2	2.1.1.3	5.2.2	3.3	3.4.6	3.2.1	8.1.5	7.1.6	7.1.6	7.1.5	3.2.1
(1)	(2)	(3a)	(3b)	(4)	(5)	(6)	(7)	(8)	(9)	(10)	(11)	(12)	(13)
1493	SILVER NITRATE	5.1	O2	II	5.1		LQ11		PP			0	
1494	SODIUM BROMATE	5.1	O2	II	5.1		LQ11		PP			0	
1495	SODIUM CHLORATE	5.1	O2	II	5.1		LQ11		PP			0	
1496	SODIUM CHLORITE	5.1	O2	II	5.1		LQ11		PP			0	
1498	SODIUM NITRATE	5.1	O2	III	5.1		LQ12	B	PP		CO02, LO04	0	CO02 and LO04 apply only when this substance is carried in bulk or without packaging
1499	SODIUM NITRATE AND POTASSIUM NITRATE MIXTURE	5.1	O2	III	5.1		LQ12	B	PP		CO02, LO04	0	CO02 and LO04 apply only when this substance is carried in bulk or without packaging
1500	SODIUM NITRITE	5.1	OT2	III	5.1+6.1	802	LQ12		PP			0	
1502	SODIUM PERCHLORATE	5.1	O2	II	5.1		LQ11		PP			0	
1503	SODIUM PERMANGANATE	5.1	O2	II	5.1		LQ11		PP			0	
1504	SODIUM PEROXIDE	5.1	O2	I	5.1		LQ0		PP			0	
1505	SODIUM PERSULPHATE	5.1	O2	III	5.1		LQ12		PP			0	
1506	STRONTIUM CHLORATE	5.1	O2	II	5.1		LQ11		PP			0	
1507	STRONTIUM NITRATE	5.1	O2	III	5.1		LQ12	B	PP		CO02, LO04	0	CO02 and LO04 apply only when this substance is carried in bulk or without packaging
1508	STRONTIUM PERCHLORATE	5.1	O2	II	5.1		LQ11		PP			0	
1509	STRONTIUM PEROXIDE	5.1	O2	II	5.1		LQ11		PP			0	
1510	TETRANITROMETHANE	5.1	OT1	I	5.1+6.1	609 802	LQ0		PP, EP, TOX, A	VE02		2	
1511	UREA HYDROGEN PEROXIDE	5.1	OC2	III	5.1+8		LQ12		PP			0	
1512	ZINC AMMONIUM NITRITE	5.1	O2	II	5.1		LQ11		PP			0	
1513	ZINC CHLORATE	5.1	O2	II	5.1		LQ11		PP			0	
1514	ZINC NITRATE	5.1	O2	II	5.1		LQ11		PP			0	
1515	ZINC PERMANGANATE	5.1	O2	II	5.1		LQ11		PP			0	
1516	ZINC PEROXIDE	5.1	O2	II	5.1		LQ11		PP			0	

UN No. or ID No. (1)	Name and description 3.1.2 (2)	Class 2.2 (3a)	Classification Code 2.2 (3b)	Packing group 2.1.1.3 (4)	Labels 5.2.2 (5)	Special provisions 3.3 (6)	Limited quantities 3.4.6 (7)	Carriage permitted 3.2.1 (8)	Equipment required 8.1.5 (9)	Ventilation 7.1.6 (10)	Provisions concerning loading, unloading and carriage 7.1.6 (11)	Number of cones, blue lights 7.1.5 (12)	Remarks 3.2.1 (13)
1517	ZIRCONIUM PICRAMATE, WETTED with not less than 20% water, by mass	4.1	D	I	4.1		LQ0		PP			1	
1541	ACETONE CYANOHYDRIN, STABILIZED	6.1	T1	I	6.1	802	LQ0	T	PP, EP, TOX, A	VE02		2	
1544	ALKALOIDS, SOLID, N.O.S. or ALKALOID SALTS, SOLID, N.O.S.	6.1	T2	I	6.1	43 274 802	LQ0		PP, EP			2	
1544	ALKALOIDS, SOLID, N.O.S. or ALKALOID SALTS, SOLID, N.O.S.	6.1	T2	II	6.1	43 274 802	LQ18		PP, EP			2	
1544	ALKALOIDS, SOLID, N.O.S. or ALKALOID SALTS, SOLID, N.O.S.	6.1	T2	III	6.1	43 274 802	LQ9		PP, EP			0	
1545	ALLYL ISOTHIOCYANATE, STABILIZED	6.1	TF1	II	6.1+3	802	LQ17	T	PP, EP, EX, TOX, A	VE01, VE02		2	
1546	AMMONIUM ARSENATE	6.1	T5	II	6.1	802	LQ18		PP, EP			2	
1547	ANILINE	6.1	T1	II	6.1	279 802	LQ17	T	PP, EP, TOX, A	VE02		2	
1548	ANILINE HYDROCHLORIDE	6.1	T2	III	6.1	802	LQ9		PP, EP			0	
1549	ANTIMONY COMPOUND, INORGANIC, SOLID, N.O.S.	6.1	T5	III	6.1	45 274 512 802	LQ9		PP, EP			0	
1550	ANTIMONY LACTATE	6.1	T5	III	6.1	802	LQ9		PP, EP			0	
1551	ANTIMONY POTASSIUM TARTRATE	6.1	T5	III	6.1	802	LQ9		PP, EP			0	
1553	ARSENIC ACID, LIQUID	6.1	T4	I	6.1	802	LQ0		PP, EP, TOX, A	VE02		2	
1554	ARSENIC ACID, SOLID	6.1	T5	II	6.1	802	LQ18		PP, EP			2	
1555	ARSENIC BROMIDE	6.1	T5	II	6.1	802	LQ18		PP, EP			2	
1556	ARSENIC COMPOUND, LIQUID, N.O.S. inorganic, including: Arsenates, n.o.s., Arsenites, n.o.s.; and Arsenic sulphides, n.o.s.	6.1	T4	I	6.1	43 274 802	LQ0		PP, EP, TOX, A	VE02		2	

(1) UN No. or ID No.	(2) Name and description	(3a) Class 2.2	(3b) Classification Code 2.2	(4) Packing group 2.1.1.3	(5) Labels 5.2.2	(6) Special provisions 3.3	(7) Limited quantities 3.4.6	(8) Carriage permitted 3.2.1	(9) Equipment required 8.1.5	(10) Ventilation 7.1.6	(11) Provisions concerning loading, unloading and carriage 7.1.6	(12) Number of cones, blue lights 7.1.5	(13) Remarks 3.2.1
1556	ARSENIC COMPOUND, LIQUID, N.O.S., inorganic, including: Arsenates, n.o.s.. Arsenites, n.o.s.; and Arsenic sulphides, n.o.s.	6.1	T4	II	6.1	43 274 802	LQ17		PP, EP, TOX, A	VE02		2	
1556	ARSENIC COMPOUND, LIQUID, N.O.S., inorganic, including: Arsenates, n.o.s.. Arsenites, n.o.s.; and Arsenic sulphides, n.o.s.	6.1	T4	III	6.1	43 274 802	LQ7		PP, EP, TOX, A	VE02		0	
1557	ARSENIC COMPOUND, SOLID, N.O.S., inorganic, including: Arsenates. n.o.s.; Arsenites, n.o.s.; and Arsenic sulphides, n.o.s.	6.1	T5	I	6.1	43 274 802	LQ0		PP, EP			2	
1557	ARSENIC COMPOUND, SOLID, N.O.S., inorganic, including: Arsenates. n.o.s.; Arsenites, n.o.s.; and Arsenic sulphides, n.o.s.	6.1	T5	II	6.1	43 274 802	LQ18		PP, EP			2	
1557	ARSENIC COMPOUND, SOLID, N.O.S., inorganic, including: Arsenates. n.o.s.; Arsenites, n.o.s.; and Arsenic sulphides, n.o.s.	6.1	T5	III	6.1	43 274 802	LQ9		PP, EP			0	
1558	ARSENIC	6.1	T5	II	6.1	802	LQ18		PP, EP			2	
1559	ARSENIC PENTOXIDE	6.1	T5	II	6.1	802	LQ18		PP, EP			2	
1560	ARSENIC TRICHLORIDE	6.1	T4	I	6.1	802	LQ0		PP, EP, TOX, A	VE02		2	
1561	ARSENIC TRIOXIDE	6.1	T5	II	6.1	802	LQ18		PP, EP			2	
1562	ARSENICAL DUST	6.1	T5	II	6.1	802	LQ18		PP, EP			2	
1564	BARIUM COMPOUND, N.O.S.	6.1	T5	II	6.1	177 274 513 587 802	LQ18		PP, EP			2	
1564	BARIUM COMPOUND, N.O.S.	6.1	T5	III	6.1	177 274 513 587 802	LQ9		PP, EP			0	
1565	BARIUM CYANIDE	6.1	T5	I	6.1	802	LQ0		PP, EP			2	

UN No. or ID No.	Name and description	Class	Classification Code	Packing group	Labels	Special provisions	Limited quantities	Carriage permitted	Equipment required	Ventilation	Provisions concerning loading, unloading and carriage	Number of cones, blue lights	Remarks
	3.1.2	2.2	2.2	2.1.1.3	5.2.2	3.3	3.4.6	3.2.1	8.1.5	7.1.6	7.1.6	7.1.5	3.2.1
(1)	(2)	(3a)	(3b)	(4)	(5)	(6)	(7)	(8)	(9)	(10)	(11)	(12)	(13)
1566	BERYLLIUM COMPOUND, N.O.S.	6.1	T5	II	6.1	274 514 802	LQ18		PP, EP			2	
1566	BERYLLIUM COMPOUND, N.O.S.	6.1	T5	III	6.1	274 514 802	LQ9		PP, EP			0	
1567	BERYLLIUM POWDER	6.1	TF3	II	6.1+4.1	802	LQ18		PP, EP			2	
1569	BROMOACETONE	6.1	TF1	II	6.1+3	802	LQ17		PP, EP, EX, TOX, A	VE01, VE02		2	
1570	BRUCINE	6.1	T2	I	6.1	43 802	LQ0		PP, EP			2	
1571	BARIUM AZIDE, WETTED with not less than 50% water, by mass	4.1	DT	I	4.1+6.1	568 802	LQ0		PP			2	
1572	CACODYLIC ACID	6.1	T5	II	6.1	802	LQ18		PP, EP			2	
1573	CALCIUM ARSENATE	6.1	T5	II	6.1	802	LQ18		PP, EP			2	
1574	CALCIUM ARSENATE AND CALCIUM ARSENITE MIXTURE, SOLID	6.1	T5	II	6.1	802	LQ18		PP, EP			2	
1575	CALCIUM CYANIDE	6.1	T5	I	6.1	802	LQ0		PP, EP			2	
1577	CHLORODINITRO-BENZENES, LIQUID	6.1	T1	II	6.1	279 802	LQ17		PP, EP, TOX, A	VE02		2	
1578	CHLORONITROBENZENES, SOLID	6.1	T1	II	6.1	279 802	LQ18	T	PP, EP, TOX, A	VE02		2	
1579	4-CHLORO-o-TOLUIDINE HYDROCHLORIDE, SOLID	6.1	T2	III	6.1	802	LQ9		PP, EP			0	
1580	CHLOROPICRIN	6.1	T1	I	6.1	802	LQ0		PP, EP, TOX, A	VE02		2	
1581	CHLOROPICRIN AND METHYL BROMIDE MIXTURE with more than 2% chloropicrin	2	2T		2.3		LQ0		PP, EP, TOX, A	VE02		2	
1582	CHLOROPICRIN AND METHYL CHLORIDE MIXTURE	2	2T		2.3		LQ0		PP, EP, TOX, A	VE02		2	
1583	CHLOROPICRIN MIXTURE, N.O.S.	6.1	T1	I	6.1	274 315 515 802	LQ0		PP, EP, TOX, A	VE02		2	

UN No. or ID No.	Name and description	Class	Classification Code	Packing group	Labels	Special provisions	Limited quantities	Carriage permitted	Equipment required	Ventilation	Provisions concerning loading, unloading and carriage	Number of cones, blue lights	Remarks
3.1.2	3.1.2	2.2	2.2	2.1.1.3	5.2.2	3.3	3.4.6	3.2.1	8.1.5	7.1.6	7.1.6	7.1.5	3.2.1
(1)	(2)	(3a)	(3b)	(4)	(5)	(6)	(7)	(8)	(9)	(10)	(11)	(12)	(13)
1583	CHLOROPICRIN MIXTURE, N.O.S.	6.1	T1	II	6.1	274 515 802	LQ17		PP, EP, TOX, A	VE02		2	
1583	CHLOROPICRIN MIXTURE, N.O.S.	6.1	T1	III	6.1	274 515 802	LQ7		PP, EP, TOX, A	VE02		0	
1585	COPPER ACETOARSENITE	6.1	T5	II	6.1	802	LQ18		PP, EP			2	
1586	COPPER ARSENITE	6.1	T5	II	6.1	802	LQ18		PP, EP			2	
1587	COPPER CYANIDE	6.1	T5	II	6.1	802	LQ18		PP, EP			2	
1588	CYANIDES, INORGANIC, SOLID, N.O.S.	6.1	T5	I	6.1	47 274 802	LQ0		PP, EP			2	
1588	CYANIDES, INORGANIC, SOLID, N.O.S.	6.1	T5	II	6.1	47 274 802	LQ18		PP, EP			2	
1588	CYANIDES, INORGANIC, SOLID, N.O.S.	6.1	T5	III	6.1	47 274 802	LQ9		PP, EP			0	
1589	CYANOGEN CHLORIDE, STABILIZED	2	2TC		2.3+8		LQ0		PP, EP, TOX, A	VE02		2	
1590	DICHLOROANILINES, LIQUID	6.1	T1	II	6.1	279 802	LQ17		PP, EP, TOX, A	VE02		2	
1591	o-DICHLOROBENZENE	6.1	T1	III	6.1	279 802	LQ7	T	PP, EP, TOX, A	VE02		0	
1593	DICHLOROMETHANE	6.1	T1	III	6.1	516 802	LQ7	T	PP, EP, TOX, A	VE02		0	
1594	DIETHYL SULPHATE	6.1	T1	II	6.1	802	LQ17	T	PP, EP, TOX, A	VE02		2	
1595	DIMETHYL SULPHATE	6.1	TC1	I	6.1+8	802	LQ0	T	PP, EP, TOX, A	VE02		2	
1596	DINITROANILINES	6.1	T2	II	6.1	802	LQ18		PP, EP			2	
1597	DINITROBENZENES, LIQUID	6.1	T1	II	6.1	802	LQ17		PP, EP, TOX, A	VE02		2	
1597	DINITROBENZENES, LIQUID	6.1	T1	III	6.1	802	LQ7		PP, EP, TOX, A	VE02		0	
1598	DINITRO-o-CRESOL	6.1	T2	II	6.1	43 802	LQ18		PP, EP			2	

UN No. or ID No.	Name and description	Class	Classifi-cation Code	Packing group	Labels	Special provisions	Limited quantities	Carriage permitted	Equipment required	Ventilation	Provisions concerning loading, unloading and carriage	Number of cones, blue lights	Remarks
3.1.1	3.1.2	2.2	2.2	2.1.1.3	5.2.2	3.3	3.4.6	3.2.1	8.1.5	7.1.6	7.1.6	7.1.5	3.2.1
(1)	(2)	(3a)	(3b)	(4)	(5)	(6)	(7)	(8)	(9)	(10)	(11)	(12)	(13)
1599	DINITROPHENOL SOLUTION	6.1	T1	II	6.1	802	LQ17		PP, EP, A			2	
1599	DINITROPHENOL SOLUTION	6.1	T1	III	6.1	802	LQ7		PP, EP, A			0	
1600	DINITROTOLUENES, MOLTEN	6.1	T1	II	6.1	802	LQ0		PP, EP, TOX, A	VE02		2	
1601	DISINFECTANT, SOLID, TOXIC, N.O.S.	6.1	T2	I	6.1	274 802	LQ0		PP, EP			2	
1601	DISINFECTANT, SOLID, TOXIC, N.O.S.	6.1	T2	II	6.1	274 802	LQ18		PP, EP			2	
1601	DISINFECTANT, SOLID, TOXIC, N.O.S.	6.1	T2	III	6.1	274 802	LQ9		PP, EP			0	
1602	DYE, LIQUID, TOXIC, N.O.S. or DYE INTERMEDIATE, LIQUID, TOXIC, N.O.S.	6.1	T1	I	6.1	274 802	LQ0		PP, EP, TOX, A	VE02		2	
1602	DYE, LIQUID, TOXIC, N.O.S. or DYE INTERMEDIATE, LIQUID, TOXIC, N.O.S.	6.1	T1	II	6.1	274 802	LQ17		PP, EP, TOX, A	VE02		2	
1602	DYE, LIQUID, TOXIC, N.O.S. or DYE INTERMEDIATE, LIQUID, TOXIC, N.O.S.	6.1	T1	III	6.1	274 802	LQ7		PP, EP, TOX, A	VE02		0	
1603	ETHYL BROMOACETATE	6.1	TF1	II	6.1+3	802	LQ17		PP, EP, EX, TOX, A	VE01, VE02		2	
1604	ETHYLENEDIAMINE	8	CF1	II	8+3		LQ22	T	PP, EP, EX, A	VE01		1	
1605	ETHYLENE DIBROMIDE	6.1	T1	I	6.1	802	LQ0	T	PP, EP, TOX, A	VE02		2	
1606	FERRIC ARSENATE	6.1	T5	II	6.1	802	LQ18		PP, EP			2	
1607	FERRIC ARSENITE	6.1	T5	II	6.1	802	LQ18		PP, EP			2	
1608	FERROUS ARSENATE	6.1	T5	II	6.1	802	LQ18		PP, EP			2	
1611	HEXAETHYL TETRAPHOSPHATE	6.1	T1	II	6.1	802	LQ17		PP, EP, TOX, A	VE02		2	
1612	HEXAETHYL TETRAPHOSPHATE AND COMPRESSED GAS MIXTURE	2	1T		2.3		LQ0		PP, EP, TOX, A	VE02		2	

UN No. or ID No.	Name and description	Class	Classification Code	Packing group	Labels	Special provisions	Limited quantities	Carriage permitted	Equipment required	Ventilation	Provisions concerning loading, unloading and carriage	Number of cones, blue lights	Remarks
3.1.2	3.1.2	2.2	2.2	2.1.1.3	5.2.2	3.3	3.4.6	3.2.1	8.1.5	7.1.6	7.1.6	7.1.5	3.2.1
(1)	(2)	(3a)	(3b)	(4)	(5)	(6)	(7)	(8)	(9)	(10)	(11)	(12)	(13)
1613	HYDROCYANIC ACID, AQUEOUS SOLUTION (HYDROGEN CYANIDE, AQUEOUS SOLUTION) with not more than 20% hydrogen cyanide	6.1	TF1	I	6.1+3	48 802	LQ0		PP, EP, EX, TOX, A	VE01, VE02		2	
1614	HYDROGEN CYANIDE, STABILIZED, containing less than 3% water and absorbed in a porous inert material	6.1	TF1	I	6.1+3	603 802	LQ0		PP, EP, EX, TOX, A	VE01, VE02		2	
1616	LEAD ACETATE	6.1	T5	III	6.1	802	LQ9		PP, EP			0	
1617	LEAD ARSENATES	6.1	T5	II	6.1	802	LQ18		PP, EP			2	
1618	LEAD ARSENITES	6.1	T5	II	6.1	802	LQ18		PP, EP			2	
1620	LEAD CYANIDE	6.1	T5	II	6.1	802	LQ18		PP, EP			2	
1621	LONDON PURPLE	6.1	T5	II	6.1	43 802	LQ18		PP, EP			2	
1622	MAGNESIUM ARSENATE	6.1	T5	II	6.1	802	LQ18		PP, EP			2	
1623	MERCURIC ARSENATE	6.1	T5	II	6.1	802	LQ18		PP, EP			2	
1624	MERCURIC CHLORIDE	6.1	T5	II	6.1	802	LQ18		PP, EP			2	
1625	MERCURIC NITRATE	6.1	T5	II	6.1	802	LQ18		PP, EP			2	
1626	MERCURIC POTASSIUM CYANIDE	6.1	T5	I	6.1	802	LQ0		PP, EP			2	
1627	MERCUROUS NITRATE	6.1	T5	II	6.1	802	LQ18		PP, EP			2	
1629	MERCURY ACETATE	6.1	T5	II	6.1	802	LQ18		PP, EP			2	
1630	MERCURY AMMONIUM CHLORIDE	6.1	T5	II	6.1	802	LQ18		PP, EP			2	
1631	MERCURY BENZOATE	6.1	T5	II	6.1	802	LQ18		PP, EP			2	
1634	MERCURY BROMIDES	6.1	T5	II	6.1	802	LQ18		PP, EP			2	
1636	MERCURY CYANIDE	6.1	T5	II	6.1	802	LQ18		PP, EP			2	
1637	MERCURY GLUCONATE	6.1	T5	II	6.1	802	LQ18		PP, EP			2	
1638	MERCURY IODIDE	6.1	T5	II	6.1	802	LQ18		PP, EP			2	
1639	MERCURY NUCLEATE	6.1	T5	II	6.1	802	LQ18		PP, EP			2	
1640	MERCURY OLEATE	6.1	T5	II	6.1	802	LQ18		PP, EP			2	
1641	MERCURY OXIDE	6.1	T5	II	6.1	802	LQ18		PP, EP			2	
1642	MERCURY OXYCYANIDE, DESENSITIZED	6.1	T5	II	6.1	802	LQ18		PP, EP			2	
1643	MERCURY POTASSIUM IODIDE	6.1	T5	II	6.1	802	LQ18		PP, EP			2	
1644	MERCURY SALICYLATE	6.1	T5	II	6.1	802	LQ18		PP, EP			2	

UN No. or ID No.	Name and description	Class	Classification Code	Packing group	Labels	Special provisions	Limited quantities	Carriage permitted	Equipment required	Ventilation	Provisions concerning loading, unloading and carriage	Number of cones, blue lights	Remarks
	3.1.2	2.2	2.2	2.1.1.3	5.2.2	3.3	3.4.6	3.2.1	8.1.5	7.1.6	7.1.6	7.1.5	3.2.1
(1)	(2)	(3a)	(3b)	(4)	(5)	(6)	(7)	(8)	(9)	(10)	(11)	(12)	(13)
1645	MERCURY SULPHATE	6.1	T5	II	6.1	802	LQ18		PP, EP			2	
1646	MERCURY THIOCYANATE	6.1	T5	II	6.1	802	LQ18		PP, EP			2	
1647	METHYL BROMIDE AND ETHYLENE DIBROMIDE MIXTURE, LIQUID	6.1	T1	I	6.1	802	LQ0		PP, EP, TOX, A	VE02		2	
1648	ACETONITRILE	3	F1	II	3		LQ4	T	PP, EX, A	VE01		1	
1649	MOTOR FUEL ANTI-KNOCK MIXTURE having a flash-point above 60 °C	6.1	T3	I	6.1	802	LQ0		PP, EP, TOX, A	VE02		2	
1649	MOTOR FUEL ANTI-KNOCK MIXTURE having a flash-point of not more than 60 °C	6.1	TF1	I	6.1 +3	802	LQ0		PP, EP, EX, TOX, A	VE01, VE02		2	
1650	beta-NAPHTHYLAMINE, SOLID	6.1	T2	II	6.1	802	LQ18		PP, EP			2	
1651	NAPHTHYLTHIOUREA	6.1	T2	II	6.1	43 802	LQ18		PP, EP			2	
1652	NAPHTHYLUREA	6.1	T2	II	6.1	802	LQ18		PP, EP			2	
1653	NICKEL CYANIDE	6.1	T5	II	6.1	802	LQ18		PP, EP			2	
1654	NICOTINE	6.1	T1	II	6.1	802	LQ17		PP, EP, TOX, A	VE02		2	
1655	NICOTINE COMPOUND, SOLID, N.O.S. or NICOTINE PREPARATION, SOLID, N.O.S.	6.1	T2	I	6.1	43 274 802	LQ0		PP, EP			2	
1655	NICOTINE COMPOUND, SOLID, N.O.S. or NICOTINE PREPARATION, SOLID, N.O.S.	6.1	T2	II	6.1	43 274 802	LQ18		PP, EP			2	
1655	NICOTINE COMPOUND, SOLID, N.O.S. or NICOTINE PREPARATION, SOLID, N.O.S.	6.1	T2	III	6.1	43 274 802	LQ9		PP, EP			0	
1656	NICOTINE HYDROCHLORIDE, LIQUID or SOLUTION	6.1	T1	II	6.1	43 802	LQ17		PP, EP, TOX, A	VE02		2	
1656	NICOTINE HYDROCHLORIDE, LIQUID or SOLUTION	6.1	T1	III	6.1	43 802	LQ7		PP, EP, TOX, A	VE02		0	
1657	NICOTINE SALICYLATE	6.1	T2	II	6.1	802	LQ18		PP, EP			2	
1658	NICOTINE SULPHATE, SOLUTION	6.1	T1	II	6.1	802	LQ17		PP, EP, TOX, A	VE02		2	
1658	NICOTINE SULPHATE, SOLUTION	6.1	T1	III	6.1	802	LQ7		PP, EP, TOX, A	VE02		0	

UN No. or ID No.	Name and description	Class	Classification Code	Packing group	Labels	Special provisions	Limited quantities	Carriage permitted	Equipment required	Ventilation	Provisions concerning loading, unloading and carriage	Number of cones, blue lights	Remarks
3.1.2	3.1.2	2.2	2.2	2.1.1.3	5.2.2	3.3	3.4.6	3.2.1	8.1.5	7.1.6	7.1.6	7.1.5	3.2.1
(1)	(2)	(3a)	(3b)	(4)	(5)	(6)	(7)	(8)	(9)	(10)	(11)	(12)	(13)
1659	NICOTINE TARTRATE	6.1	T2	II	6.1	802	LQ18		PP, EP			2	
1660	NITRIC OXIDE, COMPRESSED	2	1TOC		2.3+5.1+8		LQ0		PP, EP, TOX, A	VE02		2	
1661	NITROANILINES (o-, m-, p-)	6.1	T2	II	6.1	279 802	LQ18		PP, EP			2	
1662	NITROBENZENE	6.1	T1	II	6.1	279 802	LQ17	T	PP, EP, TOX, A	VE02		2	
1663	NITROPHENOLS (o-, m-, p-)	6.1	T2	III	6.1	279 802	LQ9	T	PP, EP			0	
1664	NITROTOLUENES, LIQUID	6.1	T1	II	6.1	802	LQ17	T	PP, EP, TOX, A	VE02		2	
1665	NITROXYLENES, LIQUID	6.1	T1	II	6.1	802	LQ17		PP, EP, TOX, A	VE02		2	
1669	PENTACHLOROETHANE	6.1	T1	II	6.1	802	LQ17		PP, EP, TOX, A	VE02		2	
1670	PERCHLOROMETHYL MERCAPTAN	6.1	T1	I	6.1	802	LQ0		PP, EP, TOX, A	VE02		2	
1671	PHENOL, SOLID	6.1	T2	II	6.1	279 802	LQ18		PP, EP			2	
1672	PHENYLCARBYLAMINE CHLORIDE	6.1	T1	I	6.1	802	LQ0		PP, EP, TOX, A	VE02		2	
1673	PHENYLENEDIAMINES (o-, m-, p-)	6.1	T2	III	6.1	279 802	LQ9		PP, EP			0	
1674	PHENYLMERCURIC ACETATE	6.1	T3	II	6.1	43 802	LQ18		PP, EP, TOX, A	VE02		2	
1677	POTASSIUM ARSENATE	6.1	T5	II	6.1	802	LQ18		PP, EP			2	
1678	POTASSIUM ARSENITE	6.1	T5	II	6.1	802	LQ18		PP, EP			2	
1679	POTASSIUM CUPROCYANIDE	6.1	T5	II	6.1	802	LQ18		PP, EP			2	
1680	POTASSIUM CYANIDE, SOLID	6.1	T5	I	6.1	802	LQ0		PP, EP			2	
1683	SILVER ARSENITE	6.1	T5	II	6.1	802	LQ18		PP, EP			2	
1684	SILVER CYANIDE	6.1	T5	II	6.1	802	LQ18		PP, EP			2	
1685	SODIUM ARSENATE	6.1	T5	II	6.1	802	LQ18		PP, EP			2	
1686	SODIUM ARSENITE, AQUEOUS SOLUTION	6.1	T4	II	6.1	43 802	LQ17		PP, EP			2	
1686	SODIUM ARSENITE, AQUEOUS SOLUTION	6.1	T4	III	6.1	43 802	LQ7		PP, EP			0	
1687	SODIUM AZIDE	6.1	T5	II	6.1	802	LQ18		PP, EP			2	

UN No. or ID No.	Name and description	Class	Classification Code	Packing group	Labels	Special provisions	Limited quantities	Carriage permitted	Equipment required	Ventilation	Provisions concerning loading, unloading and carriage	Number of cones, blue lights	Remarks
3.1.2		2.2	2.2	2.1.1.3	5.2.2	3.3	3.4.6	3.2.1	8.1.5	7.1.6	7.1.6	7.1.5	3.2.1
(1)	(2)	(3a)	(3b)	(4)	(5)	(6)	(7)	(8)	(9)	(10)	(11)	(12)	(13)
1688	SODIUM CACODYLATE	6.1	T5	II	6.1	802	LQ18		PP, EP			2	
1689	SODIUM CYANIDE, SOLID	6.1	T5	I	6.1	802	LQ0		PP, EP			2	
1690	SODIUM FLUORIDE, SOLID	6.1	T5	III	6.1	802	LQ9	B	PP, EP			0	
1691	STRONTIUM ARSENITE	6.1	T5	II	6.1	802	LQ18		PP, EP			2	
1692	STRYCHNINE or STRYCHNINE SALTS	6.1	T2	I	6.1	802	LQ0		PP, EP			2	
1693	TEAR GAS SUBSTANCE, LIQUID, N.O.S.	6.1	T1	I	6.1	274 802	LQ0		PP, EP, TOX, A	VE02		2	
1693	TEAR GAS SUBSTANCE, LIQUID, N.O.S.	6.1	T1	II	6.1	274 802	LQ17		PP, EP, TOX, A	VE02		2	
1694	BROMOBENZYL CYANIDES, LIQUID	6.1	T1	I	6.1	138 802	LQ0		PP, EP, TOX, A	VE02		2	
1695	CHLOROACETONE, STABILIZED	6.1	TFC	I	6.1+3+8	802	LQ0		PP, EP, EX, TOX, A	VE01, VE02		2	
1697	CHLOROACETOPHENONE, SOLID	6.1	T2	II	6.1	802	LQ18		PP, EP, TOX, A	VE02		2	
1698	DIPHENYLAMINE CHLOROARSINE	6.1	T3	I	6.1	802	LQ0		PP, EP, TOX, A	VE02		2	
1699	DIPHENYLCHLOROARSINE, LIQUID	6.1	T3	I	6.1	802	LQ0		PP, EP, TOX, A	VE02		2	
1700	TEAR GAS CANDLES	6.1	TF3	II	6.1+4.1	802	LQ18		PP, EP			2	
1701	XYLYL BROMIDE, LIQUID	6.1	T1	II	6.1	802	LQ17		PP, EP, TOX, A	VE02		2	
1702	1.1.2.2-TETRACHLOROETHANE	6.1	T1	II	6.1	802	LQ17		PP, EP, TOX, A	VE02		2	
1704	TETRAETHYL DITHIOPYROPHOSPHATE	6.1	T2	II	6.1	43 802	LQ18		PP, EP			2	
1707	THALLIUM COMPOUND, N.O.S.	6.1	T5	II	6.1	43 274 802	LQ18		PP, EP			2	
1708	TOLUIDINES, LIQUID	6.1	T1	II	6.1	279 802	LQ17	T	PP, EP, TOX, A	VE02		2	
1709	2.4-TOLUYLENEDIAMINE, SOLID	6.1	T2	III	6.1	802	LQ9		PP, EP			0	
1710	TRICHLOROETHYLENE	6.1	T1	III	6.1	802	LQ7	T	PP, EP, TOX, A	VE02		0	

UN No. or ID No.	Name and description	Class	Classification Code	Packing group	Labels	Special provisions	Limited quantities	Carriage permitted	Equipment required	Ventilation	Provisions concerning loading, unloading and carriage	Number of cones, blue lights	Remarks
3.1.2	3.1.2	2.2	2.2	2.1.1.3	5.2.2	3.3	3.4.6	3.2.1	8.1.5	7.1.6	7.1.6	7.1.5	3.2.1
(1)	(2)	(3a)	(3b)	(4)	(5)	(6)	(7)	(8)	(9)	(10)	(11)	(12)	(13)
1711	XYLIDINES, LIQUID	6.1	T1	II	6.1	802	LQ17		PP, EP, TOX, A	VE02		2	
1712	ZINC ARSENATE, ZINC ARSENATE or ZINC ARSENITE AND ZINC ARSENITE MIXTURE	6.1	T5	II	6.1	802	LQ18		PP, EP			2	
1713	ZINC CYANIDE	6.1	T5	I	6.1	802	LQ0		PP, EP			2	
1714	ZINC PHOSPHIDE	4.3	WT2	I	4.3+6.1	802	LQ0		PP, EP, TOX, A	VE01, VE02	HA08	2	
1715	ACETIC ANHYDRIDE	8	CF1	II	8+3		LQ22	T	PP, EP, EX, A	VE01		1	
1716	ACETYL BROMIDE	8	C3	II	8		LQ22		PP, EP			0	
1717	ACETYL CHLORIDE	3	FC	II	3+8		LQ4	T	PP, EP, EX, A	VE01		1	
1718	BUTYL ACID PHOSPHATE	8	C3	III	8		LQ7	T	PP, EP			0	
1719	CAUSTIC ALKALI LIQUID, N.O.S.	8	C5	II	8	274	LQ22	T	PP, EP			0	
1719	CAUSTIC ALKALI LIQUID, N.O.S.	8	C5	III	8	274	LQ7	T	PP, EP			0	
1722	ALLYL CHLOROFORMATE	6.1	TFC	I	6.1+3+8	802	LQ0		PP, EP, EX, TOX, A	VE01, VE02		2	
1723	ALLYL IODIDE	3	FC	II	3+8		LQ4		PP, EP, EX, A	VE01		1	
1724	ALLYLTRICHLOROSILANE, STABILIZED	8	CF1	II	8+3		LQ22		PP, EP, EX, A	VE01		1	
1725	ALUMINIUM BROMIDE, ANHYDROUS	8	C2	II	8	588	LQ23		PP, EP			0	
1726	ALUMINIUM CHLORIDE, ANHYDROUS	8	C2	II	8	588	LQ23		PP, EP			0	
1727	AMMONIUM HYDROGENDIFLUORIDE, SOLID	8	C2	II	8		LQ23		PP, EP			0	
1728	AMYLTRICHLOROSILANE	8	C3	II	8		LQ22		PP, EP			0	
1729	ANISOYL CHLORIDE	8	C4	II	8		LQ23		PP, EP			0	
1730	ANTIMONY PENTACHLORIDE, LIQUID	8	C1	II	8		LQ22		PP, EP			0	

UN No. or ID No.	Name and description	Class	Classi-fi-cation Code	Packing group	Labels	Special provisions	Limited quantities	Carriage permitted	Equipment required	Ventilation	Provisions concerning loading, unloading and carriage	Number of cones, blue lights	Remarks
	3.1.2	2.2	2.2	2.1.1.3	5.2.2	3.3	3.4.6	3.2.1	8.1.5	7.1.6	7.1.6	7.1.5	3.2.1
(1)	(2)	(3a)	(3b)	(4)	(5)	(6)	(7)	(8)	(9)	(10)	(11)	(12)	(13)
1731	ANTIMONY PENTACHLORIDE SOLUTION	8	C1	II	8		LQ22		PP, EP			0	
1731	ANTIMONY PENTACHLORIDE SOLUTION	8	C1	III	8		LQ7		PP, EP			0	
1732	ANTIMONY PENTAFLUORIDE	8	CT1	II	8+6.1	802	LQ22		PP, EP, TOX, A	VE02		2	
1733	ANTIMONY TRICHLORIDE	8	C2	II	8		LQ23		PP, EP			0	
1736	BENZOYL CHLORIDE	8	C3	II	8		LQ22		PP, EP			0	
1737	BENZYL BROMIDE	6.1	TC1	II	6.1+8	802	LQ17		PP, EP, TOX, A	VE02		2	
1738	BENZYL CHLORIDE	6.1	TC1	II	6.1+8	802	LQ17	T	PP, EP, TOX, A	VE02		2	
1739	BENZYL CHLOROFORMATE	8	C9	I	8		LQ0		PP, EP			0	
1740	HYDROGENDIFLUORIDES, SOLID. N.O.S.	8	C2	II	8	274 517	LQ23		PP, EP			0	
1740	HYDROGENDIFLUORIDES, SOLID. N.O.S.	8	C2	III	8	274 517	LQ24		PP, EP			0	
1741	BORON TRICHLORIDE	2	2TC		2.3+8		LQ0		PP, EP, TOX, A	VE02		2	
1742	BORON TRIFLUORIDE ACETIC ACID COMPLEX, LIQUID	8	C3	II	8		LQ22	T	PP, EP			0	
1743	BORON TRIFLUORIDE PROPIONIC ACID COMPLEX, LIQUID	8	C3	II	8		LQ22		PP, EP			0	
1744	BROMINE or BROMINE SOLUTION	8	CT1	I	8+6.1	802	LQ0		PP, EP, TOX, A	VE02		2	
1745	BROMINE PENTAFLUORIDE	5.1	OTC	I	5.1+6.1+8	802	LQ0		PP, EP, TOX, A	VE02		2	
1746	BROMINE TRIFLUORIDE	5.1	OTC	I	5.1+6.1+8	802	LQ0		PP, EP, TOX, A	VE02		2	
1747	BUTYLTRICHLOROSILANE	8	CF1	II	8+3		LQ22		PP, EP, EX, A	VE01		1	
1748	CALCIUM HYPOCHLORITE, DRY or CALCIUM HYPOCHLORITE MIXTURE, DRY with more than 39% available chlorine (8.8% available oxygen)	5.1	O2	II	5.1	313 314 589	LQ11		PP			0	

UN No. or ID No.	Name and description	Class	Classification Code	Packing group	Labels	Special provisions	Limited quantities	Carriage permitted	Equipment required	Ventilation	Provisions concerning loading, unloading and carriage	Number of cones, blue lights	Remarks
(1)	(2)	(3a)	(3b)	(4)	(5)	(6)	(7)	(8)	(9)	(10)	(11)	(12)	(13)
		2.2	2.2	2.1.1.3	5.2.2	3.3	3.4.6	3.2.1	8.1.5	7.1.6	7.1.6	7.1.5	3.2.1
1748	CALCIUM HYPOCHLORITE, DRY or CALCIUM HYPOCHLORITE MIXTURE, DRY with more than 39% available chlorine (8.8% available oxygen)	5.1	O2	III	5.1	316 589	LQ12		PP			0	
1749	CHLORINE TRIFLUORIDE	2	2TOC		2.3+5.1+8		LQ0		PP, EP, TOX, A	VE02		2	
1750	CHLOROACETIC ACID SOLUTION	6.1	TC1	II	6.1+8	802	LQ17	T	PP, EP, TOX, A	VE02		2	
1751	CHLOROACETIC ACID, SOLID	6.1	TC2	II	6.1+8	802	LQ18		PP, EP			2	
1752	CHLOROACETYL CHLORIDE	6.1	TC1	I	6.1+8	802	LQ0		PP, EP, TOX, A	VE02		2	
1753	CHLOROPHENYL-TRICHLOROSILANE	8	C3	II	8		LQ22		PP, EP			0	
1754	CHLOROSULPHONIC ACID (with or without sulphur trioxide)	8	C1	I	8		LQ0		PP, EP			0	
1755	CHROMIC ACID SOLUTION	8	C1	II	8	518	LQ22		PP, EP			0	
1755	CHROMIC ACID SOLUTION	8	C1	III	8	518	LQ7		PP, EP			0	
1756	CHROMIC FLUORIDE, SOLID	8	C2	II	8		LQ23		PP, EP			0	
1757	CHROMIC FLUORIDE SOLUTION	8	C1	II	8		LQ22		PP, EP			0	
1757	CHROMIC FLUORIDE SOLUTION	8	C1	III	8		LQ7		PP, EP			0	
1758	CHROMIUM OXYCHLORIDE	8	C1	I	8		LQ0		PP, EP			0	
1759	CORROSIVE SOLID, N.O.S.	8	C10	I	8	274	LQ0		PP, EP			0	
1759	CORROSIVE SOLID, N.O.S.	8	C10	II	8	274	LQ23		PP, EP			0	
1759	CORROSIVE SOLID, N.O.S.	8	C10	III	8	274	LQ24		PP, EP			0	
1760	CORROSIVE LIQUID, N.O.S.	8	C9	I	8	274	LQ0	T	PP, EP			0	
1760	CORROSIVE LIQUID, N.O.S.	8	C9	II	8	274	LQ22	T	PP, EP			0	
1760	CORROSIVE LIQUID, N.O.S.	8	C9	III	8	274	LQ7	T	PP, EP			0	
1761	CUPRIETHYLENEDIAMINE SOLUTION	8	CT1	II	8+6.1	802	LQ22		PP, EP, A			2	
1761	CUPRIETHYLENEDIAMINE SOLUTION	8	CT1	III	8+6.1	802	LQ7		PP, EP, A			0	
1762	CYCLOHEXENYLTRICHLORO-SILANE	8	C3	II	8		LQ22		PP, EP			0	

UN No. or ID No. (1) 3.1.2	Name and description (2) 3.1.2	Class (3a) 2.2	Classification Code (3b) 2.2	Packing group (4) 2.1.1.3	Labels (5) 5.2.2	Special provisions (6) 3.3	Limited quantities (7) 3.4.6	Carriage permitted (8) 3.2.1	Equipment required (9) 8.1.5	Ventilation (10) 7.1.6	Provisions concerning loading, unloading and carriage (11) 7.1.6	Number of cones, blue lights (12) 7.1.5	Remarks (13) 3.2.1
1763	CYCLOHEXYLTRICHLORO-SILANE	8	C3	II	8		LQ22		PP, EP			0	
1764	DICHLOROACETIC ACID	8	C3	II	8		LQ22	T	PP, EP			0	
1765	DICHLOROACETYL CHLORIDE	8	C3	II	8		LQ22		PP, EP			0	
1766	DICHLOROPHENYL-TRICHLOROSILANE	8	C3	II	8		LQ22		PP, EP			0	
1767	DIETHYLDICHLOROSILANE	8	CF1	II	8+3		LQ22		PP, EP, EX, A	VE01		1	
1768	DIFLUOROPHOSPHORIC ACID, ANHYDROUS	8	C1	II	8		LQ22		PP, EP			0	
1769	DIPHENYLDICHLORO-SILANE	8	C3	II	8		LQ22		PP, EP			0	
1770	DIPHENYLMETHYL BROMIDE	8	C10	II	8		LQ23		PP, EP			0	
1771	DODECYLTRICHLORO-SILANE	8	C3	II	8		LQ22		PP, EP			0	
1773	FERRIC CHLORIDE, ANHYDROUS	8	C2	III	8	590	LQ24		PP, EP			0	
1774	FIRE EXTINGUISHER CHARGES, corrosive liquid	8	C11	II	8		LQ22		PP, EP			0	
1775	FLUOROBORIC ACID	8	C1	II	8		LQ22		PP, EP			0	
1776	FLUOROPHOSPHORIC ACID, ANHYDROUS	8	C1	II	8		LQ22		PP, EP			0	
1777	FLUOROSULPHONIC ACID	8	C1	I	8		LQ0		PP, EP			0	
1778	FLUOROSILICIC ACID	8	C1	II	8		LQ22	T	PP, EP			0	
1779	FORMIC ACID with more than 85% acid by mass	8	CF1	II	8+3		LQ22	T	PP, EP, EX, A			0	
1780	FUMARYL CHLORIDE	8	C3	II	8		LQ22	T	PP, EP			0	
1781	HEXADECYLTRICHLORO-SILANE	8	C3	II	8		LQ22		PP, EP			0	
1782	HEXAFLUORO-PHOSPHORIC ACID	8	C1	II	8		LQ22		PP, EP			0	
1783	HEXAMETHYLENE-DIAMINE SOLUTION	8	C7	II	8		LQ22	T	PP, EP			0	
1783	HEXAMETHYLENE-DIAMINE SOLUTION	8	C7	III	8		LQ7	T	PP, EP			0	
1784	HEXYLTRICHLOROSILANE	8	C3	II	8		LQ22		PP, EP			0	
1786	HYDROFLUORIC ACID AND SULPHURIC ACID MIXTURE	8	CT1	I	8+6.1	802	LQ0		PP, EP, TOX, A	VE02		2	
1787	HYDRIODIC ACID	8	C1	II	8		LQ22		PP, EP			0	

UN No. or ID No. (1)	Name and description 3.1.2 (2)	Class 2.2 (3a)	Classification Code 2.2 (3b)	Packing group 2.1.1.3 (4)	Labels 5.2.2 (5)	Special provisions 3.3 (6)	Limited quantities 3.4.6 (7)	Carriage permitted 3.2.1 (8)	Equipment required 8.1.5 (9)	Ventilation 7.1.6 (10)	Provisions concerning loading, unloading and carriage 7.1.6 (11)	Number of cones, blue lights 7.1.5 (12)	Remarks 3.2.1 (13)
1787	HYDRIODIC ACID	8	C1	III	8		LQ7		PP, EP			0	
1788	HYDROBROMIC ACID	8	C1	II	8	519	LQ22		PP, EP			0	
1788	HYDROBROMIC ACID	8	C1	III	8	519	LQ7		PP, EP			0	
1789	HYDROCHLORIC ACID	8	C1	II	8	520	LQ22	T	PP, EP			0	
1789	HYDROCHLORIC ACID	8	C1	III	8	520	LQ7	T	PP, EP			0	
1790	HYDROFLUORIC ACID with more than 85% hydrofluoric acid	8	CT1	I	8+6.1	640I 802	LQ0		PP, EP, TOX, A	VE02		2	
1790	HYDROFLUORIC ACID with more than 60% but not more than 85% hydrofluoric acid	8	CT1	I	8+6.1	640J 802	LQ0		PP, EP, TOX, A	VE02		2	
1790	HYDROFLUORIC ACID with not more than 60% hydrofluoric acid	8	CT1	II	8+6.1	802	LQ22		PP, EP, TOX, A	VE02		2	
1791	HYPOCHLORITE SOLUTION	8	C9	II	8	521	LQ22		PP, EP			0	
1791	HYPOCHLORITE SOLUTION	8	C9	III	8	521	LQ7		PP, EP			0	
1792	IODINE MONOCHLORIDE	8	C1	II	8		LQ22		PP, EP			0	
1793	ISOPROPYL ACID PHOSPHATE	8	C3	III	8		LQ7		PP, EP			0	
1794	LEAD SULPHATE with more than 3% free acid	8	C2	II	8	591	LQ23		PP, EP			0	
1796	NITRATING ACID MIXTURE with more than 50% nitric acid	8	CO1	I	8+5.1		LQ0		PP, EP			0	
1796	NITRATING ACID MIXTURE with not more than 50% nitric acid	8	C1	II	8		LQ22		PP, EP			0	
1798	NITROHYDROCHLORIC ACID	8	COT					CARRIAGE PROHIBITED					
1799	NONYLTRICHLOROSILANE	8	C3	II	8		LQ22		PP, EP			0	
1800	OCTADECYLTRICHLORO-SILANE	8	C3	II	8		LQ22		PP, EP			0	
1801	OCTYLTRICHLOROSILANE	8	C3	II	8		LQ22		PP, EP			0	
1802	PERCHLORIC ACID with not more than 50% acid, by mass	8	CO1	II	8+5.1	522	LQ22	T	PP, EP			0	
1803	PHENOLSULPHONIC ACID, LIQUID	8	C3	II	8		LQ22		PP, EP			0	
1804	PHENYLTRICHLORO-SILANE	8	C3	II	8		LQ22		PP, EP			0	
1805	PHOSPHORIC ACID, SOLUTION	8	C1	III	8		LQ7		PP, EP			0	
1806	PHOSPHORUS PENTACHLORIDE	8	C2	II	8		LQ23		PP, EP			0	
1807	PHOSPHORUS PENTOXIDE	8	C2	II	8		LQ23		PP, EP			0	
1808	PHOSPHORUS TRIBROMIDE	8	C1	II	8		LQ22		PP, EP			0	

UN No. or ID No. (1)	Name and description 3.1.2 (2)	Class 2.2 (3a)	Classification Code 2.2 (3b)	Packing group 2.1.1.3 (4)	Labels 5.2.2 (5)	Special provisions 3.3 (6)	Limited quantities 3.4.6 (7)	Carriage permitted 3.2.1 (8)	Equipment required 8.1.5 (9)	Ventilation 7.1.6 (10)	Provisions concerning loading, unloading and carriage 7.1.6 (11)	Number of cones, blue lights 7.1.5 (12)	Remarks 3.2.1 (13)
1809	PHOSPHORUS TRICHLORIDE	6.1	TC3	I	6.1+8	802	LQ0		PP, EP, TOX, A	VE02		2	
1810	PHOSPHORUS OXYCHLORIDE	8	C1	II	8		LQ22		PP, EP			0	
1811	POTASSIUM HYDROGENDIFLUORIDE, SOLID	8	CT2	II	8+6.1	802	LQ23		PP, EP			2	
1812	POTASSIUM FLUORIDE, SOLID	6.1	T5	III	6.1	802	LQ9	B	PP, EP			0	
1813	POTASSIUM HYDROXIDE, SOLID	8	C6	II	8		LQ23		PP, EP			0	
1814	POTASSIUM HYDROXIDE SOLUTION	8	C5	II	8		LQ22	T	PP, EP			0	
1814	POTASSIUM HYDROXIDE SOLUTION	8	C5	III	8		LQ7	T	PP, EP			0	
1815	PROPIONYL CHLORIDE	3	FC	II	3+8		LQ4		PP, EP, EX, A	VE01		1	
1816	PROPYLTRICHLOROSILANE	8	CF1	II	8+3		LQ22		PP, EP, EX, A	VE01		1	
1817	PYROSULPHURYL CHLORIDE	8	C1	II	8		LQ22		PP, EP			0	
1818	SILICON TETRACHLORIDE	8	C1	II	8		LQ22		PP, EP			0	
1819	SODIUM ALUMINATE SOLUTION	8	C5	II	8		LQ22		PP, EP			0	
1819	SODIUM ALUMINATE SOLUTION	8	C5	III	8		LQ7		PP, EP			0	
1823	SODIUM HYDROXIDE, SOLID	8	C6	II	8		LQ23	T	PP, EP			0	
1824	SODIUM HYDROXIDE SOLUTION	8	C5	II	8		LQ22	T	PP, EP			0	
1824	SODIUM HYDROXIDE SOLUTION	8	C5	III	8		LQ7	T	PP, EP			0	
1825	SODIUM MONOXIDE	8	C6	II	8		LQ23		PP, EP			0	
1826	NITRATING ACID MIXTURE, SPENT, with more than 50% nitric acid	8	CO1	I	8+5.1	113	LQ0		PP, EP			0	
1826	NITRATING ACID MIXTURE, SPENT, with not more than 50% nitric acid	8	C1	II	8	113	LQ22		PP, EP			0	
1827	STANNIC CHLORIDE, ANHYDROUS	8	C1	II	8		LQ22		PP, EP			0	

(1)	(2)	(3a)	(3b)	(4)	(5)	(6)	(7)	(8)	(9)	(10)	(11)	(12)	(13)
UN No. or ID No.	Name and description	Class	Classification Code	Packing group	Labels	Special provisions	Limited quantities	Carriage permitted	Equipment required	Ventilation	Provisions concerning loading, unloading and carriage	Number of cones, blue lights	Remarks
3.1.2	3.1.2	2.2	2.2	2.1.1.3	5.2.2	3.3	3.4.6	3.2.1	8.1.5	7.1.6	7.1.6	7.1.5	3.2.1
1828	SULPHUR CHLORIDES	8	C1	I	8		LQ0		PP, EP			0	
1829	SULPHUR TRIOXIDE, STABILIZED	8	C1	I	8	623	LQ0		PP, EP			0	
1830	SULPHURIC ACID with more than 51% acid	8	C1	II	8		LQ22	T	PP, EP			0	
1831	SULPHURIC ACID, FUMING	8	CT1	I	8+6.1	802	LQ0	T	PP, EP, TOX, A	VE02		2	
1832	SULPHURIC ACID, SPENT	8	C1	II	8	113	LQ22	T	PP, EP			0	
1833	SULPHUROUS ACID	8	C1	II	8		LQ22		PP, EP			0	
1834	SULPHURYL CHLORIDE	8	C1	I	8		LQ0		PP, EP			0	
1835	TETRAMETHYL-AMMONIUM HYDROXIDE. SOLUTION	8	C7	II	8		LQ22		PP, EP			0	
1835	TETRAMETHYLAMMONIUM HYDROXIDE SOLUTION	8	C7	III	8		LQ7		PP, EP			0	
1836	THIONYL CHLORIDE	8	C1	I	8		LQ0		PP, EP			0	
1837	THIOPHOSPHORYL CHLORIDE	8	C1	II	8		LQ22		PP, EP			0	
1838	TITANIUM TETRACHLORIDE	8	C1	II	8		LQ22		PP, EP			0	
1839	TRICHLOROACETIC ACID	8	C4	II	8		LQ23		PP, EP			0	
1840	ZINC CHLORIDE SOLUTION	8	C1	III	8		LQ7		PP, EP			0	
1841	ACETALDEHYDE AMMONIA	9	M11	III	9		LQ27		PP			0	
1843	AMMONIUM DINITRO-o-CRESOLATE. SOLID	6.1	T2	II	6.1	802	LQ18		PP, EP			2	
1845	Carbon dioxide, solid (Dry ice)	9	M11				NOT SUBJECT TO ADN						
1846	CARBON TETRACHLORIDE	6.1	T1	II	6.1	802	LQ17	T	PP, EP, TOX, A	VE02		2	
1847	POTASSIUM SULPHIDE, HYDRATED with not less than 30% water of crystallization	8	C6	II	8	523	LQ23		PP, EP			0	
1848	PROPIONIC ACID with not less than 10% and less than 90% acid by mass	8	C3	III	8		LQ7	T	PP, EP			0	
1849	SODIUM SULPHIDE, HYDRATED with not less than 30% water	8	C6	II	8	523	LQ23		PP, EP			0	

UN No. or ID No.	Name and description	Class	Classification Code	Packing group	Labels	Special provisions	Limited quantities	Carriage permitted	Equipment required	Ventilation	Provisions concerning loading, unloading and carriage	Number of cones, blue lights	Remarks
3.1.1	3.1.2	2.2	2.2	2.1.1.3	5.2.2	3.3	3.4.6	3.2.1	8.1.5	7.1.6	7.1.6	7.1.5	3.2.1
(1)	(2)	(3a)	(3b)	(4)	(5)	(6)	(7)	(8)	(9)	(10)	(11)	(12)	(13)
1851	MEDICINE, LIQUID, TOXIC, N.O.S.	6.1	T1	II	6.1	221 274 601 802	LQ17		PP, EP, TOX, A	VE02		2	
1851	MEDICINE, LIQUID, TOXIC, N.O.S.	6.1	T1	III	6.1	221 274 601 802	LQ7		PP, EP, TOX, A	VE02		0	
1854	BARIUM ALLOYS, PYROPHORIC	4.2	S4	I	4.2		LQ0		PP			0	
1855	CALCIUM, PYROPHORIC or CALCIUM ALLOYS, PYROPHORIC	4.2	S4	I	4.2		LQ0		PP			0	
1856	Rags, oily	4.2	S2	NOT SUBJECT TO ADN									
1857	Textile waste, wet	4.2	S2	NOT SUBJECT TO ADN									
1858	HEXAFLUOROPROPYLENE (REFRIGERANT GAS R 1216)	2	2A		2.2		LQ1		PP			0	
1859	SILICON TETRAFLUORIDE	2	2TC		2.3+8		LQ0		PP, EP, TOX, A	VE02		2	
1860	VINYL FLUORIDE, STABILIZED	2	2F		2.1		LQ0		PP, EX, A	VE01		1	
1862	ETHYL CROTONATE	3	F1	II	3		LQ4		PP, EX, A	VE01		1	
1863	FUEL, AVIATION, TURBINE ENGINE	3	F1	I	3		LQ3	T	PP, EX, A	VE01		1	
1863	FUEL, AVIATION, TURBINE ENGINE (vapour pressure at 50 °C more than 110 kPa)	3	F1	II	3	640C	LQ4	T	PP, EX, A	VE01		1	
1863	FUEL, AVIATION, TURBINE ENGINE (vapour pressure at 50 °C not more than 110 kPa)	3	F1	II	3	640D	LQ4	T	PP, EX, A	VE01		1	
1863	FUEL, AVIATION, TURBINE ENGINE	3	F1	III	3		LQ7	T	PP, EX, A	VE01		0	
1865	n-PROPYL NITRATE	3	F1	II	3		LQ4		PP, EX, A	VE01		1	
1866	RESIN SOLUTION, flammable	3	F1	I	3		LQ3		PP, EX, A	VE01		1	
1866	RESIN SOLUTION, flammable (vapour pressure at 50 °C more than 110 kPa)	3	F1	II	3	640C	LQ6		PP, EX, A	VE01		1	

- 292 -

(1) UN No. or ID No.	(2) Name and description	(3a) Class	(3b) Classification Code	(4) Packing group	(5) Labels	(6) Special provisions	(7) Limited quantities	(8) Carriage permitted	(9) Equipment required	(10) Ventilation	(11) Provisions concerning loading, unloading and carriage	(12) Number of cones, blue lights	(13) Remarks
		2.2	2.2	2.1.1.3	5.2.2	3.3	3.4.6	3.2.1	8.1.5	7.1.6	7.1.6	7.1.5	3.2.1
1866	RESIN SOLUTION, flammable (vapour pressure at 50 °C not more than 110 kPa)	3	F1	II	3	640D	LQ6		PP, EX, A	VE01		1	
1866	RESIN SOLUTION, flammable	3	F1	III	3	640E	LQ7		PP, EX, A	VE01		0	
1866	RESIN SOLUTION, flammable (having a flash-point below 23 °C and viscous according to 2.2.3.1.4) (boiling point not more than 35° C)	3	F1	III	3	640F	LQ7		PP, EX, A	VE01		0	
1866	RESIN SOLUTION, flammable (having a flash-point below 23 °C and viscous according to 2.2.3.1.4) (vapour pressure at 50 °C more than 110 kPa , boiling point of more than 35° C)	3	F1	III	3	640G	LQ7		PP, EX, A	VE01		0	
1866	RESIN SOLUTION, flammable (having a flash-point below 23 °C and viscous according to 2.2.3.1.4) (vapour pressure at 50 °C not more than 110 kPa)	3	F1	III	3	640H	LQ7		PP, EX, A	VE01		0	
1868	DECABORANE	4.1	FT2	II	4.1+6.1	802	LQ0		PP			2	
1869	MAGNESIUM or MAGNESIUM ALLOYS with more than 50% magnesium in pellets, turnings or ribbons	4.1	F3	III	4.1	59	LQ9		PP			0	
1870	POTASSIUM BOROHYDRIDE	4.3	W2	I	4.3		LQ0		PP, EX, A	VE01	HA08	0	
1871	TITANIUM HYDRIDE	4.1	F3	II	4.1		LQ8		PP			1	
1872	LEAD DIOXIDE	5.1	OT2	III	5.1+6.1	802	LQ12		PP			0	
1873	PERCHLORIC ACID with more than 50% but not more than 72% acid, by mass	5.1	OC1	I	5.1+8	60	LQ0		PP, EP			0	
1884	BARIUM OXIDE	6.1	T5	III	6.1	802	LQ9		PP, EP			0	
1885	BENZIDINE	6.1	T2	II	6.1	802	LQ18		PP, EP			2	
1886	BENZYLIDENE CHLORIDE	6.1	T1	II	6.1	802	LQ17		PP, EP, TOX, A	VE02		2	

UN No. or ID No.	Name and description	Class	Classification Code	Packing group	Labels	Special provisions	Limited quantities	Carriage permitted	Equipment required	Ventilation	Provisions concerning loading, unloading and carriage	Number of cones, blue lights	Remarks
	3.1.2	2.2	2.2	2.1.1.3	5.2.2	3.3	3.4.6	3.2.1	8.1.5	7.1.6	7.1.6	7.1.5	3.2.1
(1)	(2)	(3a)	(3b)	(4)	(5)	(6)	(7)	(8)	(9)	(10)	(11)	(12)	(13)
1887	BROMOCHLOROMETHANE	6.1	T1	III	6.1	802	LQ7		PP, EP, TOX, A	VE02		0	
1888	CHLOROFORM	6.1	T1	III	6.1	802	LQ7	T	PP, EP, TOX, A	VE02		0	
1889	CYANOGEN BROMIDE	6.1	TC2	I	6.1+8	802	LQ0		PP, EP			2	
1891	ETHYL BROMIDE	6.1	T1	II	6.1	802	LQ17		PP, EP, TOX, A	VE02		2	
1892	ETHYLDICHLOROARSINE	6.1	T3	I	6.1	802	LQ0		PP, EP, TOX, A	VE02		2	
1894	PHENYLMERCURIC HYDROXIDE	6.1	T3	II	6.1	802	LQ18		PP, EP, TOX, A	VE02		2	
1895	PHENYLMERCURIC NITRATE	6.1	T3	II	6.1	802	LQ18		PP, EP, TOX, A	VE02		2	
1897	TETRACHLOROETHYLENE	6.1	T1	III	6.1	802	LQ7	T	PP, EP, TOX, A	VE02		0	
1898	ACETYL IODIDE	8	C3	II	8		LQ22		PP, EP			0	
1902	DIISOOCTYL ACID PHOSPHATE	8	C3	III	8		LQ7		PP, EP			0	
1903	DISINFECTANT, LIQUID, CORROSIVE, N.O.S.	8	C9	I	8	274	LQ0		PP, EP			0	
1903	DISINFECTANT, LIQUID, CORROSIVE, N.O.S.	8	C9	II	8	274	LQ22		PP, EP			0	
1903	DISINFECTANT, LIQUID, CORROSIVE, N.O.S.	8	C9	III	8	274	LQ7		PP, EP			0	
1905	SELENIC ACID	8	C2	I	8		LQ0		PP, EP			0	
1906	SLUDGE ACID	8	C1	II	8		LQ22		PP, EP			0	
1907	SODA LIME with more than 4% sodium hydroxide	8	C6	III	8	62	LQ24		PP, EP			0	
1908	CHLORITE SOLUTION	8	C9	II	8	521	LQ22		PP, EP			0	
1908	CHLORITE SOLUTION	8	C9	III	8	521	LQ7		PP, EP			0	
1910	Calcium oxide	8	C6					NOT SUBJECT TO ADN					
1911	DIBORANE	2	2TF		2.3+2.1		LQ0		PP, EP, EX, TOX, A	VE01, VE02		2	
1912	METHYL CHLORIDE AND METHYLENE CHLORIDE MIXTURE	2	2F		2.1	228	LQ0	T	PP, EX, A	VE01		1	
1913	NEON. REFRIGERATED LIQUID	2	3A		2.2	593	LQ1		PP			0	

UN No. or ID No.	Name and description	Class	Classification Code	Packing group	Labels	Special provisions	Limited quantities	Carriage permitted	Equipment required	Ventilation	Provisions concerning loading, unloading and carriage	Number of cones, blue lights	Remarks
3.1.2 (1)	3.1.2 (2)	2.2 (3a)	2.2 (3b)	2.1.1.3 (4)	5.2.2 (5)	3.3 (6)	3.4.6 (7)	3.2.1 (8)	8.1.5 (9)	7.1.6 (10)	7.1.6 (11)	7.1.5 (12)	3.2.1 (13)
1914	BUTYL PROPIONATES	3	F1	III	3		LQ7		PP, EX, A	VE01		0	
1915	CYCLOHEXANONE	3	F1	III	3		LQ7	T	PP, EX, A	VE01		0	
1916	2.2'-DICHLORODIETHYL ETHER	6.1	TF1	II	6.1+3	802	LQ17		PP, EP, EX, TOX, A	VE01, VE02		2	
1917	ETHYL ACRYLATE, STABILIZED	3	F1	II	3		LQ4	T	PP, EX, A	VE01		1	
1918	ISOPROPYLBENZENE	3	F1	III	3		LQ7	T	PP, EX, A	VE01		0	
1919	METHYL ACRYLATE, STABILIZED	3	F1	II	3		LQ4	T	PP, EX, A	VE01		1	
1920	NONANES	3	F1	III	3		LQ7	T	PP, EX, A	VE01		0	
1921	PROPYLENEIMINE, STABILIZED	3	FT1	I	3+6.1	802	LQ0		PP, EP, EX, TOX, A	VE01, VE02		2	
1922	PYRROLIDINE	3	FC	II	3+8		LQ4	T	PP, EP, EX, A	VE01		1	
1923	CALCIUM DITHIONITE (CALCIUM HYDROSULPHITE)	4.2	S4	II	4.2		LQ0		PP			0	
1928	METHYL MAGNESIUM BROMIDE IN ETHYL ETHER	4.3	WF1	I	4.3+3		LQ0		PP, EX, A	VE01	HA08	1	
1929	POTASSIUM DITHIONITE (POTASSIUM HYDROSULPHITE)	4.2	S4	II	4.2		LQ0		PP			0	
1931	ZINC DITHIONITE (ZINC HYDROSULPHITE)	9	M11	III	9		LQ27		PP			0	
1932	ZIRCONIUM SCRAP	4.2	S4	III	4.2	524 592	LQ0		PP			0	
1935	CYANIDE SOLUTION, N.O.S.	6.1	T4	I	6.1	274 525 802	LQ0		PP, EP, TOX, A	VE02		2	
1935	CYANIDE SOLUTION, N.O.S.	6.1	T4	II	6.1	274 525 802	LQ17		PP, EP, TOX, A	VE02		2	
1935	CYANIDE SOLUTION, N.O.S.	6.1	T4	III	6.1	274 525 802	LQ7		PP, EP, TOX, A	VE02		0	
1938	BROMOACETIC ACID, SOLUTION	8	C3	II	8		LQ22		PP, EP			0	

UN No. or ID No.	Name and description	Class	Classification Code	Packing group	Labels	Special provisions	Limited quantities	Carriage permitted	Equipment required	Ventilation	Provisions concerning loading, unloading and carriage	Number of cones, blue lights	Remarks
3.1.2	3.1.2	2.2	2.2	2.1.1.3	5.2.2	3.3	3.4.6	3.2.1	8.1.5	7.1.6	7.1.6	7.1.5	3.2.1
(1)	(2)	(3a)	(3b)	(4)	(5)	(6)	(7)	(8)	(9)	(10)	(11)	(12)	(13)
1938	BROMOACETIC ACID SOLUTION	8	C3	III	8		LQ7		PP, EP			0	
1939	PHOSPHORUS OXYBROMIDE	8	C2	II	8		LQ23		PP, EP			0	
1940	THIOGLYCOLIC ACID	8	C3	II	8		LQ22		PP, EP			0	
1941	DIBROMODIFLUORO-METHANE	9	M11	III	9		LQ28		PP			0	
1942	AMMONIUM NITRATE with not more than 0.2% total combustible material, including any organic substance calculated as carbon, to the exclusion of any other added substance	5.1	O2	III	5.1	306 611	LQ12	B	PP		ST01, CO02 HA09	0	CO02 and HA09 apply only when this substance is carried in bulk or without packaging
1944	MATCHES, SAFETY (book, card or strike on box)	4.1	F1	III	4.1	293	LQ9		PP			0	
1945	MATCHES, WAX 'VESTA'	4.1	F1	III	4.1	293	LQ9		PP			0	
1950	AEROSOLS, asphyxiant	2	5A		2.2	190 327 625	LQ2		PP	VE04		0	
1950	AEROSOLS, corrosive	2	5C		2.2+8	190 327 625	LQ2		PP, EP	VE04		0	
1950	AEROSOLS, corrosive, oxidizing	2	5CO		2.2+5.1+8	190 327 625	LQ2		PP, EP	VE04		0	
1950	AEROSOLS, flammable	2	5F		2.1	190 327 625	LQ2		PP, EX, A	VE01, VE04		1	
1950	AEROSOLS, flammable, corrosive	2	5FC		2.1+8	190 327 625	LQ2		PP, EX, A	VE01, VE04		1	
1950	AEROSOLS, oxidizing	2	5O		2.2+5.1	190 327 625	LQ2		PP	VE04		0	
1950	AEROSOLS, toxic	2	5T		2.2+6.1	190 327 625	LQ1		PP, EP, TOX, A	VE02, VE04		2	

UN No. or ID No.	Name and description	Class	Classification Code	Packing group	Labels	Special provisions	Limited quantities	Carriage permitted	Equipment required	Ventilation	Provisions concerning loading, unloading and carriage	Number of cones, blue lights	Remarks
3.1.2	3.1.2	2.2	2.2	2.1.1.3	5.2.2	3.3	3.4.6	3.2.1	8.1.5	7.1.6	7.1.6	7.1.5	3.2.1
(1)	(2)	(3a)	(3b)	(4)	(5)	(6)	(7)	(8)	(9)	(10)	(11)	(12)	(13)
1950	AEROSOLS, toxic, corrosive	2	5TC		2.2+6.1+8	190 327 625	LQ1		PP, EP, TOX, A	VE02, VE04		2	
1950	AEROSOLS, toxic, flammable	2	5TF		2.1+6.1	190 327 625	LQ1		PP, EP, EX, TOX, A	VE01, VE02, VE04		2	
1950	AEROSOLS, toxic, flammable, corrosive	2	5TFC		2.1+6.1+8	190 327 625	LQ1		PP, EP, EX, TOX, A	VE01, VE02		2	
1950	AEROSOLS, toxic, oxidizing	2	5TO		2.2+6.1+5.1	190 327 625	LQ1		PP, EP, TOX, A	VE02, VE04		2	
1950	AEROSOLS, toxic, oxidizing, corrosive	2	5TOC		2.2+6.1+5.1 +8	190 327 625	LQ1		PP, EP, TOX, A	VE02, VE04		2	
1951	ARGON, REFRIGERATED LIQUID	2	3A		2.2	593	LQ1		PP			0	
1952	ETHYLENE OXIDE AND CARBON DIOXIDE MIXTURE with not more than 9% ethylene oxide	2	2A		2.2		LQ1		PP			0	
1953	COMPRESSED GAS, TOXIC, FLAMMABLE, N.O.S.	2	1TF		2.3+2.1	274	LQ0		PP, EP, EX, TOX, A	VE01, VE02		2	
1954	COMPRESSED GAS, FLAMMABLE, N.O.S.	2	1F		2.1	274	LQ0		PP, EX, A	VE01		1	
1955	COMPRESSED GAS, TOXIC, N.O.S.	2	1T		2.3	274	LQ0		PP, EP, TOX, A	VE02		2	
1956	COMPRESSED GAS, N.O.S.	2	1A		2.2	274 292 567	LQ1		PP			0	
1957	DEUTERIUM, COMPRESSED	2	1F		2.1		LQ0		PP, EX, A	VE01		1	
1958	1,2-DICHLORO-1,1,2,2-TETRAFLUOROETHANE (REFRIGERANT GAS R 114)	2	2A		2.2		LQ1		PP			0	
1959	1,1-DIFLUOROETHYLENE (REFRIGERANT GAS R 1132a)	2	2F		2.1		LQ0		PP, EX, A	VE01		1	

UN No. or ID No.	Name and description	Class	Classifi- cation Code	Packing group	Labels	Special provisions	Limited quantities	Carriage permitted	Equipment required	Ventilation	Provisions concerning loading, unloading and carriage	Number of cones, blue lights	Remarks
3.1.2	3.1.2	2.2	2.2	2.1.1.3	5.2.2	3.3	3.4.6	3.2.1	8.1.5	7.1.6	7.1.6	7.1.5	3.2.1
(1)	(2)	(3a)	(3b)	(4)	(5)	(6)	(7)	(8)	(9)	(10)	(11)	(12)	(13)
1961	ETHANE, REFRIGERATED LIQUID	2	3F		2.1		LQ0		PP, EX, A	VE01		1	
1962	ETHYLENE	2	2F		2.1		LQ0		PP, EX, A	VE01		1	
1963	HELIUM, REFRIGERATED LIQUID	2	3A		2.2	593	LQ1		PP			0	
1964	HYDROCARBON GAS MIXTURE, COMPRESSED, N.O.S.	2	1F		2.1	274	LQ0		PP, EX, A	VE01		1	
1965	HYDROCARBON GAS MIXTURE, LIQUEFIED, N.O.S. such as mixtures A, A01, A02, A0, A1, B1, B2, B or C	2	2F		2.1	274 583	LQ0	T	PP, EX, A	VE01		1	
1966	HYDROGEN, REFRIGERATED LIQUID	2	3F		2.1		LQ0		PP, EX, A	VE01		1	
1967	INSECTICIDE GAS, TOXIC, N.O.S.	2	2T		2.3	274	LQ0		PP, EP, TOX, A	VE02		2	
1968	INSECTICIDE GAS, N.O.S.	2	2A		2.2	274	LQ1		PP			0	
1969	ISOBUTANE	2	2F		2.1		LQ0	T	PP, EX, A	VE01		1	
1970	KRYPTON, REFRIGERATED LIQUID	2	3A		2.2	593	LQ1		PP			0	
1971	METHANE, COMPRESSED or NATURAL GAS, COMPRESSED with high methane content	2	1F		2.1		LQ0		PP, EX, A	VE01		1	
1972	METHANE, REFRIGERATED LIQUID or NATURAL GAS, REFRIGERATED LIQUID with high methane content	2	3F		2.1		LQ0		PP, EX, A	VE01		1	
1973	CHLORODIFLUORO-METHANE AND CHLOROPENTAFLUORO-ETHANE MIXTURE with fixed boiling point, with approximately 49% chlorodifluoromethane (REFRIGERANT GAS R 502)	2	2A		2.2		LQ1		PP			0	
1974	CHLORODIFLUOROBROMO-METHANE (REFRIGERANT GAS R 12B1)	2	2A		2.2		LQ1		PP			0	

UN No. or ID No.	Name and description	Class	Classification Code	Packing group	Labels	Special provisions	Limited quantities	Carriage permitted	Equipment required	Ventilation	Provisions concerning loading, unloading and carriage	Number of cones, blue lights	Remarks
	3.1.2	2.2	2.2	2.1.1.3	5.2.2	3.3	3.4.6	3.2.1	8.1.5	7.1.6	7.1.6	7.1.5	3.2.1
(1)	(2)	(3a)	(3b)	(4)	(5)	(6)	(7)	(8)	(9)	(10)	(11)	(12)	(13)
1975	NITRIC OXIDE AND DINITROGEN TETROXIDE MIXTURE (NITRIC OXIDE AND NITROGEN DIOXIDE MIXTURE)	2	2TOC		2.3+5.1+8		LQ0		PP, EP, TOX, A	VE02		2	
1976	OCTAFLUOROCYCLO-BUTANE (REFRIGERANT GAS RC 318)	2	2A		2.2		LQ1		PP			0	
1977	NITROGEN, REFRIGERATED LIQUID	2	3A		2.2	593	LQ1		PP			0	
1978	PROPANE	2	2F		2.1		LQ0		PP, EX, A	VE01		1	
1982	TETRAFLUOROMETHANE (REFRIGERANT GAS R 14)	2	2A		2.2		LQ1		PP			0	
1983	1-CHLORO-2,2,2-TRIFLUOROETHANE (REFRIGERANT GAS R 133a)	2	2A		2.2		LQ1		PP			0	
1984	TRIFLUOROMETHANE (REFRIGERANT GAS R 23)	2	2A		2.2		LQ1		PP			0	
1986	ALCOHOLS, FLAMMABLE, TOXIC, N.O.S.	3	FT1	I	3+6.1	274 802	LQ0	T	PP, EP, EX, TOX, A	VE01, VE02		2	
1986	ALCOHOLS, FLAMMABLE, TOXIC, N.O.S.	3	FT1	II	3+6.1	274 802	LQ0	T	PP, EP, EX, TOX, A	VE01, VE02		2	
1986	ALCOHOLS, FLAMMABLE, TOXIC, N.O.S.	3	FT1	III	3+6.1	274 802	LQ7	T	PP, EP, EX, TOX, A	VE01, VE02		0	
1987	ALCOHOLS, N.O.S. (vapour pressure at 50 °C more than 110 kPa)	3	F1	II	3	274 330 601 640C	LQ4	T	PP, EX, A	VE01		1	
1987	ALCOHOLS, N.O.S. (vapour pressure at 50 °C not more than 110 kPa)	3	F1	II	3	274 330 601 640D	LQ4	T	PP, EX, A	VE01		1	
1987	ALCOHOLS, N.O.S.	3	F1	III	3	274 330 601	LQ7	T	PP, EX, A	VE01		0	
1988	ALDEHYDES, FLAMMABLE, TOXIC, N.O.S.	3	FT1	I	3+6.1	274 802	LQ0	T	PP, EP, EX, TOX, A	VE01, VE02		2	

UN No. or ID No.	Name and description	Class	Classification Code	Packing group	Labels	Special provisions	Limited quantities	Carriage permitted	Equipment required	Ventilation	Provisions concerning loading, unloading and carriage	Number of cones, blue lights	Remarks
3.1.2	3.1.2	2.2	2.2	2.1.1.3	5.2.2	3.3	3.4.6	3.2.1	8.1.5	7.1.6	7.1.6	7.1.5	3.2.1
(1)	(2)	(3a)	(3b)	(4)	(5)	(6)	(7)	(8)	(9)	(10)	(11)	(12)	(13)
1988	ALDEHYDES, FLAMMABLE, TOXIC, N.O.S.	3	FT1	II	3+6.1	274 802	LQ0		PP, EP, EX, TOX, A	VE01, VE02		2	
1988	ALDEHYDES, FLAMMABLE, TOXIC, N.O.S.	3	FT1	III	3+6.1	274 802	LQ7		PP, EP, EX, TOX, A	VE01, VE02		0	
1989	ALDEHYDES, N.O.S.	3	F1	I	3	274	LQ3		PP, EX, A	VE01		1	
1989	ALDEHYDES, N.O.S. (vapour pressure at 50 °C more than 110 kPa)	3	F1	II	3	274 640C	LQ4	T	PP, EX, A	VE01		1	
1989	ALDEHYDES, N.O.S. (vapour pressure at 50 °C not more than 110 kPa)	3	F1	II	3	274 640D	LQ4	T	PP, EX, A	VE01		1	
1989	ALDEHYDES, N.O.S.	3	F1	III	3	274	LQ7	T	PP, EX, A	VE01		0	
1990	BENZALDEHYDE	9	M11	III	9		LQ28		PP			0	
1991	CHLOROPRENE, STABILIZED	3	FT1	I	3+6.1	802	LQ0	T	PP, EP, EX, TOX, A	VE01, VE02		2	
1992	FLAMMABLE LIQUID, TOXIC, N.O.S.	3	FT1	I	3+6.1	274 802	LQ0	T	PP, EP, EX, TOX, A	VE01, VE02		2	
1992	FLAMMABLE LIQUID, TOXIC, N.O.S.	3	FT1	II	3+6.1	274 802	LQ0	T	PP, EP, EX, TOX, A	VE01, VE02		2	
1992	FLAMMABLE LIQUID, TOXIC, N.O.S.	3	FT1	III	3+6.1	274 802	LQ7	T	PP, EP, EX, TOX, A	VE01, VE02		0	
1993	FLAMMABLE LIQUID, N.O.S.	3	F1	I	3	274 330	LQ3	T	PP, EX, A	VE01		1	
1993	FLAMMABLE LIQUID, N.O.S. (vapour pressure at 50 °C more than 110 kPa)	3	F1	II	3	274 330 601 640C	LQ4	T	PP, EX, A	VE01		1	
1993	FLAMMABLE LIQUID, N.O.S. (vapour pressure at 50 °C not more than 110 kPa)	3	F1	II	3	274 330 601 640D	LQ4	T	PP, EX, A	VE01		1	
1993	FLAMMABLE LIQUID, N.O.S.	3	F1	III	3	274 330 601 640E	LQ7	T	PP, EX, A	VE01		0	

UN No. or ID No.	Name and description	Class	Classification Code	Packing group	Labels	Special provisions	Limited quantities	Carriage permitted	Equipment required	Ventilation	Provisions concerning loading, unloading and carriage	Number of cones, blue lights	Remarks
	3.1.2	2.2	2.2	2.1.1.3	5.2.2	3.3	3.4.6	3.2.1	8.1.5	7.1.6	7.1.6	7.1.5	3.2.1
(1)	(2)	(3a)	(3b)	(4)	(5)	(6)	(7)	(8)	(9)	(10)	(11)	(12)	(13)
1993	FLAMMABLE LIQUID, N.O.S. (having a flash-point below 23 °C and viscous according to 2.2.3.1.4) (boiling point not more than 35° C)	3	F1	III	3	274 330 601 640F	LQ7	T	PP, EX, A	VE01		0	
1993	FLAMMABLE LIQUID, N.O.S. (having a flash-point below 23 °C and viscous according to 2.2.3.1.4) (vapour pressure at 50 °C more than 110 kPa , boiling point of more than 35° C)	3	F1	III	3	274 330 601 640G	LQ7	T	PP, EX, A	VE01		0	
1993	FLAMMABLE LIQUID, N.O.S. (having a flash-point below 23 °C and viscous according to 2.2.3.1.4) (vapour pressure at 50 °C not more than 110 kPa)	3	F1	III	3	274 330 601 640H	LQ7	T	PP, EX, A	VE01		0	
1994	IRON PENTACARBONYL	6.1	TF1	I	6.1+3	802	LQ0		PP, EP, EX, TOX, A	VE01, VE02		2	
1999	TARS, LIQUID, including road asphalt and oils, bitumen and cut backs (vapour pressure at 50 °C more than 110 kPa)	3	F1	II	3	640C	LQ6		PP, EX, A	VE01		1	
1999	TARS, LIQUID, including road asphalt and oils, bitumen and cut backs (vapour pressure at 50 °C not more than 110 kPa)	3	F1	II	3	640D	LQ6		PP, EX, A	VE01		1	
1999	TARS, LIQUID, including road asphalt and oils, bitumen and cut backs	3	F1	III	3	640E	LQ7	T	PP, EX, A	VE01		0	
1999	TARS, LIQUID, including road asphalt and oils, bitumen and cut backs (having a flash-point below 23 °C and viscous according to 2.2.3.1.4) (boiling point not more than 35° C)	3	F1	III	3	640F	LQ7		PP, EX, A	VE01		0	

UN No. or ID No.	Name and description	Class	Classification Code	Packing group	Labels	Special provisions	Limited quantities	Carriage permitted	Equipment required	Ventilation	Provisions concerning loading, unloading and carriage	Number of cones, blue lights	Remarks
3.1.2	3.1.2	2.2	2.2	2.1.1.3	5.2.2	3.3	3.4.6	3.2.1	8.1.5	7.1.6	7.1.6	7.1.5	3.2.1
(1)	(2)	(3a)	(3b)	(4)	(5)	(6)	(7)	(8)	(9)	(10)	(11)	(12)	(13)
1999	TARS, LIQUID, including road asphalt and oils, bitumen and cut backs (having a flash-point below 23 °C and viscous according to 2.2.3.1.4) (vapour pressure at 50 °C not more than 110 kPa) (vapour pressure at 50 °C more than 110 kPa , boiling point of more than 35° C)	3	F1	III	3	640G	LQ7		PP, EX, A	VE01		0	
1999	TARS, LIQUID, including road asphalt and oils, bitumen and cut backs (having a flash-point below 23 °C and viscous according to 2.2.3.1.4) (vapour pressure at 50 °C not more than 110 kPa) (vapour pressure at 50 °C not more than 110 kPa)	3	F1	III	3	640H	LQ7		PP, EX, A	VE01		0	
2000	CELLULOID in block, rods, rolls, sheets, tubes, etc., except scrap	4.1	F1	III	4.1	502	LQ9		PP			0	
2001	COBALT NAPHTHENATES, POWDER	4.1	F3	III	4.1		LQ9		PP			0	
2002	CELLULOID, SCRAP	4.2	S2	III	4.2	526 592	LQ0		PP			0	
2004	MAGNESIUM DIAMIDE	4.2	S4	II	4.2		LQ0		PP			0	
2006	PLASTICS, NITROCELLULOSE-BASED, SELF-HEATING, N.O.S.	4.2	S2	III	4.2	274 528	LQ0		PP			0	
2008	ZIRCONIUM POWDER, DRY	4.2	S4	I	4.2	524 540	LQ0		PP			0	
2008	ZIRCONIUM POWDER, DRY	4.2	S4	II	4.2	524 540	LQ0		PP			0	
2008	ZIRCONIUM POWDER, DRY	4.2	S4	III	4.2	540	LQ0		PP			0	
2009	ZIRCONIUM, DRY, finished sheets, strip or coiled wire	4.2	S4	III	4.2	524 592	LQ0		PP			0	
2010	MAGNESIUM HYDRIDE	4.3	W2	I	4.3		LQ0		PP, EX, A	VE01	HA08	0	
2011	MAGNESIUM PHOSPHIDE	4.3	WT2	I	4.3+6.1	802	LQ0		PP, EP, EX, TOX, A	VE01, VE02	HA08	2	

UN No. or ID No.	Name and description	Class	Classification Code	Packing group	Labels	Special provisions	Limited quantities	Carriage permitted	Equipment required	Ventilation	Provisions concerning loading, unloading and carriage	Number of cones, blue lights	Remarks
	3.1.2	2.2	2.2	2.1.1.3	5.2.2	3.3	3.4.6	3.2.1	8.1.5	7.1.6	7.1.6	7.1.5	3.2.1
(1)	(2)	(3a)	(3b)	(4)	(5)	(6)	(7)	(8)	(9)	(10)	(11)	(12)	(13)
2012	POTASSIUM PHOSPHIDE	4.3	WT2	I	4.3+6.1	802	LQ0		PP, EP, EX, TOX, A	VE01, VE02	HA08	2	
2013	STRONTIUM PHOSPHIDE	4.3	WT2	I	4.3+6.1	802	LQ0		PP, EP, EX, TOX, A	VE01, VE02	HA08	2	
2014	HYDROGEN PEROXIDE, AQUEOUS SOLUTION with not less than 20% but not more than 60% hydrogen peroxide (stabilized as necessary)	5.1	OC1	II	5.1+8		LQ10	T	PP, EP			0	
2015	HYDROGEN PEROXIDE, AQUEOUS SOLUTION, STABILIZED with more than 70% hydrogen peroxide	5.1	OC1	I	5.1+8	640N	LQ0		PP, EP			0	
2015	HYDROGEN PEROXIDE, AQUEOUS SOLUTION, STABILIZED with more than 60% hydrogen peroxide and not more than 70% hydrogen peroxide	5.1	OC1	I	5.1+8	640O	LQ0		PP, EP			0	
2016	AMMUNITION, TOXIC, NON-EXPLOSIVE without burster or expelling charge, non-fuzed	6.1	T2	II	6.1	802	LQ0		PP, EP			2	
2017	AMMUNITION, TEAR-PRODUCING, NON-EXPLOSIVE without burster or expelling charge, non-fuzed	6.1	TC2	II	6.1+8	802	LQ0		PP, EP			2	
2018	CHLOROANILINES, SOLID	6.1	T2	II	6.1	802	LQ18		PP, EP			2	
2019	CHLOROANILINES, LIQUID	6.1	T1	II	6.1	802	LQ17		PP, EP, TOX, A	VE02		2	
2020	CHLOROPHENOLS, SOLID	6.1	T2	III	6.1	205 802	LQ9		PP, EP			0	
2021	CHLOROPHENOLS, LIQUID	6.1	T1	III	6.1	802	LQ7	T	PP, EP, TOX, A	VE02		0	
2022	CRESYLIC ACID	6.1	TC1	II	6.1+8	802	LQ17	T	PP, EP, TOX, A	VE02		2	
2023	EPICHLOROHYDRIN	6.1	TF1	II	6.1+3	279 802	LQ17	T	PP, EP, EX, TOX, A	VE01, VE02		2	

UN No. or ID No. (1)	Name and description 3.1.2 (2)	Class 2.2 (3a)	Classification Code 2.2 (3b)	Packing group 2.1.1.3 (4)	Labels 5.2.2 (5)	Special provisions 3.3 (6)	Limited quantities 3.4.6 (7)	Carriage permitted 3.2.1 (8)	Equipment required 8.1.5 (9)	Ventilation 7.1.6 (10)	Provisions concerning loading, unloading and carriage 7.1.6 (11)	Number of cones, blue lights 7.1.5 (12)	Remarks 3.2.1 (13)
2024	MERCURY COMPOUND, LIQUID, N.O.S.	6.1	T4	I	6.1	43 274 802	LQ0		PP, EP, TOX, A	VE02		2	
2024	MERCURY COMPOUND, LIQUID, N.O.S.	6.1	T4	II	6.1	43 274 802	LQ17		PP, EP, TOX, A	VE02		2	
2024	MERCURY COMPOUND, LIQUID, N.O.S.	6.1	T4	III	6.1	43 274 802	LQ7		PP, EP, TOX, A	VE02		0	
2025	MERCURY COMPOUND, SOLID, N.O.S.	6.1	T5	I	6.1	43 274 529 585 802	LQ0		PP, EP			2	
2025	MERCURY COMPOUND, SOLID, N.O.S.	6.1	T5	II	6.1	43 274 529 585 802	LQ18		PP, EP			2	
2025	MERCURY COMPOUND, SOLID, N.O.S.	6.1	T5	III	6.1	43 274 529 585 802	LQ9		PP, EP			0	
2026	PHENYLMERCURIC COMPOUND, N.O.S.	6.1	T3	I	6.1	43 274 802	LQ0		PP, EP, TOX, A	VE02		2	
2026	PHENYLMERCURIC COMPOUND, N.O.S.	6.1	T3	II	6.1	43 274 802	LQ18		PP, EP, TOX, A	VE02		2	
2026	PHENYLMERCURIC COMPOUND, N.O.S.	6.1	T3	III	6.1	43 274 802	LQ9		PP, EP, TOX, A	VE02		0	
2027	SODIUM ARSENITE, SOLID	6.1	T5	II	6.1	43 802	LQ18		PP, EP			2	

UN No. or ID No.	Name and description	Class	Classification Code	Packing group	Labels	Special provisions	Limited quantities	Carriage permitted	Equipment required	Ventilation	Provisions concerning loading, unloading and carriage	Number of cones, blue lights	Remarks
3.1.2	3.1.2	2.2	2.2	2.1.1.3	5.2.2	3.3	3.4.6	3.2.1	8.1.5	7.1.6	7.1.6	7.1.5	3.2.1
(1)	(2)	(3a)	(3b)	(4)	(5)	(6)	(7)	(8)	(9)	(10)	(11)	(12)	(13)
2028	BOMBS, SMOKE, NON-EXPLOSIVE with corrosive liquid, without initiating device	8	C11	II	8		LQ0		PP, EP			0	
2029	HYDRAZINE, ANHYDROUS	8	CFT	I	8+3+6.1	802	LQ0		PP, EP, EX, TOX, A	VE01, VE02		2	
2030	HYDRAZINE AQUEOUS SOLUTION, with more than 37% hydrazine by mass, having a flash-point above 60 °C	8	CT1	I	8+6.1	530 802	LQ0		PP, EP, TOX, A	VE02		2	
2030	HYDRAZINE AQUEOUS SOLUTION, with more than 37% hydrazine by mass having a flash-point of not more than 60 °C	8	CFT	I	8 +3 +6.1	530 802	LQ0		PP, EP, EX, TOX, A	VE01, VE02		2	
2030	HYDRAZINE AQUEOUS SOLUTION, with more than 37% hydrazine by mass	8	CT1	II	8+6.1	530 802	LQ22		PP, EP, TOX, A	VE02		2	
2030	HYDRAZINE AQUEOUS SOLUTION, with more than 37% hydrazine by mass	8	CT1	III	8+6.1	530 802	LQ7		PP, EP, TOX, A	VE02		0	
2031	NITRIC ACID, other than red fuming, with more than 70% nitric acid	8	CO1	I	8+5.1		LQ0	T	PP, EP			0	
2031	NITRIC ACID, other than red fuming, with not more than 70% nitric acid	8	CO1	II	8		LQ22	T	PP, EP			0	
2032	NITRIC ACID, RED FUMING	8	COT	I	8+5.1+6.1	802	LQ0	T	PP, EP, TOX, A	VE02		2	
2033	POTASSIUM MONOXIDE	8	C6	II	8		LQ23		PP, EP			0	
2034	HYDROGEN AND METHANE MIXTURE, COMPRESSED	2	1F		2.1		LQ0		PP, EX, A	VE01		1	
2035	1,1,1-TRIFLUOROETHANE (REFRIGERANT GAS R 143a)	2	2F		2.1		LQ0		PP, EX, A	VE01		1	
2036	XENON	2	2A		2.2		LQ1		PP			0	
2037	RECEPTACLES, SMALL, CONTAINING GAS (GAS CARTRIDGES) without a release device, non-refillable	2	5A		2.2	191 303	LQ2		PP			0	

UN No. or ID No.	Name and description	Class	Classification Code	Packing group	Labels	Special provisions	Limited quantities	Carriage permitted	Equipment required	Ventilation	Provisions concerning loading, unloading and carriage	Number of cones, blue lights	Remarks
	3.1.2	2.2	2.2	2.1.1.3	5.2.2	3.3	3.4.6	3.2.1	8.1.5	7.1.6	7.1.6	7.1.5	3.2.1
(1)	(2)	(3a)	(3b)	(4)	(5)	(6)	(7)	(8)	(9)	(10)	(11)	(12)	(13)
2037	RECEPTACLES, SMALL, CONTAINING GAS (GAS CARTRIDGES) without a release device, non-refillable	2	5F		2.1	191 303	LQ2		PP, EX, A	VE01		1	
2037	RECEPTACLES, SMALL, CONTAINING GAS (GAS CARTRIDGES) without a release device, non-refillable	2	5O		2.2+5.1	191 303	LQ2		PP			0	
2037	RECEPTACLES, SMALL, CONTAINING GAS (GAS CARTRIDGES) without a release device, non-refillable	2	5T		2.3	303	LQ1		PP, EP, TOX, A	VE02		2	
2037	RECEPTACLES, SMALL, CONTAINING GAS (GAS CARTRIDGES) without a release device, non-refillable	2	5TC		2.3+8	303	LQ1		PP, EP, TOX, A	VE02		2	
2037	RECEPTACLES, SMALL, CONTAINING GAS (GAS CARTRIDGES) without a release device, non-refillable	2	5TF		2.3+2.1	303	LQ1		PP, EP, EX, TOX, A	VE01, VE02		2	
2037	RECEPTACLES, SMALL, CONTAINING GAS (GAS CARTRIDGES) without a release device, non-refillable	2	5TFC		2.3+2.1+8	303	LQ1		PP, EP, EX, TOX, A	VE01, VE02		2	
2037	RECEPTACLES, SMALL, CONTAINING GAS (GAS CARTRIDGES) without a release device, non-refillable	2	5TO		2.3+5.1	303	LQ1		PP, EP, TOX, A	VE02		2	
2037	RECEPTACLES, SMALL, CONTAINING GAS (GAS CARTRIDGES) without a release device, non-refillable	2	5TOC		2.3+5.1+8	303	LQ1		PP, EP, TOX, A	VE02		2	
2038	DINITROTOLUENES, LIQUID	6.1	T1	II	6.1	802	LQ17		PP, EP, TOX, A	VE02		2	
2044	2,2-DIMETHYLPROPANE	2	2F		2.1		LQ0		PP, EX, A	VE01		1	
2045	ISOBUTYRALDEHYDE (ISOBUTYL ALDEHYDE)	3	F1	II	3		LQ4	T	PP, EX, A	VE01		1	

UN No. or ID No.	Name and description	Class	Classification Code	Packing group	Labels	Special provisions	Limited quantities	Carriage permitted	Equipment required	Ventilation	Provisions concerning loading, unloading and carriage	Number of cones, blue lights	Remarks
3.1.2	3.1.2	2.2	2.2	2.1.1.3	5.2.2	3.3	3.4.6	3.2.1	8.1.5	7.1.6	7.1.6	7.1.5	3.2.1
(1)	(2)	(3a)	(3b)	(4)	(5)	(6)	(7)	(8)	(9)	(10)	(11)	(12)	(13)
2046	CYMENES	3	F1	III	3		LQ7	T	PP, EX, A	VE01		0	
2047	DICHLOROPROPENES	3	F1	II	3		LQ4	T	PP, EX, A	VE01		1	
2047	DICHLOROPROPENES	3	F1	III	3		LQ7	T	PP, EX, A	VE01		0	
2048	DICYCLOPENTADIENE	3	F1	III	3		LQ7	T	PP, EX, A	VE01		0	
2049	DIETHYLBENZENE	3	F1	III	3		LQ7	T	PP, EX, A	VE01		0	
2050	DIISOBUTYLENE, ISOMERIC COMPOUNDS	3	F1	II	3		LQ4	T	PP, EX, A	VE01		1	
2051	2-DIMETHYLAMINO-ETHANOL	8	CF1	II	8+3		LQ22	T	PP, EP, EX, A	VE01		1	
2052	DIPENTENE	3	F1	III	3		LQ7		PP, EX, A	VE01		0	
2053	METHYL ISOBUTYL CARBINOL	3	F1	III	3		LQ7	T	PP, EX, A	VE01		0	
2054	MORPHOLINE	8	CF1	I	8+3		LQ0	T	PP, EP, EX, A	VE01		1	
2055	STYRENE MONOMER, STABILIZED	3	F1	III	3		LQ7	T	PP, EX, A	VE01		0	
2056	TETRAHYDROFURAN	3	F1	II	3		LQ4	T	PP, EX, A	VE01		1	
2057	TRIPROPYLENE	3	F1	II	3		LQ4	T	PP, EX, A	VE01		1	
2057	TRIPROPYLENE	3	F1	III	3		LQ7	T	PP, EX, A	VE01		0	
2058	VALERALDEHYDE	3	F1	II	3		LQ4		PP, EX, A	VE01		1	
2059	NITROCELLULOSE SOLUTION, FLAMMABLE with not more than 12.6% nitrogen, by dry mass, and not more than 55% nitrocellulose	3	D	I	3	198 531	LQ3		PP, EX, A	VE01		1	
2059	NITROCELLULOSE SOLUTION, FLAMMABLE with not more than 12.6% nitrogen, by dry mass, and not more than 55% nitrocellulose (vapour pressure at 50 °C more than 110 kPa)	3	D	II	3	198 531 640C	LQ4		PP, EX, A	VE01		1	
2059	NITROCELLULOSE SOLUTION, FLAMMABLE with not more than 12.6% nitrogen, by dry mass, and not more than 55% nitrocellulose (vapour pressure at 50 °C not more than 110 kPa)	3	D	II	3	198 531 640D	LQ4		PP, EX, A	VE01		1	

UN No. or ID No.	Name and description	Class	Classification Code	Packing group	Labels	Special provisions	Limited quantities	Carriage permitted	Equipment required	Ventilation	Provisions concerning loading, unloading and carriage	Number of cones, blue lights	Remarks
3.1.2	3.1.2	2.2	2.2	2.1.1.3	5.2.2	3.3	3.4.6	3.2.1	8.1.5	7.1.6	7.1.6	7.1.5	3.2.1
(1)	(2)	(3a)	(3b)	(4)	(5)	(6)	(7)	(8)	(9)	(10)	(11)	(12)	(13)
2059	NITROCELLULOSE SOLUTION, FLAMMABLE with not more than 12.6% nitrogen, by dry mass, and not more than 55% nitrocellulose	3	D	III	3	198 531	LQ7		PP, EX, A	VE01		0	
2067	AMMONIUM NITRATE BASED FERTILIZER	5.1	O2	III	5.1	186 306 307	LQ12	B	PP		CO02, ST01, LO04	HA09	0	CO02, LO04 and HA09 apply only when this substance is carried in bulk or without packaging
2071	AMMONIUM NITRATE BASED FERTILIZERS, uniform mixtures of the nitrogen/phosphate, nitrogen/potash or nitrogen/phosphate/potash type, containing not more than 70% ammonium nitrate and not more than 0.4% total combustible/organic material calculated as carbon or with not more than 45% ammonium nitrate and unrestricted combustible material	9	M11			186 193		B	PP		CO02, ST02	HA09	0	Dangerous only in bulk or without packaging. CO02, ST02 and HA09 apply only when this substance is carried in bulk or without packaging
2073	AMMONIA SOLUTION, relative density less than 0.880 at 15 °C in water, with more than 35% but not more than 50% ammonia	2	4A		2.2	532	LQ1		PP				0	
2074	ACRYLAMIDE, SOLID	6.1	T2	III	6.1	802	LQ9	T	PP, EP				0	
2075	CHLORAL, ANHYDROUS, STABILIZED	6.1	T1	II	6.1	802	LQ17		PP, EP, TOX, A	VE02			2	
2076	CRESOLS, LIQUID	6.1	TC1	II	6.1+8	802	LQ17		PP, EP, TOX, A	VE02			2	
2077	alpha-NAPHTHYLAMINE	6.1	T2	III	6.1	802	LQ9		PP, EP				0	
2078	TOLUENE DIISOCYANATE	6.1	T1	II	6.1	279 802	LQ17		PP, EP, TOX, A	VE02			2	
2078	TOLUENE DIISOCYANATE (2,4-TOLUENE DIISOCYANATE)	6.1	T1	II	6.1	279 802	LQ17	T	PP, EP, TOX, A	VE02			2	
2079	DIETHYLENETRIAMINE	8	C7	II	8		LQ22	T	PP, EP				0	

UN No. or ID No.	Name and description	Class	Classification Code	Packing group	Labels	Special provisions	Limited quantities	Carriage permitted	Equipment required	Ventilation	Provisions concerning loading, unloading and carriage	Number of cones, blue lights	Remarks
(1)	3.1.2 (2)	2.2 (3a)	2.2 (3b)	2.1.1.3 (4)	5.2.2 (5)	3.3 (6)	3.4.6 (7)	3.2.1 (8)	8.1.5 (9)	7.1.6 (10)	7.1.6 (11)	7.1.5 (12)	3.2.1 (13)
2186	HYDROGEN CHLORIDE, REFRIGERATED LIQUID	2	3TC										CARRIAGE PROHIBITED
2187	CARBON DIOXIDE, REFRIGERATED LIQUID	2	3A		2.2	593	LQ1		PP			0	
2188	ARSINE	2	2TF		2.3+2.1		LQ0		PP, EP, EX, TOX, A	VE01, VE02		2	
2189	DICHLOROSILANE	2	2TFC		2.3+2.1+8		LQ0		PP, EP, EX, TOX, A	VE01, VE02		2	
2190	OXYGEN DIFLUORIDE, COMPRESSED	2	1TOC		2.3+5.1+8		LQ0		PP, EP, TOX, A	VE02		2	
2191	SULPHURYL FLUORIDE	2	2T		2.3		LQ0		PP, EP, TOX, A	VE02		2	
2192	GERMANE	2	2TF		2.3+2.1	632	LQ0		PP, EP, EX, TOX, A	VE01, VE02		2	
2193	HEXAFLUOROETHANE (REFRIGERANT GAS R 116)	2	2A		2.2		LQ1		PP			0	
2194	SELENIUM HEXAFLUORIDE	2	2TC		2.3+8		LQ0		PP, EP, TOX, A	VE02		2	
2195	TELLURIUM HEXAFLUORIDE	2	2TC		2.3+8		LQ0		PP, EP, TOX, A	VE02		2	
2196	TUNGSTEN HEXAFLUORIDE	2	2TC		2.3+8		LQ0		PP, EP, TOX, A	VE02		2	
2197	HYDROGEN IODIDE, ANHYDROUS	2	2TC		2.3+8		LQ0		PP, EP, TOX, A	VE02		2	
2198	PHOSPHORUS PENTAFLUORIDE	2	2TC		2.3+8		LQ0		PP, EP, TOX, A	VE02		2	
2199	PHOSPHINE	2	2TF		2.3+2.1	632	LQ0		PP, EP, EX, TOX, A	VE01, VE02		2	
2200	PROPADIENE, STABILIZED	2	2F		2.1		LQ0		PP, EX, A	VE01		1	
2201	NITROUS OXIDE, REFRIGERATED LIQUID	2	3O		2.2+5.1		LQ0		PP			0	
2202	HYDROGEN SELENIDE, ANHYDROUS	2	2TF		2.3+2.1		LQ0		PP, EP, EX, TOX, A	VE01, VE02		2	
2203	SILANE	2	2F		2.1	632	LQ0		PP, EX, A	VE01		1	
2204	CARBONYL SULPHIDE	2	2TF		2.3+2.1		LQ0		PP, EP, EX, TOX, A	VE01, VE02		2	

UN No. or ID No.	Name and description	Class	Classification Code	Packing group	Labels	Special provisions	Limited quantities	Carriage permitted	Equipment required	Ventilation	Provisions concerning loading, unloading and carriage	Number of cones, blue lights	Remarks
	3.1.2	2.2	2.2	2.1.1.3	5.2.2	3.3	3.4.6	3.2.1	8.1.5	7.1.6	7.1.6	7.1.5	3.2.1
(1)	(2)	(3a)	(3b)	(4)	(5)	(6)	(7)	(8)	(9)	(10)	(11)	(12)	(13)
2205	ADIPONITRILE	6.1	T1	III	6.1	802	LQ7	T	PP, EP, TOX, A	VE02		0	
2206	ISOCYANATES, TOXIC. N.O.S. or ISOCYANATE SOLUTION, TOXIC. N.O.S.	6.1	T1	II	6.1	274 551 802	LQ17	T	PP, EP, TOX, A	VE02		2	
2206	ISOCYANATES, TOXIC. N.O.S. or ISOCYANATE SOLUTION, TOXIC. N.O.S.	6.1	T1	III	6.1	274 551 802	LQ7		PP, EP, TOX, A	VE02		0	
2208	CALCIUM HYPOCHLORITE MIXTURE, DRY with more than 10% but not more than 39% available chlorine	5.1	O2	III	5.1	313 314	LQ12		PP			0	
2209	FORMALDEHYDE SOLUTION with not less than 25% formaldehyde	8	C9	III	8	533	LQ7	T	PP, EP			0	
2210	MANEB or MANEB PREPARATION with not less than 60% maneb	4.2	SW	III	4.2+4.3	273	LQ0	B	PP, EX, A	VE01, VE03	IN01, IN03	0	VE03, IN01 and IN03 apply only when this substance is carried in bulk or without packaging
2211	POLYMERIC BEADS, EXPANDABLE, evolving flammable vapour	9	M3	III	none	207 633	LQ27		PP, EX, EP, A	VE01, VE03	IN01	0	VE03 and IN01 apply only when this substance is carried in bulk or without packaging
2212	BLUE ASBESTOS (crocidolite) or BROWN ASBESTOS (amosite, mysorite)	9	M1	II	9	168 802	LQ25		PP			0	
2213	PARAFORMALDEHYDE	4.1	F1	III	4.1		LQ9		PP			0	
2214	PHTHALIC ANHYDRIDE with more than 0.05% of maleic anhydride	8	C4	III	8	169	LQ24		PP, EP			0	
2215	MALEIC ANHYDRIDE, MOLTEN	8	C3	III	8		LQ0	T	PP, EP			0	
2215	MALEIC ANHYDRIDE	8	C4	III	8		LQ24		PP, EP			0	

UN No. or ID No.	Name and description	Class	Classification Code	Packing group	Labels	Special provisions	Limited quantities	Carriage permitted	Equipment required	Ventilation	Provisions concerning loading, unloading and carriage	Number of cones, blue lights	Remarks
3.1.2	3.1.2	2.2	2.2	2.1.1.3	5.2.2	3.3	3.4.6	3.2.1	8.1.5	7.1.6	7.1.6	7.1.5	3.2.1
(1)	(2)	(3a)	(3b)	(4)	(5)	(6)	(7)	(8)	(9)	(10)	(11)	(12)	(13)
2216	FISH MEAL, STABILISED or FISH SCRAP, STABILISED	9	M11					B	PP			0	
2217	SEED CAKE with not more than 1.5% oil and not more than 11% moisture	4.2	S2	III	4.2	142 800	LQ0	B	PP		IN01	0	IN01 applies only when this substance is carried in bulk or without packaging
2218	ACRYLIC ACID, STABILIZED	8	CF1	II	8+3		LQ22	T	PP, EP, EX, A	VE01		1	
2219	ALLYL GLYCIDYL ETHER	3	F1	III	3		LQ7		PP, EX, A	VE01		0	
2222	ANISOLE	3	F1	III	3		LQ7		PP, EX, A	VE01		0	
2224	BENZONITRILE	6.1	T1	II	6.1	802	LQ17		PP, EP, TOX, A	VE02		2	
2225	BENZENESULPHONYL CHLORIDE	8	C3	III	8		LQ7		PP, EP			0	
2226	BENZOTRICHLORIDE	8	C9	II	8		LQ22		PP, EP			0	
2227	n-BUTYL METHACRYLATE, STABILIZED	3	F1	III	3		LQ7	T	PP, EX, A	VE01		0	
2232	2-CHLOROETHANAL	6.1	T1	I	6.1	802	LQ0		PP, EP, TOX, A	VE02		2	
2233	CHLOROANISIDINES	6.1	T2	III	6.1	802	LQ9		PP, EP	VE01		0	
2234	CHLOROBENZOTRIFLUORIDES	3	F1	III	3		LQ7		PP, EX, A	VE01		0	
2235	CHLOROBENZYL CHLORIDES, LIQUID	6.1	T1	III	6.1	802	LQ7		PP, EP			0	
2236	3-CHLORO-4-METHYLPHENYL ISOCYANATE, LIQUID	6.1	T1	II	6.1	802	LQ17		PP, EP			2	
2237	CHLORONITROANILINES	6.1	T2	III	6.1	802	LQ9		PP, EP			0	
2238	CHLOROTOLUENES	3	F1	III	3	802	LQ7	T	PP, EX, A	VE01		0	
2239	CHLOROTOLUIDINES, SOLID	6.1	T2	III	6.1	802	LQ9		PP, EP			0	
2240	CHROMOSULPHURIC ACID	8	C1	I	8		LQ0		PP, EP			0	
2241	CYCLOHEPTANE	3	F1	II	3		LQ4		PP, EX, A	VE01		1	
2242	CYCLOHEPTENE	3	F1	II	3		LQ4		PP, EX, A	VE01		1	
2243	CYCLOHEXYL ACETATE	3	F1	III	3		LQ7		PP, EX, A	VE01		0	
2244	CYCLOPENTANOL	3	F1	III	3		LQ7		PP, EX, A	VE01		0	
2245	CYCLOPENTANONE	3	F1	III	3		LQ7		PP, EX, A	VE01		0	
2246	CYCLOPENTENE	3	F1	II	3		LQ4		PP, EX, A	VE01		1	

UN No. or ID No.	Name and description	Class	Classification Code	Packing group	Labels	Special provisions	Limited quantities	Carriage permitted	Equipment required	Ventilation	Provisions concerning loading, unloading and carriage	Number of cones, blue lights	Remarks
		2.2	2.2	2.1.1.3	5.2.2	3.3	3.4.6	3.2.1	8.1.5	7.1.6	7.1.6	7.1.5	3.2.1
(1)	(2)	(3a)	(3b)	(4)	(5)	(6)	(7)	(8)	(9)	(10)	(11)	(12)	(13)
2247	n-DECANE	3	F1	III	3		LQ7	T	PP, EX, A	VE01		0	
2248	DI-n-BUTYLAMINE	8	CF1	II	8+3		LQ22	T	PP, EP, EX, A	VE01		1	
2249	DICHLORODIMETHYL ETHER, SYMMETRICAL	6.1	TF1					CARRIAGE PROHIBITED					
2250	DICHLOROPHENYL ISOCYANATES	6.1	T2	II	6.1	802	LQ17		PP, EP			2	
2251	BICYCLO[2.2.1]HEPTA-2,5-DIENE, STABILIZED (2,5-NORBORNADIENE, STABILIZED)	3	F1	II	3		LQ4		PP, EX, A	VE01		1	
2252	1,2-DIMETHOXYETHANE	3	F1	II	3		LQ4		PP, EX, A	VE01		1	
2253	N,N-DIMETHYLANILINE	6.1	T1	II	6.1	802	LQ17		PP, EP, TOX, A	VE02		2	
2254	MATCHES, FUSEE	4.1	F1	III	4.1	293	LQ9		PP			0	
2256	CYCLOHEXENE	3	F1	II	3		LQ4		PP, EX, A	VE01		1	
2257	POTASSIUM	4.3	W2	I	4.3		LQ0		PP, EX, A	VE01	HA08	0	
2258	1,2-PROPYLENEDIAMINE	8	CF1	II	8+3		LQ22		PP, EP, EX, A	VE01		1	
2259	TRIETHYLENETETRAMINE	8	C7	II	8		LQ22	T	PP, EP	VE01		0	
2260	TRIPROPYLAMINE	3	FC	III	3+8		LQ7		PP, EP, EX, A			0	
2261	XYLENOLS, SOLID	6.1	T2	II	6.1	802	LQ18		PP, EP			2	
2262	DIMETHYLCARBAMOYL CHLORIDE	8	C3	II	8		LQ22		PP, EP			0	
2263	DIMETHYLCYCLOHEXANES	3	F1	II	3		LQ4	T	PP, EX, A	VE01		1	
2264	N,N-DIMETHYL-CYCLOHEXYLAMINE	8	CF1	II	8+3		LQ22	T	PP, EP, EX, A	VE01		1	
2265	N,N-DIMETHYL-FORMAMIDE	3	F1	III	3		LQ7	T	PP, EX, A	VE01		0	
2266	DIMETHYL-N-PROPYLAMINE	3	FC	II	3+8		LQ4	T	PP, EP, EX, A	VE01		1	
2267	DIMETHYL THIOPHOSPHORYL CHLORIDE	6.1	TC1	II	6.1+8	802	LQ17		PP, EP, TOX, A	VE02		2	
2269	3,3'-IMINODIPROPYLAMINE	8	C7	III	8		LQ7		PP, EP			0	
2270	ETHYLAMINE, AQUEOUS SOLUTION with not less than 50% but not more than 70% ethylamine	3	FC	II	3+8		LQ4		PP, EP, EX, A	VE01		1	

UN No. or ID No.	Name and description	Class	Classification Code	Packing group	Labels	Special provisions	Limited quantities	Carriage permitted	Equipment required	Ventilation	Provisions concerning loading, unloading and carriage	Number of cones, blue lights	Remarks
	3.1.2	2.2	2.2	2.1.1.3	5.2.2	3.3	3.4.6	3.2.1	8.1.5	7.1.6	7.1.6	7.1.5	3.2.1
(1)	(2)	(3a)	(3b)	(4)	(5)	(6)	(7)	(8)	(9)	(10)	(11)	(12)	(13)
2271	ETHYL AMYL KETONE	3	F1	III	3		LQ7		PP, EX, A	VE01		0	
2272	N-ETHYLANILINE	6.1	T1	III	6.1	802	LQ7		PP, EP, TOX, A	VE02		0	
2273	2-ETHYLANILINE	6.1	T1	III	6.1	802	LQ7		PP, EP, TOX, A	VE02		0	
2274	N-ETHYL-N-BENZYLANILINE	6.1	T1	III	6.1	802	LQ7		PP, EP, TOX, A	VE02		0	
2275	2-ETHYLBUTANOL	3	F1	III	3		LQ7		PP, EX, A	VE01		0	
2276	2-ETHYLHEXYLAMINE	3	FC	III	3+8		LQ7	T	PP, EP, EX, A	VE01		0	
2277	ETHYL METHACRYLATE, STABILIZED	3	F1	II	3		LQ4		PP, EX, A	VE01		1	
2278	n-HEPTENE	3	F1	II	3		LQ4	T	PP, EX, A	VE01		1	
2279	HEXACHLOROBUTADIENE	6.1	T1	III	6.1	802	LQ7		PP, EP, TOX, A	VE02		0	
2280	HEXAMETHYLENE-DIAMINE, SOLID	8	C8	III	8		LQ24	T	PP, EP			0	
2281	HEXAMETHYLENE DIISOCYANATE	6.1	T1	II	6.1	802	LQ17		PP, EP, TOX, A	VE02		2	
2282	HEXANOLS	3	F1	III	3		LQ7		PP, EX, A	VE01		0	
2283	ISOBUTYL METHACRYLATE, STABILIZED	3	F1	III	3	802	LQ7		PP, EP, TOX, A	VE01		0	
2284	ISOBUTYRONITRILE	3	FT1	II	3+6.1	802	LQ0	T	PP, EP, EX, TOX, A	VE01, VE02		2	
2285	ISOCYANATOBENZO-TRIFLUORIDES	6.1	TF1	II	6.1+3	802	LQ17		PP, EP, EX, TOX, A	VE01, VE02		2	
2286	PENTAMETHYLHEPTANE	3	F1	III	3		LQ7	T	PP, EX, A	VE01		0	
2287	ISOHEPTENES	3	F1	II	3		LQ4		PP, EX, A	VE01		1	
2288	ISOHEXENES	3	F1	II	3		LQ4		PP, EX, A	VE01		1	
2289	ISOPHORONEDIAMINE	8	C7	III	8		LQ7		PP, EP			0	
2290	ISOPHORONE DIISOCYANATE	6.1	T1	III	6.1	802	LQ7	T	PP, EP, TOX, A	VE02		0	
2291	LEAD COMPOUND, SOLUBLE, N.O.S.	6.1	T5	III	6.1	199 274 535 802	LQ9		PP, EP			0	

UN No. or ID No. (1)	Name and description (2) 3.1.2	Class (3a) 2.2	Classification Code (3b) 2.2	Packing group (4) 2.1.1.3	Labels (5) 5.2.2	Special provisions (6) 3.3	Limited quantities (7) 3.4.6	Carriage permitted (8) 3.2.1	Equipment required (9) 8.1.5	Ventilation (10) 7.1.6	Provisions concerning loading, unloading and carriage (11) 7.1.6	Number of cones, blue lights (12) 7.1.5	Remarks (13) 3.2.1
2293	4-METHOXY-4-METHYLPENTAN-2-ONE	3	F1	III	3		LQ7		PP, EX, A	VE01		0	
2294	N-METHYLANILINE	6.1	T1	III	6.1	802	LQ7		PP, EP, TOX, A	VE02		0	
2295	METHYL CHLOROACETATE	6.1	TF1	I	6.1+3	802	LQ0		PP, EP, EX, TOX, A	VE01, VE02		2	
2296	METHYLCYCLOHEXANE	3	F1	II	3		LQ4		PP, EX, A	VE01		1	
2297	METHYLCYCLOHEXANONE	3	F1	III	3		LQ7		PP, EX, A	VE01		0	
2298	METHYLCYCLOPENTANE	3	F1	II	3		LQ4		PP, EX, A	VE01		1	
2299	METHYL DICHLOROACETATE	6.1	T1	III	6.1	802	LQ7		PP, EP, TOX, A	VE02		0	
2300	2-METHYL-5-ETHYLPYRIDINE	6.1	T1	III	6.1	802	LQ7		PP, EP, TOX, A	VE02		0	
2301	2-METHYLFURAN	3	F1	II	3		LQ4		PP, EX, A	VE01		1	
2302	5-METHYLHEXAN-2-ONE	3	F1	III	3		LQ7	T	PP, EX, A	VE01		0	
2303	ISOPROPENYLBENZENE	3	F1	III	3		LQ7	T	PP, EX, A	VE01		0	
2304	NAPHTHALENE, MOLTEN	4.1	F2	III	4.1	536	LQ0		PP			0	
2305	NITROBENZENE-SULPHONIC ACID	8	C4	II	8		LQ23		PP, EP			0	
2306	NITROBENZO-TRIFLUORIDES, LIQUID	6.1	T1	II	6.1	802	LQ17		PP, EP, TOX, A	VE02		2	
2307	3-NITRO-4-CHLORO-BENZOTRIFLUORIDE	6.1	T1	II	6.1	802	LQ17		PP, EP, TOX, A	VE02		2	
2308	NITROSYLSULPHURIC ACID, LIQUID	8	C1	II	8		LQ22		PP, EP			0	
2309	OCTADIENE	3	F1	II	3		LQ4	T	PP, EX, A	VE01		1	
2310	PENTANE-2,4-DIONE	3	FT1	III	3+6.1	802	LQ7		PP, EP, EX, TOX, A	VE01, VE02		0	
2311	PHENETIDINES	6.1	T1	III	6.1	279 802	LQ7	T	PP, EP, TOX, A	VE02		0	
2312	PHENOL, MOLTEN	6.1	T1	II	6.1	802	LQ0		PP, EP, TOX, A	VE02		2	
2313	PICOLINES	3	F1	III	3		LQ7		PP, EX, A	VE01		0	
2315	POLYCHLORINATED BIPHENYLS, LIQUID	9	M2	II	9	305 802	LQ26		PP, EP			0	
2316	SODIUM CUPROCYANIDE, SOLID	6.1	T5	I	6.1	802	LQ0		PP, EP			2	

UN No. or ID No.	Name and description	Class	Classification Code	Packing group	Labels	Special provisions	Limited quantities	Carriage permitted	Equipment required	Ventilation	Provisions concerning loading, unloading and carriage	Number of cones, blue lights	Remarks
3.1.2	3.1.2	2.2	2.2	2.1.1.3	5.2.2	3.3	3.4.6	3.2.1	8.1.5	7.1.6	7.1.6	7.1.5	3.2.1
(1)	(2)	(3a)	(3b)	(4)	(5)	(6)	(7)	(8)	(9)	(10)	(11)	(12)	(13)
2317	SODIUM CUPROCYANIDE SOLUTION	6.1	T4	I	6.1	802	LQ0		PP, EP			2	
2318	SODIUM HYDROSULPHIDE with less than 25% water of crystallization	4.2	S4	II	4.2	504	LQ0		PP			0	
2319	TERPENE HYDROCARBONS, N.O.S.	3	F1	III	3		LQ7		PP, EX, A	VE01		0	
2320	TETRAETHYLENE-PENTAMINE	8	C7	III	8		LQ7	T	PP, EP			0	
2321	TRICHLOROBENZENES, LIQUID	6.1	T1	III	6.1	802	LQ7	T	PP, EP, TOX, A	VE02		0	
2322	TRICHLOROBUTENE	6.1	T1	II	6.1	802	LQ17		PP, EP, TOX, A	VE02		2	
2323	TRIETHYL PHOSPHITE	3	F1	III	3		LQ7	T	PP, EX, A	VE01		0	
2324	TRIISOBUTYLENE	3	F1	III	3		LQ7	T	PP, EX, A	VE01		0	
2325	1,3,5-TRIMETHYLBENZENE	3	F1	III	3		LQ7	T	PP, EX, A	VE01		0	
2326	TRIMETHYLCYCLO-HEXYLAMINE	8	C7	III	8		LQ7		PP, EP			0	
2327	TRIMETHYLHEXA-METHYLENEDIAMINES	8	C7	III	8		LQ7		PP, EP			0	
2328	TRIMETHYL-HEXAMETHYLENE DIISOCYANATE	6.1	T1	III	6.1	802	LQ7		PP, EP, TOX, A	VE02		0	
2329	TRIMETHYL PHOSPHITE	3	F1	III	3		LQ7		PP, EX, A	VE01		0	
2330	UNDECANE	3	F1	III	3		LQ7		PP, EX, A	VE01		0	
2331	ZINC CHLORIDE, ANHYDROUS	8	C2	III	8		LQ24		PP, EP			0	
2332	ACETALDEHYDE OXIME	3	F1	III	3		LQ7		PP, EX, A	VE01		0	
2333	ALLYL ACETATE	3	FT1	II	3+6.1	802	LQ0	T	PP, EP, EX, TOX, A	VE01, VE02		2	
2334	ALLYLAMINE	6.1	TF1	I	6.1+3	802	LQ0		PP, EP, EX, TOX, A	VE01, VE02		2	
2335	ALLYL ETHYL ETHER	3	FT1	II	3+6.1	802	LQ0		PP, EP, EX, TOX, A	VE01, VE02		2	
2336	ALLYL FORMATE	3	FT1	I	3+6.1	802	LQ0		PP, EP, EX, TOX, A	VE01, VE02		2	
2337	PHENYL MERCAPTAN	6.1	TF1	I	6.1+3	802	LQ0		PP, EP, EX, TOX, A	VE01, VE02		2	
2338	BENZOTRIFLUORIDE	3	F1	II	3		LQ4		PP, EX, A	VE01		1	

UN No. or ID No. (1) 3.1.2	Name and description (2) 3.1.2	Class (3a) 2.2	Classification Code (3b) 2.2	Packing group (4) 2.1.1.3	Labels (5) 5.2.2	Special provisions (6) 3.3	Limited quantities (7) 3.4.6	Carriage permitted (8) 3.2.1	Equipment required (9) 8.1.5	Ventilation (10) 7.1.6	Provisions concerning loading, unloading and carriage (11) 7.1.6	Number of cones, blue lights (12) 7.1.5	Remarks (13) 3.2.1
2339	2-BROMOBUTANE	3	F1	II	3		LQ4		PP, EX, A	VE01		1	
2340	2-BROMOETHYL ETHYL ETHER	3	F1	II	3		LQ4		PP, EX, A	VE01		1	
2341	1-BROMO-3-METHYLBUTANE	3	F1	III	3		LQ7		PP, EX, A	VE01		0	
2342	BROMOMETHYLPROPANES	3	F1	II	3		LQ4		PP, EX, A	VE01		1	
2343	2-BROMOPENTANE	3	F1	II	3		LQ4		PP, EX, A	VE01		1	
2344	BROMOPROPANES	3	F1	II	3		LQ4		PP, EX, A	VE01		1	
2344	BROMOPROPANES	3	F1	III	3		LQ7		PP, EX, A	VE01		0	
2345	3-BROMOPROPYNE	3	F1	II	3		LQ4		PP, EX, A	VE01		1	
2346	BUTANEDIONE	3	F1	II	3		LQ4		PP, EX, A	VE01		1	
2347	BUTYL MERCAPTAN	3	F1	II	3		LQ4		PP, EX, A	VE01		1	
2348	BUTYL ACRYLATES, STABILIZED	3	F1	III	3		LQ7	T	PP, EX, A	VE01		0	
2350	BUTYL METHYL ETHER	3	F1	II	3		LQ4	T	PP, EX, A	VE01		1	
2351	BUTYL NITRITES	3	F1	II	3		LQ4		PP, EX, A	VE01		1	
2351	BUTYL NITRITES	3	F1	III	3		LQ7		PP, EX, A	VE01		0	
2352	BUTYL VINYL ETHER, STABILIZED	3	F1	II	3		LQ4		PP, EX, A	VE01		1	
2353	BUTYRYL CHLORIDE	3	FC	II	3+8		LQ4		PP, EP, EX, A	VE01		1	
2354	CHLOROMETHYL ETHYL ETHER	3	FT1	II	3+6.1	802	LQ0		PP, EP, EX, TOX, A	VE01, VE02		2	
2356	2-CHLOROPROPANE	3	F1	I	3		LQ3	T	PP, EX, A	VE01		1	
2357	CYCLOHEXYLAMINE	8	CF1	II	8+3		LQ22	T	PP, EP, EX, A	VE01		1	
2358	CYCLOOCTATETRAENE	3	F1	II	3		LQ4		PP, EX, A	VE01		1	
2359	DIALLYLAMINE	3	FTC	II	3+6.1+8	802	LQ0		PP, EP, EX, TOX, A	VE01, VE02		2	
2360	DIALLYL ETHER	3	FT1	II	3+6.1	802	LQ0		PP, EP, EX, TOX, A	VE01, VE02		2	
2361	DIISOBUTYLAMINE	3	FC	III	3+8		LQ7		PP, EP, EX, A	VE01		0	
2362	1,1-DICHLOROETHANE	3	F1	II	3		LQ4	T	PP, EX, A	VE01		1	
2363	ETHYL MERCAPTAN	3	F1	I	3		LQ3		PP, EX, A	VE01		1	
2364	n-PROPYLBENZENE	3	F1	III	3		LQ7		PP, EX, A	VE01		0	
2366	DIETHYL CARBONATE	3	F1	III	3		LQ7		PP, EX, A	VE01		0	

(1)	(2)	(3a)	(3b)	(4)	(5)	(6)	(7)	(8)	(9)	(10)	(11)	(12)	(13)
UN No. or ID No.	Name and description	Class	Classification Code	Packing group	Labels	Special provisions	Limited quantities	Carriage permitted	Equipment required	Ventilation	Provisions concerning loading, unloading and carriage	Number of cones, blue lights	Remarks
3.1.2	3.1.2	2.2	2.2	2.1.1.3	5.2.2	3.3	3.4.6	3.2.1	8.1.5	7.1.6	7.1.6	7.1.5	3.2.1
2367	alpha-METHYLVALERALDEHYDE	3	F1	II	3		LQ4		PP, EX, A	VE01		1	
2368	alpha-PINENE	3	F1	III	3		LQ7		PP, EX, A	VE01		0	
2370	1-HEXENE	3	F1	II	3		LQ4	T	PP, EX, A	VE01		1	
2371	ISOPENTENES	3	F1	I	3		LQ3		PP, EX, A	VE01		1	
2372	1,2-DI-(DIMETHYLAMINO)ETHANE	3	F1	II	3		LQ4		PP, EX, A	VE01		1	
2373	DIETHOXYMETHANE	3	F1	II	3		LQ4		PP, EX, A	VE01		1	
2374	3,3-DIETHOXYPROPENE	3	F1	II	3		LQ4		PP, EX, A	VE01		1	
2375	DIETHYL SULPHIDE	3	F1	II	3		LQ4		PP, EX, A	VE01		1	
2376	2,3-DIHYDROPYRAN	3	F1	II	3		LQ4		PP, EX, A	VE01		1	
2377	1,1-DIMETHOXYETHANE	3	F1	II	3		LQ4		PP, EX, A	VE01		1	
2378	2-DIMETHYLAMINO-ACETONITRILE	3	FT1	II	3+6.1	802	LQ0		PP, EP, EX, TOX, A	VE01, VE02		2	
2379	1,3-DIMETHYLBUTYLAMINE	3	FC	II	3+8		LQ4		PP, EP, EX, A	VE01		1	
2380	DIMETHYLDIETHOXY-SILANE	3	F1	II	3		LQ4		PP, EX, A	VE01		1	
2381	DIMETHYL DISULPHIDE	3	F1	II	3		LQ4		PP, EX, A	VE01		1	
2382	DIMETHYLHYDRAZINE, SYMMETRICAL	6.1	TF1	I	6.1+3	802	LQ0	T	PP, EP, EX, TOX, A	VE01, VE02		2	
2383	DIPROPYLAMINE	3	FC	II	3+8		LQ4	T	PP, EP, EX, A	VE01		1	
2384	DI-n-PROPYL ETHER	3	F1	II	3		LQ4		PP, EX, A	VE01		1	
2385	ETHYL ISOBUTYRATE	3	F1	II	3		LQ4		PP, EX, A	VE01		1	
2386	1-ETHYLPIPERIDINE	3	FC	II	3+8		LQ4		PP, EP, EX, A	VE01		1	
2387	FLUOROBENZENE	3	F1	II	3		LQ4		PP, EX, A	VE01		1	
2388	FLUOROTOLUENES	3	F1	II	3		LQ4		PP, EX, A	VE01		1	
2389	FURAN	3	F1	I	3		LQ3		PP, EX, A	VE01		1	
2390	2-IODOBUTANE	3	F1	II	3		LQ4		PP, EX, A	VE01		1	
2391	IODOMETHYLPROPANES	3	F1	II	3		LQ4		PP, EX, A	VE01		1	
2392	IODOPROPANES	3	F1	III	3		LQ7		PP, EX, A	VE01		0	
2393	ISOBUTYL FORMATE	3	F1	II	3		LQ4		PP, EX, A	VE01		1	
2394	ISOBUTYL PROPIONATE	3	F1	III	3		LQ7		PP, EX, A	VE01		0	
2395	ISOBUTYRYL CHLORIDE	3	FC	II	3+8		LQ4		PP, EP, EX, A	VE01		1	

UN No. or ID No.	Name and description	Class	Classification Code	Packing group	Labels	Special provisions	Limited quantities	Carriage permitted	Equipment required	Ventilation	Provisions concerning loading, unloading and carriage	Number of cones, blue lights	Remarks
	3.1.2	2.2	2.2	2.1.1.3	5.2.2	3.3	3.4.6	3.2.1	8.1.5	7.1.6	7.1.6	7.1.5	3.2.1
(1)	(2)	(3a)	(3b)	(4)	(5)	(6)	(7)	(8)	(9)	(10)	(11)	(12)	(13)
2396	METHACRYLALDEHYDE, STABILIZED	3	FT1	II	3+6.1	802	LQ0		PP, EP, EX, TOX, A	VE01, VE02		2	
2397	3-METHYLBUTAN-2-ONE	3	F1	II	3		LQ4	T	PP, EX, A	VE01		1	
2398	METHYL tert-BUTYL ETHER	3	F1	II	3		LQ4	T	PP, EX, A	VE01		1	
2399	1-METHYLPIPERIDINE	3	FC	II	3+8		LQ4		PP, EP, EX, A	VE01		1	
2400	METHYL ISOVALERATE	3	F1	II	3		LQ4		PP, EX, A	VE01		1	
2401	PIPERIDINE	8	CF1	I	8+3		LQ0		PP, EP, EX, A	VE01		1	
2402	PROPANETHIOLS	3	F1	II	3		LQ4		PP, EX, A	VE01		1	
2403	ISOPROPENYL ACETATE	3	F1	II	3		LQ4		PP, EX, A	VE01		1	
2404	PROPIONITRILE	3	FT1	II	3+6.1	802	LQ0	T	PP, EP, EX, TOX, A	VE01, VE02		2	
2405	ISOPROPYL BUTYRATE	3	F1	III	3		LQ7		PP, EX, A	VE01		0	
2406	ISOPROPYL ISOBUTYRATE	3	F1	II	3		LQ4		PP, EX, A	VE01		1	
2407	ISOPROPYL CHLOROFORMATE	6.1	TFC	I	6.1+3+8	802	LQ0		PP, EP, EX, TOX, A	VE01, VE02		2	
2409	ISOPROPYL PROPIONATE	3	F1	II	3		LQ4		PP, EX, A	VE01		1	
2410	1,2,3,6-TETRAHYDROPYRIDINE	3	F1	II	3		LQ4		PP, EX, A	VE01		1	
2411	BUTYRONITRILE	3	FT1	II	3+6.1	802	LQ0		PP, EP, EX, TOX, A	VE01, VE02		2	
2412	TETRAHYDROTHIOPHENE	3	F1	II	3		LQ4	T	PP, EX, A	VE01		1	
2413	TETRAPROPYL ORTHOTITANATE	3	F1	III	3		LQ7		PP, EX, A	VE01		0	
2414	THIOPHENE	3	F1	II	3		LQ4		PP, EX, A	VE01		1	
2416	TRIMETHYL BORATE	3	F1	II	3		LQ4		PP, EX, A	VE01		1	
2417	CARBONYL FLUORIDE	2	2TC		2.3+8		LQ0		PP, EP, TOX, A	VE02		2	
2418	SULPHUR TETRAFLUORIDE	2	2TC		2.3+8		LQ0		PP, EP, TOX, A	VE02		2	
2419	BROMOTRIFLUORO-ETHYLENE	2	2F		2.1		LQ0		PP, EP, EX, A	VE01		1	
2420	HEXAFLUOROACETONE	2	2TC		2.3+8		LQ0		PP, EP, TOX, A	VE02		2	
2421	NITROGEN TRIOXIDE	2	2TOC					CARRIAGE PROHIBITED					
2422	OCTAFLUOROBUT-2-ENE (REFRIGERANT GAS R 1318)	2	2A		2.2		LQ1		PP			0	

UN No. or ID No.	Name and description	Class	Classification Code	Packing group	Labels	Special provisions	Limited quantities	Carriage permitted	Equipment required	Ventilation	Provisions concerning loading, unloading and carriage	Number of cones, blue lights	Remarks
(1)	3.1.2 (2)	2.2 (3a)	2.2 (3b)	2.1.1.3 (4)	5.2.2 (5)	3.3 (6)	3.4.6 (7)	3.2.1 (8)	8.1.5 (9)	7.1.6 (10)	7.1.6 (11)	7.1.5 (12)	3.2.1 (13)
2424	OCTAFLUOROPROPANE (REFRIGERANT GAS R 218)	2	2A		2.2		LQ1		PP			0	
2426	AMMONIUM NITRATE, LIQUID, hot concentrated solution, in a concentration of more than 80% but not more than 93%	5.1	O1		5.1	252 644	LQ0		PP			0	
2427	POTASSIUM CHLORATE, AQUEOUS SOLUTION	5.1	O1	II	5.1		LQ10		PP			0	
2427	POTASSIUM CHLORATE, AQUEOUS SOLUTION	5.1	O1	III	5.1		LQ13		PP			0	
2428	SODIUM CHLORATE, AQUEOUS SOLUTION	5.1	O1	II	5.1		LQ10		PP			0	
2428	SODIUM CHLORATE, AQUEOUS SOLUTION	5.1	O1	III	5.1		LQ13		PP			0	
2429	CALCIUM CHLORATE, AQUEOUS SOLUTION	5.1	O1	II	5.1		LQ10		PP			0	
2429	CALCIUM CHLORATE, AQUEOUS SOLUTION	5.1	O1	III	5.1		LQ13		PP			0	
2430	ALKYLPHENOLS, SOLID, N.O.S. (including C$_2$-C$_{12}$ homologues)	8	C4	I	8	274	LQ0		PP, EP			0	
2430	ALKYLPHENOLS, SOLID, N.O.S. (including C$_2$-C$_{12}$ homologues)	8	C4	II	8	274	LQ23	T	PP, EP			0	
2430	ALKYLPHENOLS, SOLID, N.O.S. (including C$_2$-C$_{12}$ homologues)	8	C4	III	8	274	LQ24		PP, EP			0	
2431	ANISIDINES	6.1	T1	III	6.1	802	LQ7		PP, EP, TOX, A	VE02		0	
2432	N,N-DIETHYLANILINE	6.1	T1	III	6.1	279 802	LQ7	T	PP, EP, TOX, A	VE02		0	
2433	CHLORONITROTOLUENES, LIQUID	6.1	T1	III	6.1	802	LQ7		PP, EP, TOX, A	VE02		0	
2434	DIBENZYLDICHLORO-SILANE	8	C3	II	8		LQ22		PP, EP			0	
2435	ETHYLPHENYLDICHLORO-SILANE	8	C3	II	8		LQ22		PP, EP			0	
2436	THIOACETIC ACID	3	F1	II	3		LQ4		PP, EX, A	VE01		1	
2437	METHYLPHENYL-DICHLOROSILANE	8	C3	II	8		LQ22		PP, EP			0	

- 319 -

UN No. or ID No. (1) 3.1.2	Name and description (2) 3.1.2	Class (3a) 2.2	Classification Code (3b) 2.2	Packing group (4) 2.1.1.3	Labels (5) 5.2.2	Special provisions (6) 3.3	Limited quantities (7) 3.4.6	Carriage permitted (8) 3.2.1	Equipment required (9) 8.1.5	Ventilation (10) 7.1.6	Provisions concerning loading, unloading and carriage (11) 7.1.6	Number of cones, blue lights (12) 7.1.5	Remarks (13) 3.2.1
2438	TRIMETHYLACETYL CHLORIDE	6.1	TFC	I	6.1+3+8	802	LQ0		PP, EP, EX, TOX, A	VE01, VE02		2	
2439	SODIUM HYDROGENDIFLUORIDE	8	C2	II	8		LQ23		PP, EP			0	
2440	STANNIC CHLORIDE PENTAHYDRATE	8	C2	III	8		LQ24		PP, EP			0	
2441	TITANIUM TRICHLORIDE, PYROPHORIC or TITANIUM TRICHLORIDE MIXTURE, PYROPHORIC	4.2	SC4	I	4.2+8	537	LQ0		PP			0	
2442	TRICHLOROACETYL CHLORIDE	8	C3	II	8		LQ22		PP, EP			0	
2443	VANADIUM OXYTRICHLORIDE	8	C1	II	8		LQ22		PP, EP			0	
2444	VANADIUM TETRACHLORIDE	8	C1	I	8		LQ0		PP, EP			0	
2446	NITROCRESOLS, SOLID	6.1	T2	III	6.1	802	LQ9		PP, EP			0	
2447	PHOSPHORUS, WHITE, MOLTEN	4.2	ST3	I	4.2+6.1	802	LQ0		PP, EP, TOX, A	VE02		2	
2448	SULPHUR, MOLTEN	4.1	F3	III	4.1	538	LQ0	T	PP			0	
2451	NITROGEN TRIFLUORIDE	2	2O		2.2+5.1		LQ0		PP			0	
2452	ETHYLACETYLENE, STABILIZED	2	2F		2.1		LQ0		PP, EX, A	VE01		1	
2453	ETHYL FLUORIDE (REFRIGERANT GAS R 161)	2	2F		2.1		LQ0		PP, EX, A	VE01		1	
2454	METHYL FLUORIDE (REFRIGERANT GAS R 41)	2	2F		2.1		LQ0		PP, EX, A	VE01		1	
2455	METHYL NITRITE	2	2A						CARRIAGE PROHIBITED				
2456	2-CHLOROPROPENE	3	F1	I	3		LQ3		PP, EX, A	VE01		1	
2457	2,3-DIMETHYLBUTANE	3	F1	II	3		LQ4		PP, EX, A	VE01		1	
2458	HEXADIENES	3	F1	II	3		LQ4	T	PP, EX, A	VE01		1	
2459	2-METHYL-1-BUTENE	3	F1	I	3		LQ3		PP, EX, A	VE01		1	
2460	2-METHYL-2-BUTENE	3	F1	II	3		LQ4		PP, EX, A	VE01		1	
2461	METHYLPENTADIENE	3	F1	II	3		LQ4		PP, EX, A	VE01		1	
2463	ALUMINIUM HYDRIDE	4.3	W2	I	4.3		LQ0		PP, EX, A	VE01	HA08	0	
2464	BERYLLIUM NITRATE	5.1	OT2	II	5.1+6.1	802	LQ11		PP			2	

UN No. or ID No. (1) 3.1.2	Name and description (2) 3.1.2	Class (3a) 2.2	Classification Code (3b) 2.2	Packing group (4) 2.1.1.3	Labels (5) 5.2.2	Special provisions (6) 3.3	Limited quantities (7) 3.4.6	Carriage permitted (8) 3.2.1	Equipment required (9) 8.1.5	Ventilation (10) 7.1.6	Provisions concerning loading, unloading and carriage (11) 7.1.6	Number of cones, blue lights (12) 7.1.5	Remarks (13) 3.2.1
2465	DICHLOROISOCYANURIC ACID, DRY or DICHLOROISOCYANURIC ACID SALTS	5.1	O2	II	5.1	135	LQ11		PP			0	
2466	POTASSIUM SUPEROXIDE	5.1	O2	I	5.1		LQ0		PP			0	
2468	TRICHLOROISOCYANURIC ACID, DRY	5.1	O2	II	5.1		LQ11		PP			0	
2469	ZINC BROMATE	5.1	O2	III	5.1		LQ12		PP			0	
2470	PHENYLACETONITRILE, LIQUID	6.1	T1	III	6.1	802	LQ7		PP, EP, TOX, A	VE02		0	
2471	OSMIUM TETROXIDE	6.1	T5	I	6.1	802	LQ0		PP, EP			2	
2473	SODIUM ARSANILATE	6.1	T3	III	6.1	802	LQ9		PP, EP, TOX, A	VE02		0	
2474	THIOPHOSGENE	6.1	T1	II	6.1	279 802	LQ17		PP, EP, TOX, A	VE02		2	
2475	VANADIUM TRICHLORIDE	8	C2	III	8		LQ24		PP, EP			0	
2477	METHYL ISOTHIOCYANATE	6.1	TF1	I	6.1+3	802	LQ0	T	PP, EP, EX, TOX, A	VE01, VE02		2	
2478	ISOCYANATES, FLAMMABLE, TOXIC, N.O.S. or ISOCYANATE SOLUTION, FLAMMABLE, TOXIC, N.O.S.	3	FT1	II	3+6.1	274 539 802	LQ0		PP, EP, EX, TOX, A	VE01, VE02		2	
2478	ISOCYANATES, FLAMMABLE, TOXIC, N.O.S. or ISOCYANATE SOLUTION, FLAMMABLE, TOXIC, N.O.S.	3	FT1	III	3+6.1	274 802	LQ7		PP, EP, EX, TOX, A	VE01, VE02		0	
2480	METHYL ISOCYANATE	6.1	TF1	I	6.1+3	802	LQ0		PP, EP, EX, TOX, A	VE01, VE02		2	
2481	ETHYL ISOCYANATE	3	FT1	I	3+6.1	802	LQ0		PP, EP, EX, TOX, A	VE01, VE02		2	
2482	n-PROPYL ISOCYANATE	6.1	TF1	I	6.1+3	802	LQ0		PP, EP, EX, TOX, A	VE01, VE02		2	
2483	ISOPROPYL ISOCYANATE	3	FT1	I	3+6.1	802	LQ0		PP, EP, EX, TOX, A	VE01, VE02		2	
2484	tert-BUTYL ISOCYANATE	6.1	TF1	I	6.1+3	802	LQ0		PP, EP, EX, TOX, A	VE01, VE02		2	

UN No. or ID No.	Name and description	Class	Classification Code	Packing group	Labels	Special provisions	Limited quantities	Carriage permitted	Equipment required	Ventilation	Provisions concerning loading, unloading and carriage	Number of cones, blue lights	Remarks
3.1.2	3.1.2	2.2	2.2	2.1.1.3	5.2.2	3.3	3.4.6	3.2.1	8.1.5	7.1.6	7.1.6	7.1.5	3.2.1
(1)	(2)	(3a)	(3b)	(4)	(5)	(6)	(7)	(8)	(9)	(10)	(11)	(12)	(13)
2485	n-BUTYL ISOCYANATE	6.1	TF1	I	6.1+3	802	LQ0	T	PP, EP, EX, TOX, A	VE01, VE02		2	
2486	ISOBUTYL ISOCYANATE	3	FT1	II	3+6.1	802	LQ0	T	PP, EP, EX, TOX, A	VE01, VE02		2	
2487	PHENYL ISOCYANATE	6.1	TF1	I	6.1+3	802	LQ0	T	PP, EP, EX, TOX, A	VE01, VE02		2	
2488	CYCLOHEXYL ISOCYANATE	6.1	TF1	I	6.1+3	802	LQ0	T	PP, EP, EX, TOX, A	VE01, VE02		2	
2490	DICHLOROISOPROPYL ETHER	6.1	T1	II	6.1	802	LQ17	T	PP, EP, TOX, A	VE02		2	
2491	ETHANOLAMINE or ETHANOLAMINE SOLUTION	8	C7	III	8		LQ7	T	PP, EP			0	
2493	HEXAMETHYLENEIMINE	3	FC	II	3+8		LQ4	T	PP, EP, EX, A	VE01		1	
2495	IODINE PENTAFLUORIDE	5.1	OTC	I	5.1+6.1+8	802	LQ0	T	PP, EP, TOX, A	VE02		2	
2496	PROPIONIC ANHYDRIDE	8	C3	III	8		LQ7		PP, EP			0	
2498	1.2.3.6-TETRAHYDRO-BENZALDEHYDE	3	F1	III	3		LQ7		PP, EX, A	VE01		0	
2501	TRIS-(1-AZIRIDINYL) PHOSPHINE OXIDE SOLUTION	6.1	T1	II	6.1	802	LQ17		PP, EP, TOX, A	VE02		2	
2501	TRIS-(1-AZIRIDINYL) PHOSPHINE OXIDE SOLUTION	6.1	T1	III	6.1	802	LQ7		PP, EP, TOX, A	VE02		0	
2502	VALERYL CHLORIDE	8	CF1	II	8+3		LQ22		PP, EP, EX, A	VE01		1	
2503	ZIRCONIUM TETRACHLORIDE	8	C2	III	8		LQ24		PP, EP			0	
2504	TETRABROMOETHANE	6.1	T1	III	6.1	802	LQ7		PP, EP, TOX, A	VE02		0	
2505	AMMONIUM FLUORIDE	6.1	T5	III	6.1	802	LQ9	B	PP, EP			0	
2506	AMMONIUM HYDROGEN SULPHATE	8	C2	II	8		LQ23	B	PP, EP		CO03	0	CO03 applies only when this substance is carried in bulk or without packaging
2507	CHLOROPLATINIC ACID, SOLID	8	C2	III	8		LQ24		PP, EP			0	

UN No. or ID No.	Name and description	Class	Classification Code	Packing group	Labels	Special provisions	Limited quantities	Carriage permitted	Equipment required	Ventilation	Provisions concerning loading, unloading and carriage	Number of cones, blue lights	Remarks
3.1.2	3.1.2	2.2	2.2	2.1.1.3	5.2.2	3.3	3.4.6	3.2.1	8.1.5	7.1.6	7.1.6	7.1.5	3.2.1
(1)	(2)	(3a)	(3b)	(4)	(5)	(6)	(7)	(8)	(9)	(10)	(11)	(12)	(13)
2508	MOLYBDENUM PENTACHLORIDE	8	C2	III	8		LQ24		PP, EP			0	
2509	POTASSIUM HYDROGEN SULPHATE	8	C2	II	8		LQ23	B	PP, EP		CO03	0	CO03 applies only when this substance is carried in bulk or without packaging
2511	2-CHLOROPROPIONIC ACID	8	C3	III	8		LQ7		PP, EP			0	
2512	AMINOPHENOLS (o-, m-, p-)	6.1	T2	III	6.1	279 802	LQ9		PP, EP			0	
2513	BROMOACETYL BROMIDE	8	C3	II	8		LQ22		PP, EP			0	
2514	BROMOBENZENE	3	F1	III	3		LQ7		PP, EX, A	VE01		0	
2515	BROMOFORM	6.1	T1	III	6.1	802	LQ7		PP, EP, TOX, A	VE02		0	
2516	CARBON TETRABROMIDE	6.1	T2	III	6.1	802	LQ9		PP, EP			0	
2517	1-CHLORO-1,1-DIFLUOROETHANE (REFRIGERANT GAS R 142b)	2	2F		2.1		LQ0		PP, EX, A	VE01		1	
2518	1,5,9-CYCLODODECATRIENE	6.1	T1	III	6.1	802	LQ7	T	PP, EP, TOX, A	VE02		0	
2520	CYCLOOCTADIENES	3	F1	III	3		LQ7		PP, EX, A	VE01		0	
2521	DIKETENE, STABILIZED	6.1	TF1	I	6.1+3	802	LQ0		PP, EP, EX, TOX, A	VE01, VE02		2	
2522	2-DIMETHYLAMINOETHYL METHACRYLATE	6.1	T1	II	6.1	802	LQ17		PP, EP, TOX, A	VE02		2	
2524	ETHYL ORTHOFORMATE	3	F1	III	3		LQ7		PP, EX, A	VE01		0	
2525	ETHYL OXALATE	6.1	T1	III	6.1	802	LQ7		PP, EP, TOX, A	VE02		0	
2526	FURFURYLAMINE	3	FC	III	3+8		LQ7		PP, EP, EX, A	VE01		0	
2527	ISOBUTYL ACRYLATE, STABILIZED	3	F1	III	3		LQ7	T	PP, EX, A	VE01		0	
2528	ISOBUTYL ISOBUTYRATE	3	F1	III	3		LQ7	T	PP, EX, A	VE01		0	
2529	ISOBUTYRIC ACID	3	FC	III	3+8		LQ7	T	PP, EP, EX, A	VE01		0	
2531	METHACRYLIC ACID, STABILIZED	8	C3	II	8		LQ22	T	PP, EP			0	

UN No. or ID No.	Name and description	Class	Classification Code	Packing group	Labels	Special provisions	Limited quantities	Carriage permitted	Equipment required	Ventilation	Provisions concerning loading, unloading and carriage	Number of cones, blue lights	Remarks
3.1.2	3.1.2	2.2	2.2	2.1.1.3	5.2.2	3.3	3.4.6	3.2.1	8.1.5	7.1.6	7.1.6	7.1.5	3.2.1
(1)	(2)	(3a)	(3b)	(4)	(5)	(6)	(7)	(8)	(9)	(10)	(11)	(12)	(13)
2533	METHYL TRICHLOROACETATE	6.1	T1	III	6.1	802	LQ7		PP, EP, TOX, A	VE02		0	
2534	METHYLCHLOROSILANE	2	2TFC		2.3+2.1+8		LQ0		PP, EP, EX, TOX, A	VE01, VE02		2	
2535	4-METHYLMORPHOLINE (N-METHYLMORPHOLINE)	3	FC	II	3+8		LQ4		PP, EP, EX, A	VE01		1	
2536	METHYL-TETRAHYDROFURAN	3	F1	II	3		LQ4		PP, EX, A	VE01		1	
2538	NITRONAPHTHALENE	4.1	F1	III	4.1		LQ9		PP			0	
2541	TERPINOLENE	3	F1	III	3		LQ7		PP, EX, A	VE01		0	
2542	TRIBUTYLAMINE	6.1	T1	II	6.1	802	LQ17		PP, EP, TOX, A	VE02		2	
2545	HAFNIUM POWDER, DRY	4.2	S4	I	4.2	540	LQ0		PP			0	
2545	HAFNIUM POWDER, DRY	4.2	S4	II	4.2	540	LQ0		PP			0	
2545	HAFNIUM POWDER, DRY	4.2	S4	III	4.2	540	LQ0		PP			0	
2546	TITANIUM POWDER, DRY	4.2	S4	I	4.2	540	LQ0		PP			0	
2546	TITANIUM POWDER, DRY	4.2	S4	II	4.2	540	LQ0		PP			0	
2546	TITANIUM POWDER, DRY	4.2	S4	III	4.2	540	LQ0		PP			0	
2547	SODIUM SUPEROXIDE	5.1	O2	I	5.1		LQ0		PP			0	
2548	CHLORINE PENTAFLUORIDE	2	2TOC		2.3+5.1+8		LQ0		PP, EP, TOX, A	VE02		2	
2552	HEXAFLUOROACETONE HYDRATE, LIQUID	6.1	T1	II	6.1	802	LQ17		PP, EP, TOX, A	VE02		2	
2554	METHYLALLYL CHLORIDE	3	F1	II	3		LQ4		PP, EX, A	VE01		1	
2555	NITROCELLULOSE WITH WATER (not less than 25% water, by mass)	4.1	D	II	4.1	541	LQ0		PP			0	
2556	NITROCELLULOSE WITH ALCOHOL (not less than 25% alcohol, by mass, and not more than 12.6% nitrogen, by dry mass)	4.1	D	II	4.1	541	LQ0		PP			0	
2557	NITROCELLULOSE, with not more than 12.6% nitrogen, by dry mass, MIXTURE WITH or WITHOUT PLASTICIZER, WITH or WITHOUT PIGMENT	4.1	D	II	4.1	241 541	LQ0		PP			0	

UN No. or ID No. (1) 3.1.2	Name and description (2) 3.1.2	Class (3a) 2.2	Classification Code (3b) 2.2	Packing group (4) 2.1.1.3	Labels (5) 5.2.2	Special provisions (6) 3.3	Limited quantities (7) 3.4.6	Carriage permitted (8) 3.2.1	Equipment required (9) 8.1.5	Ventilation (10) 7.1.6	Provisions concerning loading, unloading and carriage (11) 7.1.6	Number of cones, blue lights (12) 7.1.5	Remarks (13) 3.2.1
2558	EPIBROMOHYDRIN	6.1	TF1	I	6.1+3	802	LQ0		PP, EP, EX, TOX, A	VE01, VE02		2	
2560	2-METHYLPENTAN-2-OL	3	F1	III	3		LQ7		PP, EX, A	VE01		0	
2561	3-METHYL-1-BUTENE	3	F1	I	3		LQ3		PP, EX, A	VE01		1	
2564	TRICHLOROACETIC ACID SOLUTION	8	C3	II	8		LQ22	T	PP, EP			0	
2564	TRICHLOROACETIC ACID SOLUTION	8	C3	III	8		LQ7	T	PP, EP			0	
2565	DICYCLOHEXYLAMINE	8	C7	III	8		LQ7		PP, EP			0	
2567	SODIUM PENTACHLOROPHENATE	6.1	T2	II	6.1	802	LQ18		PP, EP			2	
2570	CADMIUM COMPOUND	6.1	T5	I	6.1	274 596 802	LQ0		PP, EP			2	
2570	CADMIUM COMPOUND	6.1	T5	II	6.1	274 596 802	LQ18		PP, EP			2	
2570	CADMIUM COMPOUND	6.1	T5	III	6.1	274 596 802	LQ9		PP, EP			0	
2571	ALKYLSULPHURIC ACIDS	8	C3	II	8		LQ22		PP, EP			0	
2572	PHENYLHYDRAZINE	6.1	T1	II	6.1	802	LQ17		PP, EP, TOX, A	VE02		2	
2573	THALLIUM CHLORATE	5.1	OT2	II	5.1+6.1	802	LQ11		PP			2	
2574	TRICRESYL PHOSPHATE with more than 3% ortho isomer	6.1	T1	II	6.1	802	LQ17	T	PP, EP, TOX, A	VE02		2	
2576	PHOSPHORUS OXYBROMIDE, MOLTEN	8	C1	II	8		LQ0		PP, EP			0	
2577	PHENYLACETYL CHLORIDE	8	C3	II	8		LQ22		PP, EP			0	
2578	PHOSPHORUS TRIOXIDE	8	C2	III	8		LQ24		PP, EP			0	
2579	PIPERAZINE	8	C8	III	8		LQ24	T	PP, EP			0	
2580	ALUMINIUM BROMIDE SOLUTION	8	C1	III	8		LQ7		PP, EP			0	
2581	ALUMINIUM CHLORIDE SOLUTION	8	C1	III	8		LQ7		PP, EP			0	
2582	FERRIC CHLORIDE SOLUTION	8	C1	III	8		LQ7		PP, EP			0	

UN No. or ID No.	Name and description	Class	Classification Code	Packing group	Labels	Special provisions	Limited quantities	Carriage permitted	Equipment required	Ventilation	Provisions concerning loading, unloading and carriage	Number of cones, blue lights	Remarks
	3.1.2	2.2	2.2	2.1.1.3	5.2.2	3.3	3.4.6	3.2.1	8.1.5	7.1.6	7.1.6	7.1.5	3.2.1
(1)	(2)	(3a)	(3b)	(4)	(5)	(6)	(7)	(8)	(9)	(10)	(11)	(12)	(13)
2583	ALKYLSULPHONIC ACIDS, SOLID or ARYLSULPHONIC ACIDS, SOLID with more than 5% free sulphuric acid	8	C2	II	8	274	LQ23		PP, EP			0	
2584	ALKYLSULPHONIC ACIDS, LIQUID or ARYLSULPHONIC ACIDS, LIQUID with more than 5% free sulphuric acid	8	C1	II	8	274	LQ22		PP, EP			0	
2585	ALKYLSULPHONIC ACIDS, SOLID or ARYLSULPHONIC ACIDS, SOLID with not more than 5% free sulphuric acid	8	C4	III	8	274	LQ24		PP, EP			0	
2586	ALKYLSULPHONIC ACIDS, LIQUID or ARYLSULPHONIC ACIDS, LIQUID with not more than 5% free sulphuric acid	8	C3	III	8	274	LQ7	T	PP, EP			0	
2587	BENZOQUINONE	6.1	T2	II	6.1	802	LQ18		PP, EP			2	
2588	PESTICIDE, SOLID, TOXIC, N.O.S.	6.1	T7	I	6.1	61 648 802	LQ0		PP, EP			2	
2588	PESTICIDE, SOLID, TOXIC, N.O.S.	6.1	T7	II	6.1	61 648 802	LQ18		PP, EP			2	
2588	PESTICIDE, SOLID, TOXIC, N.O.S.	6.1	T7	III	6.1	61 648 802	LQ9		PP, EP			0	
2589	VINYL CHLOROACETATE	6.1	TF1	II	6.1+3	802	LQ17		PP, EP, EX, TOX, A			2	
2590	WHITE ASBESTOS (chrysotile, actinolite, anthophyllite, tremolite)	9	M1	III	9	168 542 802	LQ27		PP	VE01, VE02		0	
2591	XENON, REFRIGERATED LIQUID	2	3A		2.2	593	LQ1		PP			0	

UN No. or ID No.	Name and description	Class	Classification Code	Packing group	Labels	Special provisions	Limited quantities	Carriage permitted	Equipment required	Ventilation	Provisions concerning loading, unloading and carriage	Number of cones, blue lights	Remarks
3.1.2	3.1.2	2.2	2.2	2.1.1.3	5.2.2	3.3	3.4.6	3.2.1	8.1.5	7.1.6	7.1.6	7.1.5	3.2.1
(1)	(2)	(3a)	(3b)	(4)	(5)	(6)	(7)	(8)	(9)	(10)	(11)	(12)	(13)
2599	CHLOROTRIFLUOROMETHANE AND TRIFLUOROMETHANE AZEOTROPIC MIXTURE with approximately 60% chlorotrifluoromethane (REFRIGERANT GAS R 503)	2	2A		2.2		LQ1		PP			0	
2601	CYCLOBUTANE	2	2F		2.1		LQ0		PP, EX, A	VE01		1	
2602	DICHLORODIFLUOROMETHANE AND 1,1-DIFLUOROETHANE AZEOTROPIC MIXTURE with approximately 74% dichlorodifluoromethane (REFRIGERANT GAS R 500)	2	2A		2.2		LQ1		PP			0	
2603	CYCLOHEPTATRIENE	3	FT1	II	3+6.1	802	LQ0		PP, EP, EX, TOX, A	VE01, VE02		2	
2604	BORON TRIFLUORIDE DIETHYL ETHERATE	8	CF1	I	8+3		LQ0		PP, EP, EX, A	VE01		1	
2605	METHOXYMETHYL ISOCYANATE	3	FT1	I	3+6.1	802	LQ0		PP, EP, EX, TOX, A	VE01, VE02		2	
2606	METHYL ORTHOSILICATE	6.1	TF1	I	6.1+3	802	LQ0		PP, EP, EX, TOX, A	VE01, VE02		2	
2607	ACROLEIN DIMER, STABILIZED	3	F1	III	3		LQ7		PP, EX, A	VE01		0	
2608	NITROPROPANES	3	F1	III	3		LQ7	T	PP, EX, A	VE01		0	
2609	TRIALLYL BORATE	6.1	T1	III	6.1	802	LQ7		PP, EP, TOX, A	VE02		0	
2610	TRIALLYLAMINE	3	FC	III	3+8		LQ7	T	PP, EP, EX, A	VE01		0	
2611	PROPYLENE CHLOROHYDRIN	6.1	TF1	II	6.1+3	802	LQ17		PP, EP, EX, TOX, A	VE01, VE02		2	
2612	METHYL PROPYL ETHER	3	F1	II	3		LQ4		PP, EX, A	VE01		1	
2614	METHALLYL ALCOHOL	3	F1	III	3		LQ7		PP, EX, A	VE01		0	
2615	ETHYL PROPYL ETHER	3	F1	II	3		LQ4		PP, EX, A	VE01		1	
2616	TRIISOPROPYL BORATE	3	F1	II	3		LQ4		PP, EX, A	VE01		1	
2616	TRIISOPROPYL BORATE	3	F1	III	3		LQ7		PP, EX, A	VE01		0	

UN No. or ID No.	Name and description	Class	Classification Code	Packing group	Labels	Special provisions	Limited quantities	Carriage permitted	Equipment required	Ventilation	Provisions concerning loading, unloading and carriage	Number of cones, blue lights	Remarks
3.1.2	3.1.2	2.2	2.2	2.1.1.3	5.2.2	3.3	3.4.6	3.2.1	8.1.5	7.1.6	7.1.6	7.1.5	3.2.1
(1)	(2)	(3a)	(3b)	(4)	(5)	(6)	(7)	(8)	(9)	(10)	(11)	(12)	(13)
2617	METHYLCYCLOHEXANOLS, flammable	3	F1	III	3		LQ7		PP, EX, A	VE01		0	
2618	VINYLTOLUENES, STABILIZED	3	F1	III	3		LQ7	T	PP, EX, A	VE01		0	
2619	BENZYLDIMETHYLAMINE	8	CF1	II	8+3		LQ22		PP, EP, EX, A	VE01		1	
2620	AMYL BUTYRATES	3	F1	III	3		LQ7		PP, EX, A	VE01		0	
2621	ACETYL METHYL CARBINOL	3	F1	III	3		LQ7		PP, EX, A	VE01		0	
2622	GLYCIDALDEHYDE	3	FT1	II	3+6.1	802	LQ0		PP, EP, EX, TOX, A	VE01, VE02		2	
2623	FIRELIGHTERS, SOLID with flammable liquid	4.1	F1	III	4.1		LQ9		PP			0	
2624	MAGNESIUM SILICIDE	4.3	W2	II	4.3		LQ11		PP, EX, A	VE01	HA08	0	
2626	CHLORIC ACID, AQUEOUS SOLUTION with not more than 10% chloric acid	5.1	O1	II	5.1	613	LQ10		PP			0	
2627	NITRITES, INORGANIC, N.O.S.	5.1	O2	II	5.1	103 274	LQ11		PP			0	
2628	POTASSIUM FLUOROACETATE	6.1	T2	I	6.1	802	LQ0		PP, EP			2	
2629	SODIUM FLUOROACETATE	6.1	T2	I	6.1	802	LQ0		PP, EP			2	
2630	SELENATES or SELENITES	6.1	T5	I	6.1	274 802	LQ0		PP, EP			2	
2642	FLUOROACETIC ACID	6.1	T2	I	6.1	802	LQ0		PP, EP	VE02		2	
2643	METHYL BROMOACETATE	6.1	T1	II	6.1	802	LQ17		PP, EP, TOX, A			2	
2644	METHYL IODIDE	6.1	T1	I	6.1	802	LQ0		PP, EP, TOX, A	VE02		2	
2645	PHENACYL BROMIDE	6.1	T2	II	6.1	802	LQ18		PP, EP			2	
2646	HEXACHLOROCYCLO-PENTADIENE	6.1	T1	I	6.1	802	LQ0		PP, EP, TOX, A	VE02		2	
2647	MALONONITRILE	6.1	T2	II	6.1	802	LQ18		PP, EP	VE02		2	
2648	1,2-DIBROMOBUTAN-3-ONE	6.1	T1	II	6.1	802	LQ17		PP, EP, TOX, A			2	
2649	1,3-DICHLOROACETONE	6.1	T2	II	6.1	802	LQ18		PP, EP			2	
2650	1,1-DICHLORO-1-NITROETHANE	6.1	T1	II	6.1	802	LQ17		PP, EP, TOX, A	VE02		2	
2651	4,4'-DIAMINODIPHENYL-METHANE	6.1	T2	III	6.1	802	LQ9	T	PP, EP			0	

UN No. or ID No.	Name and description	Class	Classification Code	Packing group	Labels	Special provisions	Limited quantities	Carriage permitted	Equipment required	Ventilation	Provisions concerning loading, unloading and carriage	Number of cones, blue lights	Remarks
	3.1.2	2.2	2.2	2.1.1.3	5.2.2	3.3	3.4.6	3.2.1	8.1.5	7.1.6	7.1.6	7.1.5	3.2.1
(1)	(2)	(3a)	(3b)	(4)	(5)	(6)	(7)	(8)	(9)	(10)	(11)	(12)	(13)
2653	BENZYL IODIDE	6.1	T1	II	6.1	802	LQ17		PP, EP, TOX, A	VE02		2	
2655	POTASSIUM FLUOROSILICATE	6.1	T5	III	6.1	802	LQ9		PP, EP			0	
2656	QUINOLINE	6.1	T1	III	6.1	802	LQ7		PP, EP, TOX, A	VE02		0	
2657	SELENIUM DISULPHIDE	6.1	T5	II	6.1	802	LQ18		PP, EP			2	
2659	SODIUM CHLOROACETATE	6.1	T2	III	6.1	807	LQ9		PP, EP			0	
2660	NITROTOLUIDINES (MONO)	6.1	T2	III	6.1	802	LQ9		PP, EP			0	
2661	HEXACHLOROACETONE	6.1	T1	III	6.1	802	LQ7		PP, EP, TOX, A	VE02		0	
2664	DIBROMOMETHANE	6.1	T1	III	6.1	802	LQ7		PP, EP, TOX, A	VE02		0	
2667	BUTYLTOLUENES	6.1	T1	III	6.1	802	LQ7		PP, EP, TOX, A	VE02		0	
2668	CHLOROACETONITRILE	6.1	TF1	II	6.1+3	802	LQ17		PP, EP, EX, TOX, A	VE01, VE02		2	
2669	CHLOROCRESOLS, SOLUTION	6.1	T1	II	6.1	802	LQ17		PP, EP, TOX, A	VE02		2	
2669	CHLOROCRESOLS, SOLUTION	6.1	T1	III	6.1	802	LQ7		PP, EP, TOX, A	VE02		0	
2670	CYANURIC CHLORIDE	8	C4	II	8		LQ23		PP, EP			0	
2671	AMINOPYRIDINES (o-, m-, p-)	6.1	T2	II	6.1	802	LQ18		PP, EP			2	
2672	AMMONIA SOLUTION, relative density between 0.880 and 0.957 at 15 °C in water, with more than 10% but not more than 35% ammonia	8	C5	III	8	543	LQ7	T	PP, EP			0	
2673	2-AMINO-4-CHLOROPHENOL	6.1	T2	II	6.1	802	LQ18		PP, EP			2	
2674	SODIUM FLUOROSILICATE	6.1	T5	III	6.1	802	LQ9		PP, EP			0	
2676	STIBINE	2	2TF		2.3+2.1		LQ0		PP, EP, EX, TOX, A	VE01, VE02		2	
2677	RUBIDIUM HYDROXIDE SOLUTION	8	C5	II	8		LQ22		PP, EP			0	
2677	RUBIDIUM HYDROXIDE SOLUTION	8	C5	III	8		LQ7		PP, EP			0	
2678	RUBIDIUM HYDROXIDE	8	C6	II	8		LQ23		PP, EP			0	
2679	LITHIUM HYDROXIDE SOLUTION	8	C5	II	8		LQ22		PP, EP			0	

UN No. or ID No.	Name and description	Class	Classification Code	Packing group	Labels	Special provisions	Limited quantities	Carriage permitted	Equipment required	Ventilation	Provisions concerning loading, unloading and carriage	Number of cones, blue lights	Remarks
	3.1.2	2.2	2.2	2.1.1.3	5.2.2	3.3	3.4.6	3.2.1	8.1.5	7.1.6	7.1.6	7.1.5	3.2.1
(1)	(2)	(3a)	(3b)	(4)	(5)	(6)	(7)	(8)	(9)	(10)	(11)	(12)	(13)
2679	LITHIUM HYDROXIDE SOLUTION	8	C5	III	8		LQ7		PP, EP			0	
2680	LITHIUM HYDROXIDE	8	C6	II	8		LQ23		PP, EP			0	
2681	CAESIUM HYDROXIDE SOLUTION	8	C5	II	8		LQ22		PP, EP			0	
2681	CAESIUM HYDROXIDE SOLUTION	8	C5	III	8		LQ7		PP, EP			0	
2682	CAESIUM HYDROXIDE	8	C6	II	8		LQ23		PP, EP			0	
2683	AMMONIUM SULPHIDE SOLUTION	8	CFT	II	8+3+6.1	802	LQ22	T	PP, EP, EX, TOX, A	VE01, VE02		2	
2684	3-DIETHYLAMINO-PROPYLAMINE	3	FC	III	3+8		LQ7		PP, EP, EX, A	VE01		0	
2685	N,N-DIETHYLETHYLENE-DIAMINE	8	CF1	II	8+3		LQ22		PP, EP, EX, A	VE01		1	
2686	2-DIETHYLAMINO-ETHANOL	8	CF1	II	8+3		LQ22		PP, EP, EX, A	VE01		1	
2687	DICYCLOHEXYL-AMMONIUM NITRITE	4.1	F3	III	4.1		LQ9		PP			0	
2688	1-BROMO-3-CHLOROPROPANE	6.1	T1	III	6.1	802	LQ7		PP, EP, TOX, A	VE02		0	
2689	GLYCEROL alpha-MONOCHLOROHYDRIN	6.1	T1	III	6.1	802	LQ7		PP, EP, TOX, A	VE02		0	
2690	N.n-BUTYLIMIDAZOLE	6.1	T1	II	6.1	802	LQ17		PP, EP, TOX, A	VE02		2	
2691	PHOSPHORUS PENTABROMIDE	8	C2	II	8		LQ23		PP, EP			0	
2692	BORON TRIBROMIDE	8	C1	I	8		LQ0		PP, EP			0	
2693	BISULPHITES, AQUEOUS SOLUTION, N.O.S.	8	C1	III	8	274	LQ7	T	PP, EP			0	
2698	TETRAHYDROPHTHALIC ANHYDRIDES with more than 0.05% of maleic anhydride	8	C4	III	8	169	LQ24		PP, EP			0	
2699	TRIFLUOROACETIC ACID	8	C3	I	8		LQ0		PP, EP			0	
2705	1-PENTOL	8	C9	II	8		LQ22		PP, EP			0	
2707	DIMETHYLDIOXANES	3	F1	II	3		LQ4		PP, EX, A	VE01		1	
2707	DIMETHYLDIOXANES	3	F1	III	3		LQ7		PP, EX, A	VE01		0	
2709	BUTYLBENZENES	3	F1	III	3		LQ7		PP, EX, A	VE01		0	
2710	DIPROPYL KETONE	3	F1	III	3		LQ7	T	PP, EX, A	VE01		0	

UN No. or ID No.	Name and description	Class	Classification Code	Packing group	Labels	Special provisions	Limited quantities	Carriage permitted	Equipment required	Ventilation	Provisions concerning loading, unloading and carriage	Number of cones, blue lights	Remarks
	3.1.2	2.2	2.2	2.1.1.3	5.2.2	3.3	3.4.6	3.2.1	8.1.5	7.1.6	7.1.6	7.1.5	3.2.1
(1)	(2)	(3a)	(3b)	(4)	(5)	(6)	(7)	(8)	(9)	(10)	(11)	(12)	(13)
2713	ACRIDINE	6.1	T2	III	6.1	802	LQ9		PP, EP			0	
2714	ZINC RESINATE	4.1	F3	III	4.1		LQ9		PP			0	
2715	ALUMINIUM RESINATE	4.1	F3	III	4.1		LQ9		PP			0	
2716	1,4-BUTYNEDIOL	6.1	T2	III	6.1	802	LQ9		PP, EP			0	
2717	CAMPHOR, synthetic	4.1	F1	III	4.1		LQ9		PP			0	
2719	BARIUM BROMATE	5.1	OT2	II	5.1+6.1	802	LQ11		PP			2	
2720	CHROMIUM NITRATE	5.1	O2	III	5.1		LQ12	B	PP		CO02, LO04	0	CO02 and LO04 apply only when this substance is carried in bulk or without packaging
2721	COPPER CHLORATE	5.1	O2	II	5.1		LQ11		PP			0	
2722	LITHIUM NITRATE	5.1	O2	III	5.1		LQ12	B	PP		CO02, LO04	0	CO02 and LO04 apply only when this substance is carried in bulk or without packaging
2723	MAGNESIUM CHLORATE	5.1	O2	II	5.1		LQ11		PP			0	
2724	MANGANESE NITRATE	5.1	O2	III	5.1		LQ12	B	PP		CO02, LO04	0	CO02 and LO04 apply only when this substance is carried in bulk or without packaging
2725	NICKEL NITRATE	5.1	O2	III	5.1		LQ12	B	PP		CO02, LO04	0	CO02 and LO04 apply only when this substance is carried in bulk or without packaging
2726	NICKEL NITRITE	5.1	O2	III	5.1		LQ12		PP			0	
2727	THALLIUM NITRATE	6.1	TO2	II	6.1+5.1	802	LQ18		PP, EP			2	
2728	ZIRCONIUM NITRATE	5.1	O2	III	5.1		LQ12	B	PP		CO02, LO04	0	CO02 and LO04 apply only when this substance is carried in bulk or without packaging
2729	HEXACHLOROBENZENE	6.1	T2	III	6.1	802	LQ9		PP, EP			0	

UN No. or ID No.	Name and description	Class	Classification Code	Packing group	Labels	Special provisions	Limited quantities	Carriage permitted	Equipment required	Ventilation	Provisions concerning loading, unloading and carriage	Number of cones, blue lights	Remarks
3.1.1	3.1.2	2.2	2.2	2.1.1.3	5.2.2	3.3	3.4.6	3.2.1	8.1.5	7.1.6	7.1.6	7.1.5	3.2.1
(1)	(2)	(3a)	(3b)	(4)	(5)	(6)	(7)	(8)	(9)	(10)	(11)	(12)	(13)
2730	NITROANISOLES, LIQUID	6.1	T1	III	6.1	279 802	LQ7		PP, EP, TOX, A	VE02		0	
2732	NITROBROMOBENZENES, LIQUID	6.1	T1	III	6.1	802	LQ7		PP, EP, TOX, A	VE02		0	
2733	AMINES, FLAMMABLE, CORROSIVE, N.O.S. or POLYAMINES, FLAMMABLE, CORROSIVE, N.O.S.	3	FC	I	3+8	274 544	LQ3		PP, EP, EX, A	VE01		1	
2733	AMINES, FLAMMABLE, CORROSIVE, N.O.S. or POLYAMINES, FLAMMABLE, CORROSIVE, N.O.S.	3	FC	II	3+8	274 544	LQ4	T	PP, EP, EX, A	VE01		1	
2733	AMINES, FLAMMABLE, CORROSIVE, N.O.S. or POLYAMINES, FLAMMABLE, CORROSIVE, N.O.S.	3	FC	III	3+8	274 544	LQ7		PP, EP, EX, A	VE01		0	
2734	AMINES, LIQUID, CORROSIVE, FLAMMABLE, N.O.S. or POLYAMINES, LIQUID, CORROSIVE, FLAMMABLE, N.O.S.	8	CF1	I	8+3	274	LQ0		PP, EP, EX, A	VE01		1	
2734	AMINES, LIQUID, CORROSIVE, FLAMMABLE, N.O.S. or POLYAMINES, LIQUID, CORROSIVE, FLAMMABLE, N.O.S.	8	CF1	II	8+3	274	LQ22		PP, EP, EX, A	VE01		1	
2735	AMINES, LIQUID, CORROSIVE, N.O.S. or POLYAMINES, LIQUID, CORROSIVE, N.O.S.	8	C7	I	8	274	LQ0	T	PP, EP			0	
2735	AMINES, LIQUID, CORROSIVE, N.O.S. or POLYAMINES, LIQUID, CORROSIVE, N.O.S.	8	C7	II	8	274	LQ22	T	PP, EP			0	
2735	AMINES, LIQUID, CORROSIVE, N.O.S. or POLYAMINES, LIQUID, CORROSIVE, N.O.S.	8	C7	III	8	274	LQ7	T	PP, EP			0	
2738	N-BUTYLANILINE	6.1	T1	II	6.1	802	LQ17		PP, EP, TOX, A	VE02		2	

(1) UN No. or ID No.	(2) Name and description	(3a) Class	(3b) Classification Code	(4) Packing group	(5) Labels	(6) Special provisions	(7) Limited quantities	(8) Carriage permitted	(9) Equipment required	(10) Ventilation	(11) Provisions concerning loading, unloading and carriage	(12) Number of cones, blue lights	(13) Remarks
		2.2	2.2	2.1.1.3	5.2.2	3.3	3.4.6	3.2.1	8.1.5	7.1.6	7.1.6	7.1.5	3.2.1
2739	BUTYRIC ANHYDRIDE	8	C3	III	8		LQ7		PP, EP			0	
2740	n-PROPYL CHLOROFORMATE	6.1	TFC	I	6.1+3+8	802	LQ0		PP, EP, EX, TOX, A	VE01, VE02		2	
2741	BARIUM HYPOCHLORITE with more than 22% available chlorine	5.1	OT2	II	5.1+6.1	802	LQ11		PP			2	
2742	CHLOROFORMATES, TOXIC, CORROSIVE, FLAMMABLE, N.O.S.	6.1	TFC	II	6.1+3+8	274 561 802	LQ17		PP, EP, EX, TOX, A	VE01, VE02		2	
2743	n-BUTYL CHLOROFORMATE	6.1	TFC	II	6.1+3+8	802	LQ17		PP, EP, EX, TOX, A	VE01, VE02		2	
2744	CYCLOBUTYL CHLOROFORMATE	6.1	TFC	II	6.1+3+8	802	LQ17		PP, EP, EX, TOX, A	VE01, VE02		2	
2745	CHLOROMETHYL CHLOROFORMATE	6.1	TC1	II	6.1+8	802	LQ17		PP, EP, TOX, A	VE02		2	
2746	PHENYL CHLOROFORMATE	6.1	TC1	II	6.1+8	802	LQ17		PP, EP, TOX, A	VE02		2	
2747	tert-BUTYLCYCLOHEXYL CHLOROFORMATE	6.1	T1	III	6.1	802	LQ7		PP, EP, TOX, A	VE02		0	
2748	2-ETHYLHEXYL CHLOROFORMATE	6.1	TC1	II	6.1+8	802	LQ17		PP, EP, TOX, A	VE02		2	
2749	TETRAMETHYLSILANE	3	F1	I	3		LQ3		PP, EX, A	VE01		1	
2750	1,3-DICHLOROPROPANOL-2	6.1	T1	II	6.1	802	LQ17		PP, EP, TOX, A	VE02		2	
2751	DIETHYLTHIOPHOSPHORYL CHLORIDE	8	C3	II	8		LQ22		PP, EP	VE02		0	
2752	1,2-EPOXY-3-ETHOXYPROPANE	3	F1	III	3		LQ7		PP, EX, A	VE01		0	
2753	N-ETHYLBENZYL-TOLUIDINES, LIQUID	6.1	T1	III	6.1	802	LQ7		PP, EP, TOX, A	VE02		0	
2754	N-ETHYLTOLUIDINES	6.1	T1	II	6.1	802	LQ17	T	PP, EP, TOX, A	VE02		2	
2757	CARBAMATE PESTICIDE, SOLID, TOXIC	6.1	T7	I	6.1	61 648 802	LQ0		PP, EP			2	
2757	CARBAMATE PESTICIDE, SOLID, TOXIC	6.1	T7	II	6.1	61 648 802	LQ18		PP, EP			2	

UN No. or ID No.	Name and description	Class	Classification Code	Packing group	Labels	Special provisions	Limited quantities	Carriage permitted	Equipment required	Ventilation	Provisions concerning loading, unloading and carriage	Number of cones, blue lights	Remarks
3.1.2	3.1.2	2.2	2.2	2.1.1.3	5.2.2	3.3	3.4.6	3.2.1	8.1.5	7.1.6	7.1.6	7.1.5	3.2.1
(1)	(2)	(3a)	(3b)	(4)	(5)	(6)	(7)	(8)	(9)	(10)	(11)	(12)	(13)
2757	CARBAMATE PESTICIDE, SOLID, TOXIC	6.1	T7	III	6.1	61 648 802	LQ9		PP, EP			0	
2758	CARBAMATE PESTICIDE, LIQUID, FLAMMABLE, TOXIC, flash-point less than 23 °C	3	FT2	I	3+6.1	61 802	LQ3		PP, EP, EX, TOX, A	VE01, VE02		2	
2758	CARBAMATE PESTICIDE, LIQUID, FLAMMABLE, TOXIC, flash-point less than 23 °C	3	FT2	II	3+6.1	61 802	LQ4		PP, EP, EX, TOX, A	VE01, VE02		2	
2759	ARSENICAL PESTICIDE, SOLID, TOXIC	6.1	T7	I	6.1	61 648 802	LQ0		PP, EP			2	
2759	ARSENICAL PESTICIDE, SOLID, TOXIC	6.1	T7	II	6.1	61 648 802	LQ18		PP, EP			2	
2759	ARSENICAL PESTICIDE, SOLID, TOXIC	6.1	T7	III	6.1	61 648 802	LQ9		PP, EP			0	
2760	ARSENICAL PESTICIDE, LIQUID, FLAMMABLE, TOXIC, flash-point less than 23 °C	3	FT2	I	3+6.1	61 802	LQ3		PP, EP, EX, TOX, A	VE01, VE02		2	
2760	ARSENICAL PESTICIDE, LIQUID, FLAMMABLE, TOXIC, flash-point less than 23 °C	3	FT2	II	3+6.1	61 802	LQ4		PP, EP, EX, TOX, A	VE01, VE02		2	
2761	ORGANOCHLORINE PESTICIDE, SOLID, TOXIC	6.1	T7	I	6.1	61 648 802	LQ0		PP, EP			2	
2761	ORGANOCHLORINE PESTICIDE, SOLID, TOXIC	6.1	T7	II	6.1	61 648 802	LQ18		PP, EP			2	
2761	ORGANOCHLORINE PESTICIDE, SOLID, TOXIC	6.1	T7	III	6.1	61 802	LQ9		PP, EP			0	
2762	ORGANOCHLORINE PESTICIDE, LIQUID, FLAMMABLE, TOXIC, flash-point less than 23 °C	3	FT2	I	3+6.1	61 802	LQ3		PP, EP, EX, TOX, A	VE01, VE02		2	

UN No. or ID No. (1)	Name and description 3.1.2 (2)	Class 2.2 (3a)	Classification Code 2.2 (3b)	Packing group 2.1.1.3 (4)	Labels 5.2.2 (5)	Special provisions 3.3 (6)	Limited quantities 3.4.6 (7)	Carriage permitted 3.2.1 (8)	Equipment required 8.1.5 (9)	Ventilation 7.1.6 (10)	Provisions concerning loading, unloading and carriage 7.1.6 (11)	Number of cones, blue lights 7.1.5 (12)	Remarks 3.2.1 (13)
2762	ORGANOCHLORINE PESTICIDE, LIQUID, FLAMMABLE, TOXIC, flash-point less than 23 °C	3	FT2	II	3+6.1	61 802	LQ4		PP, EP, EX, TOX, A	VE01, VE02		2	
2763	TRIAZINE PESTICIDE, SOLID, TOXIC	6.1	T7	I	6.1	61 648 802	LQ0		PP, EP			2	
2763	TRIAZINE PESTICIDE, SOLID, TOXIC	6.1	T7	II	6.1	61 648 802	LQ18		PP, EP			2	
2763	TRIAZINE PESTICIDE, SOLID, TOXIC	6.1	T7	III	6.1	61 648 802	LQ9		PP, EP			0	
2764	TRIAZINE PESTICIDE, LIQUID, FLAMMABLE, TOXIC, flash-point less than 23 °C	3	FT2	I	3+6.1	61 802	LQ3		PP, EP, EX, TOX, A	VE01, VE02		2	
2764	TRIAZINE PESTICIDE, LIQUID, FLAMMABLE, TOXIC, flash-point less than 23 °C	3	FT2	II	3+6.1	61 802	LQ4		PP, EP, EX, TOX, A	VE01, VE02		2	
2771	THIOCARBAMATE PESTICIDE, SOLID, TOXIC	6.1	T7	I	6.1	61 648 802	LQ0		PP, EP			2	
2771	THIOCARBAMATE PESTICIDE, SOLID, TOXIC	6.1	T7	II	6.1	61 648 802	LQ18		PP, EP			2	
2771	THIOCARBAMATE PESTICIDE, SOLID, TOXIC	6.1	T7	III	6.1	61 648 802	LQ9		PP, EP			0	
2772	THIOCARBAMATE PESTICIDE, LIQUID, FLAMMABLE, TOXIC, flash-point less than 23 °C	3	FT2	I	3+6.1	61 802	LQ3		PP, EP, EX, TOX, A	VE01, VE02		2	
2772	THIOCARBAMATE PESTICIDE, LIQUID, FLAMMABLE, TOXIC, flash-point less than 23 °C	3	FT2	II	3+6.1	61 802	LQ4		PP, EP, EX, TOX, A	VE01, VE02		2	
2775	COPPER BASED PESTICIDE, SOLID, TOXIC	6.1	T7	I	6.1	61 648 802	LQ0		PP, EP			2	

UN No. or ID No.	Name and description	Class	Classification Code	Packing group	Labels	Special provisions	Limited quantities	Carriage permitted	Equipment required	Ventilation	Provisions concerning loading, unloading and carriage	Number of cones, blue lights	Remarks
	3.1.2	2.2	2.2	2.1.1.3	5.2.2	3.3	3.4.6	3.2.1	8.1.5	7.1.6	7.1.6	7.1.5	3.2.1
(1)	(2)	(3a)	(3b)	(4)	(5)	(6)	(7)	(8)	(9)	(10)	(11)	(12)	(13)
2775	COPPER BASED PESTICIDE, SOLID, TOXIC	6.1	T7	II	6.1	61 648 802	LQ18		PP, EP			2	
2775	COPPER BASED PESTICIDE, SOLID, TOXIC	6.1	T7	III	6.1	61 648 802	LQ9		PP, EP			0	
2776	COPPER BASED PESTICIDE, LIQUID, FLAMMABLE, TOXIC, flash-point less than 23 °C	3	FT2	I	3+6.1	61 802	LQ3		PP, EP, EX, TOX, A	VE01, VE02		2	
2776	COPPER BASED PESTICIDE, LIQUID, FLAMMABLE, TOXIC, flash-point less than 23 °C	3	FT2	II	3+6.1	61 802	LQ4		PP, EP, EX, TOX, A	VE01, VE02		2	
2777	MERCURY BASED PESTICIDE, SOLID, TOXIC	6.1	T7	I	6.1	61 648 802	LQ0		PP, EP			2	
2777	MERCURY BASED PESTICIDE, SOLID, TOXIC	6.1	T7	II	6.1	61 648 802	LQ18		PP, EP			2	
2777	MERCURY BASED PESTICIDE, SOLID, TOXIC	6.1	T7	III	6.1	61 648 802	LQ9		PP, EP			0	
2778	MERCURY BASED PESTICIDE, LIQUID, FLAMMABLE, TOXIC, flash-point less than 23 °C	3	FT2	I	3+6.1	61 802	LQ3		PP, EP, EX, TOX, A	VE01, VE02		2	
2778	MERCURY BASED PESTICIDE, LIQUID, FLAMMABLE, TOXIC, flash-point less than 23 °C	3	FT2	II	3+6.1	61 802	LQ4		PP, EP, EX, TOX, A	VE01, VE02		2	
2779	SUBSTITUTED NITROPHENOL PESTICIDE, SOLID, TOXIC	6.1	T7	I	6.1	61 648 802	LQ0		PP, EP			2	
2779	SUBSTITUTED NITROPHENOL PESTICIDE, SOLID, TOXIC	6.1	T7	II	6.1	61 648 802	LQ18		PP, EP			2	
2779	SUBSTITUTED NITROPHENOL PESTICIDE, SOLID, TOXIC	6.1	T7	III	6.1	61 648 802	LQ9		PP, EP			0	

UN No. or ID No.	Name and description	Class	Classification Code	Packing group	Labels	Special provisions	Limited quantities	Carriage permitted	Equipment required	Ventilation	Provisions concerning loading, unloading and carriage	Number of cones, blue lights	Remarks
	3.1.2	2.2	2.2	2.1.1.3	5.2.2	3.3	3.4.6	3.2.1	8.1.5	7.1.6	7.1.6	7.1.5	3.2.1
(1)	(2)	(3a)	(3b)	(4)	(5)	(6)	(7)	(8)	(9)	(10)	(11)	(12)	(13)
2780	SUBSTITUTED NITROPHENOL PESTICIDE, LIQUID, FLAMMABLE, TOXIC, flash-point less than 23 °C	3	FT2	I	3+6.1	61 802	LQ3		PP, EP, EX, TOX, A	VE01, VE02		2	
2780	SUBSTITUTED NITROPHENOL PESTICIDE, LIQUID, FLAMMABLE, TOXIC, flash-point less than 23 °C	3	FT2	II	3+6.1	61 802	LQ4		PP, EP, EX, TOX, A	VE01, VE02		2	
2781	BIPYRIDILIUM PESTICIDE, SOLID, TOXIC	6.1	T7	I	6.1	61 648 802	LQ0		PP, EP			2	
2781	BIPYRIDILIUM PESTICIDE, SOLID, TOXIC	6.1	T7	II	6.1	61 648 802	LQ18		PP, EP			2	
2781	BIPYRIDILIUM PESTICIDE, SOLID, TOXIC	6.1	T7	III	6.1	61 648 802	LQ9		PP, EP			0	
2782	BIPYRIDILIUM PESTICIDE, LIQUID, FLAMMABLE, TOXIC, flash-point less than 23 °C	3	FT2	I	3+6.1	61 802	LQ3		PP, EP, EX, TOX, A	VE01, VE02		2	
2782	BIPYRIDILIUM PESTICIDE, LIQUID, FLAMMABLE, TOXIC, flash-point less than 23 °C	3	FT2	II	3+6.1	61 802	LQ4		PP, EP, EX, TOX, A	VE01, VE02		2	
2783	ORGANOPHOSPHORUS PESTICIDE, SOLID, TOXIC	6.1	T7	I	6.1	61 648 802	LQ0		PP, EP			2	
2783	ORGANOPHOSPHORUS PESTICIDE, SOLID, TOXIC	6.1	T7	II	6.1	61 648 802	LQ18		PP, EP			2	
2783	ORGANOPHOSPHORUS PESTICIDE, SOLID, TOXIC	6.1	T7	III	6.1	61 648 802	LQ9		PP, EP			0	
2784	ORGANOPHOSPHORUS PESTICIDE, LIQUID, FLAMMABLE, TOXIC, flash-point less than 23 °C	3	FT2	I	3+6.1	61 802	LQ3		PP, EP, EX, TOX, A	VE01, VE02		2	

UN No. or ID No.	Name and description	Class	Classification Code	Packing group	Labels	Special provisions	Limited quantities	Carriage permitted	Equipment required	Ventilation	Provisions concerning loading, unloading and carriage	Number of cones, blue lights	Remarks
3.1.2	3.1.2	2.2	2.2	2.1.1.3	5.2.2	3.3	3.4.6	3.2.1	8.1.5	7.1.6	7.1.6	7.1.5	3.2.1
(1)	(2)	(3a)	(3b)	(4)	(5)	(6)	(7)	(8)	(9)	(10)	(11)	(12)	(13)
2784	ORGANOPHOSPHORUS PESTICIDE, LIQUID, FLAMMABLE, TOXIC, flash-point less than 23 °C	3	FT2	II	3+6.1	61 802	LQ4		PP, EP, EX, TOX, A	VE01, VE02		2	
2785	4-THIAPENTANAL	6.1	T1	III	6.1	802	LQ7		PP, EP, TOX, A	VE02		0	
2786	ORGANOTIN PESTICIDE, SOLID, TOXIC	6.1	T7	I	6.1	61 648 802	LQ0		PP, EP			2	
2786	ORGANOTIN PESTICIDE, SOLID, TOXIC	6.1	T7	II	6.1	61 648 802	LQ18		PP, EP			2	
2786	ORGANOTIN PESTICIDE, SOLID, TOXIC	6.1	T7	III	6.1	61 648 802	LQ9		PP, EP			0	
2787	ORGANOTIN PESTICIDE, LIQUID, FLAMMABLE, TOXIC, flash-point less than 23 °C	3	FT2	I	3+6.1	61 802	LQ3		PP, EP, EX, TOX, A	VE01, VE02		2	
2787	ORGANOTIN PESTICIDE, LIQUID, FLAMMABLE, TOXIC, flash-point less than 23 °C	3	FT2	II	3+6.1	61 802	LQ4		PP, EP, EX, TOX, A	VE01, VE02		2	
2788	ORGANOTIN COMPOUND, LIQUID, N.O.S.	6.1	T3	I	6.1	43 274 802	LQ0		PP, EP, TOX, A	VE02		2	
2788	ORGANOTIN COMPOUND, LIQUID, N.O.S.	6.1	T3	II	6.1	43 274 802	LQ17		PP, EP, TOX, A	VE02		2	
2788	ORGANOTIN COMPOUND, LIQUID, N.O.S.	6.1	T3	III	6.1	43 274 802	LQ7		PP, EP, TOX, A	VE02		0	
2789	ACETIC ACID, GLACIAL or ACETIC ACID SOLUTION, more than 80% acid, by mass	8	CF1	II	8+3		LQ22	T	PP, EP, EX, A	VE01		1	
2790	ACETIC ACID SOLUTION, not less than 50% but not more than 80% acid, by mass	8	C3	II	8		LQ22	T	PP, EP			0	

(1)	(2)	(3a)	(3b)	(4)	(5)	(6)	(7)	(8)	(9)	(10)	(11)	(12)	(13)
UN No. or ID No.	Name and description	Class	Classification Code	Packing group	Labels	Special provisions	Limited quantities	Carriage permitted	Equipment required	Ventilation	Provisions concerning loading, unloading and carriage	Number of cones, blue lights	Remarks
	3.1.2	2.2	2.2	2.1.1.3	5.2.2	3.3	3.4.6	3.2.1	8.1.5	7.1.6	7.1.6	7.1.5	3.2.1
2790	ACETIC ACID SOLUTION, more than 10% and less than 50% acid, by mass	8	C3	III	8	597 647	LQ7	T	PP, EP			0	
2793	FERROUS METAL BORINGS, SHAVINGS, TURNINGS or CUTTINGS in a form liable to self-heating	4.2	S4	III	4.2	592	LQ0	B	PP		LO02	0	LO02 applies only when this substance is carried in bulk or without packaging
2794	BATTERIES, WET, FILLED WITH ACID, electric storage	8	C11		8	295 598	LQ0		PP, EP			0	
2795	BATTERIES, WET, FILLED WITH ALKALI, electric storage	8	C11		8	295 598	LQ0		PP, EP			0	
2796	SULPHURIC ACID with not more than 51% acid or BATTERY FLUID, ACID	8	C1	II	8		LQ22	T	PP, EP			0	
2797	BATTERY FLUID, ALKALI	8	C5	II	8		LQ22	T	PP, EP			0	
2798	PHENYLPHOSPHORUS DICHLORIDE	8	C3	II	8		LQ22		PP, EP			0	
2799	PHENYLPHOSPHORUS THIODICHLORIDE	8	C3	II	8		LQ22		PP, EP			0	
2800	BATTERIES, WET, NON-SPILLABLE, electric storage	8	C11	II	8	238 295 598	LQ0		PP, EP			0	
2801	DYE, LIQUID, CORROSIVE, N.O.S. or DYE INTERMEDIATE, LIQUID, CORROSIVE, N.O.S.	8	C9	I	8	274	LQ0		PP, EP			0	
2801	DYE, LIQUID, CORROSIVE, N.O.S. or DYE INTERMEDIATE, LIQUID, CORROSIVE, N.O.S.	8	C9	II	8	274	LQ22		PP, EP			0	
2801	DYE, LIQUID, CORROSIVE, N.O.S. or DYE INTERMEDIATE, LIQUID, CORROSIVE, N.O.S.	8	C9	III	8	274	LQ7		PP, EP			0	
2802	COPPER CHLORIDE	8	C2	III	8		LQ24		PP, EP			0	
2803	GALLIUM	8	C10	III	8		LQ24		PP, EP			0	
2805	LITHIUM HYDRIDE, FUSED SOLID	4.3	W2	II	4.3		LQ11		PP, EX, A	VE01	HA08	0	
2806	LITHIUM NITRIDE	4.3	W2	I	4.3		LQ0		PP, EX, A	VE01	HA08	0	

UN No. or ID No.	Name and description	Class	Classification Code	Packing group	Labels	Special provisions	Limited quantities	Carriage permitted	Equipment required	Ventilation	Provisions concerning loading, unloading and carriage	Number of cones, blue lights	Remarks
3.1.2	3.1.2	2.2	2.2	2.1.1.3	5.2.2	3.3	3.4.6	3.2.1	8.1.5	7.1.6	7.1.6	7.1.5	3.2.1
(1)	(2)	(3a)	(3b)	(4)	(5)	(6)	(7)	(8)	(9)	(10)	(11)	(12)	(13)
2807	Magnetized material	9	M11	NOT SUBJECT TO ADN									
2809	MERCURY	8	C9	III	8	599	LQ19		PP, EP			0	
2810	TOXIC LIQUID, ORGANIC, N.O.S.	6.1	T1	I	6.1	274 315 614 802	LQ0	T	PP, EP, TOX, A	VE02		2	
2810	TOXIC LIQUID, ORGANIC, N.O.S.	6.1	T1	II	6.1	274 614 802	LQ17	T	PP, EP, TOX, A	VE02		2	
2810	TOXIC LIQUID, ORGANIC, N.O.S.	6.1	T1	III	6.1	274 614 802	LQ7	T	PP, EP, TOX, A	VE02		0	
2811	TOXIC SOLID, ORGANIC, N.O.S.	6.1	T2	I	6.1	274 614 802	LQ0		PP, EP			2	
2811	TOXIC SOLID, ORGANIC, N.O.S.	6.1	T2	II	6.1	274 614 802	LQ18		PP, EP			2	
2811	TOXIC SOLID, ORGANIC, N.O.S.	6.1	T2	III	6.1	274 614 802	LQ9	T	PP, EP			0	
2812	Sodium aluminate, solid	8	C6	NOT SUBJECT TO ADN									
2813	WATER-REACTIVE SOLID, N.O.S.	4.3	W2	I	4.3	274	LQ0		PP, EX, A	VE01	HA08	0	
2813	WATER-REACTIVE SOLID, N.O.S.	4.3	W2	II	4.3	274	LQ11		PP, EX, A	VE01	HA08	0	
2813	WATER-REACTIVE SOLID, N.O.S.	4.3	W2	III	4.3	274	LQ12		PP, EX, A	VE01	HA08	0	
2814	INFECTIOUS SUBSTANCE, AFFECTING HUMANS	6.2	I1		6.2	318 802	LQ0		PP			0	
2814	INFECTIOUS SUBSTANCE, AFFECTING HUMANS, in refrigerated liquid nitrogen	6.2	I1		6.2 +2.2	318 802	LQ0		PP			0	
2814	INFECTIOUS SUBSTANCE, AFFECTING HUMANS (animal carcasses only)	6.2	I1		6.2	318 802	LQ0		PP			0	

UN No. or ID No.	Name and description	Class	Classification Code	Packing group	Labels	Special provisions	Limited quantities	Carriage permitted	Equipment required	Ventilation	Provisions concerning loading, unloading and carriage	Number of cones, blue lights	Remarks
3.1.2	3.1.2	2.2	2.2	2.1.1.3	5.2.2	3.3	3.4.6	3.2.1	8.1.5	7.1.6	7.1.6	7.1.5	3.2.1
(1)	(2)	(3a)	(3b)	(4)	(5)	(6)	(7)	(8)	(9)	(10)	(11)	(12)	(13)
2815	N-AMINOETHYL-PIPERAZINE	8	C7	III	8		LQ7	T	PP, EP			0	
2817	AMMONIUM HYDROGENDIFLUORIDE SOLUTION	8	CT1	II	8+6.1	802	LQ22		PP, EP			2	
2817	AMMONIUM HYDROGENDIFLUORIDE SOLUTION	8	CT1	III	8+6.1	802	LQ7		PP, EP			0	
2818	AMMONIUM POLYSULPHIDE SOLUTION	8	CT1	II	8+6.1	802	LQ22		PP, EP			2	
2818	AMMONIUM POLYSULPHIDE SOLUTION	8	CT1	III	8+6.1	802	LQ7		PP, EP			0	
2819	AMYL ACID PHOSPHATE	8	C3	III	8		LQ7		PP, EP			0	
2820	BUTYRIC ACID	8	C3	III	8		LQ7	T	PP, EP			0	
2821	PHENOL SOLUTION	6.1	T1	II	6.1	802	LQ17		PP, EP, TOX, A	VE02		2	
2821	PHENOL SOLUTION	6.1	T1	III	6.1	802	LQ7		PP, EP, TOX, A	VE02		0	
2822	2-CHLOROPYRIDINE	6.1	T1	II	6.1	802	LQ17		PP, EP, TOX, A	VE02		2	
2823	CROTONIC ACID, SOLID	8	C4	III	8		LQ24		PP, EP			0	
2826	ETHYL CHLOROTHIOFORMATE	8	CF1	II	8+3		LQ22		PP, EP, EX, A	VE01		1	
2829	CAPROIC ACID	8	C3	III	8		LQ7	T	PP, EP			0	
2830	LITHIUM FERROSILICON	4.3	W2	II	4.3		LQ11		PP, EX, A	VE01	HA08	0	
2831	1,1,1-TRICHLOROETHANE	6.1	T1	III	6.1	802	LQ7	T	PP, EP, TOX, A	VE02		0	
2834	PHOSPHOROUS ACID	8	C2	III	8		LQ24		PP, EP			0	
2835	SODIUM ALUMINIUM HYDRIDE	4.3	W2	II	4.3		LQ11		PP, EX, A	VE01	HA08	0	
2837	BISULPHATES, AQUEOUS SOLUTION	8	C1	II	8	274	LQ22		PP, EP			0	
2837	BISULPHATES, AQUEOUS SOLUTION	8	C1	III	8	274	LQ7		PP, EP			0	
2838	VINYL BUTYRATE, STABILIZED	3	F1	II	3		LQ4		PP, EX, A	VE01		1	
2839	ALDOL	6.1	T1	II	6.1	802	LQ17		PP, EP, TOX, A	VE02		2	

UN No. or ID No.	Name and description	Class	Classification Code	Packing group	Labels	Special provisions	Limited quantities	Carriage permitted	Equipment required	Ventilation	Provisions concerning loading, unloading and carriage	Number of cones, blue lights	Remarks
		2.2	2.2	2.1.1.3	5.2.2	3.3	3.4.6	3.2.1	8.1.5	7.1.6	7.1.6	7.1.5	3.2.1
(1)	(2) 3.1.2	(3a)	(3b)	(4)	(5)	(6)	(7)	(8)	(9)	(10)	(11)	(12)	(13)
2840	BUTYRALDOXIME	3	F1	III	3		LQ7		PP, EX, A	VE01		0	
2841	DI-n-AMYLAMINE	3	FT1	III	3+6.1	802	LQ7		PP, EP, EX, TOX, A	VE01, VE02		2	
2842	NITROETHANE	3	F1	III	3		LQ7		PP, EX, A	VE01		0	
2844	CALCIUM MANGANESE SILICON	4.3	W2	III	4.3		LQ12		PP, EX, A	VE01	HA08	0	
2845	PYROPHORIC LIQUID, ORGANIC, N.O.S.	4.2	S1	I	4.2	274	LQ0		PP			0	
2846	PYROPHORIC SOLID, ORGANIC, N.O.S.	4.2	S2	I	4.2	274	LQ0		PP			0	
2849	3-CHLOROPROPANOL-1	6.1	T1	III	6.1	802	LQ7		PP, EP, TOX, A	VE02		0	
2850	PROPYLENE TETRAMER	3	F1	III	3		LQ7		PP, EX, A	VE01		0	
2851	BORON TRIFLUORIDE DIHYDRATE	8	C1	II	8		LQ22	T	PP, EP			0	
2852	DIPICRYL SULPHIDE, WETTED with not less than 10% water, by mass	4.1	D	I	4.1	545	LQ0		PP			1	
2853	MAGNESIUM FLUOROSILICATE	6.1	T5	III	6.1	802	LQ9		PP, EP			0	
2854	AMMONIUM FLUOROSILICATE	6.1	T5	III	6.1	802	LQ9		PP, EP			0	
2855	ZINC FLUOROSILICATE	6.1	T5	III	6.1	802	LQ9		PP, EP			0	
2856	FLUOROSILICATES, N.O.S.	6.1	T5	III	6.1	274 802	LQ9		PP, EP			0	
2857	REFRIGERATING MACHINES containing non-flammable, non-toxic gases or ammonia solutions (UN 2672)	2	6A		2.2	119	LQ0		PP			0	
2858	ZIRCONIUM, DRY, coiled wire, finished metal sheets, strip (thinner than 254 microns but not thinner than 18 microns)	4.1	F3	III	4.1	546	LQ9		PP			0	
2859	AMMONIUM METAVANADATE	6.1	T5	II	6.1	802	LQ18		PP, EP			2	
2861	AMMONIUM POLYVANADATE	6.1	T5	II	6.1	802	LQ18		PP, EP			2	
2862	VANADIUM PENTOXIDE, non-fused form	6.1	T5	III	6.1	600 802	LQ9		PP, EP			0	

UN No. or ID No. (1) 3.1.2	Name and description (2) 3.1.2	Class (3a) 2.2	Classification Code (3b) 2.2	Packing group (4) 2.1.1.3	Labels (5) 5.2.2	Special provisions (6) 3.3	Limited quantities (7) 3.4.6	Carriage permitted (8) 3.2.1	Equipment required (9) 8.1.5	Ventilation (10) 7.1.6	Provisions concerning loading, unloading and carriage (11) 7.1.6	Number of cones, blue lights (12) 7.1.5	Remarks (13) 3.2.1
2863	SODIUM AMMONIUM VANADATE	6.1	T5	II	6.1	802	LQ18		PP, EP			2	
2864	POTASSIUM METAVANADATE	6.1	T5	II	6.1	802	LQ18		PP, EP			2	
2865	HYDROXYLAMINE SULPHATE	8	C2	III	8		LQ24		PP, EP			0	
2869	TITANIUM TRICHLORIDE MIXTURE	8	C2	II	8		LQ23		PP, EP			0	
2869	TITANIUM TRICHLORIDE MIXTURE	8	C2	III	8		LQ24		PP, EP			0	
2870	ALUMINIUM BOROHYDRIDE	4.2	SW	I	4.2+4.3		LQ0		PP, EX, A	VE01		0	
2870	ALUMINIUM BOROHYDRIDE IN DEVICES	4.2	SW	I	4.2+4.3		LQ0		PP, EX, A	VE01		0	
2871	ANTIMONY POWDER	6.1	T5	III	6.1	802	LQ9		PP, EP			0	
2872	DIBROMOCHLORO-PROPANES	6.1	T1	II	6.1	802	LQ17		PP, EP, TOX, A	VE02		2	
2872	DIBROMOCHLORO-PROPANES	6.1	T1	III	6.1	802	LQ7		PP, EP, TOX, A	VE02		0	
2873	DIBUTYLAMINOETHANOL	6.1	T1	III	6.1	802	LQ7		PP, EP, TOX, A	VE02		0	
2874	FURFURYL ALCOHOL	6.1	T1	III	6.1	802	LQ7	T	PP, EP, TOX, A	VE02		0	
2875	HEXACHLOROPHENE	6.1	T2	III	6.1	802	LQ9		PP, EP			0	
2876	RESORCINOL	6.1	T2	III	6.1	802	LQ9		PP, EP			0	
2878	TITANIUM SPONGE GRANULES or TITANIUM SPONGE POWDERS	4.1	F3	III	4.1		LQ9		PP			0	
2879	SELENIUM OXYCHLORIDE	8	CT1	I	8+6.1	802	LQ0		PP, EP, TOX, A	VE02		2	
2880	CALCIUM HYPOCHLORITE, HYDRATED, or CALCIUM HYPOCHLORITE, HYDRATED MIXTURE, with not less than 5.5% but not more than 16% water	5.1	O2	II	5.1	313 314 322	LQ11		PP			0	
2880	CALCIUM HYPOCHLORITE, HYDRATED or CALCIUM HYPOCHLORITE HYDRATED MIXTURE, with not less than 5.5% but not more than 16% water	5.1	O2	III	5.1	223 313 314	LQ12		PP			0	

UN No. or ID No.	Name and description	Class	Classification Code	Packing group	Labels	Special provisions	Limited quantities	Carriage permitted	Equipment required	Ventilation	Provisions concerning loading, unloading and carriage	Number of cones, blue lights	Remarks
	3.1.2	2.2	2.2	2.1.1.3	5.2.2	3.3	3.4.6	3.2.1	8.1.5	7.1.6	7.1.6	7.1.5	3.2.1
(1)	(2)	(3a)	(3b)	(4)	(5)	(6)	(7)	(8)	(9)	(10)	(11)	(12)	(13)
2881	METAL CATALYST, DRY	4.2	S4	I	4.2	274	LQ0		PP			0	
2881	METAL CATALYST, DRY	4.2	S4	II	4.2	274	LQ0		PP			0	
2881	METAL CATALYST, DRY	4.2	S4	III	4.2	274	LQ0		PP			0	
2900	INFECTIOUS SUBSTANCE, AFFECTING ANIMALS only	6.2	I2		6.2	318 802	LQ0		PP			0	
2900	INFECTIOUS SUBSTANCE, AFFECTING ANIMALS only, in refrigerated liquid nitrogen	6.2	I2		6.2 +2.2	318 802	LQ0		PP			0	
2900	INFECTIOUS SUBSTANCE, AFFECTING ANIMALS only (animal carcasses and wastes only)	6.2	I2		6.2	318 802	LQ0		PP			0	
2901	BROMINE CHLORIDE	2	2TOC		2.3+5.1+8		LQ0		PP, EP, TOX, A	VE02		2	
2902	PESTICIDE, LIQUID, TOXIC, N.O.S.	6.1	T6	I	6.1	61 648 802	LQ0		PP, EP, TOX, A	VE02		2	
2902	PESTICIDE, LIQUID, TOXIC, N.O.S.	6.1	T6	II	6.1	61 648 802	LQ17		PP, EP, TOX, A	VE02		2	
2902	PESTICIDE, LIQUID, TOXIC, N.O.S.	6.1	T6	III	6.1	61 648 802	LQ7		PP, EP, TOX, A	VE02		0	
2903	PESTICIDE, LIQUID, TOXIC, FLAMMABLE, N.O.S., flash-point not less than 23 °C	6.1	TF2	I	6.1+3	61 802	LQ0		PP, EP, EX, TOX, A	VE01, VE02		2	
2903	PESTICIDE, LIQUID, TOXIC, FLAMMABLE, N.O.S., flash-point not less than 23 °C	6.1	TF2	II	6.1+3	61 802	LQ17		PP, EP, EX, TOX, A	VE01, VE02		2	
2903	PESTICIDE, LIQUID, TOXIC, FLAMMABLE, N.O.S., flash-point not less than 23 °C	6.1	TF2	III	6.1+3	61 802	LQ7		PP, EP, EX, TOX, A	VE01, VE02		0	
2904	CHLOROPHENOLATES, LIQUID or PHENOLATES, LIQUID	8	C9	III	8		LQ7	T	PP, EP			0	* applies only to phenolates but not to chlorophenolates
2905	CHLOROPHENOLATES, SOLID or PHENOLATES, SOLID	8	C10	III	8		LQ24		PP, EP			0	

UN No. or ID No.	Name and description	Class	Classification Code	Packing group	Labels	Special provisions	Limited quantities	Carriage permitted	Equipment required	Ventilation	Provisions concerning loading, unloading and carriage	Number of cones, blue lights	Remarks
	3.1.2	2.2	2.2	2.1.1.3	5.2.2	3.3	3.4.6	3.2.1	8.1.5	7.1.6	7.1.6	7.1.5	3.2.1
(1)	(2)	(3a)	(3b)	(4)	(5)	(6)	(7)	(8)	(9)	(10)	(11)	(12)	(13)
2907	ISOSORBIDE DINITRATE MIXTURE with not less than 60% lactose, mannose, starch or calcium hydrogen phosphate	4.1	D	II	4.1	127	LQ8		PP			0	
2908	RADIOACTIVE MATERIAL, EXCEPTED PACKAGE - EMPTY PACKAGING	7				290	LQ0		PP			0	
2909	RADIOACTIVE MATERIAL, EXCEPTED PACKAGE - ARTICLES MANUFACTURED FROM NATURAL URANIUM or DEPLETED URANIUM or NATURAL THORIUM	7				290	LQ0		PP			0	
2910	RADIOACTIVE MATERIAL, EXCEPTED PACKAGE - LIMITED QUANTITY OF MATERIAL	7				290	LQ0		PP			0	
2911	RADIOACTIVE MATERIAL, EXCEPTED PACKAGE - INSTRUMENTS or ARTICLES	7				290	LQ0		PP			0	
2912	RADIOACTIVE MATERIAL, LOW SPECIFIC ACTIVITY (LSA-I), non fissile or fissile-excepted	7			7X	172 317 325	LQ0	B	PP		RA01	2	
2913	RADIOACTIVE MATERIAL, SURFACE CONTAMINATED OBJECTS (SCO-I or SCO-II), non fissile or fissile-excepted	7			7X	172 317	LQ0	B	PP		RA02, RA03	2	
2915	RADIOACTIVE MATERIAL, TYPE A PACKAGE, non-special form, non fissile or fissile-excepted	7			7X	172 317 325	LQ0		PP			2	
2916	RADIOACTIVE MATERIAL, TYPE B(U) PACKAGE, non fissile or fissile-excepted	7			7X	172 317	LQ0		PP			2	
2917	RADIOACTIVE MATERIAL, TYPE B(M) PACKAGE, non fissile or fissile-excepted	7			7X	172 317	LQ0		PP			2	

UN No. or ID No. (1)	Name and description 3.1.2 (2)	Class 2.2 (3a)	Classification Code 2.2 (3b)	Packing group 2.1.1.3 (4)	Labels 5.2.2 (5)	Special provisions 3.3 (6)	Limited quantities 3.4.6 (7)	Carriage permitted 3.2.1 (8)	Equipment required 8.1.5 (9)	Ventilation 7.1.6 (10)	Provisions concerning loading, unloading and carriage 7.1.6 (11)	Number of cones, blue lights 7.1.5 (12)	Remarks 3.2.1 (13)
2919	RADIOACTIVE MATERIAL, TRANSPORTED UNDER SPECIAL ARRANGEMENT, non fissile or fissile-excepted	7			7X	172 317	LQ0		PP			2	
2920	CORROSIVE LIQUID, FLAMMABLE, N.O.S.	8	CF1	I	8+3	274	LQ0		PP, EP, EX, A	VE01		1	
2920	CORROSIVE LIQUID, FLAMMABLE, N.O.S.	8	CF1	II	8+3	274	LQ22	T	PP, EP, EX, A	VE01		1	
2921	CORROSIVE SOLID, FLAMMABLE, N.O.S.	8	CF2	I	8+4.1	274	LQ0		PP, EP			1	
2921	CORROSIVE SOLID, FLAMMABLE, N.O.S.	8	CF2	II	8+4.1	274	LQ23		PP, EP			1	
2922	CORROSIVE LIQUID, TOXIC, N.O.S.	8	CT1	I	8+6.1	274 802	LQ0	T	PP, EP, TOX, A	VE02		2	
2922	CORROSIVE LIQUID, TOXIC, N.O.S.	8	CT1	II	8+6.1	274 802	LQ22	T	PP, EP, TOX, A	VE02		2	
2922	CORROSIVE LIQUID, TOXIC, N.O.S.	8	CT1	III	8+6.1	274 802	LQ7	T	PP, EP, TOX, A	VE02		0	
2923	CORROSIVE SOLID, TOXIC, N.O.S.	8	CT2	I	8+6.1	274 802	LQ0		PP, EP			2	
2923	CORROSIVE SOLID, TOXIC, N.O.S.	8	CT2	II	8+6.1	274 802	LQ23		PP, EP			2	
2923	CORROSIVE SOLID, TOXIC, N.O.S.	8	CT2	III	8+6.1	274 802	LQ24		PP, EP			0	
2924	FLAMMABLE LIQUID, CORROSIVE, N.O.S.	3	FC	I	3+8	274	LQ3	T	PP, EP, EX, A	VE01		1	
2924	FLAMMABLE LIQUID, CORROSIVE, N.O.S.	3	FC	II	3+8	274	LQ4	T	PP, EP, EX, A	VE01		1	
2924	FLAMMABLE LIQUID, CORROSIVE, N.O.S.	3	FC	III	3+8	274	LQ7	T	PP, EP, EX, A	VE01		0	
2925	FLAMMABLE SOLID, CORROSIVE, ORGANIC, N.O.S.	4.1	FC1	II	4.1+8	274	LQ0		PP			1	
2925	FLAMMABLE SOLID, CORROSIVE, ORGANIC, N.O.S.	4.1	FC1	III	4.1+8	274	LQ0		PP			0	
2926	FLAMMABLE SOLID, TOXIC, ORGANIC, N.O.S.	4.1	FT1	II	4.1+6.1	274 802	LQ0		PP			2	

UN No. or ID No.	Name and description	Class	Classification Code	Packing group	Labels	Special provisions	Limited quantities	Carriage permitted	Equipment required	Ventilation	Provisions concerning loading, unloading and carriage	Number of cones, blue lights	Remarks
	3.1.2	2.2	2.2	2.1.1.3	5.2.2	3.3	3.4.6	3.2.1	8.1.5	7.1.6	7.1.6	7.1.5	3.2.1
(1)	(2)	(3a)	(3b)	(4)	(5)	(6)	(7)	(8)	(9)	(10)	(11)	(12)	(13)
2926	FLAMMABLE SOLID, TOXIC, ORGANIC, N.O.S.	4.1	FT1	III	4.1+6.1	274 802	LQ0		PP			0	
2927	TOXIC LIQUID, CORROSIVE, ORGANIC, N.O.S.	6.1	TC1	I	6.1+8	274 315 802	LQ0	T	PP, EP, TOX, A	VE02		2	
2927	TOXIC LIQUID, CORROSIVE, ORGANIC, N.O.S.	6.1	TC1	II	6.1+8	274 802	LQ17	T	PP, EP, TOX, A	VE02		2	
2928	TOXIC SOLID, CORROSIVE, ORGANIC, N.O.S.	6.1	TC2	I	6.1+8	274 802	LQ0		PP, EP			2	
2928	TOXIC SOLID, CORROSIVE, ORGANIC, N.O.S.	6.1	TC2	II	6.1+8	274 802	LQ18		PP, EP			2	
2929	TOXIC LIQUID, FLAMMABLE, ORGANIC, N.O.S.	6.1	TF1	I	6.1+3	274 315 802	LQ0	T	PP, EP, EX, TOX, A	VE01, VE02		2	
2929	TOXIC LIQUID, FLAMMABLE, ORGANIC, N.O.S.	6.1	TF1	II	6.1+3	274 802	LQ17	T	PP, EP, EX, TOX, A	VE01, VE02		2	
2930	TOXIC SOLID, FLAMMABLE, ORGANIC, N.O.S.	6.1	TF3	I	6.1+4.1	274 802	LQ0		PP, EP			2	
2930	TOXIC SOLID, FLAMMABLE, ORGANIC, N.O.S.	6.1	TF3	II	6.1+4.1	274 802	LQ18		PP, EP			2	
2931	VANADYL SULPHATE	6.1	T5	II	6.1	802	LQ18		PP, EP			2	
2933	METHYL 2-CHLOROPROPIONATE	3	F1	III	3		LQ7		PP, EX, A	VE01		0	
2934	ISOPROPYL 2-CHLOROPROPIONATE	3	F1	III	3		LQ7		PP, EX, A	VE01		0	
2935	ETHYL 2-CHLOROPROPIONATE	3	F1	III	3		LQ7	T	PP, EX, A	VE01		0	
2936	THIOLACTIC ACID	6.1	T1	II	6.1	802	LQ17		PP, EP, TOX, A	VE02		2	
2937	alpha-METHYLBENZYL ALCOHOL, LIQUID	6.1	T1	III	6.1	802	LQ7		PP, EP, TOX, A	VE02		0	
2940	9-PHOSPHABICYCLO-NONANES (CYCLOOCTADIENE PHOSPHINES)	4.2	S2	II	4.2		LQ0		PP			0	
2941	FLUOROANILINES	6.1	T1	III	6.1	802	LQ7		PP, EP, TOX, A	VE02		0	

UN No. or ID No. (1)	Name and description (2) 3.1.2	Class (3a) 2.2	Classification Code (3b) 2.2	Packing group (4) 2.1.1.3	Labels (5) 5.2.2	Special provisions (6) 3.3	Limited quantities (7) 3.4.6	Carriage permitted (8) 3.2.1	Equipment required (9) 8.1.5	Ventilation (10) 7.1.6	Provisions concerning loading, unloading and carriage (11) 7.1.6	Number of cones, blue lights (12) 7.1.5	Remarks (13) 3.2.1
2942	2-TRIFLUOROMETHYL-ANILINE	6.1	T1	III	6.1	802	LQ7		PP, EP, TOX, A	VE02		0	
2943	TETRAHYDRO-FURFURYLAMINE	3	F1	III	3		LQ7		PP, EX, A	VE01		0	
2945	N-METHYLBUTYLAMINE	3	FC	II	3+8		LQ4		PP, EP, EX, A	VE01		1	
2946	2-AMINO-5-DIETHYL-AMINOPENTANE	6.1	T1	III	6.1	802	LQ7		PP, EP, TOX, A	VE02		0	
2947	ISOPROPYL CHLOROACETATE	3	F1	III	3		LQ7	T	PP, EX, A	VE01		0	
2948	3-TRIFLUOROMETHYL-ANILINE	6.1	T1	II	6.1	802	LQ17		PP, EP, TOX, A	VE02		2	
2949	SODIUM HYDROSULPHIDE, HYDRATED with not less than 25% water of crystallization	8	C6	II	8	523	LQ23		PP, EP			0	
2950	MAGNESIUM GRANULES, COATED, particle size not less than 149 microns	4.3	W2	III	4.3		LQ12		PP, EX, A	VE01	HA08	0	
2956	5-tert-BUTYL-2,4,6-TRINITRO-m-XYLENE (MUSK XYLENE)	4.1	SR1	III	4.1	638	LQ0		PP			0	
2965	BORON TRIFLUORIDE DIMETHYL ETHERATE	4.3	WFC	I	4.3+3+8		LQ0		PP, EP, EX, A	VE01	HA08	1	
2966	THIOGLYCOL	6.1	T1	II	6.1	802	LQ17	T	PP, EP, TOX, A	VE02		2	
2967	SULPHAMIC ACID	8	C2	III	8		LQ24		PP, EP			0	
2968	MANEB, STABILIZED or MANEB PREPARATION, STABILIZED against self-heating	4.3	W2	III	4.3	547	LQ12		PP, EX, A	VE01	HA08	0	
2969	CASTOR BEANS or CASTOR MEAL or CASTOR POMACE or CASTOR FLAKE	9	M11	II	9	141	LQ25	B	PP			0	
2977	RADIOACTIVE MATERIAL, URANIUM HEXAFLUORIDE, FISSILE	7			7X+7E+8	172	LQ0		PP			2	
2978	RADIOACTIVE MATERIAL, URANIUM HEXAFLUORIDE, non fissile or fissile-excepted	7			7X+8	172 317	LQ0	B	PP		RA01	2	

UN No. or ID No. (1)	Name and description 3.1.2 (2)	Class 2.2 (3a)	Classification Code 2.2 (3b)	Packing group 2.1.1.3 (4)	Labels 5.2.2 (5)	Special provisions 3.3 (6)	Limited quantities 3.4.6 (7)	Carriage permitted 3.2.1 (8)	Equipment required 8.1.5 (9)	Ventilation 7.1.6 (10)	Provisions concerning loading, unloading and carriage 7.1.6 (11)	Number of cones, blue lights 7.1.5 (12)	Remarks 3.2.1 (13)
2983	ETHYLENE OXIDE AND PROPYLENE OXIDE MIXTURE, not more than 30% ethylene oxide	3	FT1	I	3+6.1	802	LQ0	T	PP, EP, EX, TOX, A	VE01, VE02		2	
2984	HYDROGEN PEROXIDE, AQUEOUS SOLUTION with not less than 8% but less than 20% hydrogen peroxide (stabilized as necessary)	5.1	O1	III	5.1	65	LQ13		PP			0	
2985	CHLOROSILANES, FLAMMABLE, CORROSIVE, N.O.S.	3	FC	II	3+8	274 548	LQ4		PP, EP, EX, A	VE01		1	
2986	CHLOROSILANES, CORROSIVE, FLAMMABLE, N.O.S.	8	CF1	II	8+3	274 548	LQ22		PP, EP, EX, A	VE01		1	
2987	CHLOROSILANES, CORROSIVE, N.O.S.	8	C3	II	8	274 548	LQ22		PP, EP			0	
2988	CHLOROSILANES, WATER-REACTIVE, FLAMMABLE, CORROSIVE, N.O.S.	4.3	WFC	I	4.3+3+8	274 549	LQ0		PP, EP, EX, A	VE01	HA08	1	
2989	LEAD PHOSPHITE, DIBASIC	4.1	F3	II	4.1		LQ8		PP			1	
2989	LEAD PHOSPHITE, DIBASIC	4.1	F3	III	4.1		LQ9		PP			0	
2990	LIFE-SAVING APPLIANCES, SELF-INFLATING	9	M5		9	296 635	LQ0		PP			0	
2991	CARBAMATE PESTICIDE, LIQUID, TOXIC, FLAMMABLE, flash-point not less than 23 °C	6.1	TF2	I	6.1+3	61 802	LQ0		PP, EP, EX, TOX, A	VE01, VE02		2	
2991	CARBAMATE PESTICIDE, LIQUID, TOXIC, FLAMMABLE, flash-point not less than 23 °C	6.1	TF2	II	6.1+3	61 802	LQ17		PP, EP, EX, TOX, A	VE01, VE02		2	
2991	CARBAMATE PESTICIDE, LIQUID, TOXIC, FLAMMABLE, flash-point not less than 23 °C	6.1	TF2	III	6.1+3	61 802	LQ7		PP, EP, EX, TOX, A	VE01, VE02		0	
2992	CARBAMATE PESTICIDE, LIQUID, TOXIC	6.1	T6	I	6.1	61 648 802	LQ0		PP, EP, TOX, A	VE02		2	

UN No. or ID No. (1)	Name and description (2)	Class (3a)	Classification Code (3b)	Packing group (4)	Labels (5)	Special provisions (6)	Limited quantities (7)	Carriage permitted (8)	Equipment required (9)	Ventilation (10)	Provisions concerning loading, unloading and carriage (11)	Number of cones, blue lights (12)	Remarks (13)
2992	CARBAMATE PESTICIDE, LIQUID, TOXIC	6.1	T6	II	6.1	61 648 802	LQ17		PP, EP, TOX, A	VE02		2	
2992	CARBAMATE PESTICIDE, LIQUID, TOXIC	6.1	T6	III	6.1	61 648 802	LQ7		PP, EP, TOX, A	VE02		0	
2993	ARSENICAL PESTICIDE, LIQUID, TOXIC, FLAMMABLE, flash-point not less than 23 °C	6.1	TF2	I	6.1+3	61 802	LQ0		PP, EP, EX, TOX, A	VE01, VE02		2	
2993	ARSENICAL PESTICIDE, LIQUID, TOXIC, FLAMMABLE, flash-point not less than 23 °C	6.1	TF2	II	6.1+3	61 802	LQ17		PP, EP, EX, TOX, A	VE01, VE02		2	
2993	ARSENICAL PESTICIDE, LIQUID, TOXIC, FLAMMABLE, flash-point not less than 23 °C	6.1	TF2	III	6.1+3	61 802	LQ7		PP, EP, EX, TOX, A	VE01, VE02		0	
2994	ARSENICAL PESTICIDE, LIQUID, TOXIC	6.1	T6	I	6.1	61 648 802	LQ0		PP, EP, TOX, A	VE02		2	
2994	ARSENICAL PESTICIDE, LIQUID, TOXIC	6.1	T6	II	6.1	61 648 802	LQ17		PP, EP, TOX, A	VE02		2	
2994	ARSENICAL PESTICIDE, LIQUID, TOXIC	6.1	T6	III	6.1	61 648 802	LQ7		PP, EP, TOX, A	VE02		0	
2995	ORGANOCHLORINE PESTICIDE, LIQUID, TOXIC, FLAMMABLE, flash-point not less than 23 °C	6.1	TF2	I	6.1+3	61 802	LQ0		PP, EP, EX, TOX, A	VE01, VE02		2	
2995	ORGANOCHLORINE PESTICIDE, LIQUID, TOXIC, FLAMMABLE, flash-point not less than 23 °C	6.1	TF2	II	6.1+3	61 802	LQ17		PP, EP, EX, TOX, A	VE01, VE02		2	
2995	ORGANOCHLORINE PESTICIDE, LIQUID, TOXIC, FLAMMABLE, flash-point not less than 23 °C	6.1	TF2	III	6.1+3	61 802	LQ7		PP, EP, EX, TOX, A	VE01, VE02		0	

UN No. or ID No.	Name and description	Class	Classification Code	Packing group	Labels	Special provisions	Limited quantities	Carriage permitted	Equipment required	Ventilation	Provisions concerning loading, unloading and carriage	Number of cones, blue lights	Remarks
	3.1.2	2.2	2.2	2.1.1.3	5.2.2	3.3	3.4.6	3.2.1	8.1.5	7.1.6	7.1.6	7.1.5	3.2.1
(1)	(2)	(3a)	(3b)	(4)	(5)	(6)	(7)	(8)	(9)	(10)	(11)	(12)	(13)
2996	ORGANOCHLORINE PESTICIDE, LIQUID, TOXIC	6.1	T6	I	6.1	61 648 802	LQ0		PP, EP, TOX, A	VE02		2	
2996	ORGANOCHLORINE PESTICIDE, LIQUID, TOXIC	6.1	T6	II	6.1	61 648 802	LQ17		PP, EP, TOX, A	VE02		2	
2996	ORGANOCHLORINE PESTICIDE, LIQUID, TOXIC	6.1	T6	III	6.1	61 648 802	LQ7		PP, EP, TOX, A	VE02		0	
2997	TRIAZINE PESTICIDE, LIQUID, TOXIC, FLAMMABLE, flash-point not less than 23 °C	6.1	TF2	I	6.1+3	61 802	LQ0		PP, EP, EX, TOX, A	VE01, VE02		2	
2997	TRIAZINE PESTICIDE, LIQUID, TOXIC, FLAMMABLE, flash-point not less than 23 °C	6.1	TF2	II	6.1+3	61 802	LQ17		PP, EP, EX, TOX, A	VE01, VE02		2	
2997	TRIAZINE PESTICIDE, LIQUID, TOXIC, FLAMMABLE, flash-point not less than 23 °C	6.1	TF2	III	6.1+3	61 802	LQ7		PP, EP, EX, TOX, A	VE01, VE02		0	
2998	TRIAZINE PESTICIDE, LIQUID, TOXIC	6.1	T6	I	6.1	61 648 802	LQ0		PP, EP, TOX, A	VE02		2	
2998	TRIAZINE PESTICIDE, LIQUID, TOXIC	6.1	T6	II	6.1	61 802	LQ17		PP, EP, TOX, A	VE02		2	
2998	TRIAZINE PESTICIDE, LIQUID, TOXIC	6.1	T6	III	6.1	61 648 802	LQ7		PP, EP, TOX, A	VE02		0	
3005	THIOCARBAMATE PESTICIDE, LIQUID, TOXIC, FLAMMABLE, flash-point not less than 23 °C	6.1	TF2	I	6.1+3	61 802	LQ0		PP, EP, EX, TOX, A	VE01, VE02		2	
3005	THIOCARBAMATE PESTICIDE, LIQUID, TOXIC, FLAMMABLE, flash-point not less than 23 °C	6.1	TF2	II	6.1+3	61 802	LQ17		PP, EP, EX, TOX, A	VE01, VE02		2	
3005	THIOCARBAMATE PESTICIDE, LIQUID, TOXIC, FLAMMABLE, flash-point not less than 23 °C	6.1	TF2	III	6.1+3	61 802	LQ7		PP, EP, EX, TOX, A	VE01, VE02		0	

UN No. or ID No.	Name and description	Class	Classification Code	Packing group	Labels	Special provisions	Limited quantities	Carriage permitted	Equipment required	Ventilation	Provisions concerning loading, unloading and carriage	Number of cones, blue lights	Remarks
3.1.2	3.1.2	2.2	2.2	2.1.1.3	5.2.2	3.3	3.4.6	3.2.1	8.1.5	7.1.6	7.1.6	7.1.5	3.2.1
(1)	(2)	(3a)	(3b)	(4)	(5)	(6)	(7)	(8)	(9)	(10)	(11)	(12)	(13)
3006	THIOCARBAMATE PESTICIDE, LIQUID, TOXIC	6.1	T6	I	6.1	61 648 802	LQ0		PP, EP, TOX, A	VE02		2	
3006	THIOCARBAMATE PESTICIDE, LIQUID, TOXIC	6.1	T6	II	6.1	61 648 802	LQ17		PP, EP, TOX, A	VE02		2	
3006	THIOCARBAMATE PESTICIDE, LIQUID, TOXIC	6.1	T6	III	6.1	61 648 802	LQ7		PP, EP, TOX, A	VE02		0	
3009	COPPER BASED PESTICIDE, LIQUID, TOXIC, FLAMMABLE, flash-point not less than 23 °C	6.1	TF2	I	6.1+3	61 802	LQ0		PP, EP, EX, TOX, A	VE01, VE02		2	
3009	COPPER BASED PESTICIDE, LIQUID, TOXIC, FLAMMABLE, flash-point not less than 23 °C	6.1	TF2	II	6.1+3	61 802	LQ17		PP, EP, EX, TOX, A	VE01, VE02		2	
3009	COPPER BASED PESTICIDE, LIQUID, TOXIC, FLAMMABLE, flash-point not less than 23 °C	6.1	TF2	III	6.1+3	61 802	LQ7		PP, EP, EX, TOX, A	VE01, VE02		0	
3010	COPPER BASED PESTICIDE, LIQUID, TOXIC	6.1	T6	I	6.1	61 648 802	LQ0		PP, EP, TOX, A	VE02		2	
3010	COPPER BASED PESTICIDE, LIQUID, TOXIC	6.1	T6	II	6.1	61 648 802	LQ17		PP, EP, TOX, A	VE02		2	
3010	COPPER BASED PESTICIDE, LIQUID, TOXIC	6.1	T6	III	6.1	61 648 802	LQ7		PP, EP, TOX, A	VE02		0	
3011	MERCURY BASED PESTICIDE, LIQUID, TOXIC, FLAMMABLE, flash-point not less than 23 °C	6.1	TF2	I	6.1+3	61 802	LQ0		PP, EP, EX, TOX, A	VE01, VE02		2	
3011	MERCURY BASED PESTICIDE, LIQUID, TOXIC, FLAMMABLE, flash-point not less than 23 °C	6.1	TF2	II	6.1+3	61 802	LQ17		PP, EP, EX, TOX, A	VE01, VE02		2	
3011	MERCURY BASED PESTICIDE, LIQUID, TOXIC, FLAMMABLE, flash-point not less than 23 °C	6.1	TF2	III	6.1+3	61 802	LQ7		PP, EP, EX, TOX, A	VE01, VE02		0	

UN No. or ID No.	Name and description	Class	Classification Code	Packing group	Labels	Special provisions	Limited quantities	Carriage permitted	Equipment required	Ventilation	Provisions concerning loading, unloading and carriage	Number of cones, blue lights	Remarks
3.1.2	3.1.2	2.2	2.2	2.1.1.3	5.2.2	3.3	3.4.6	3.2.1	8.1.5	7.1.6	7.1.6	7.1.5	3.2.1
(1)	(2)	(3a)	(3b)	(4)	(5)	(6)	(7)	(8)	(9)	(10)	(11)	(12)	(13)
3012	MERCURY BASED PESTICIDE, LIQUID, TOXIC	6.1	T6	I	6.1	61 648 802	LQ0		PP, EP, TOX, A	VE02		2	
3012	MERCURY BASED PESTICIDE, LIQUID, TOXIC	6.1	T6	II	6.1	61 648 802	LQ17		PP, EP, TOX, A	VE02		2	
3012	MERCURY BASED PESTICIDE, LIQUID, TOXIC	6.1	T6	III	6.1	61 648 802	LQ7		PP, EP, TOX, A	VE02		0	
3013	SUBSTITUTED NITROPHENOL PESTICIDE, LIQUID, TOXIC, FLAMMABLE, flash-point not less than 23 °C	6.1	TF2	I	6.1+3	61 802	LQ0		PP, EP, EX, TOX, A	VE01, VE02		2	
3013	SUBSTITUTED NITROPHENOL PESTICIDE, LIQUID, TOXIC, FLAMMABLE, flash-point not less than 23 °C	6.1	TF2	II	6.1+3	61 802	LQ17		PP, EP, EX, TOX, A	VE01, VE02		2	
3013	SUBSTITUTED NITROPHENOL PESTICIDE, LIQUID, TOXIC, FLAMMABLE, flash-point not less than 23 °C	6.1	TF2	III	6.1+3	61 802	LQ7		PP, EP, EX, TOX, A	VE01, VE02		0	
3014	SUBSTITUTED NITROPHENOL PESTICIDE, LIQUID, TOXIC	6.1	T6	I	6.1	61 648 802	LQ0		PP, EP, TOX, A	VE02		2	
3014	SUBSTITUTED NITROPHENOL PESTICIDE, LIQUID, TOXIC	6.1	T6	II	6.1	61 648 802	LQ17		PP, EP, TOX, A	VE02		2	
3014	SUBSTITUTED NITROPHENOL PESTICIDE, LIQUID, TOXIC	6.1	T6	III	6.1	61 648 802	LQ7		PP, EP, TOX, A	VE02		0	
3015	BIPYRIDILIUM PESTICIDE, LIQUID, TOXIC, FLAMMABLE, flash-point not less than 23 °C	6.1	TF2	I	6.1+3	61 802	LQ0		PP, EP, EX, TOX, A	VE01, VE02		2	
3015	BIPYRIDILIUM PESTICIDE, LIQUID, TOXIC, FLAMMABLE, flash-point not less than 23 °C	6.1	TF2	II	6.1+3	61 802	LQ17		PP, EP, EX, TOX, A	VE01, VE02		2	

UN No. or ID No.	Name and description	Class	Classification Code	Packing group	Labels	Special provisions	Limited quantities	Carriage permitted	Equipment required	Ventilation	Provisions concerning loading, unloading and carriage	Number of cones, blue lights	Remarks
3.1.1	3.1.2	2.2	2.2	2.1.1.3	5.2.2	3.3	3.4.6	3.2.1	8.1.5	7.1.6	7.1.6	7.1.5	3.2.1
(1)	(2)	(3a)	(3b)	(4)	(5)	(6)	(7)	(8)	(9)	(10)	(11)	(12)	(13)
3015	BIPYRIDILIUM PESTICIDE, LIQUID, TOXIC, FLAMMABLE, flash-point not less than 23 °C	6.1	TF2	III	6.1+3	61 802	LQ7		PP, EP, EX, TOX, A	VE01, VE02		0	
3016	BIPYRIDILIUM PESTICIDE, LIQUID, TOXIC	6.1	T6	I	6.1	61 648 802	LQ0		PP, EP, TOX, A	VE02		2	
3016	BIPYRIDILIUM PESTICIDE, LIQUID, TOXIC	6.1	T6	II	6.1	61 648 802	LQ17		PP, EP, TOX, A	VE02		2	
3016	BIPYRIDILIUM PESTICIDE, LIQUID, TOXIC	6.1	T6	III	6.1	61 648 802	LQ7		PP, EP, TOX, A	VE02		0	
3017	ORGANOPHOSPHORUS PESTICIDE, LIQUID, TOXIC, FLAMMABLE, flash-point not less than 23 °C	6.1	TF2	I	6.1+3	61 802	LQ0		PP, EP, EX, TOX, A	VE01, VE02		2	
3017	ORGANOPHOSPHORUS PESTICIDE, LIQUID, TOXIC, FLAMMABLE, flash-point not less than 23 °C	6.1	TF2	II	6.1+3	61 802	LQ17		PP, EP, EX, TOX, A	VE01, VE02		2	
3017	ORGANOPHOSPHORUS PESTICIDE, LIQUID, TOXIC, FLAMMABLE, flash-point not less than 23 °C	6.1	TF2	III	6.1+3	61 802	LQ7		PP, EP, EX, TOX, A	VE01, VE02		0	
3018	ORGANOPHOSPHORUS PESTICIDE, LIQUID, TOXIC	6.1	T6	I	6.1	61 648 802	LQ0		PP, EP, TOX, A	VE02		2	
3018	ORGANOPHOSPHORUS PESTICIDE, LIQUID, TOXIC	6.1	T6	II	6.1	61 648 802	LQ17		PP, EP, TOX, A	VE02		2	
3018	ORGANOPHOSPHORUS PESTICIDE, LIQUID, TOXIC	6.1	T6	III	6.1	61 648 802	LQ7		PP, EP, TOX, A	VE02		0	
3019	ORGANOTIN PESTICIDE, LIQUID, TOXIC, FLAMMABLE, flash-point not less than 23 °C	6.1	TF2	I	6.1+3	61 802	LQ0		PP, EP, EX, TOX, A	VE01, VE02		2	

UN No. or ID No. (1)	Name and description 3.1.2 (2)	Class 2.2 (3a)	Classification Code 2.2 (3b)	Packing group 2.1.1.3 (4)	Labels 5.2.2 (5)	Special provisions 3.3 (6)	Limited quantities 3.4.6 (7)	Carriage permitted 3.2.1 (8)	Equipment required 8.1.5 (9)	Ventilation 7.1.6 (10)	Provisions concerning loading, unloading and carriage 7.1.6 (11)	Number of cones, blue lights 7.1.5 (12)	Remarks 3.2.1 (13)
3019	ORGANOTIN PESTICIDE, LIQUID, TOXIC, FLAMMABLE, flash-point not less than 23 °C	6.1	TF2	II	6.1+3	61 802	LQ17		PP, EP, EX, TOX, A	VE01, VE02		2	
3019	ORGANOTIN PESTICIDE, LIQUID, TOXIC, FLAMMABLE, flash-point not less than 23 °C	6.1	TF2	III	6.1+3	61 802	LQ7		PP, EP, EX, TOX, A	VE01, VE02		0	
3020	ORGANOTIN PESTICIDE, LIQUID, TOXIC	6.1	T6	I	6.1	61 648 802	LQ0		PP, EP, TOX, A	VE02		2	
3020	ORGANOTIN PESTICIDE, LIQUID, TOXIC	6.1	T6	II	6.1	61 648 802	LQ17		PP, EP, TOX, A	VE02		2	
3020	ORGANOTIN PESTICIDE, LIQUID, TOXIC	6.1	T6	III	6.1	61 648 802	LQ7		PP, EP, TOX, A	VE02		0	
3021	PESTICIDE, LIQUID, FLAMMABLE, TOXIC, N.O.S., flash-point less than 23 °C	3	FT2	I	3+6.1	61 802	LQ3		PP, EP, EX, TOX, A	VE01, VE02		2	
3021	PESTICIDE, LIQUID, FLAMMABLE, TOXIC, N.O.S., flash-point less than 23 °C	3	FT2	II	3+6.1	61 802	LQ4		PP, EP, EX, TOX, A	VE01, VE02		2	
3022	1,2-BUTYLENE OXIDE, STABILIZED	3	F1	II	3		LQ4		PP, EX, A	VE01		1	
3023	2-METHYL-2-HEPTANETHIOL	6.1	TF1	I	6.1+3	802	LQ0		PP, EP, EX, TOX, A	VE01, VE02		2	
3024	COUMARIN DERIVATIVE PESTICIDE, LIQUID, FLAMMABLE, TOXIC, flash-point less than 23 °C	3	FT2	I	3+6.1	61 802	LQ3		PP, EP, EX, TOX, A	VE01, VE02		2	
3024	COUMARIN DERIVATIVE PESTICIDE, LIQUID, FLAMMABLE, TOXIC, flash-point less than 23 °C	3	FT2	II	3+6.1	61 802	LQ4		PP, EP, EX, TOX, A	VE01, VE02		2	
3025	COUMARIN DERIVATIVE PESTICIDE, LIQUID, TOXIC, FLAMMABLE, flash-point not less than 23 °C	6.1	TF2	I	6.1+3	61 802	LQ0		PP, EP, EX, TOX, A	VE01, VE02		2	

UN No. or ID No. (1)	Name and description (2)	Class (3a)	Classification Code (3b)	Packing group (4)	Labels (5)	Special provisions (6)	Limited quantities (7)	Carriage permitted (8)	Equipment required (9)	Ventilation (10)	Provisions concerning loading, unloading and carriage (11)	Number of cones, blue lights (12)	Remarks (13)
3025	COUMARIN DERIVATIVE PESTICIDE, LIQUID, TOXIC, FLAMMABLE, flash-point not less than 23 °C	6.1	TF2	II	6.1+3	61 802	LQ17		PP, EP, EX, TOX, A	VE01, VE02		2	
3025	COUMARIN DERIVATIVE PESTICIDE, LIQUID, TOXIC, FLAMMABLE, flash-point not less than 23 °C	6.1	TF2	III	6.1+3	61 802	LQ7		PP, EP, EX, TOX, A	VE01, VE02		0	
3026	COUMARIN DERIVATIVE PESTICIDE, LIQUID, TOXIC	6.1	T6	I	6.1	61 648 802	LQ0		PP, EP, TOX, A	VE02		2	
3026	COUMARIN DERIVATIVE PESTICIDE, LIQUID, TOXIC	6.1	T6	II	6.1	61 648 802	LQ17		PP, EP, TOX, A	VE02		2	
3026	COUMARIN DERIVATIVE PESTICIDE, LIQUID, TOXIC	6.1	T6	III	6.1	61 648 802	LQ7		PP, EP, TOX, A	VE02		0	
3027	COUMARIN DERIVATIVE PESTICIDE, SOLID, TOXIC	6.1	T7	I	6.1	61 648 802	LQ0		PP, EP			2	
3027	COUMARIN DERIVATIVE PESTICIDE, SOLID, TOXIC	6.1	T7	II	6.1	61 648 802	LQ18		PP, EP			2	
3027	COUMARIN DERIVATIVE PESTICIDE, SOLID, TOXIC	6.1	T7	III	6.1	61 648 802	LQ9		PP, EP			0	
3028	BATTERIES, DRY, CONTAINING POTASSIUM HYDROXIDE SOLID, electric storage	8	C11		8	295 304 598	LQ0		PP, EP			0	
3048	ALUMINIUM PHOSPHIDE PESTICIDE	6.1	T7	I	6.1	61 648 153 802	LQ0		PP, EP			2	
3054	CYCLOHEXYL MERCAPTAN	3	F1	III	3		LQ7		PP, EX, A	VE01		0	
3055	2-(2-AMINOETHOXY)ETHANOL	8	C7	III	8		LQ7		PP, EP	VE01		0	
3056	n-HEPTALDEHYDE	3	F1	III	3		LQ7		PP, EX, A	VE01		0	

UN No. or ID No.	Name and description	Class	Classification Code	Packing group	Labels	Special provisions	Limited quantities	Carriage permitted	Equipment required	Ventilation	Provisions concerning loading, unloading and carriage	Number of cones, blue lights	Remarks
3.1.2	3.1.2	2.2	2.2	2.1.1.3	5.2.2	3.3	3.4.6	3.2.1	8.1.5	7.1.6	7.1.6	7.1.5	3.2.1
(1)	(2)	(3a)	(3b)	(4)	(5)	(6)	(7)	(8)	(9)	(10)	(11)	(12)	(13)
3057	TRIFLUOROACETYL CHLORIDE	2	2TC		2.3+8		LQ0		PP, EP, TOX, A	VE02		2	
3064	NITROGLYCERIN, SOLUTION IN ALCOHOL with more than 1% but not more than 5% nitroglycerin	3	D	II	3		LQ0		PP, EX, A	VE01		1	
3065	ALCOHOLIC BEVERAGES, with more than 70% alcohol by volume	3	F1	II	3		LQ5		PP, EX, A	VE01		1	
3065	ALCOHOLIC BEVERAGES, with more than 24% but not more than 70% alcohol by volume	3	F1	III	3	144 145 247	LQ7		PP, EX, A	VE01		0	
3066	PAINT (including paint, lacquer, enamel. stain. shellac, varnish, polish, liquid filler and liquid lacquer base) or PAINT RELATED MATERIAL (including paint thinning and reducing compound)	8	C9	II	8	163	LQ22		PP, EP			0	
3066	PAINT (including paint, lacquer, enamel. stain. shellac, varnish, polish, liquid filler and liquid lacquer base) or PAINT RELATED MATERIAL (including paint thinning and reducing compound)	8	C9	III	8	163	LQ7		PP, EP			0	
3070	ETHYLENE OXIDE AND DICHLORODIFLUORO-METHANE MIXTURE with not more than 12.5% ethylene oxide	2	2A		2.2		LQ1		PP			0	
3071	MERCAPTANS, LIQUID, TOXIC, FLAMMABLE, N.O.S. or MERCAPTAN MIXTURE, LIQUID, TOXIC, FLAMMABLE, N.O.S.	6.1	TF1	II	6.1+3	274 802	LQ17		PP, EP, EX, TOX, A	VE01, VE02		2	
3072	LIFE-SAVING APPLIANCES NOT SELF-INFLATING containing dangerous goods as equipment	9	M5		9	296 635	LQ0		PP			0	
3073	VINYLPYRIDINES, STABILIZED	6.1	TFC	II	6.1+3+8	802	LQ17		PP, EP, EX, TOX, A	VE01, VE02		2	

UN No. or ID No.	Name and description	Class	Classification Code	Packing group	Labels	Special provisions	Limited quantities	Carriage permitted	Equipment required	Ventilation	Provisions concerning loading, unloading and carriage	Number of cones, blue lights	Remarks
3.1.2	3.1.2	2.2	2.2	2.1.1.3	5.2.2	3.3	3.4.6	3.2.1	8.1.5	7.1.6	7.1.6	7.1.5	3.2.1
(1)	(2)	(3a)	(3b)	(4)	(5)	(6)	(7)	(8)	(9)	(10)	(11)	(12)	(13)
3077	ENVIRONMENTALLY HAZARDOUS SUBSTANCE, SOLID. N.O.S.	9	M7	III	9	274 601	LQ27	T	PP			0	
3078	CERIUM. turnings or gritty powder	4.3	W2	II	4.3	550	LQ11		PP, EX, A	VE01	HA08	0	
3079	METHACRYLONITRILE, STABILIZED	3	FT1	I	3+6.1	802	LQ0	T	PP, EP, EX, TOX, A	VE01, VE02		2	
3080	ISOCYANATES, TOXIC, FLAMMABLE, N.O.S. or ISOCYANATE SOLUTION, TOXIC. FLAMMABLE, N.O.S.	6.1	TF1	II	6.1+3	274 551 802	LQ17		PP, EP, EX, TOX, A	VE01, VE02		2	
3082	ENVIRONMENTALLY HAZARDOUS SUBSTANCE, LIQUID. N.O.S.	9	M6	III	9	274 601	LQ7	T	PP			0	
3083	PERCHLORYL FLUORIDE	2	2TO		2.3+5.1		LQ0		PP, EP, TOX, A	VE02		2	
3084	CORROSIVE SOLID, OXIDIZING. N.O.S.	8	CO2	I	8+5.1	274	LQ0		PP, EP			0	
3084	CORROSIVE SOLID, OXIDIZING. N.O.S.	8	CO2	II	8+5.1	274	LQ23		PP, EP			0	
3085	OXIDIZING SOLID, CORROSIVE. N.O.S.	5.1	OC2	I	5.1+8	274	LQ0		PP			0	
3085	OXIDIZING SOLID. CORROSIVE. N.O.S.	5.1	OC2	II	5.1+8	274	LQ11		PP			0	
3085	OXIDIZING SOLID, CORROSIVE. N.O.S.	5.1	OC2	III	5.1+8	274	LQ12		PP			0	
3086	TOXIC SOLID. OXIDIZING, N.O.S.	6.1	TO2	I	6.1+5.1	274 802	LQ0		PP, EP			2	
3086	TOXIC SOLID. OXIDIZING, N.O.S.	6.1	TO2	II	6.1+5.1	274 802	LQ18		PP, EP			2	
3087	OXIDIZING SOLID, TOXIC, N.O.S.	5.1	OT2	I	5.1+6.1	274 802	LQ0		PP			2	
3087	OXIDIZING SOLID, TOXIC, N.O.S.	5.1	OT2	II	5.1+6.1	274 802	LQ11		PP			2	
3087	OXIDIZING SOLID, TOXIC, N.O.S.	5.1	OT2	III	5.1+6.1	274 802	LQ12		PP			0	
3088	SELF-HEATING SOLID, ORGANIC. N.O.S.	4.2	S2	II	4.2	274	LQ0		PP			0	

UN No. or ID No.	Name and description	Class	Classification Code	Packing group	Labels	Special provisions	Limited quantities	Carriage permitted	Equipment required	Ventilation	Provisions concerning loading, unloading and carriage	Number of cones, blue lights	Remarks
(1)	3.1.2 (2)	2.2 (3a)	2.2 (3b)	2.1.1.3 (4)	5.2.2 (5)	3.3 (6)	3.4.6 (7)	3.2.1 (8)	8.1.5 (9)	7.1.6 (10)	7.1.6 (11)	7.1.5 (12)	3.2.1 (13)
3088	SELF-HEATING SOLID, ORGANIC, N.O.S.	4.2	S2	III	4.2	274	LQ0		PP			0	
3089	METAL POWDER, FLAMMABLE, N.O.S.	4.1	F3	II	4.1	274 552	LQ8		PP			1	
3089	METAL POWDER, FLAMMABLE, N.O.S.	4.1	F3	III	4.1	274 552	LQ9		PP			0	
3090	LITHIUM BATTERIES	9	M4	II	9	188 230 310 636	LQ0		PP			0	
3091	LITHIUM BATTERIES CONTAINED IN EQUIPMENT or LITHIUM BATTERIES PACKED WITH EQUIPMENT	9	M4	II	9	188 230 636	LQ0		PP			0	
3092	1-METHOXY-2-PROPANOL	3	F1	III	3		LQ7	T	PP, EX, A	VE01		0	
3093	CORROSIVE LIQUID, OXIDIZING, N.O.S.	8	CO1	I	8+5.1	274	LQ0		PP, EP			0	
3093	CORROSIVE LIQUID, OXIDIZING, N.O.S.	8	CO1	II	8+5.1	274	LQ22		PP, EP			0	
3094	CORROSIVE LIQUID, WATER-REACTIVE, N.O.S.	8	CW1	I	8+4.3	274	LQ0		PP, EP			0	
3094	CORROSIVE LIQUID, WATER-REACTIVE, N.O.S.	8	CW1	II	8+4.3	274	LQ22		PP, EP			0	
3095	CORROSIVE SOLID, SELF-HEATING, N.O.S.	8	CS2	I	8+4.2	274	LQ0		PP, EP			0	
3095	CORROSIVE SOLID, SELF-HEATING, N.O.S.	8	CS2	II	8+4.2	274	LQ23		PP, EP			0	
3096	CORROSIVE SOLID, WATER-REACTIVE, N.O.S.	8	CW2	I	8+4.3	274	LQ0		PP, EP			0	
3096	CORROSIVE SOLID, WATER-REACTIVE, N.O.S.	8	CW2	II	8+4.3	274	LQ23		PP, EP			0	
3097	FLAMMABLE SOLID, OXIDIZING, N.O.S.	4.1	FO					CARRIAGE PROHIBITED					
3098	OXIDIZING LIQUID, CORROSIVE, N.O.S.	5.1	OC1	I	5.1+8	274	LQ0		PP, EP			0	
3098	OXIDIZING LIQUID, CORROSIVE, N.O.S.	5.1	OC1	II	5.1+8	274	LQ10		PP, EP			0	

UN No. or ID No.	Name and description	Class	Classification Code	Packing group	Labels	Special provisions	Limited quantities	Carriage permitted	Equipment required	Ventilation	Provisions concerning loading, unloading and carriage	Number of cones, blue lights	Remarks
	3.1.2	2.2	2.2	2.1.1.3	5.2.2	3.3	3.4.6	3.2.1	8.1.5	7.1.6	7.1.6	7.1.5	3.2.1
(1)	(2)	(3a)	(3b)	(4)	(5)	(6)	(7)	(8)	(9)	(10)	(11)	(12)	(13)
3098	OXIDIZING LIQUID, CORROSIVE, N.O.S.	5.1	OC1	III	5.1+8	274	LQ13		PP, EP			0	
3099	OXIDIZING LIQUID, TOXIC, N.O.S.	5.1	OT1	I	5.1+6.1	274 802	LQ0		PP, EP, TOX, A	VE02		2	
3099	OXIDIZING LIQUID, TOXIC, N.O.S.	5.1	OT1	II	5.1+6.1	274 802	LQ10		PP, EP, TOX, A	VE02		2	
3099	OXIDIZING LIQUID, TOXIC, N.O.S.	5.1	OT1	III	5.1+6.1	274 802	LQ13		PP, EP, TOX, A	VE02		0	
3100	OXIDIZING SOLID, SELF-HEATING, N.O.S.	5.1	OS					CARRIAGE PROHIBITED					
3101	ORGANIC PEROXIDE TYPE B, LIQUID	5.2	P1		5.2+1	122 181 274	LQ14		PP, EX, A	VE01	HA01, HA10	3	
3102	ORGANIC PEROXIDE TYPE B, SOLID	5.2	P1		5.2+1	122 181 274	LQ15		PP, EX, A	VE01	HA01, HA10	3	
3103	ORGANIC PEROXIDE TYPE C, LIQUID	5.2	P1		5.2	122 274	LQ14		PP, EX, A	VE01		0	
3104	ORGANIC PEROXIDE TYPE C, SOLID	5.2	P1		5.2	122 274	LQ15		PP, EX, A	VE01		0	
3105	ORGANIC PEROXIDE TYPE D, LIQUID	5.2	P1		5.2	122 274	LQ16		PP, EX, A	VE01		0	
3106	ORGANIC PEROXIDE TYPE D, SOLID	5.2	P1		5.2	122 274	LQ11		PP, EX, A	VE01		0	
3107	ORGANIC PEROXIDE TYPE E, LIQUID	5.2	P1		5.2	122 274	LQ16		PP, EX, A	VE01		0	
3108	ORGANIC PEROXIDE TYPE E, SOLID	5.2	P1		5.2	122 274	LQ11		PP, EX, A	VE01		0	
3109	ORGANIC PEROXIDE TYPE F, LIQUID	5.2	P1		5.2	122 274	LQ16		PP, EX, A	VE01		0	
3110	ORGANIC PEROXIDE TYPE F, SOLID	5.2	P1		5.2	122 274	LQ11		PP, EX, A	VE01		0	
3111	ORGANIC PEROXIDE TYPE B, LIQUID, TEMPERATURE CONTROLLED	5.2	P2		5.2+1	122 181 274	LQ0		PP, EX, A	VE01	HA01, HA10	3	

UN No. or ID No. (1)	Name and description 3.1.2 (2)	Class 2.2 (3a)	Classification Code 2.2 (3b)	Packing group 2.1.1.3 (4)	Labels 5.2.2 (5)	Special provisions 3.3 (6)	Limited quantities 3.4.6 (7)	Carriage permitted 3.2.1 (8)	Equipment required 8.1.5 (9)	Ventilation 7.1.6 (10)	Provisions concerning loading, unloading and carriage 7.1.6 (11)	Number of cones, blue lights 7.1.5 (12)	Remarks 3.2.1 (13)
3112	ORGANIC PEROXIDE TYPE B, SOLID, TEMPERATURE CONTROLLED	5.2	P2		5.2+1	122 181 274	LQ0		PP, EX, A	VE01	HA01, HA10	3	
3113	ORGANIC PEROXIDE TYPE C, LIQUID, TEMPERATURE CONTROLLED	5.2	P2		5.2	122 274	LQ0		PP, EX, A	VE01		0	
3114	ORGANIC PEROXIDE TYPE C, SOLID, TEMPERATURE CONTROLLED	5.2	P2		5.2	122 274	LQ0		PP, EX, A	VE01		0	
3115	ORGANIC PEROXIDE TYPE D, LIQUID, TEMPERATURE CONTROLLED	5.2	P2		5.2	122 274	LQ0		PP, EX, A	VE01		0	
3116	ORGANIC PEROXIDE TYPE D, SOLID, TEMPERATURE CONTROLLED	5.2	P2		5.2	122 274	LQ0		PP, EX, A	VE01		0	
3117	ORGANIC PEROXIDE TYPE E, LIQUID, TEMPERATURE CONTROLLED	5.2	P2		5.2	122 274	LQ0		PP, EX, A	VE01		0	
3118	ORGANIC PEROXIDE TYPE E, SOLID, TEMPERATURE CONTROLLED	5.2	P2		5.2	122 274	LQ0		PP, EX, A	VE01		0	
3119	ORGANIC PEROXIDE TYPE F, LIQUID, TEMPERATURE CONTROLLED	5.2	P2		5.2	122 274	LQ0		PP, EX, A	VE01		0	
3120	ORGANIC PEROXIDE TYPE F, SOLID, TEMPERATURE CONTROLLED	5.2	P2		5.2	122 274	LQ0		PP, EX, A	VE01		0	
3121	OXIDIZING SOLID, WATER-REACTIVE, N.O.S.	5.1	OW					CARRIAGE PROHIBITED					
3122	TOXIC LIQUID, OXIDIZING, N.O.S.	6.1	TO1	I	6.1+5.1	274 315 802	LQ0		PP, EP, TOX, A	VE02		2	
3122	TOXIC LIQUID, OXIDIZING, N.O.S.	6.1	TO1	II	6.1+5.1	274 802	LQ17		PP, EP, TOX, A	VE02		2	
3123	TOXIC LIQUID, WATER-REACTIVE, N.O.S.	6.1	TW1	I	6.1+4.3	274 315 802	LQ0		PP, EP, TOX, A	VE02		2	

UN No. or ID No. (1)	Name and description 3.1.2 (2)	Class 2.2 (3a)	Classification Code 2.2 (3b)	Packing group 2.1.1.3 (4)	Labels 5.2.2 (5)	Special provisions 3.3 (6)	Limited quantities 3.4.6 (7)	Carriage permitted 3.2.1 (8)	Equipment required 8.1.5 (9)	Ventilation 7.1.6 (10)	Provisions concerning loading, unloading and carriage 7.1.6 (11)	Number of cones, blue lights 7.1.5 (12)	Remarks 3.2.1 (13)
3123	TOXIC LIQUID, WATER-REACTIVE. N.O.S.	6.1	TW1	II	6.1+4.3	274 802	LQ17		PP, EP, TOX, A	VE02		2	
3124	TOXIC SOLID, SELF-HEATING, N.O.S.	6.1	TS	I	6.1+4.2	274 802	LQ0		PP, EP			2	
3124	TOXIC SOLID, SELF-HEATING. N.O.S.	6.1	TS	II	6.1+4.2	274 802	LQ18		PP, EP			2	
3125	TOXIC SOLID, WATER-REACTIVE. N.O.S.	6.1	TW2	I	6.1+4.3	274 802	LQ0		PP, EP			2	
3125	TOXIC SOLID, WATER-REACTIVE, N.O.S.	6.1	TW2	II	6.1+4.3	274 802	LQ18		PP, EP			2	
3126	SELF-HEATING SOLID, CORROSIVE, ORGANIC, N.O.S.	4.2	SC2	II	4.2+8	274	LQ0		PP			0	
3126	SELF-HEATING SOLID, CORROSIVE, ORGANIC, N.O.S.	4.2	SC2	III	4.2+8	274	LQ0		PP			0	
3127	SELF-HEATING SOLID, OXIDIZING, N.O.S	4.2	SO					CARRIAGE PROHIBITED					
3128	SELF-HEATING SOLID, TOXIC, ORGANIC, N.O.S.	4.2	ST2	II	4.2+6.1	274 802	LQ0		PP			2	
3128	SELF-HEATING SOLID, TOXIC, ORGANIC, N.O.S.	4.2	ST2	III	4.2+6.1	274 802	LQ0		PP			0	
3129	WATER-REACTIVE LIQUID, CORROSIVE, N.O.S.	4.3	WC1	I	4.3+8	274	LQ0		PP, EP, EX, A	VE01	HA08	0	
3129	WATER-REACTIVE LIQUID, CORROSIVE, N.O.S.	4.3	WC1	II	4.3+8	274	LQ10		PP, EP, EX, A	VE01	HA08	0	
3129	WATER-REACTIVE LIQUID, CORROSIVE, N.O.S.	4.3	WC1	III	4.3+8	274	LQ13		PP, EP, EX, A	VE01	HA08	0	
3130	WATER-REACTIVE LIQUID, TOXIC. N.O.S.	4.3	WT1	I	4.3+6.1	274 802	LQ0		PP, EP, EX, TOX, A	VE01, VE02	HA08	2	
3130	WATER-REACTIVE LIQUID, TOXIC. N.O.S.	4.3	WT1	II	4.3+6.1	274 802	LQ10		PP, EP, EX, TOX, A	VE01, VE02	HA08	2	
3130	WATER-REACTIVE LIQUID, TOXIC. N.O.S.	4.3	WT1	III	4.3+6.1	274 802	LQ13		PP, EP, EX, TOX, A	VE01, VE02	HA08	0	
3131	WATER-REACTIVE SOLID, CORROSIVE. N.O.S.	4.3	WC2	I	4.3+8	274	LQ0		PP, EP, EX, A	VE01	HA08	0	
3131	WATER-REACTIVE SOLID, CORROSIVE, N.O.S.	4.3	WC2	II	4.3+8	274	LQ11		PP, EP, EX, A	VE01	HA08	0	

UN No. or ID No.	Name and description	Class	Classification Code	Packing group	Labels	Special provisions	Limited quantities	Carriage permitted	Equipment required	Ventilation	Provisions concerning loading, unloading and carriage	Number of cones, blue lights	Remarks
	3.1.2	2.2	2.2	2.1.1.3	5.2.2	3.3	3.4.6	3.2.1	8.1.5	7.1.6	7.1.6	7.1.5	3.2.1
(1)	(2)	(3a)	(3b)	(4)	(5)	(6)	(7)	(8)	(9)	(10)	(11)	(12)	(13)
3131	WATER-REACTIVE SOLID, CORROSIVE, N.O.S.	4.3	WC2	III	4.3+8	274	LQ12		PP, EP, EX, A	VE01	HA08	0	
3132	WATER-REACTIVE SOLID, FLAMMABLE, N.O.S.	4.3	WF2	CARRIAGE PROHIBITED									
3133	WATER-REACTIVE SOLID, OXIDIZING. N.O.S.	4.3	WO	CARRIAGE PROHIBITED									
3134	WATER-REACTIVE SOLID, TOXIC. N.O.S.	4.3	WT2	I	4.3+6.1	274 802	LQ0		PP, EP, EX, A	VE01	HA08	2	
3134	WATER-REACTIVE SOLID, TOXIC. N.O.S.	4.3	WT2	II	4.3+6.1	274 802	LQ11		PP, EP, EX, A	VE01	HA08	2	
3134	WATER-REACTIVE SOLID, TOXIC. N.O.S.	4.3	WT2	III	4.3+6.1	274 802	LQ12		PP, EP, EX, A	VE01	HA08	0	
3135	WATER-REACTIVE SOLID, SELF-HEATING, N.O.S.	4.3	WS	CARRIAGE PROHIBITED									
3136	TRIFLUOROMETHANE, REFRIGERATED LIQUID	2	3A		2.2	593	LQ1		PP			0	
3137	OXIDIZING SOLID, FLAMMABLE, N.O.S.	5.1	OF	CARRIAGE PROHIBITED									
3138	ETHYLENE, ACETYLENE AND PROPYLENE MIXTURE, REFRIGERATED LIQUID containing at least 71.5% ethylene with not more than 22.5% acetylene and not more than 6% propylene	2	3F		2.1		LQ0		PP, EX, A	VE01		1	
3139	OXIDIZING LIQUID. N.O.S.	5.1	O1	I	5.1	274	LQ0		PP			0	
3139	OXIDIZING LIQUID. N.O.S.	5.1	O1	II	5.1	274	LQ10		PP			0	
3139	OXIDIZING LIQUID. N.O.S.	5.1	O1	III	5.1	274	LQ13		PP			0	
3140	ALKALOIDS, LIQUID, N.O.S. or ALKALOID SALTS, LIQUID, N.O.S.	6.1	T1	I	6.1	43 274 802	LQ0		PP, EP, TOX, A	VE02		2	
3140	ALKALOIDS, LIQUID, N.O.S. or ALKALOID SALTS, LIQUID, N.O.S.	6.1	T1	II	6.1	43 274 802	LQ17		PP, EP, TOX, A	VE02		2	
3140	ALKALOIDS, LIQUID, N.O.S. or ALKALOID SALTS, LIQUID, N.O.S.	6.1	T1	III	6.1	43 274 802	LQ7		PP, EP, TOX, A	VE02		0	

UN No. or ID No. (1) 3.1.2	Name and description (2) 3.1.2	Class (3a) 2.2	Classification Code (3b) 2.2	Packing group (4) 2.1.1.3	Labels (5) 5.2.2	Special provisions (6) 3.3	Limited quantities (7) 3.4.6	Carriage permitted (8) 3.2.1	Equipment required (9) 8.1.5	Ventilation (10) 7.1.6	Provisions concerning loading, unloading and carriage (11) 7.1.6	Number of cones, blue lights (12) 7.1.5	Remarks (13) 3.2.1
3141	ANTIMONY COMPOUND, INORGANIC, LIQUID, N.O.S.	6.1	T4	III	6.1	45 274 512 802	LQ7		PP, EP, TOX, A	VE02		0	
3142	DISINFECTANT, LIQUID, TOXIC, N.O.S.	6.1	T1	I	6.1	274 802	LQ0		PP, EP, TOX, A	VE02		2	
3142	DISINFECTANT, LIQUID, TOXIC, N.O.S.	6.1	T1	II	6.1	274 802	LQ17		PP, EP, TOX, A	VE02		2	
3142	DISINFECTANT, LIQUID, TOXIC, N.O.S.	6.1	T1	III	6.1	274 802	LQ7		PP, EP, TOX, A	VE02		0	
3143	DYE, SOLID, TOXIC, N.O.S. or DYE INTERMEDIATE, SOLID, TOXIC, N.O.S.	6.1	T2	I	6.1	274 802	LQ0		PP, EP			2	
3143	DYE, SOLID, TOXIC, N.O.S. or DYE INTERMEDIATE, SOLID, TOXIC, N.O.S.	6.1	T2	II	6.1	274 802	LQ18		PP, EP			2	
3143	DYE, SOLID, TOXIC, N.O.S. or DYE INTERMEDIATE, SOLID, TOXIC, N.O.S.	6.1	T2	III	6.1	274 802	LQ9		PP, EP			0	
3144	NICOTINE COMPOUND, LIQUID, N.O.S. or NICOTINE PREPARATION, LIQUID, N.O.S.	6.1	T1	I	6.1	43 274 802	LQ0		PP, EP, TOX, A	VE02		2	
3144	NICOTINE COMPOUND, LIQUID, N.O.S. or NICOTINE PREPARATION, LIQUID, N.O.S.	6.1	T1	II	6.1	43 274 802	LQ17		PP, EP, TOX, A	VE02		2	
3144	NICOTINE COMPOUND, LIQUID, N.O.S. or NICOTINE PREPARATION, LIQUID, N.O.S.	6.1	T1	III	6.1	43 274 802	LQ7		PP, EP, TOX, A	VE02		0	
3145	ALKYLPHENOLS, LIQUID, N.O.S. (including C_2-C_{12} homologues)	8	C3	I	8	274	LQ0		PP, EP			0	
3145	ALKYLPHENOLS, LIQUID, N.O.S. (including C_2-C_{12} homologues)	8	C3	II	8	274	LQ22	T	PP, EP			0	
3145	ALKYLPHENOLS, LIQUID, N.O.S. (including C_2-C_{12} homologues)	8	C3	III	8	274	LQ7	T	PP, EP			0	

UN No. or ID No.	Name and description	Class	Classification Code	Packing group	Labels	Special provisions	Limited quantities	Carriage permitted	Equipment required	Ventilation	Provisions concerning loading, unloading and carriage	Number of cones, blue lights	Remarks
3.1.2	3.1.2	2.2	2.2	2.1.1.3	5.2.2	3.3	3.4.6	3.2.1	8.1.5	7.1.6	7.1.6	7.1.5	3.2.1
(1)	(2)	(3a)	(3b)	(4)	(5)	(6)	(7)	(8)	(9)	(10)	(11)	(12)	(13)
3146	ORGANOTIN COMPOUND, SOLID, N.O.S.	6.1	T3	I	6.1	43 274 802	LQ0		PP, EP			2	
3146	ORGANOTIN COMPOUND, SOLID, N.O.S.	6.1	T3	II	6.1	43 274 802	LQ18		PP, EP			2	
3146	ORGANOTIN COMPOUND, SOLID, N.O.S.	6.1	T3	III	6.1	43 274 802	LQ9		PP, EP			0	
3147	DYE, SOLID, CORROSIVE, N.O.S. or DYE INTERMEDIATE, SOLID, CORROSIVE, N.O.S.	8	C10	I	8	274	LQ0		PP, EP			0	
3147	DYE, SOLID, CORROSIVE, N.O.S. or DYE INTERMEDIATE, SOLID, CORROSIVE, N.O.S.	8	C10	II	8	274	LQ23		PP, EP			0	
3147	DYE, SOLID, CORROSIVE, N.O.S. or DYE INTERMEDIATE, SOLID, CORROSIVE, N.O.S.	8	C10	III	8	274	LQ24		PP, EP			0	
3148	WATER-REACTIVE LIQUID, N.O.S.	4.3	W1	I	4.3	274	LQ0		PP, EX, A	VE01	HA08	0	
3148	WATER-REACTIVE LIQUID, N.O.S.	4.3	W1	II	4.3	274	LQ10		PP, EX, A	VE01	HA08	0	
3148	WATER-REACTIVE LIQUID, N.O.S.	4.3	W1	III	4.3	274	LQ13		PP, EX, A	VE01	HA08	0	
3149	HYDROGEN PEROXIDE AND PEROXYACETIC ACID MIXTURE with acid(s), water and not more than 5% peroxyacetic acid, STABILIZED	5.1	OC1	II	5.1+8	196 553	LQ10		PP, EP			0	
3150	DEVICES, SMALL, HYDROCARBON GAS POWERED or HYDROCARBON GAS REFILLS FOR SMALL DEVICES with release device	2	6F		2.1		LQ0		PP, EX, A	VE01		1	

- 365 -

UN No. or ID No. 3.1.2 (1)	Name and description 3.1.2 (2)	Class 2.2 (3a)	Classification Code 2.2 (3b)	Packing group 2.1.1.3 (4)	Labels 5.2.2 (5)	Special provisions 3.3 (6)	Limited quantities 3.4.6 (7)	Carriage permitted 3.2.1 (8)	Equipment required 8.1.5 (9)	Ventilation 7.1.6 (10)	Provisions concerning loading, unloading and carriage 7.1.6 (11)	Number of cones, blue lights 7.1.5 (12)	Remarks 3.2.1 (13)
3151	POLYHALOGENATED BIPHENYLS, LIQUID or POLYHALOGENATED TERPHENYLS, LIQUID	9	M2	II	9	203 305 802	LQ26		PP, EP			0	
3152	POLYHALOGENATED BIPHENYLS, SOLID or POLYHALOGENATED TERPHENYLS, SOLID	9	M2	II	9	203 305 802	LQ25		PP, EP			0	
3153	PERFLUORO(METHYL VINYL ETHER)	2	2F		2.1		LQ0		PP, EX, A	VE01		1	
3154	PERFLUORO(ETHYL VINYL ETHER)	2	2F		2.1		LQ0		PP, EX, A	VE01		1	
3155	PENTACHLOROPHENOL	6.1	T2	II	6.1	43 802	LQ18		PP, EP			2	
3156	COMPRESSED GAS, OXIDIZING, N.O.S.	2	1O		2.2+5.1	274	LQ0		PP			0	
3157	LIQUEFIED GAS, OXIDIZING, N.O.S.	2	2O		2.2+5.1	274	LQ0		PP			0	
3158	GAS, REFRIGERATED LIQUID, N.O.S.	2	3A		2.2	274 593	LQ1		PP			0	
3159	1,1,1,2-TETRAFLUOROETHANE (REFRIGERANT GAS R 134a)	2	2A		2.2		LQ1		PP			0	
3160	LIQUEFIED GAS, TOXIC, FLAMMABLE, N.O.S.	2	2TF		2.3+2.1	274	LQ0		PP, EP, EX, TOX, A	VE01, VE02		2	
3161	LIQUEFIED GAS, FLAMMABLE, N.O.S.	2	2F		2.1	274	LQ0		PP, EX, A	VE01		1	
3162	LIQUEFIED GAS, TOXIC, N.O.S.	2	2T		2.3	274	LQ0		PP, EP, TOX, A	VE02		2	
3163	LIQUEFIED GAS, N.O.S.	2	2A		2.2	274	LQ1		PP			0	
3164	ARTICLES, PRESSURIZED, PNEUMATIC or HYDRAULIC (containing non-flammable gas)	2	6A		2.2	283 594	LQ0		PP			0	
3165	AIRCRAFT HYDRAULIC POWER UNIT FUEL TANK (containing a mixture of anhydrous hydrazine and methylhydrazine) (M86 fuel)	3	FTC	I	3+6.1+8	802	LQ0		PP, EP, EX, TOX, A	VE01, VE02		2	

UN No. or ID No.	Name and description	Class	Classification Code	Packing group	Labels	Special provisions	Limited quantities	Carriage permitted	Equipment required	Ventilation	Provisions concerning loading, unloading and carriage	Number of cones, blue lights	Remarks
3.1.2	3.1.2	2.2	2.2	2.1.1.3	5.2.2	3.3	3.4.6	3.2.1	8.1.5	7.1.6	7.1.6	7.1.5	3.2.1
(1)	(2)	(3a)	(3b)	(4)	(5)	(6)	(7)	(8)	(9)	(10)	(11)	(12)	(13)
3166	Engine, internal combustion or vehicle, flammable gas powered or vehicle, flammable liquid powered	9	M11					NOT SUBJECT TO ADN					
3167	GAS SAMPLE, NON-PRESSURIZED, FLAMMABLE, N.O.S., not refrigerated liquid	2	7F		2.1	274	LQ0		PP, EX, A	VE01		1	
3168	GAS SAMPLE, NON-PRESSURIZED, TOXIC, FLAMMABLE, N.O.S., not refrigerated liquid	2	7TF		2.3+2.1	274	LQ0		PP, EP, EX, TOX, A	VE01, VE02		2	
3169	GAS SAMPLE, NON-PRESSURIZED, TOXIC, N.O.S., not refrigerated liquid	2	7T		2.3	274	LQ0		PP, EP, TOX, A	VE02		2	
3170	ALUMINIUM SMELTING BY-PRODUCTS or ALUMINIUM REMELTING BY-PRODUCTS	4.3	W2	II	4.3	244	LQ11		PP, EX, A	VE01	HA08	0	
3170	ALUMINIUM SMELTING BY-PRODUCTS or ALUMINIUM REMELTING BY-PRODUCTS	4.3	W2	III	4.3	244	LQ12	B	PP, EX, A	VE01, VE03	LO03 HA07, HA08 IN01, IN02, IN03	0	VE03, LO03, HA07, IN01, IN02 and IN03 apply only when this substance is carried in bulk or without packaging
3171	Battery-powered vehicle or Battery-powered equipment	9	M11					NOT SUBJECT TO ADN					
3172	TOXINS, EXTRACTED FROM LIVING SOURCES, LIQUID, N.O.S.	6.1	T1	I	6.1	210 274 802	LQ0		PP, EP, TOX, A	VE02		2	
3172	TOXINS, EXTRACTED FROM LIVING SOURCES, LIQUID, N.O.S.	6.1	T1	II	6.1	210 274 802	LQ17		PP, EP, TOX, A	VE02		2	
3172	TOXINS, EXTRACTED FROM LIVING SOURCES, LIQUID, N.O.S.	6.1	T1	III	6.1	210 274 802	LQ7		PP, EP, TOX, A	VE02		0	
3174	TITANIUM DISULPHIDE	4.2	S4	III	4.2		LQ0		PP			0	

UN No. or ID No.	Name and description	Class	Classification Code	Packing group	Labels	Special provisions	Limited quantities	Carriage permitted	Equipment required	Ventilation	Provisions concerning loading, unloading and carriage	Number of cones, blue lights	Remarks
(1)	3.1.2 (2)	2.2 (3a)	2.2 (3b)	2.1.1.3 (4)	5.2.2 (5)	3.3 (6)	3.4.6 (7)	3.2.1 (8)	8.1.5 (9)	7.1.6 (10)	7.1.6 (11)	7.1.5 (12)	3.2.1 (13)
3175	SOLIDS or mixtures of solids (such as preparations and wastes) CONTAINING FLAMMABLE LIQUID, N.O.S. having a flash-point up to 60°C	4.1	F1	II	4.1	216 274 800	LQ8	B	PP, EX, A	VE01, VE03	IN01, IN02	1	VE03, IN01and IN02 apply only when this substance is carried in bulk or without packaging
3175	SOLIDS CONTAINING FLAMMABLE LIQUID, MOLTEN, having a flash-point up to 60°C (DIALKYL-(C₁₂-C₁₈)-DIMETHYL-AMMONIUM and 2-PROPANOL)	4.1	F1	II	4.1	216 274 800	LQ8	T	PP, EX, A	VE01, VE03	IN01, IN02	1	VE03, IN01and IN02 apply only when this substance is carried in bulk or without packaging
3176	FLAMMABLE SOLID, ORGANIC, MOLTEN, N.O.S.	4.1	F2	II	4.1	274	LQ0		PP			0	
3176	FLAMMABLE SOLID, ORGANIC, MOLTEN, N.O.S.	4.1	F2	III	4.1	274	LQ0		PP			0	
3178	FLAMMABLE SOLID, INORGANIC, N.O.S.	4.1	F3	II	4.1	274	LQ8		PP			1	
3178	FLAMMABLE SOLID, INORGANIC, N.O.S.	4.1	F3	III	4.1	274	LQ9		PP			0	
3179	FLAMMABLE SOLID, TOXIC, INORGANIC, N.O.S.	4.1	FT2	II	4.1+6.1	274 802	LQ0		PP			2	
3179	FLAMMABLE SOLID, TOXIC, INORGANIC, N.O.S.	4.1	FT2	III	4.1+6.1	274 802	LQ0		PP			0	
3180	FLAMMABLE SOLID, CORROSIVE, INORGANIC, N.O.S.	4.1	FC2	II	4.1+8	274	LQ0		PP			1	
3180	FLAMMABLE SOLID, CORROSIVE, INORGANIC, N.O.S.	4.1	FC2	III	4.1+8	274	LQ0		PP			0	
3181	METAL SALTS OF ORGANIC COMPOUNDS, FLAMMABLE, N.O.S.	4.1	F3	II	4.1	274	LQ8		PP			1	
3181	METAL SALTS OF ORGANIC COMPOUNDS, FLAMMABLE, N.O.S.	4.1	F3	III	4.1	274	LQ9		PP			0	

UN No. or ID No.	Name and description	Class	Classification Code	Packing group	Labels	Special provisions	Limited quantities	Carriage permitted	Equipment required	Ventilation	Provisions concerning loading, unloading and carriage	Number of cones, blue lights	Remarks
3.1.2	3.1.2	2.2	2.2	2.1.1.3	5.2.2	3.3	3.4.6	3.2.1	8.1.5	7.1.6	7.1.6	7.1.5	3.2.1
(1)	(2)	(3a)	(3b)	(4)	(5)	(6)	(7)	(8)	(9)	(10)	(11)	(12)	(13)
3182	METAL HYDRIDES, FLAMMABLE, N.O.S.	4.1	F3	II	4.1	274 554	LQ8		PP			1	
3182	METAL HYDRIDES, FLAMMABLE, N.O.S.	4.1	F3	III	4.1	274 554	LQ9		PP			0	
3183	SELF-HEATING LIQUID, ORGANIC. N.O.S.	4.2	S1	II	4.2	274	LQ0		PP			0	
3183	SELF-HEATING LIQUID, ORGANIC. N.O.S.	4.2	S1	III	4.2	274	LQ0		PP			0	
3184	SELF-HEATING LIQUID, TOXIC, ORGANIC. N.O.S.	4.2	ST1	II	4.2+6.1	274 802	LQ0		PP, EP, TOX, A	VE02		2	
3184	SELF-HEATING LIQUID, TOXIC, ORGANIC. N.O.S.	4.2	ST1	III	4.2+6.1	274 802	LQ0		PP, EP, TOX, A	VE02		0	
3185	SELF-HEATING LIQUID, CORROSIVE, ORGANIC, N.O.S.	4.2	SC1	II	4.2+8	274	LQ0		PP, EP			0	
3185	SELF-HEATING LIQUID, CORROSIVE, ORGANIC, N.O.S.	4.2	SC1	III	4.2+8	274	LQ0		PP, EP			0	
3186	SELF-HEATING LIQUID, INORGANIC. N.O.S.	4.2	S3	II	4.2	274	LQ0		PP			0	
3186	SELF-HEATING LIQUID, INORGANIC. N.O.S.	4.2	S3	III	4.2	274	LQ0		PP			0	
3187	SELF-HEATING LIQUID, TOXIC, INORGANIC. N.O.S.	4.2	ST3	II	4.2+6.1	274 802	LQ0		PP, EP, TOX, A	VE02		2	
3187	SELF-HEATING LIQUID, TOXIC, INORGANIC. N.O.S.	4.2	ST3	III	4.2+6.1	274 802	LQ0		PP, EP, TOX, A	VE02		0	
3188	SELF-HEATING LIQUID, CORROSIVE, INORGANIC. N.O.S.	4.2	SC3	II	4.2+8	274	LQ0		PP, EP			0	
3188	SELF-HEATING LIQUID, CORROSIVE, INORGANIC. N.O.S.	4.2	SC3	III	4.2+8	274	LQ0		PP, EP			0	
3189	METAL POWDER, SELF-HEATING. N.O.S.	4.2	S4	II	4.2	274 555	LQ0		PP			0	
3189	METAL POWDER, SELF-HEATING. N.O.S.	4.2	S4	III	4.2	274 555	LQ0		PP			0	
3190	SELF-HEATING SOLID, INORGANIC. N.O.S.	4.2	S4	II	4.2	274	LQ0		PP			0	

UN No. or ID No.	Name and description	Class	Classification Code	Packing group	Labels	Special provisions	Limited quantities	Carriage permitted	Equipment required	Ventilation	Provisions concerning loading, unloading and carriage	Number of cones, blue lights	Remarks
3.1.2	3.1.2	2.2	2.2	2.1.1.3	5.2.2	3.3	3.4.6	3.2.1	8.1.5	7.1.6	7.1.6	7.1.5	3.2.1
(1)	(2)	(3a)	(3b)	(4)	(5)	(6)	(7)	(8)	(9)	(10)	(11)	(12)	(13)
3190	SELF-HEATING SOLID, INORGANIC, N.O.S.	4.2	S4	III	4.2	274	LQ0	B	PP			0	
3191	SELF-HEATING SOLID, TOXIC, INORGANIC, N.O.S.	4.2	ST4	II	4.2+6.1	274 802	LQ0		PP			2	
3191	SELF-HEATING SOLID, TOXIC, INORGANIC, N.O.S.	4.2	ST4	III	4.2+6.1	274 802	LQ0		PP			0	
3192	SELF-HEATING SOLID, CORROSIVE, INORGANIC, N.O.S.	4.2	SC4	II	4.2+8	274	LQ0		PP			0	
3192	SELF-HEATING SOLID, CORROSIVE, INORGANIC, N.O.S.	4.2	SC4	III	4.2+8	274	LQ0		PP			0	
3194	PYROPHORIC LIQUID, INORGANIC, N.O.S.	4.2	S3	I	4.2	274	LQ0		PP			0	
3200	PYROPHORIC SOLID, INORGANIC, N.O.S.	4.2	S4	I	4.2	274	LQ0		PP			0	
3205	ALKALINE EARTH METAL ALCOHOLATES, N.O.S.	4.2	S4	II	4.2	183 274	LQ0		PP			0	
3205	ALKALINE EARTH METAL ALCOHOLATES, N.O.S.	4.2	S4	III	4.2	183 274	LQ0		PP			0	
3206	ALKALI METAL ALCOHOLATES, SELF-HEATING, CORROSIVE, N.O.S.	4.2	SC4	II	4.2+8	182 274	LQ0		PP			0	
3206	ALKALI METAL ALCOHOLATES, SELF-HEATING, CORROSIVE, N.O.S.	4.2	SC4	III	4.2+8	183 274	LQ0		PP			0	
3208	METALLIC SUBSTANCE, WATER-REACTIVE, N.O.S.	4.3	W2	I	4.3	274 557	LQ0		PP, EX, A	VE01	HA08	0	
3208	METALLIC SUBSTANCE, WATER-REACTIVE, N.O.S.	4.3	W2	II	4.3	274 557	LQ11		PP, EX, A	VE01	HA08	0	
3208	METALLIC SUBSTANCE, WATER-REACTIVE, N.O.S.	4.3	W2	III	4.3	274 557	LQ12		PP, EX, A	VE01	HA08	0	
3209	METALLIC SUBSTANCE, WATER-REACTIVE, SELF-HEATING, N.O.S.	4.3	WS	I	4.3+4.2	274 558	LQ0		PP, EX, A	VE01	HA08	0	

UN No. or ID No.	Name and description	Class	Classification Code	Packing group	Labels	Special provisions	Limited quantities	Carriage permitted	Equipment required	Ventilation	Provisions concerning loading, unloading and carriage	Number of cones, blue lights	Remarks
	3.1.2	2.2	2.2	2.1.1.3	5.2.2	3.3	3.4.6	3.2.1	8.1.5	7.1.6	7.1.6	7.1.5	3.2.1
(1)	(2)	(3a)	(3b)	(4)	(5)	(6)	(7)	(8)	(9)	(10)	(11)	(12)	(13)
3209	METALLIC SUBSTANCE, WATER-REACTIVE, SELF-HEATING, N.O.S.	4.3	WS	II	4.3+4.2	274 558	LQ11		PP, EX, A	VE01	HA08	0	
3209	METALLIC SUBSTANCE, WATER-REACTIVE, SELF-HEATING, N.O.S.	4.3	WS	III	4.3+4.2	274 558	LQ12		PP, EX, A	VE01	HA08	0	
3210	CHLORATES, INORGANIC, AQUEOUS SOLUTION, N.O.S.	5.1	O1	II	5.1	274 605	LQ10		PP			0	
3210	CHLORATES, INORGANIC, AQUEOUS SOLUTION, N.O.S.	5.1	O1	III	5.1	274 605	LQ13		PP			0	
3211	PERCHLORATES, INORGANIC, AQUEOUS SOLUTION, N.O.S.	5.1	O1	II	5.1	274	LQ10		PP			0	
3211	PERCHLORATES, INORGANIC, AQUEOUS SOLUTION, N.O.S.	5.1	O1	III	5.1	274	LQ13		PP			0	
3212	HYPOCHLORITES, INORGANIC, N.O.S.	5.1	O2	II	5.1	274 559	LQ11		PP			0	
3213	BROMATES, INORGANIC, AQUEOUS SOLUTION, N.O.S.	5.1	O1	II	5.1	274 604	LQ10		PP			0	
3213	BROMATES, INORGANIC, AQUEOUS SOLUTION, N.O.S.	5.1	O1	III	5.1	274 604	LQ13		PP			0	
3214	PERMANGANATES, INORGANIC, AQUEOUS SOLUTION, N.O.S.	5.1	O1	II	5.1	274 608	LQ10		PP			0	
3215	PERSULPHATES, INORGANIC, N.O.S.	5.1	O2	III	5.1	274	LQ12		PP			0	
3216	PERSULPHATES, INORGANIC, AQUEOUS SOLUTION, N.O.S.	5.1	O1	III	5.1	274	LQ13		PP			0	
3218	NITRATES, INORGANIC, AQUEOUS SOLUTION, N.O.S.	5.1	O1	II	5.1	270 274 511	LQ10		PP			0	
3218	NITRATES, INORGANIC, AQUEOUS SOLUTION, N.O.S.	5.1	O1	III	5.1	270 274 511	LQ13		PP			0	
3219	NITRITES, INORGANIC, AQUEOUS SOLUTION, N.O.S.	5.1	O1	II	5.1	103 274	LQ10		PP			0	
3219	NITRITES, INORGANIC, AQUEOUS SOLUTION, N.O.S.	5.1	O1	III	5.1	103 274	LQ13		PP			0	

UN No. or ID No.	Name and description	Class	Classification Code	Packing group	Labels	Special provisions	Limited quantities	Carriage permitted	Equipment required	Ventilation	Provisions concerning loading, unloading and carriage	Number of cones, blue lights	Remarks
	3.1.2	2.2	2.2	2.1.1.3	5.2.2	3.3	3.4.6	3.2.1	8.1.5	7.1.6	7.1.6	7.1.5	3.2.1
(1)	(2)	(3a)	(3b)	(4)	(5)	(6)	(7)	(8)	(9)	(10)	(11)	(12)	(13)
3220	PENTAFLUOROETHANE (REFRIGERANT GAS R 125)	2	2A		2.2		LQ1		PP			0	
3221	SELF-REACTIVE LIQUID TYPE B	4.1	SR1		4.1+1	181 194 274	LQ14		PP		HA01, HA10	3	
3222	SELF-REACTIVE SOLID TYPE B	4.1	SR1		4.1+1	181 194 274	LQ15		PP		HA01, HA10	3	
3223	SELF-REACTIVE LIQUID TYPE C	4.1	SR1		4.1	194 274	LQ14		PP			0	
3224	SELF-REACTIVE SOLID TYPE C	4.1	SR1		4.1	194 274	LQ15		PP			0	
3225	SELF-REACTIVE LIQUID TYPE D	4.1	SR1		4.1	194 274	LQ16		PP			0	
3226	SELF-REACTIVE SOLID TYPE D	4.1	SR1		4.1	194 274	LQ11		PP			0	
3227	SELF-REACTIVE LIQUID TYPE E	4.1	SR1		4.1	194 274	LQ16		PP			0	
3228	SELF-REACTIVE SOLID TYPE E	4.1	SR1		4.1	194 274	LQ11		PP			0	
3229	SELF-REACTIVE LIQUID TYPE F	4.1	SR1		4.1	194 274	LQ16		PP			0	
3230	SELF-REACTIVE SOLID TYPE F	4.1	SR1		4.1	194 274	LQ11		PP			0	
3231	SELF-REACTIVE LIQUID TYPE B, TEMPERATURE CONTROLLED	4.1	SR2		4.1+1	181 194 274	LQ0		PP		HA01, HA10	3	
3232	SELF-REACTIVE SOLID TYPE B, TEMPERATURE CONTROLLED	4.1	SR2		4.1+1	181 194 274	LQ0		PP		HA01, HA10	3	
3233	SELF-REACTIVE LIQUID TYPE C, TEMPERATURE CONTROLLED	4.1	SR2		4.1	194 274	LQ0		PP			0	
3234	SELF-REACTIVE SOLID TYPE C, TEMPERATURE CONTROLLED	4.1	SR2		4.1	194 274	LQ0		PP			0	

UN No. or ID No.	Name and description	Class	Classification Code	Packing group	Labels	Special provisions	Limited quantities	Carriage permitted	Equipment required	Ventilation	Provisions concerning loading, unloading and carriage	Number of cones, blue lights	Remarks
3.1.2	3.1.2	2.2	2.2	2.1.1.3	5.2.2	3.3	3.4.6	3.2.1	8.1.5	7.1.6	7.1.6	7.1.5	3.2.1
(1)	(2)	(3a)	(3b)	(4)	(5)	(6)	(7)	(8)	(9)	(10)	(11)	(12)	(13)
3235	SELF-REACTIVE LIQUID TYPE D, TEMPERATURE CONTROLLED	4.1	SR2		4.1	194 274	LQ0		PP			0	
3236	SELF-REACTIVE SOLID TYPE D, TEMPERATURE CONTROLLED	4.1	SR2		4.1	194 274	LQ0		PP			0	
3237	SELF-REACTIVE LIQUID TYPE E, TEMPERATURE CONTROLLED	4.1	SR2		4.1	194 274	LQ0		PP			0	
3238	SELF-REACTIVE SOLID TYPE E, TEMPERATURE CONTROLLED	4.1	SR2		4.1	194 274	LQ0		PP			0	
3239	SELF-REACTIVE LIQUID TYPE F, TEMPERATURE CONTROLLED	4.1	SR2		4.1	194 274	LQ0		PP			0	
3240	SELF-REACTIVE SOLID TYPE F, TEMPERATURE CONTROLLED	4.1	SR2		4.1	194 274	LQ0		PP			0	
3241	2-BROMO-2-NITROPROPANE-1,3-DIOL	4.1	SR1	III	4.1	638	LQ0		PP			0	
3242	AZODICARBONAMIDE	4.1	SR1	II	4.1	215 638	LQ0		PP			0	
3243	SOLIDS CONTAINING TOXIC LIQUID, N.O.S.	6.1	T9	II	6.1	217 274 802	LQ18		PP, EP, TOX, A	VE02		2	
3244	SOLIDS CONTAINING CORROSIVE LIQUID, N.O.S.	8	C10	II	8	218 274	LQ23		PP, EP			0	
3245	GENETICALLY MODIFIED MICROORGANISMS or GENETICALLY MODIFIED ORGANISMS	9	M8		9	219 637 802	LQ0		PP			0	
3245	GENETICALLY MODIFIED MICRO-ORGANISMS or GENETICALLY MODIFIED ORGANISMS, in refrigerated liquid nitrogen	9	M8		9 +2.2	219 637 802	LQ0		PP			0	
3246	METHANESULPHONYL CHLORIDE	6.1	TC1	I	6.1+8	802	LQ0		PP, EP, TOX, A	VE02		2	

UN No. or ID No.	Name and description	Class	Classification Code	Packing group	Labels	Special provisions	Limited quantities	Carriage permitted	Equipment required	Ventilation	Provisions concerning loading, unloading and carriage	Number of cones, blue lights	Remarks
3.1.1	3.1.2	2.2	2.2	2.1.1.3	5.2.2	3.3	3.4.6	3.2.1	8.1.5	7.1.6	7.1.6	7.1.5	3.2.1
(1)	(2)	(3a)	(3b)	(4)	(5)	(6)	(7)	(8)	(9)	(10)	(11)	(12)	(13)
3247	SODIUM PEROXOBORATE, ANHYDROUS	5.1	O2	II	5.1		LQ11		PP			0	
3248	MEDICINE, LIQUID, FLAMMABLE, TOXIC, N.O.S.	3	FT1	II	3+6.1	220 221 274 601 802	LQ0		PP, EP, EX, TOX, A	VE01, VE02		2	
3248	MEDICINE, LIQUID, FLAMMABLE, TOXIC, N.O.S.	3	FT1	III	3+6.1	220 221 274 601 802	LQ7		PP, EP, EX, TOX, A	VE01, VE02		0	
3249	MEDICINE, SOLID, TOXIC, N.O.S.	6.1	T2	II	6.1	221 274 601 802	LQ18		PP, EP			2	
3249	MEDICINE, SOLID, TOXIC, N.O.S.	6.1	T2	III	6.1	221 274 601 802	LQ9		PP, EP			0	
3250	CHLOROACETIC ACID, MOLTEN	6.1	TC1	II	6.1+8	802	LQ0		PP, EP, TOX, A	VE02		2	
3251	ISOSORBIDE-5-MONONITRATE	4.1	SR1	III	4.1	226 638	LQ0		PP			0	
3252	DIFLUOROMETHANE (REFRIGERANT GAS R 32)	2	2F		2.1		LQ0		PP, EX, A	VE01		1	
3253	DISODIUM TRIOXOSILICATE	8	C6	III	8		LQ24		PP, EP			0	
3254	TRIBUTYLPHOSPHANE	4.2	S1	I	4.2		LQ0		PP			0	
3255	tert-BUTYL HYPOCHLORITE	4.2	SC1				CARRIAGE PROHIBITED							
3256	ELEVATED TEMPERATURE LIQUID, FLAMMABLE, N.O.S. with flash-point above 60 °C, at or above its flash-point	3	F2	III	3	274 560	LQ0	T	PP, EX, A	VE01		0	

UN No. or ID No. (1)	Name and description (2)	Class (3a)	Classification Code (3b)	Packing group (4)	Labels (5)	Special provisions (6)	Limited quantities (7)	Carriage permitted (8)	Equipment required (9)	Ventilation (10)	Provisions concerning loading, unloading and carriage (11)	Number of cones, blue lights (12)	Remarks (13)
	3.1.2	2.2	2.2	2.1.1.3	5.2.2	3.3	3.4.6	3.2.1	8.1.5	7.1.6	7.1.6	7.1.5	3.2.1
3257	ELEVATED TEMPERATURE LIQUID, N.O.S., at or above 100 °C and below its flash-point (including molten metals, molten salts, etc.)	9	M9	III	9	274 580 643	LQ0	T	PP			0	
3258	ELEVATED TEMPERATURE SOLID, N.O.S., at or above 240 °C	9	M10	III	9	274 580 643	LQ0		PP			0	
3259	AMINES, SOLID, CORROSIVE, N.O.S. or POLYAMINES, SOLID, CORROSIVE, N.O.S.	8	C8	I	8	274	LQ0		PP, EP			0	
3259	AMINES, SOLID, CORROSIVE, N.O.S. or POLYAMINES, SOLID, CORROSIVE, N.O.S.	8	C8	II	8	274	LQ23		PP, EP			0	
3259	AMINES, SOLID, CORROSIVE, N.O.S. or POLYAMINES, SOLID, CORROSIVE, N.O.S.	8	C8	III	8	274	LQ24	T	PP, EP			0	
3260	CORROSIVE SOLID, ACIDIC, INORGANIC, N.O.S.	8	C2	I	8	274	LQ0		PP, EP			0	
3260	CORROSIVE SOLID, ACIDIC, INORGANIC, N.O.S.	8	C2	II	8	274	LQ23		PP, EP			0	
3260	CORROSIVE SOLID, ACIDIC, INORGANIC, N.O.S.	8	C2	III	8	274	LQ24		PP, EP			0	
3261	CORROSIVE SOLID, ACIDIC, ORGANIC, N.O.S.	8	C4	I	8	274	LQ0		PP, EP			0	
3261	CORROSIVE SOLID, ACIDIC, ORGANIC, N.O.S.	8	C4	II	8	274	LQ23		PP, EP			0	
3261	CORROSIVE SOLID, ACIDIC, ORGANIC, N.O.S.	8	C4	III	8	274	LQ24		PP, EP			0	
3262	CORROSIVE SOLID, BASIC, INORGANIC, N.O.S.	8	C6	I	8	274	LQ0		PP, EP			0	
3262	CORROSIVE SOLID, BASIC, INORGANIC, N.O.S.	8	C6	II	8	274	LQ23		PP, EP			0	
3262	CORROSIVE SOLID, BASIC, INORGANIC, N.O.S.	8	C6	III	8	274	LQ24		PP, EP			0	
3263	CORROSIVE SOLID, BASIC, ORGANIC, N.O.S.	8	C8	I	8	274	LQ0		PP, EP			0	

UN No. or ID No.	Name and description	Class	Classification Code	Packing group	Labels	Special provisions	Limited quantities	Carriage permitted	Equipment required	Ventilation	Provisions concerning loading, unloading and carriage	Number of cones, blue lights	Remarks
	3.1.2	2.2	2.2	2.1.1.3	5.2.2	3.3	3.4.6	3.2.1	8.1.5	7.1.6	7.1.6	7.1.5	3.2.1
(1)	(2)	(3a)	(3b)	(4)	(5)	(6)	(7)	(8)	(9)	(10)	(11)	(12)	(13)
3263	CORROSIVE SOLID, BASIC, ORGANIC, N.O.S.	8	C8	II	8	274	LQ23		PP, EP			0	
3263	CORROSIVE SOLID, BASIC, ORGANIC, N.O.S.	8	C8	III	8	274	LQ24		PP, EP			0	
3264	CORROSIVE LIQUID, ACIDIC, INORGANIC. N.O.S.	8	C1	I	8	274	LQ0	T	PP, EP			0	
3264	CORROSIVE LIQUID, ACIDIC, INORGANIC. N.O.S.	8	C1	II	8	274	LQ22	T	PP, EP			0	
3264	CORROSIVE LIQUID, ACIDIC, INORGANIC. N.O.S.	8	C1	III	8	274	LQ7	T	PP, EP			0	
3265	CORROSIVE LIQUID, ACIDIC, ORGANIC. N.O.S.	8	C3	I	8	274	LQ0	T	PP, EP			0	
3265	CORROSIVE LIQUID, ACIDIC, ORGANIC. N.O.S.	8	C3	II	8	274	LQ22	T	PP, EP			0	
3265	CORROSIVE LIQUID, ACIDIC, ORGANIC. N.O.S.	8	C3	III	8	274	LQ7	T	PP, EP			0	
3266	CORROSIVE LIQUID, BASIC, INORGANIC. N.O.S.	8	C5	I	8	274	LQ0	T	PP, EP			0	
3266	CORROSIVE LIQUID, BASIC, INORGANIC. N.O.S.	8	C5	II	8	274	LQ22	T	PP, EP			0	
3266	CORROSIVE LIQUID, BASIC, INORGANIC. N.O.S.	8	C5	III	8	274	LQ7	T	PP, EP			0	
3267	CORROSIVE LIQUID, BASIC, ORGANIC. N.O.S.	8	C7	I	8	274	LQ0	T	PP, EP			0	
3267	CORROSIVE LIQUID, BASIC, ORGANIC. N.O.S.	8	C7	II	8	274	LQ22	T	PP, EP			0	
3267	CORROSIVE LIQUID, BASIC, ORGANIC. N.O.S.	8	C7	III	8	274	LQ7	T	PP, EP			0	
3268	AIR BAG INFLATORS or AIR BAG MODULES or SEAT-BELT PRETENSIONERS	9	M5	III	9	280 289	LQ0		PP			0	
3269	POLYESTER RESIN KIT	3	F1	II	3	236	LQ6		PP, EX, A	VE01		1	
3269	POLYESTER RESIN KIT	3	F1	III	3	236	LQ7		PP, EX, A	VE01		0	
3270	NITROCELLULOSE MEMBRANE FILTERS, with not more than 12.6% nitrogen, by dry mass	4.1	F1	II	4.1	237 286	LQ8		PP			1	

UN No. or ID No.	Name and description	Class	Classification Code	Packing group	Labels	Special provisions	Limited quantities	Carriage permitted	Equipment required	Ventilation	Provisions concerning loading, unloading and carriage	Number of cones, blue lights	Remarks
(1)	3.1.2 (2)	2.2 (3a)	2.2 (3b)	2.1.1.3 (4)	5.2.2 (5)	3.3 (6)	3.4.6 (7)	3.2.1 (8)	8.1.5 (9)	7.1.6 (10)	7.1.6 (11)	7.1.5 (12)	3.2.1 (13)
3271	ETHERS, N.O.S.	3	F1	II	3	274	LQ4	T	PP, EX, A	VE01		1	
3271	ETHERS, N.O.S.	3	F1	III	3	274	LQ7	T	PP, EX, A	VE01		0	
3272	ESTERS, N.O.S.	3	F1	II	3	274 601	LQ4	T	PP, EX, A	VE01		1	
3272	ESTERS, N.O.S.	3	F1	III	3	274 601	LQ7	T	PP, EX, A	VE01		0	
3273	NITRILES, FLAMMABLE, TOXIC, N.O.S.	3	FT1	I	3+6.1	274 802	LQ0		PP, EP, EX, TOX, A	VE01, VE02		2	
3273	NITRILES, FLAMMABLE, TOXIC, N.O.S.	3	FT1	II	3+6.1	274 802	LQ0		PP, EP, EX, TOX, A	VE01, VE02		2	
3274	ALCOHOLATES SOLUTION, N.O.S.. in alcohol	3	FC	II	3+8	274	LQ4		PP, EP, EX, A	VE01		1	
3275	NITRILES, TOXIC, FLAMMABLE, N.O.S.	6.1	TF1	I	6.1+3	274 315 802	LQ0		PP, EP, EX, TOX, A	VE01, VE02		2	
3275	NITRILES, TOXIC, FLAMMABLE, N.O.S.	6.1	TF1	II	6.1+3	274 802	LQ17		PP, EP, EX, TOX, A	VE01, VE02		2	
3276	NITRILES, TOXIC, LIQUID, N.O.S.	6.1	T1	I	6.1	274 315 802	LQ0		PP, EP, TOX, A	VE02		2	
3276	NITRILES, TOXIC, LIQUID, N.O.S.	6.1	T1	II	6.1	274 802	LQ17	T	PP, EP, TOX, A	VE02		2	
3276	NITRILES, TOXIC, LIQUID, N.O.S.	6.1	T1	III	6.1	274 802	LQ7		PP, EP, TOX, A	VE02		0	
3277	CHLOROFORMATES, TOXIC, CORROSIVE, N.O.S.	6.1	TC1	II	6.1+8	274 561 802	LQ17		PP, EP, TOX, A	VE02		2	
3278	ORGANOPHOSPHORUS COMPOUND, TOXIC, LIQUID, N.O.S.	6.1	T1	I	6.1	43 274 315 802	LQ0		PP, EP, TOX, A	VE02		2	
3278	ORGANOPHOSPHORUS COMPOUND, TOXIC, LIQUID, N.O.S.	6.1	T1	II	6.1	43 274 802	LQ17		PP, EP, TOX, A	VE02		2	
3278	ORGANOPHOSPHORUS COMPOUND, TOXIC, LIQUID, N.O.S.	6.1	T1	III	6.1	43 274 802	LQ7		PP, EP, TOX, A	VE02		0	

UN No. or ID No.	Name and description	Class	Classification Code	Packing group	Labels	Special provisions	Limited quantities	Carriage permitted	Equipment required	Ventilation	Provisions concerning loading, unloading and carriage	Number of cones, blue lights	Remarks
3.1.2	3.1.2	2.2	2.2	2.1.1.3	5.2.2	3.3	3.4.6	3.2.1	8.1.5	7.1.6	7.1.6	7.1.5	3.2.1
(1)	(2)	(3a)	(3b)	(4)	(5)	(6)	(7)	(8)	(9)	(10)	(11)	(12)	(13)
3279	ORGANOPHOSPHORUS COMPOUND, TOXIC, FLAMMABLE, N.O.S.	6.1	TF1	I	6.1+3	43 274 315 802	LQ0		PP, EP, EX, TOX, A	VE01, VE02		2	
3279	ORGANOPHOSPHORUS COMPOUND, TOXIC, FLAMMABLE, N.O.S.	6.1	TF1	II	6.1+3	43 274 802	LQ17		PP, EP, EX, TOX, A	VE01, VE02		2	
3280	ORGANOARSENIC COMPOUND, LIQUID, N.O.S.	6.1	T3	I	6.1	274 315 802	LQ0		PP, EP, TOX, A	VE02		2	
3280	ORGANOARSENIC COMPOUND, LIQUID, N.O.S.	6.1	T3	II	6.1	274 802	LQ17		PP, EP, TOX, A	VE02		2	
3280	ORGANOARSENIC COMPOUND, LIQUID, N.O.S.	6.1	T3	III	6.1	274 802	LQ7		PP, EP, TOX, A	VE02		0	
3281	METAL CARBONYLS, LIQUID, N.O.S.	6.1	T3	I	6.1	274 315 562 802	LQ0		PP, EP, TOX, A	VE02		2	
3281	METAL CARBONYLS, LIQUID, N.O.S.	6.1	T3	II	6.1	274 562 802	LQ17		PP, EP, TOX, A	VE02		2	
3281	METAL CARBONYLS, LIQUID, N.O.S.	6.1	T3	III	6.1	274 562 802	LQ7		PP, EP, TOX, A	VE02		0	
3282	ORGANOMETALLIC COMPOUND, TOXIC, LIQUID, N.O.S.	6.1	T3	I	6.1	274 562 802	LQ0		PP, EP, TOX, A	VE02		2	
3282	ORGANOMETALLIC COMPOUND, TOXIC, LIQUID, N.O.S.	6.1	T3	II	6.1	274 562 802	LQ17		PP, EP, TOX, A	VE02		2	
3282	ORGANOMETALLIC COMPOUND, TOXIC, LIQUID, N.O.S.	6.1	T3	III	6.1	274 562 802	LQ7		PP, EP, TOX, A	VE02		0	
3283	SELENIUM COMPOUND, SOLID, N.O.S.	6.1	T5	I	6.1	274 563 802	LQ0		PP, EP			2	

UN No. or ID No.	Name and description	Class	Classification Code	Packing group	Labels	Special provisions	Limited quantities	Carriage permitted	Equipment required	Ventilation	Provisions concerning loading, unloading and carriage	Number of cones, blue lights	Remarks
3.1.2	3.1.2	2.2	2.2	2.1.1.3	5.2.2	3.3	3.4.6	3.2.1	8.1.5	7.1.6	7.1.6	7.1.5	3.2.1
(1)	(2)	(3a)	(3b)	(4)	(5)	(6)	(7)	(8)	(9)	(10)	(11)	(12)	(13)
3283	SELENIUM COMPOUND, SOLID. N.O.S.	6.1	T5	II	6.1	274 563 802	LQ18		PP, EP			2	
3283	SELENIUM COMPOUND, SOLID. N.O.S.	6.1	T5	III	6.1	274 563 802	LQ9		PP, EP			0	
3284	TELLURIUM COMPOUND, N.O.S.	6.1	T5	I	6.1	274 802	LQ0		PP, EP			2	
3284	TELLURIUM COMPOUND, N.O.S.	6.1	T5	II	6.1	274 802	LQ18		PP, EP			2	
3284	TELLURIUM COMPOUND, N.O.S.	6.1	T5	III	6.1	274 802	LQ9		PP, EP			0	
3285	VANADIUM COMPOUND, N.O.S.	6.1	T5	I	6.1	274 564 802	LQ0		PP, EP			2	
3285	VANADIUM COMPOUND, N.O.S.	6.1	T5	II	6.1	274 564 802	LQ18		PP, EP			2	
3285	VANADIUM COMPOUND, N.O.S.	6.1	T5	III	6.1	274 564 802	LQ9		PP, EP			0	
3286	FLAMMABLE LIQUID, TOXIC, CORROSIVE. N.O.S.	3	FTC	I	3+6.1+8	274 802	LQ0	T	PP, EP, EX, TOX, A	VE01, VE02		2	
3286	FLAMMABLE LIQUID, TOXIC, CORROSIVE. N.O.S.	3	FTC	II	3+6.1+8	274 802	LQ0	T	PP, EP, EX, TOX, A	VE01, VE02		2	
3287	TOXIC LIQUID, INORGANIC, N.O.S.	6.1	T4	I	6.1	274 315 802	LQ0	T	PP, EP, TOX, A	VE02		2	
3287	TOXIC LIQUID, INORGANIC, N.O.S.	6.1	T4	II	6.1	274 802	LQ17	T	PP, EP, TOX, A	VE02		2	
3287	TOXIC LIQUID, INORGANIC, N.O.S.	6.1	T4	III	6.1	274 802	LQ7	T	PP, EP, TOX, A	VE02		0	
3288	TOXIC SOLID, INORGANIC, N.O.S.	6.1	T5	I	6.1	274 802	LQ0		PP, EP			2	
3288	TOXIC SOLID, INORGANIC, N.O.S.	6.1	T5	II	6.1	274 802	LQ18		PP, EP			2	

UN No. or ID No. (1)	Name and description 3.1.2 (2)	Class 2.2 (3a)	Classification Code 2.2 (3b)	Packing group 2.1.1.3 (4)	Labels 5.2.2 (5)	Special provisions 3.3 (6)	Limited quantities 3.4.6 (7)	Carriage permitted 3.2.1 (8)	Equipment required 8.1.5 (9)	Ventilation 7.1.6 (10)	Provisions concerning loading, unloading and carriage 7.1.6 (11)	Number of cones, blue lights 7.1.5 (12)	Remarks 3.2.1 (13)
3288	TOXIC SOLID, INORGANIC, N.O.S.	6.1	T5	III	6.1	274 802	LQ9		PP, EP			0	
3289	TOXIC LIQUID, CORROSIVE, INORGANIC, N.O.S.	6.1	TC3	I	6.1+8	274 315 802	LQ0	T	PP, EP, TOX, A	VE02		2	
3289	TOXIC LIQUID, CORROSIVE, INORGANIC, N.O.S.	6.1	TC3	II	6.1+8	274 802	LQ17	T	PP, EP, TOX, A	VE02		2	
3290	TOXIC SOLID, CORROSIVE, INORGANIC, N.O.S.	6.1	TC4	I	6.1+8	274 802	LQ0		PP, EP			2	
3290	TOXIC SOLID, CORROSIVE, INORGANIC, N.O.S.	6.1	TC4	II	6.1+8	274 802	LQ18		PP, EP			2	
3291	CLINICAL WASTE, UNSPECIFIED, N.O.S. or (BIO) MEDICAL WASTE, N.O.S. or REGULATED MEDICAL WASTE, N.O.S.	6.2	I3	II	6.2	565 802	LQ0		PP			0	
3291	CLINICAL WASTE, UNSPECIFIED, N.O.S. or (BIO) MEDICAL WASTE, N.O.S. or REGULATED MEDICAL WASTE, N.O.S.. in refrigerated liquid nitrogen	6.2	I3	II	6.2 +2.2	565 802	LQ0		PP			0	
3292	BATTERIES, CONTAINING SODIUM, or CELLS, CONTAINING SODIUM	4.3	W3	II	4.3	239 295	LQ0		PP, EX, A	VE01	HA08	0	
3293	HYDRAZINE, AQUEOUS SOLUTION with not more than 37% hydrazine, by mass	6.1	T4	III	6.1	566 802	LQ7		PP, EP, TOX, A	VE02		0	
3294	HYDROGEN CYANIDE, SOLUTION IN ALCOHOL with not more than 45% hydrogen cyanide	6.1	TF1	I	6.1+3	610 802	LQ0		PP, EP, EX, TOX, A	VE01, VE02		2	
3295	HYDROCARBONS, LIQUID, N.O.S.	3	F1	I	3	649	LQ3	T	PP, EX, A	VE01		1	
3295	HYDROCARBONS, LIQUID, N.O.S. (vapour pressure at 50 °C more than 110 kPa)	3	F1	II	3	640C 649	LQ4	T	PP, EX, A	VE01		1	

UN No. or ID No.	Name and description	Class	Classification Code	Packing group	Labels	Special provisions	Limited quantities	Carriage permitted	Equipment required	Ventilation	Provisions concerning loading, unloading and carriage	Number of cones, blue lights	Remarks
3.1.2	3.1.2	2.2	2.2	2.1.1.3	5.2.2	3.3	3.4.6	3.2.1	8.1.5	7.1.6	7.1.6	7.1.5	3.2.1
(1)	(2)	(3a)	(3b)	(4)	(5)	(6)	(7)	(8)	(9)	(10)	(11)	(12)	(13)
3295	HYDROCARBONS, LIQUID, N.O.S. (vapour pressure at 50 °C not more than 110 kPa)	3	F1	II	3	640D 649	LQ4	T	PP, EX, A	VE01		1	
3295	HYDROCARBONS, LIQUID, N.O.S.	3	F1	III	3		LQ7	T	PP, EX, A	VE01		0	
3296	HEPTAFLUOROPROPANE (REFRIGERANT GAS R 227)	2	2A		2.2		LQ1		PP			0	
3297	ETHYLENE OXIDE AND CHLOROTETRAFLUORO-ETHANE MIXTURE with not more than 8.8% ethylene oxide	2	2A		2.2		LQ1		PP			0	
3298	ETHYLENE OXIDE AND PENTAFLUOROETHANE MIXTURE with not more than 7.9% ethylene oxide	2	2A		2.2		LQ1		PP			0	
3299	ETHYLENE OXIDE AND TETRAFLUOROETHANE MIXTURE with not more than 5.6% ethylene oxide	2	2A		2.2		LQ1		PP			0	
3300	ETHYLENE OXIDE AND CARBON DIOXIDE MIXTURE with more than 87% ethylene oxide	2	2TF		2.3+2.1		LQ0		PP, EP, EX, TOX, A	VE01, VE02		2	
3301	CORROSIVE LIQUID, SELF-HEATING, N.O.S.	8	CS1	I	8+4.2	274	LQ0		PP, EP			0	
3301	CORROSIVE LIQUID, SELF-HEATING, N.O.S.	8	CS1	II	8+4.2	274	LQ22		PP, EP			0	
3302	2-DIMETHYLAMINOETHYL ACRYLATE	6.1	T1	II	6.1	802	LQ17		PP, EP, TOX, A			2	
3303	COMPRESSED GAS, TOXIC, OXIDIZING, N.O.S.	2	1TO		2.3+5.1	274	LQ0		PP, EP, TOX, A	VE02		2	
3304	COMPRESSED GAS, TOXIC, CORROSIVE, N.O.S.	2	1TC		2.3+8	274	LQ0		PP, EP, TOX, A	VE02		2	
3305	COMPRESSED GAS, TOXIC, FLAMMABLE, CORROSIVE, N.O.S.	2	1TFC		2.3+2.1+8	274	LQ0		PP, EP, EX, TOX, A	VE01, VE02		2	

UN No. or ID No.	Name and description	Class	Classification Code	Packing group	Labels	Special provisions	Limited quantities	Carriage permitted	Equipment required	Ventilation	Provisions concerning loading, unloading and carriage	Number of cones, blue lights	Remarks
3.1.2	3.1.2	2.2	2.2	2.1.1.3	5.2.2	3.3	3.4.6	3.2.1	8.1.5	7.1.6	7.1.6	7.1.5	3.2.1
(1)	(2)	(3a)	(3b)	(4)	(5)	(6)	(7)	(8)	(9)	(10)	(11)	(12)	(13)
3306	COMPRESSED GAS, TOXIC, OXIDIZING, CORROSIVE, N.O.S.	2	1TOC		2.3+5.1+8	274	LQ0		PP, EP, TOX, A	VE02		2	
3307	LIQUEFIED GAS, TOXIC, OXIDIZING, N.O.S.	2	2TO		2.3+5.1	274	LQ0		PP, EP, TOX, A	VE02		2	
3308	LIQUEFIED GAS, TOXIC, CORROSIVE, N.O.S.	2	2TC		2.3+8	274	LQ0		PP, EP, TOX, A	VE02		2	
3309	LIQUEFIED GAS, TOXIC, FLAMMABLE, CORROSIVE, N.O.S.	2	2TFC		2.3+2.1+8	274	LQ0		PP, EP, EX, TOX, A	VE01, VE02		2	
3310	LIQUEFIED GAS, TOXIC, OXIDIZING, CORROSIVE, N.O.S.	2	2TOC		2.3+5.1+8	274	LQ0		PP, EP, TOX, A	VE02		2	
3311	GAS, REFRIGERATED LIQUID, OXIDIZING, N.O.S.	2	3O		2.2+5.1	274	LQ0		PP			0	
3312	GAS, REFRIGERATED LIQUID, FLAMMABLE, N.O.S.	2	3F		2.1	274	LQ0		PP, EX, A	VE01		1	
3313	ORGANIC PIGMENTS, SELF-HEATING	4.2	S2	II	4.2		LQ0		PP			0	
3313	ORGANIC PIGMENTS, SELF-HEATING	4.2	S2	III	4.2		LQ0		PP			0	
3314	PLASTICS MOULDING COMPOUND in dough, sheet or extruded rope form evolving flammable vapour	9	M3	III	none	207 633	LQ27		PP, EP, EX, A	VE01		0	
3315	CHEMICAL SAMPLE, TOXIC	6.1	T8	I	6.1	250 802	LQ0		PP, EP, TOX, A	VE02		2	
3316	CHEMICAL KIT or FIRST AID KIT	9	M11	II	9	251	LQ0		PP			0	
3316	CHEMICAL KIT or FIRST AID KIT	9	M11	III	9	251	LQ0		PP			0	
3317	2-AMINO-4,6-DINITROPHENOL, WETTED with not less than 20% water, by mass	4.1	D	I	4.1		LQ0		PP			1	

UN No. or ID No.	Name and description	Class	Classification Code	Packing group	Labels	Special provisions	Limited quantities	Carriage permitted	Equipment required	Ventilation	Provisions concerning loading, unloading and carriage	Number of cones, blue lights	Remarks
3.1.2	3.1.2	2.2	2.2	2.1.1.3	5.2.2	3.3	3.4.6	3.2.1	8.1.5	7.1.6	7.1.6	7.1.5	3.2.1
(1)	(2)	(3a)	(3b)	(4)	(5)	(6)	(7)	(8)	(9)	(10)	(11)	(12)	(13)
3318	AMMONIA SOLUTION, relative density less than 0.880 at 15 °C in water, with more than 50% ammonia	2	4TC		2.3+8	23	LQ0		PP, EP, TOX, A	VE02		2	
3319	NITROGLYCERIN MIXTURE, DESENSITIZED, SOLID, N.O.S. with more than 2% but not more than 10% nitroglycerin, by mass	4.1	D	II	4.1	272 274	LQ0		PP			0	
3320	SODIUM BOROHYDRIDE AND SODIUM HYDROXIDE SOLUTION, with not more than 12% sodium borohydride and not more than 40% sodium hydroxide by mass	8	C5	II	8		LQ22		PP, EP			0	
3320	SODIUM BOROHYDRIDE AND SODIUM HYDROXIDE SOLUTION, with not more than 12% sodium borohydride and not more than 40% sodium hydroxide by mass	8	C5	III	8		LQ7		PP, EP			0	
3321	RADIOACTIVE MATERIAL, LOW SPECIFIC ACTIVITY (LSA-II), non fissile or fissile-excepted	7			7X	172 317 325	LQ0		PP			2	
3322	RADIOACTIVE MATERIAL, LOW SPECIFIC ACTIVITY (LSA-III), non fissile or fissile-excepted	7			7X	172 317 325	LQ0		PP			2	
3323	RADIOACTIVE MATERIAL, TYPE C PACKAGE, non fissile or fissile-excepted	7			7X	172 317	LQ0		PP			2	
3324	RADIOACTIVE MATERIAL, LOW SPECIFIC ACTIVITY (LSA-II), FISSILE	7			7X+7E	172 326	LQ0		PP			2	
3325	RADIOACTIVE MATERIAL, LOW SPECIFIC ACTIVITY, (LSA-III), FISSILE	7			7X+7E	172 326	LQ0		PP			2	

UN No. or ID No.	Name and description	Class	Classification Code	Packing group	Labels	Special provisions	Limited quantities	Carriage permitted	Equipment required	Ventilation	Provisions concerning loading, unloading and carriage	Number of cones, blue lights	Remarks
3.1.2	3.1.2	2.2	2.2	2.1.1.3	5.2.2	3.3	3.4.6	3.2.1	8.1.5	7.1.6	7.1.6	7.1.5	3.2.1
(1)	(2)	(3a)	(3b)	(4)	(5)	(6)	(7)	(8)	(9)	(10)	(11)	(12)	(13)
3326	RADIOACTIVE MATERIAL, SURFACE CONTAMINATED OBJECTS (SCO-I or SCO-II), FISSILE	7			7X+7E	172	LQ0		PP			2	
3327	RADIOACTIVE MATERIAL, TYPE A PACKAGE, FISSILE, non-special form	7			7X+7E	172 326	LQ0		PP			2	
3328	RADIOACTIVE MATERIAL, TYPE B(U) PACKAGE, FISSILE	7			7X+7E	172	LQ0		PP			2	
3329	RADIOACTIVE MATERIAL, TYPE B(M) PACKAGE, FISSILE	7			7X+7E	172	LQ0		PP			2	
3330	RADIOACTIVE MATERIAL, TYPE C PACKAGE, FISSILE	7			7X+7E	172	LQ0		PP			2	
3331	RADIOACTIVE MATERIAL, TRANSPORTED UNDER SPECIAL ARRANGEMENT, FISSILE	7			7X+7E	172	LQ0		PP			2	
3332	RADIOACTIVE MATERIAL, TYPE A PACKAGE, SPECIAL FORM, non fissile or fissile-excepted	7			7X	172 317	LQ0		PP			2	
3333	RADIOACTIVE MATERIAL, TYPE A PACKAGE, SPECIAL FORM, FISSILE	7			7X+7E	172	LQ0		PP			2	
3334	Aviation regulated liquid, n.o.s.	9	M11					NOT SUBJECT TO ADN					
3335	Aviation regulated solid, n.o.s.	9	M11					NOT SUBJECT TO ADN					
3336	MERCAPTANS, LIQUID, FLAMMABLE, N.O.S. or MERCAPTAN MIXTURE, LIQUID, FLAMMABLE, N.O.S.	3	F1	I	3	274	LQ3		PP, EX, A	VE01		1	

UN No. or ID No.	Name and description	Class	Classification Code	Packing group	Labels	Special provisions	Limited quantities	Carriage permitted	Equipment required	Ventilation	Provisions concerning loading, unloading and carriage	Number of cones, blue lights	Remarks
	3.1.2	2.2	2.2	2.1.1.3	5.2.2	3.3	3.4.6	3.2.1	8.1.5	7.1.6	7.1.6	7.1.5	3.2.1
(1)	(2)	(3a)	(3b)	(4)	(5)	(6)	(7)	(8)	(9)	(10)	(11)	(12)	(13)
3336	MERCAPTANS, LIQUID, FLAMMABLE, N.O.S. or MERCAPTAN MIXTURE, LIQUID, FLAMMABLE, N.O.S. (vapour pressure at 50 °C more than 110 kPa)	3	F1	II	3	274 640C	LQ4		PP, EX, A	VE01		1	
3336	MERCAPTANS, LIQUID, FLAMMABLE, N.O.S. or MERCAPTAN MIXTURE, LIQUID, FLAMMABLE, N.O.S. (vapour pressure at 50 °C not more than 110 kPa)	3	F1	II	3	274 640D	LQ4		PP, EX, A	VE01		1	
3336	MERCAPTANS, LIQUID, FLAMMABLE, N.O.S. or MERCAPTAN MIXTURE, LIQUID, FLAMMABLE, N.O.S.	3	F1	III	3	274	LQ7		PP, EX, A	VE01		0	
3337	REFRIGERANT GAS R 404A (Pentafluoroethane, 1,1,1-trifluoroethane, and 1,1,1,2-tetrafluoroethane zeotropic mixture with approximately 44% pentafluoroethane and 52% 1,1,1-trifluoroethane)	2	2A		2.2		LQ1		PP			0	
3338	REFRIGERANT GAS R 407A (Difluoromethane, pentafluoroethane, and 1,1,1,2-tetrafluoroethane zeotropic mixture with approximately 20% difluoromethane and 40% pentafluoroethane)	2	2A		2.2		LQ1		PP			0	
3339	REFRIGERANT GAS R 407B (Difluoromethane, pentafluoroethane, and 1,1,1,2-tetrafluoroethane zeotropic mixture with approximately 10% difluoromethane and 70% pentafluoroethane)	2	2A		2.2		LQ1		PP			0	

UN No. or ID No.	Name and description	Class	Classification Code	Packing group	Labels	Special provisions	Limited quantities	Carriage permitted	Equipment required	Ventilation	Provisions concerning loading, unloading and carriage	Number of cones, blue lights	Remarks
3.1.2	3.1.2	2.2	2.2	2.1.1.3	5.2.2	3.3	3.4.6	3.2.1	8.1.5	7.1.6	7.1.6	7.1.5	3.2.1
(1)	(2)	(3a)	(3b)	(4)	(5)	(6)	(7)	(8)	(9)	(10)	(11)	(12)	(13)
3340	REFRIGERANT GAS R 407C (Difluoromethane, pentafluoroethane, and 1,1,1,2-tetrafluoroethane zeotropic mixture with approximately 23%difluoromethane and 25% pentafluoroethane)	2	2A		2.2		LQ1		PP			0	
3341	THIOUREA DIOXIDE	4.2	S2	II	4.2		LQ0		PP			0	
3341	THIOUREA DIOXIDE	4.2	S2	III	4.2		LQ0		PP			0	
3342	XANTHATES	4.2	S2	II	4.2		LQ0		PP			0	
3342	XANTHATES	4.2	S2	III	4.2		LQ0		PP			0	
3343	NITROGLYCERIN MIXTURE, DESENSITIZED, LIQUID, FLAMMABLE, N.O.S. with not more than 30% nitroglycerin, by mass	3	D		3	274 278	LQ0		PP, EX, A	VE01		0	
3344	PENTAERYTHRITE TETRANITRATE MIXTURE, DESENSITIZED, SOLID, N.O.S. with more than 10% but not more than 20% PETN, by mass	4.1	D	II	4.1	272 274	LQ0		PP			1	
3345	PHENOXYACETIC ACID DERIVATIVE PESTICIDE, SOLID, TOXIC	6.1	T7	I	6.1	61 648 802	LQ0		PP, EP			2	
3345	PHENOXYACETIC ACID DERIVATIVE PESTICIDE, SOLID, TOXIC	6.1	T7	II	6.1	61 648 802	LQ18		PP, EP			2	
3345	PHENOXYACETIC ACID DERIVATIVE PESTICIDE, SOLID, TOXIC	6.1	T7	III	6.1	61 648 802	LQ9		PP, EP			0	
3346	PHENOXYACETIC ACID DERIVATIVE PESTICIDE, LIQUID, FLAMMABLE, TOXIC, flash-point less than 23 °C	3	FT2	I	3+6.1	61 802	LQ3		PP, EP, EX, TOX, A	VE01, VE02		2	

UN No. or ID No.	Name and description	Class	Classification Code	Packing group	Labels	Special provisions	Limited quantities	Carriage permitted	Equipment required	Ventilation	Provisions concerning loading, unloading and carriage	Number of cones, blue lights	Remarks
3.1.2	3.1.2	2.2	2.2	2.1.1.3	5.2.2	3.3	3.4.6	3.2.1	8.1.5	7.1.6	7.1.6	7.1.5	3.2.1
(1)	(2)	(3a)	(3b)	(4)	(5)	(6)	(7)	(8)	(9)	(10)	(11)	(12)	(13)
3346	PHENOXYACETIC ACID DERIVATIVE PESTICIDE, LIQUID, FLAMMABLE, TOXIC, flash-point less than 23 °C	3	FT2	II	3+6.1	61 802	LQ4		PP, EP, EX, TOX, A	VE01, VE02		2	
3347	PHENOXYACETIC ACID DERIVATIVE PESTICIDE, LIQUID, TOXIC, FLAMMABLE, flash-point not less than 23 °C	6.1	TF2	I	6.1+3	61 802	LQ0		PP, EP, EX, TOX, A	VE01, VE02		2	
3347	PHENOXYACETIC ACID DERIVATIVE PESTICIDE, LIQUID, TOXIC, FLAMMABLE, flash-point not less than 23 °C	6.1	TF2	II	6.1+3	61 802	LQ17		PP, EP, EX, TOX, A	VE01, VE02		2	
3347	PHENOXYACETIC ACID DERIVATIVE PESTICIDE, LIQUID, TOXIC, FLAMMABLE, flash-point not less than 23 °C	6.1	TF2	III	6.1+3	61 802	LQ7		PP, EP, EX, TOX, A	VE01, VE02		0	
3348	PHENOXYACETIC ACID DERIVATIVE PESTICIDE, LIQUID, TOXIC	6.1	T6	I	6.1	61 648 802	LQ0		PP, EP, TOX, A	VE02		2	
3348	PHENOXYACETIC ACID DERIVATIVE PESTICIDE, LIQUID, TOXIC	6.1	T6	II	6.1	61 648 802	LQ17		PP, EP, TOX, A	VE02		2	
3348	PHENOXYACETIC ACID DERIVATIVE PESTICIDE, LIQUID, TOXIC	6.1	T6	III	6.1	61 648 802	LQ7		PP, EP, TOX, A	VE02		0	
3349	PYRETHROID PESTICIDE, SOLID, TOXIC	6.1	T7	I	6.1	61 648 802	LQ0		PP, EP			2	
3349	PYRETHROID PESTICIDE, SOLID, TOXIC	6.1	T7	II	6.1	61 648 802	LQ18		PP, EP			2	
3349	PYRETHROID PESTICIDE, SOLID, TOXIC	6.1	T7	III	6.1	61 802	LQ9		PP, EP			0	
3350	PYRETHROID PESTICIDE, LIQUID, FLAMMABLE, TOXIC, flash-point less than 23 °C	3	FT2	I	3+6.1	61 802	LQ3		PP, EP, EX, TOX, A	VE01, VE02		2	

UN No. or ID No.	Name and description	Class	Classification Code	Packing group	Labels	Special provisions	Limited quantities	Carriage permitted	Equipment required	Ventilation	Provisions concerning loading, unloading and carriage	Number of cones, blue lights	Remarks
3.1.2	3.1.2	2.2	2.2	2.1.1.3	5.2.2	3.3	3.4.6	3.2.1	8.1.5	7.1.6	7.1.6	7.1.5	3.2.1
(1)	(2)	(3a)	(3b)	(4)	(5)	(6)	(7)	(8)	(9)	(10)	(11)	(12)	(13)
3350	PYRETHROID PESTICIDE, LIQUID, FLAMMABLE, TOXIC, flash-point less than 23 °C	3	FT2	II	3+6.1	61 802	LQ4		PP, EP, EX, TOX, A	VE01, VE02		2	
3351	PYRETHROID PESTICIDE, LIQUID, TOXIC, FLAMMABLE, flash-point not less than 23 °C	6.1	TF2	I	6.1+3	61 802	LQ0		PP, EP, EX, TOX, A	VE01, VE02		2	
3351	PYRETHROID PESTICIDE, LIQUID, TOXIC, FLAMMABLE, flash-point not less than 23 °C	6.1	TF2	II	6.1+3	61 802	LQ17		PP, EP, EX, TOX, A	VE01, VE02		2	
3351	PYRETHROID PESTICIDE, LIQUID, TOXIC, FLAMMABLE, flash-point not less than 23 °C	6.1	TF2	III	6.1+3	61 802	LQ7		PP, EP, EX, TOX, A	VE01, VE02		0	
3352	PYRETHROID PESTICIDE, LIQUID, TOXIC	6.1	T6	I	6.1	61 648 802	LQ0		PP, EP, TOX, A	VE02		2	
3352	PYRETHROID PESTICIDE, LIQUID, TOXIC	6.1	T6	II	6.1	61 648 802	LQ17		PP, EP, TOX, A	VE02		2	
3352	PYRETHROID PESTICIDE, LIQUID, TOXIC	6.1	T6	III	6.1	61 648 802	LQ7		PP, EP, TOX, A	VE02		0	
3354	INSECTICIDE GAS, FLAMMABLE, N.O.S.	2	2F		2.1	274	LQ0		PP, EX, A	VE01		1	
3355	INSECTICIDE GAS, TOXIC, FLAMMABLE, N.O.S.	2	2TF		2.3+2.1	274	LQ0		PP, EP, EX, TOX, A	VE01, VE02		2	
3356	OXYGEN GENERATOR, CHEMICAL	5.1	O3	II	5.1	284	LQ0		PP			0	
3357	NITROGLYCERIN MIXTURE, DESENSITIZED, LIQUID, N.O.S. with not more than 30% nitroglycerin, by mass	3	D	II	3	274 288	LQ4		PP, EX, A	VE01		1	
3358	REFRIGERATING MACHINES containing flammable, non-toxic, liquefied gas	2	6F		2.1	291	LQ0		PP, EX, A	VE01		1	
3359	FUMIGATED UNIT	9	M11			302			PP				
3360	Fibres, vegetable, dry	4.1	F1			NOT SUBJECT TO ADN							

- 388 -

UN No. or ID No.	Name and description	Class	Classification Code	Packing group	Labels	Special provisions	Limited quantities	Carriage permitted	Equipment required	Ventilation	Provisions concerning loading, unloading and carriage	Number of cones, blue lights	Remarks
3.1.2	3.1.2	2.2	2.2	2.1.1.3	5.2.2	3.3	3.4.6	3.2.1	8.1.5	7.1.6	7.1.6	7.1.5	3.2.1
(1)	(2)	(3a)	(3b)	(4)	(5)	(6)	(7)	(8)	(9)	(10)	(11)	(12)	(13)
3361	CHLOROSILANES, TOXIC, CORROSIVE, N.O.S.	6.1	TC1	II	6.1+8	274 802	LQ0		PP, EP, TOX, A	VE02		2	
3362	CHLOROSILANES, TOXIC, CORROSIVE, FLAMMABLE, N.O.S.	6.1	TFC	II	6.1+3+8	274	LQ0		PP, EP, EX, TOX, A	VE01 VE02		2	
3363	Dangerous goods in machinery or dangerous goods in apparatus	9	M11		NOT SUBJECT TO ADN [see also 1.1.3.1 (b)]								
3364	TRINITROPHENOL (PICRIC ACID) WETTED with not less than 10% water, by mass	4.1	D	I	4.1		LQ0		PP			1	
3365	TRINITROCHLOROBENZENE (PICRYL CHLORIDE) WETTED with not less than 10% water, by mass	4.1	D	I	4.1		LQ0		PP			1	
3366	TRINITROTOLUENE (TNT), WETTED with not less than 10% water, by mass	4.1	D	I	4.1		LQ0		PP			1	
3367	TRINITROBENZENE, WETTED with not less than 10% water, by mass	4.1	D	I	4.1		LQ0		PP			1	
3368	TRINITROBENZOIC ACID, WETTED with not less than 10% water, by mass	4.1	D	I	4.1		LQ0		PP			1	
3369	SODIUM DINITRO-o-CRESOLATE, WETTED with not less than 10% water, by mass	4.1	DT	I	4.1+6.1	802	LQ0		PP			2	
3370	UREA NITRATE, WETTED with not less than 10% water, by mass	4.1	D	I	4.1		LQ0		PP			1	
3371	2-METHYLBUTANAL	3	F1	II	3		LQ4		PP, EX, A	VE01		1	
3373	BIOLOGICAL SUBSTANCE, CATEGORY B	6.2	I4		6.2	319	LQ0		PP			0	
3374	ACETYLENE, SOLVENT FREE	2	2F		2.1		LQ0		PP, EX, A	VE01		1	
3375	AMMONIUM NITRATE EMULSION, or SUSPENSION or GEL, intermediate for blasting explosives, liquid	5.1	O1	II	5.1	309	LQ0		PP			0	

UN No. or ID No.	Name and description	Class	Classification Code	Packing group	Labels	Special provisions	Limited quantities	Carriage permitted	Equipment required	Ventilation	Provisions concerning loading, unloading and carriage	Number of cones, blue lights	Remarks
3.1.2	3.1.2	2.2	2.2	2.1.1.3	5.2.2	3.3	3.4.6	3.2.1	8.1.5	7.1.6	7.1.6	7.1.5	3.2.1
(1)	(2)	(3a)	(3b)	(4)	(5)	(6)	(7)	(8)	(9)	(10)	(11)	(12)	(13)
3375	AMMONIUM NITRATE EMULSION, or SUSPENSION or GEL, intermediate for blasting explosives, solid	5.1	O2	II	5.1	309	LQ0		PP			0	
3376	4-NITROPHENYLHYDRAZINE, with not less than 30% water, by mass	4.1	D	I	4.1		LQ0		PP			1	
3377	SODIUM PERBORATE MONOHYDRATE	5.1	O2	III	5.1		LQ12		PP			0	
3378	SODIUM CARBONATE PEROXYHYDRATE	5.1	O2	II	5.1		LQ11		PP			0	
3378	SODIUM CARBONATE PEROXYHYDRATE	5.1	O2	III	5.1		LQ12		PP			0	
3379	DESENSITIZED EXPLOSIVE, LIQUID, N.O.S.	3	D	I	3	274 311	LQ0		PP, EX, A	VE01		1	
3380	DESENSITIZED EXPLOSIVE, SOLID, N.O.S.	4.1	D	I	4.1	274 311	LQ0		PP			1	
3381	TOXIC BY INHALATION LIQUID, N.O.S. with an inhalation toxicity lower than or equal to 200 ml/m^3 and saturated vapour concentration greater than or equal to 500 LC$_{50}$	6.1	T1 or T4	I	6.1	274 802	LQ0		PP, EP, TOX, A	VE02		2	
3382	TOXIC BY INHALATION LIQUID, N.O.S. with an inhalation toxicity lower than or equal to 1000 ml/m^3 and saturated vapour concentration greater than or equal to 10 LC$_{50}$	6.1	T1 or T4	I	6.1	274 802	LQ0		PP, EP, TOX, A	VE02		2	

UN No. or ID No.	Name and description	Class	Classification Code	Packing group	Labels	Special provisions	Limited quantities	Carriage permitted	Equipment required	Ventilation	Provisions concerning loading, unloading and carriage	Number of cones, blue lights	Remarks
3.1.2	3.1.2	2.2	2.2	2.1.1.3	5.2.2	3.3	3.4.6	3.2.1	8.1.5	7.1.6	7.1.6	7.1.5	3.2.1
(1)	(2)	(3a)	(3b)	(4)	(5)	(6)	(7)	(8)	(9)	(10)	(11)	(12)	(13)
3383	TOXIC BY INHALATION LIQUID, FLAMMABLE, N.O.S. with an inhalation toxicity lower than or equal to 200 ml/m^3 and saturated vapour concentration greater than or equal to 500 LC$_{50}$	6.1	TF1	I	6.1 +3	274 802	LQ0		PP, EP, EX, TOX, A	VE01, VE02		2	
3384	TOXIC BY INHALATION LIQUID, FLAMMABLE, N.O.S. with an inhalation toxicity lower than or equal to 1000 ml/m^3 and saturated vapour concentration greater than or equal to 10 LC$_{50}$	6.1	TF1	I	6.1 +3	274 802	LQ0		PP, EP, EX, TOX, A	VE01, VE02		2	
3385	TOXIC BY INHALATION LIQUID, WATER-REACTIVE, N.O.S. with an inhalation toxicity lower than or equal to 200 ml/m^3 and saturated vapour concentration greater than or equal to 500 LC$_{50}$	6.1	TW1	I	6.1 +4.3	274 802	LQ0		PP, EP, TOX, A	VE02		2	
3386	TOXIC BY INHALATION LIQUID, WATER-REACTIVE, N.O.S. with an inhalation toxicity lower than or equal to 1000 ml/m^3 and saturated vapour concentration greater than or equal to 10 LC$_{50}$	6.1	TW1	I	6.1 +4.3	274 802	LQ0		PP, EP, TOX, A	VE02		2	
3387	TOXIC BY INHALATION LIQUID, OXIDIZING, N.O.S. with an inhalation toxicity lower than or equal to 200 ml/m^3 and saturated vapour concentration greater than or equal to 500 LC$_{50}$	6.1	TO1	I	6.1 +5.1	274 802	LQ0		PP, EP, TOX, A	VE02		2	
3388	TOXIC BY INHALATION LIQUID, OXIDIZING, N.O.S. with an inhalation toxicity lower than or equal to 1000 ml/m^3 and saturated vapour concentration greater than or equal to 10 LC$_{50}$	6.1	TO1	I	6.1 +5.1	274 802	LQ0		PP, EP, TOX, A	VE02		2	

UN No. or ID No.	Name and description	Class	Classification Code	Packing group	Labels	Special provisions	Limited quantities	Carriage permitted	Equipment required	Ventilation	Provisions concerning loading, unloading and carriage	Number of cones, blue lights	Remarks
	3.1.2	2.2	2.2	2.1.1.3	5.2.2	3.3	3.4.6	3.2.1	8.1.5	7.1.6	7.1.6	7.1.5	3.2.1
(1)	(2)	(3a)	(3b)	(4)	(5)	(6)	(7)	(8)	(9)	(10)	(11)	(12)	(13)
3389	TOXIC BY INHALATION LIQUID, CORROSIVE, N.O.S. with an inhalation toxicity lower than or equal to 200 ml/m³ and saturated vapour concentration greater than or equal to 500 LC_{50}	6.1	TC1 or TC3	I	6.1 +8	274 802	LQ0		PP, EP, TOX, A	VE02		2	
3390	TOXIC BY INHALATION LIQUID, CORROSIVE, N.O.S. with an inhalation toxicity lower than or equal to 1000 ml/m³ and saturated vapour concentration greater than or equal to 10 LC_{50}	6.1	TC1 or TC3	I	6.1 +8	274 802	LQ0		PP, EP, TOX, A	VE02		2	
3391	ORGANOMETALLIC SUBSTANCE, SOLID, PYROPHORIC	4.2	S5	I	4.2	274	LQ0		PP			0	
3392	ORGANOMETALLIC SUBSTANCE, LIQUID, PYROPHORIC	4.2	S5	I	4.2	274	LQ0		PP			0	
3393	ORGANOMETALLIC SUBSTANCE, SOLID, PYROPHORIC, WATER REACTIVE	4.2	SW	I	4.2 +4.3	274	LQ0		PP, EX, A	VE01		0	
3394	ORGANOMETALLIC SUBSTANCE, LIQUID, PYROPHORIC, WATER REACTIVE	4.2	SW	I	4.2 +4.3	274	LQ0		PP, EX, A	VE01		0	
3395	ORGANOMETALLIC SUBSTANCE, SOLID, WATER REACTIVE	4.3	W2	I	4.3	274	LQ0		PP, EX, A	VE01	HA08	0	
3395	ORGANOMETALLIC SUBSTANCE, SOLID, WATER REACTIVE	4.3	W2	II	4.3	274	LQ11		PP, EX, A	VE01	HA08	0	
3395	ORGANOMETALLIC SUBSTANCE, SOLID, WATER REACTIVE	4.3	W2	III	4.3	274	LQ12		PP, EX, A	VE01	HA08	0	

UN No. or ID No.	Name and description	Class	Classification Code	Packing group	Labels	Special provisions	Limited quantities	Carriage permitted	Equipment required	Ventilation	Provisions concerning loading, unloading and carriage	Number of cones, blue lights	Remarks
3.1.2	3.1.2	2.2	2.2	2.1.1.3	5.2.2	3.3	3.4.6	3.2.1	8.1.5	7.1.6	7.1.6	7.1.5	3.2.1
(1)	(2)	(3a)	(3b)	(4)	(5)	(6)	(7)	(8)	(9)	(10)	(11)	(12)	(13)
3396	ORGANOMETALLIC SUBSTANCE, SOLID, WATER REACTIVE, FLAMMABLE	4.3	WF2	I	4.3 +4.1	274	LQ0		PP, EX, A	VE01		1	
3396	ORGANOMETALLIC SUBSTANCE, SOLID, WATER REACTIVE, FLAMMABLE	4.3	WF2	II	4.3 +4.1	274	LQ11		PP, EX, A	VE01		1	
3396	ORGANOMETALLIC SUBSTANCE, SOLID, WATER REACTIVE, FLAMMABLE	4.3	WF2	III	4.3 +4.1	274	LQ12		PP, EX, A	VE01		0	
3397	ORGANOMETALLIC SUBSTANCE, SOLID, WATER REACTIVE, SELF-HEATING	4.3	WS	I	4.3 +4.2	274	LQ0		PP, EX, A	VE01	HA08	0	
3397	ORGANOMETALLIC SUBSTANCE, SOLID, WATER REACTIVE, SELF-HEATING	4.3	WS	II	4.3 +4.2	274	LQ11		PP, EX, A	VE01	HA08	0	
3397	ORGANOMETALLIC SUBSTANCE, SOLID, WATER REACTIVE, SELF-HEATING	4.3	WS	III	4.3 +4.2	274	LQ12		PP, EX, A	VE01	HA08	0	
3398	ORGANOMETALLIC SUBSTANCE, LIQUID, WATER REACTIVE	4.3	W1	I	4.3	274	LQ0		PP, EX, A	VE01	HA08	0	
3398	ORGANOMETALLIC SUBSTANCE, LIQUID, WATER REACTIVE	4.3	W1	II	4.3	274	LQ10		PP, EX, A	VE01	HA08	0	
3398	ORGANOMETALLIC SUBSTANCE, LIQUID, WATER REACTIVE	4.3	W1	III	4.3	274	LQ13		PP, EX, A	VE01	HA08	0	
3399	ORGANOMETALLIC SUBSTANCE, LIQUID, WATER REACTIVE, FLAMMABLE	4.3	WF1	I	4.3 +3	274	LQ0		PP, EX, A	VE01	HA08	1	
3399	ORGANOMETALLIC SUBSTANCE, LIQUID, WATER REACTIVE, FLAMMABLE	4.3	WF1	II	4.3 +3	274	LQ10		PP, EX, A	VE01	HA08	1	
3399	ORGANOMETALLIC SUBSTANCE, LIQUID, WATER REACTIVE, FLAMMABLE	4.3	WF1	III	4.3 +3	274	LQ13		PP, EX, A	VE01	HA08	0	

UN No. or ID No. (1)	Name and description 3.1.2 (2)	Class 2.2 (3a)	Classification Code 2.2 (3b)	Packing group 2.1.1.3 (4)	Labels 5.2.2 (5)	Special provisions 3.3 (6)	Limited quantities 3.4.6 (7)	Carriage permitted 3.2.1 (8)	Equipment required 8.1.5 (9)	Ventilation 7.1.6 (10)	Provisions concerning loading, unloading and carriage 7.1.6 (11)	Number of cones, blue lights 7.1.5 (12)	Remarks 3.2.1 (13)
3400	ORGANOMETALLIC SUBSTANCE, SOLID, SELF-HEATING	4.2	S5	II	4.2	274	LQ18		PP			0	
3400	ORGANOMETALLIC SUBSTANCE, SOLID, SELF-HEATING	4.2	S5	III	4.2	274	LQ11		PP			0	
3401	ALKALI METAL AMALGAM, SOLID	4.3	W2	I	4.3	182 274	LQ0		PP, EX, A	VE01	HA08	0	
3402	ALKALINE EARTH METAL AMALGAM, SOLID	4.3	W2	I	4.3	183 274	LQ0		PP, EX, A	VE01	HA08	0	
3403	POTASSIUM METAL ALLOYS, SOLID	4.3	W2	I	4.3		LQ0		PP, EX, A	VE01	HA08	0	
3404	POTASSIUM SODIUM ALLOYS, SOLID	4.3	W2	I	4.3		LQ0		PP, EX, A	VE01	HA08	0	
3405	BARIUM CHLORATE SOLUTION	5.1	OT1	II	5.1 +6.1	802	LQ10		PP, EP, TOX, A	VE02		2	
3405	BARIUM CHLORATE SOLUTION	5.1	OT1	III	5.1 +6.1	802	LQ13		PP, EP, TOX, A	VE02		0	
3406	BARIUM PERCHLORATE SOLUTION	5.1	OT1	II	5.1 +6.1	802	LQ10		PP, EP, TOX, A	VE02		2	
3406	BARIUM PERCHLORATE SOLUTION	5.1	OT1	III	5.1 +6.1	802	LQ13		PP, EP, TOX, A	VE02		0	
3407	CHLORATE AND MAGNESIUM CHLORIDE MIXTURE SOLUTION	5.1	O1	II	5.1		LQ10		PP			0	
3407	CHLORATE AND MAGNESIUM CHLORIDE MIXTURE SOLUTION	5.1	O1	III	5.1		LQ13		PP			0	
3408	LEAD PERCHLORATE SOLUTION	5.1	OT1	II	5.1 +6.1		LQ10		PP			0	
3408	LEAD PERCHLORATE SOLUTION	5.1	OT1	III	5.1 +6.1		LQ13		PP			0	
3409	CHLORONITROBENZENES, LIQUID	6.1	T1	II	6.1	279 802	LQ17		PP, EP			2	
3410	4-CHLORO-o-TOLUIDINE HYDROCHLORIDE SOLUTION	6.1	T1	III	6.1	802	LQ7		PP, EP, TOX, A	VE02		0	

UN No. or ID No.	Name and description	Class	Classification Code	Packing group	Labels	Special provisions	Limited quantities	Carriage permitted	Equipment required	Ventilation	Provisions concerning loading, unloading and carriage	Number of cones, blue lights	Remarks
	3.1.2	2.2	2.2	2.1.1.3	5.2.2	3.3	3.4.6	3.2.1	8.1.5	7.1.6	7.1.6	7.1.5	3.2.1
(1)	(2)	(3a)	(3b)	(4)	(5)	(6)	(7)	(8)	(9)	(10)	(11)	(12)	(13)
3411	beta-NAPHTHYLAMINE SOLUTION	6.1	T1	II	6.1	802	LQ17		PP, EP, TOX, A	VE02		2	
3411	beta-NAPHTHYLAMINE SOLUTION	6.1	T1	III	6.1	802	LQ7		PP, EP, TOX, A	VE02		0	
3412	FORMIC ACID with not less than 10% but not more than 85% acid by mass	8	C3	II	8		LQ22	T	PP, EP			0	
3412	FORMIC ACID with not less than 5% but less than 10% acid by mass	8	C3	III	8		LQ7	T	PP, EP			0	
3413	POTASSIUM CYANIDE SOLUTION	6.1	T4	I	6.1	802	LQ0		PP, EP, TOX, A	VE02		2	
3413	POTASSIUM CYANIDE SOLUTION	6.1	T4	II	6.1	802	LQ17		PP, EP, TOX, A	VE02		2	
3413	POTASSIUM CYANIDE SOLUTION	6.1	T4	III	6.1	802	LQ7		PP, EP, TOX, A	VE02		0	
3414	SODIUM CYANIDE SOLUTION	6.1	T4	I	6.1	802	LQ0		PP, EP, TOX, A	VE02		2	
3414	SODIUM CYANIDE SOLUTION	6.1	T4	II	6.1	802	LQ17		PP, EP, TOX, A	VE02		2	
3414	SODIUM CYANIDE SOLUTION	6.1	T4	III	6.1	802	LQ7		PP, EP, TOX, A	VE02		0	
3415	SODIUM FLUORIDE SOLUTION	6.1	T4	III	6.1	802	LQ7		PP, EP, TOX, A	VE02		0	
3416	CHLOROACETOPHENONE, LIQUID	6.1	T1	II	6.1	802	LQ17		PP, EP, TOX, A	VE02		2	
3417	XYLYL BROMIDE, SOLID	6.1	T2	II	6.1	802	LQ18		PP, EP			2	
3418	2,4-TOLUYLENEDIAMINE SOLUTION	6.1	T1	III	6.1	802	LQ7		PP, EP, TOX, A	VE02		0	
3419	BORON TRIFLUORIDE ACETIC ACID COMPLEX, SOLID	8	C4	II	8		LQ23		PP, EP			0	
3420	BORON TRIFLUORIDE PROPIONIC ACID COMPLEX, SOLID	8	C4	II	8		LQ23		PP, EP			0	
3421	POTASSIUM HYDROGENDI-FLUORIDE SOLUTION	8	CT1	II	8 +6.1	802	LQ22		PP, EP, TOX, A	VE02		2	
3421	POTASSIUM HYDROGENDI-FLUORIDE SOLUTION	8	CT1	III	8 +6.1	802	LQ7		PP, EP, TOX, A	VE02		0	

UN No. or ID No.	Name and description	Class	Classification Code	Packing group	Labels	Special provisions	Limited quantities	Carriage permitted	Equipment required	Ventilation	Provisions concerning loading, unloading and carriage	Number of cones, blue lights	Remarks
	3.1.2	2.2	2.2	2.1.1.3	5.2.2	3.3	3.4.6	3.2.1	8.1.5	7.1.6	7.1.6	7.1.5	3.2.1
(1)	(2)	(3a)	(3b)	(4)	(5)	(6)	(7)	(8)	(9)	(10)	(11)	(12)	(13)
3422	POTASSIUM FLUORIDE SOLUTION	6.1	T4	III	6.1	802	LQ7		PP, EP, TOX, A	VE02		0	
3423	TETRAMETHYLAMMONIUM HYDROXIDE, SOLID	8	C8	II	8		LQ24		PP, EP			0	
3424	AMMONIUM DINITRO-o-CRESOLATE SOLUTION	6.1	T1	II	6.1	802	LQ17		PP, EP, TOX, A	VE02		2	
3424	AMMONIUM DINITRO-o-CRESOLATE SOLUTION	6.1	T1	III	6.1	802	LQ7		PP, EP, TOX, A	VE02		0	
3425	BROMOACETIC ACID, SOLID	8	C4	II	8		LQ23		PP, EP			0	
3426	ACRYLAMIDE SOLUTION	6.1	T1	III	6.1		LQ7	T	PP, EP, TOX, A	VE02		0	
3427	CHLOROBENZYL CHLORIDES, SOLID	6.1	T2	III	6.1	802	LQ9		PP, EP			0	
3428	3-CHLORO-4-METHYLPHENYL ISOCYANATE, SOLID	6.1	T2	II	6.1	802	LQ18		PP, EP			2	
3429	CHLOROTOLUIDINES, LIQUID	6.1	T1	III	6.1	802	LQ7		PP, EP, TOX, A	VE02		0	
3430	XYLENOLS, LIQUID	6.1	T1	II	6.1	802	LQ17		PP, EP, TOX, A	VE02		2	
3431	NITROBENZOTRI-FLUORIDES, SOLID	6.1	T2	II	6.1	802	LQ18		PP, EP			2	
3432	POLYCHLORINATED BIPHENYLS, SOLID	9	M2	II	9	305 802	LQ25		PP, EP			0	
3434	NITROCRESOLS, LIQUID	6.1	T1	III	6.1	802	LQ7		PP, EP, TOX, A	VE02		0	
3436	HEXAFLUOROACETONE HYDRATE, SOLID	6.1	T2	II	6.1	802	LQ18		PP, EP			2	
3437	CHLOROCRESOLS, SOLID	6.1	T2	II	6.1	802	LQ18		PP, EP			2	
3438	alpha-METHYLBENZYL ALCOHOL, SOLID	6.1	T2	III	6.1	802	LQ9		PP, EP			0	
3439	NITRILES, TOXIC, SOLID, N.O.S.	6.1	T2	I	6.1	274 802	LQ0		PP, EP			2	
3439	NITRILES, TOXIC, SOLID, N.O.S.	6.1	T2	II	6.1	274 802	LQ18		PP, EP			2	
3439	NITRILES, TOXIC, SOLID, N.O.S.	6.1	T2	III	6.1	274 802	LQ9		PP, EP			0	

UN No. or ID No.	Name and description	Class	Classification Code	Packing group	Labels	Special provisions	Limited quantities	Carriage permitted	Equipment required	Ventilation	Provisions concerning loading, unloading and carriage	Number of cones, blue lights	Remarks
	3.1.2	2.2	2.2	2.1.1.3	5.2.2	3.3	3.4.6	3.2.1	8.1.5	7.1.6	7.1.6	7.1.5	3.2.1
(1)	(2)	(3a)	(3b)	(4)	(5)	(6)	(7)	(8)	(9)	(10)	(11)	(12)	(13)
3440	SELENIUM COMPOUND, LIQUID, N.O.S.	6.1	T4	I	6.1	274 802	LQ0		PP, EP, TOX, A	VE02		2	
3440	SELENIUM COMPOUND, LIQUID, N.O.S.	6.1	T4	II	6.1	274 802	LQ17		PP, EP, TOX, A	VE02		2	
3440	SELENIUM COMPOUND, LIQUID, N.O.S.	6.1	T4	III	6.1	274 802	LQ7		PP, EP, TOX, A	VE02		0	
3441	CHLORODINITROBENZENES, SOLID	6.1	T2	II	6.1	279 802	LQ18		PP, EP			2	
3442	DICHLOROANILINES, SOLID	6.1	T2	II	6.1	279 802	LQ18		PP, EP			2	
3443	DINITROBENZENES, SOLID	6.1	T2	II	6.1	802	LQ18		PP, EP			2	
3444	NICOTINE HYDROCHLORIDE, SOLID	6.1	T2	II	6.1	43 802	LQ18		PP, EP			2	
3445	NICOTINE SULPHATE, SOLID	6.1	T2	II	6.1	802	LQ18		PP, EP			2	
3446	NITROTOLUENES, SOLID	6.1	T2	II	6.1	802	LQ18	T	PP, EP			2	
3447	NITROXYLENES, SOLID	6.1	T2	II	6.1	802	LQ18		PP, EP			2	
3448	TEAR GAS SUBSTANCE, SOLID, N.O.S.	6.1	T2	I	6.1	274 802	LQ0		PP, EP			2	
3448	TEAR GAS SUBSTANCE, SOLID, N.O.S.	6.1	T2	II	6.1	274 802	LQ18		PP, EP			2	
3449	BROMOBENZYL CYANIDES, SOLID	6.1	T2	I	6.1	138 802	LQ0		PP, EP			2	
3450	DIPHENYLCHLOROARSINE, SOLID	6.1	T3	I	6.1	802	LQ0		PP, EP			2	
3451	TOLUIDINES, SOLID	6.1	T2	II	6.1	279 802	LQ18	T	PP, EP			2	
3452	XYLIDINES, SOLID	6.1	T2	II	6.1	802	LQ18		PP, EP			2	
3453	PHOSPHORIC ACID, SOLID	8	C2	III	8		LQ24		PP, EP			0	
3454	DINITROTOLUENES, SOLID	6.1	T2	II	6.1	802	LQ18		PP, EP			2	
3455	CRESOLS, SOLID	6.1	TC2	II	6.1+8	802	LQ18	T	PP, EP			2	
3456	NITROSYLSULPHURIC ACID, SOLID	8	C2	II	8		LQ23	T3	PP, EP			0	
3457	CHLORONITROTOLUENES, SOLID	6.1	T2	III	6.1	802	LQ9		PP, EP			0	
3458	NITROANISOLES, SOLID	6.1	T2	III	6.1	279 802	LQ9		PP, EP			0	

UN No. or ID No. (1)	Name and description 3.1.2 (2)	Class 2.2 (3a)	Classification Code 2.2 (3b)	Packing group 2.1.1.3 (4)	Labels 5.2.2 (5)	Special provisions 3.3 (6)	Limited quantities 3.4.6 (7)	Carriage permitted 3.2.1 (8)	Equipment required 8.1.5 (9)	Ventilation 7.1.6 (10)	Provisions concerning loading, unloading and carriage 7.1.6 (11)	Number of cones, blue lights 7.1.5 (12)	Remarks 3.2.1 (13)
3459	NITROBROMOBENZENES, SOLID	6.1	T2	III	6.1	802	LQ9		PP, EP			0	
3460	N-ETHYLBENZYL-TOLUIDINES, SOLID	6.1	T2	III	6.1	802	LQ9		PP, EP			0	
3462	TOXINS, EXTRACTED FROM LIVING SOURCES, SOLID, N.O.S.	6.1	T2	I	6.1	210 274 802	LQ0		PP, EP			2	
3462	TOXINS, EXTRACTED FROM LIVING SOURCES, SOLID, N.O.S.	6.1	T2	II	6.1	210 274 802	LQ18		PP, EP			2	
3462	TOXINS, EXTRACTED FROM LIVING SOURCES, SOLID, N.O.S.	6.1	T2	III	6.1	210 274 802	LQ9		PP, EP			0	
3463	PROPIONIC ACID with not less than 90% acid by mass	8	CF1	II	8 +3		LQ22	T	PP, EP, EX, A			0	
3464	ORGANOPHOSPHORUS COMPOUND, TOXIC, SOLID, N.O.S.	6.1	T2	I	6.1	43 274 802	LQ0		PP, EP			2	
3464	ORGANOPHOSPHORUS COMPOUND, TOXIC, SOLID, N.O.S.	6.1	T2	II	6.1	43 274 802	LQ18		PP, EP			2	
3464	ORGANOPHOSPHORUS COMPOUND, TOXIC, SOLID, N.O.S.	6.1	T2	III	6.1	43 274 802	LQ9		PP, EP			0	
3465	ORGANOARSENIC COMPOUND, SOLID, N.O.S.	6.1	T3	I	6.1	274 802	LQ0		PP, EP			2	
3465	ORGANOARSENIC COMPOUND, SOLID, N.O.S.	6.1	T3	II	6.1	274 802	LQ18		PP, EP			2	
3465	ORGANOARSENIC COMPOUND, SOLID, N.O.S.	6.1	T3	III	6.1	274 802	LQ9		PP, EP			0	
3466	METAL CARBONYLS, SOLID, N.O.S	6.1	T3	I	6.1	274 562 802	LQ0		PP, EP			2	
3466	METAL CARBONYLS, SOLID, N.O.S	6.1	T3	II	6.1	274 562 802	LQ18		PP, EP			2	

UN No. or ID No.	Name and description	Class	Classification Code	Packing group	Labels	Special provisions	Limited quantities	Carriage permitted	Equipment required	Ventilation	Provisions concerning loading, unloading and carriage	Number of cones, blue lights	Remarks
3.1.2	3.1.2	2.2	2.2	2.1.1.3	5.2.2	3.3	3.4.6	3.2.1	8.1.5	7.1.6	7.1.6	7.1.5	3.2.1
(1)	(2)	(3a)	(3b)	(4)	(5)	(6)	(7)	(8)	(9)	(10)	(11)	(12)	(13)
3466	METAL CARBONYLS, SOLID, N.O.S	6.1	T3	III	6.1	274 562 802	LQ9		PP, EP			0	
3467	ORGANOMETALLIC COMPOUND, TOXIC, SOLID, N.O.S.	6.1	T3	I	6.1	274 562 802	LQ0		PP, EP			2	
3467	ORGANOMETALLIC COMPOUND, TOXIC, SOLID, N.O.S.	6.1	T3	II	6.1	274 562 802	LQ18		PP, EP			2	
3467	ORGANOMETALLIC COMPOUND, TOXIC, SOLID, N.O.S	6.1	T3	III	6.1	274 562 802	LQ9		PP, EP			0	
3468	HYDROGEN IN A METAL HYDRIDE STORAGE SYSTEM	2	2F		2.1	321	LQ0	T	PP, EX, A	VE01		1	
3469	PAINT, FLAMMABLE, CORROSIVE (including paint, lacquer, enamel, stain, shellac, varnish, polish, liquid filler and liquid lacquer base) or PAINT RELATED MATERIAL, FLAMMABLE, CORROSIVE (including paint thinning or reducing compound)	3	FC	I	3 +8	163	LQ3		PP, EX, A	VE01		1	
3469	PAINT, FLAMMABLE, CORROSIVE (including paint, lacquer, enamel, stain, shellac, varnish, polish, liquid filler and liquid lacquer base) or PAINT RELATED MATERIAL, FLAMMABLE, CORROSIVE (including paint thinning or reducing compound)	3	FC	II	3 +8	163	LQ4		PP, EX, A	VE01		1	

UN No. or ID No.	Name and description	Class	Classification Code	Packing group	Labels	Special provisions	Limited quantities	Carriage permitted	Equipment required	Ventilation	Provisions concerning loading, unloading and carriage	Number of cones, blue lights	Remarks
3.1.2	3.1.2	2.2	2.2	2.1.1.3	5.2.2	3.3	3.4.6	3.2.1	8.1.5	7.1.6	7.1.6	7.1.5	3.2.1
(1)	(2)	(3a)	(3b)	(4)	(5)	(6)	(7)	(8)	(9)	(10)	(11)	(12)	(13)
3469	PAINT, FLAMMABLE, CORROSIVE (including paint, lacquer, enamel, stain, shellac, varnish, polish, liquid filler and liquid lacquer base) or PAINT RELATED MATERIAL, FLAMMABLE, CORROSIVE (including paint thinning or reducing compound)	3	FC	III	3 +8	163	LQ7		PP, EX, A	VE01		0	
3470	PAINT, CORROSIVE, FLAMMABLE (including paint, lacquer, enamel, stain, shellac, varnish, polish, liquid filler and liquid lacquer base) or PAINT RELATED MATERIAL, CORROSIVE, FLAMMABLE (including paint thinning or reducing compound)	8	CF1	II	8 +3	163	LQ22		PP, EP, EX, A	VE01		0	
3471	HYDROGENDIFLUORIDES SOLUTION, N.O.S.	8	CT1	II	8 +6.1		LQ22		PP, EP			0	
3471	HYDROGENDIFLUORIDES SOLUTION, N.O.S.	8	CT1	III	8 +6.1		LQ7		PP, EP			0	
3472	CROTONIC ACID, LIQUID	8	C3	III	8		LQ7		PP, EP			0	
3473	FUEL CELL CARTRIDGES containing flammable liquids	3	F1		3	328	LQ13						
9000	AMMONIA, DEEPLY REFRIGERATED	2	3TC		2.3+8			T	PP			2	Admitted only for carriage in tank vessel
9001	SUBSTANCES WITH A FLASH-POINT ABOVE 60 °C which are carried heated within a limiting range of 15K below their flash-point	3	F3		none			T	PP			0	Dangerous only when carried in tank vessel

UN No. or ID No.	Name and description	Class	Classification Code	Packing group	Labels	Special provisions	Limited quantities	Carriage permitted	Equipment required	Ventilation	Provisions concerning loading, unloading and carriage	Number of cones, blue lights	Remarks
	3.1.2	2.2	2.2	2.1.1.3	5.2.2	3.3	3.4.6	3.2.1	8.1.5	7.1.6	7.1.6	7.1.5	3.2.1
(1)	(2)	(3a)	(3b)	(4)	(5)	(6)	(7)	(8)	(9)	(10)	(11)	(12)	(13)
9002	SUBSTANCES WITH A SELF-IGNITION TEMPERATURE OF 200 °C AND BELOW, n.o.s.	3	F4		none			T	PP			0	Dangerous only when carried in tank vessel
9003	SUBSTANCES WITH A FLASH-POINT ABOVE 60 °C AND NOT MORE THAN 100 °C, which do not belong to another Class	9			none			T	PP			0	Dangerous only when carried in tank vessel
9004	DIPHENYLMETHANE-4, 4'-DIISOCYANATE	9			none			T	PP			0	Dangerous only when carried in tank vessel

3.2.2   **Table B: List of dangerous goods in alphabetical order**

This index is an alphabetical list of the substances and articles which are listed in the UN numerical order in Table A of 3.2.1. It does not form an integral part of ADN. It has been prepared, with all necessary care by the Secretariat of the United Nations Economic Commission for Europe, in order to facilitate the consultation of Annexes A and B, but it cannot be relied upon as a substitute for the careful study and observance of the actual provisions of those annexed Regulations which, in case of conflict, are deemed to be authoritative.

NOTE 1: For the purpose of determining the alphabetical order the following information has been ignored, even when it forms part of the proper shipping name: numbers; Greek letters; the abbreviations "sec" and "tert"; and the letters "N" (nitrogen), "n" (normal), "o" (ortho) "m" (meta), "p" (para) and "N.O.S." (not otherwise specified).

NOTE 2: The name of a substance or article in block capital letters indicates a proper shipping name (see 3.1.2).

NOTE 3: The name of a substance or article in block capital letters followed by the word "see" indicates an alternative proper shipping name or part of a proper shipping name (except for PCBs) (see 3.1.2.1).

NOTE 4: An entry in lower case letters followed by the word "see" indicates that the entry is not a proper shipping name; it is a synonym.

NOTE 5: Where an entry is partly in block capital letters and partly in lower case letters, the latter part is considered not to be part of the proper shipping name (see 3.1.2.1).

NOTE 6: A proper shipping name may be used in the singular or plural, as appropriate, for the purposes of documentation and package marking (see 3.1.2.3).

NOTE 7: For the exact determination of a proper shipping name, see 3.1.2.

Name and description	UN No.	Class	Remarks	Name and description	UN No.	Class	Remarks
Accumulators, electric, see	2794	8		ACETYL METHYL CARBINOL	2621	3	
	2795	8					
	2800	8		Acid butyl phosphate, see	1718	8	
	3028	8					
	3292	4.3		Acid mixture, hydrofluoric and sulphuric, see	1786	8	
ACETAL	1088	3					
				Acid mixture, nitrating acid, see	1796	8	
ACETALDEHYDE	1089	3					
ACETALDEHYDE AMMONIA	1841	9		Acid mixture, spent, nitrating acid, see	1826	8	
ACETALDEHYDE OXIME	2332	3					
				Acraldehyde, inhibited, see	1092	6.1	
ACETIC ACID, GLACIAL	2789	8					
				ACRIDINE	2713	6.1	
ACETIC ACID SOLUTION, more than 10% but not more than 80% acid, by mass	2790	8		ACROLEIN DIMER, STABILIZED	2607	3	
ACETIC ACID SOLUTION, more than 80% acid, by mass	2789	8		ACROLEIN, STABILIZED	1092	6.1	
ACETIC ANHYDRIDE	1715	8		ACRYLAMIDE, SOLID	2074	6.1	
Acetoin, see	2621	3		ACRYLAMIDE, SOLUTION	3426	6.1	
				ACRYLIC ACID, STABILIZED	2218	8	
ACETONE	1090	3					
ACETONE CYANOHYDRIN, STABILIZED	1541	6.1		ACRYLONITRILE, STABILIZED	1093	3	
				Actinolite, see	2590	9	
ACETONE OILS	1091	3		Activated carbon, see	1362	4.2	
ACETONITRILE	1648	3		Activated charcoal, see	1362	4.2	
ACETYL BROMIDE	1716	8		ADHESIVES containing flammable liquid	1133	3	
ACETYL CHLORIDE	1717	3					
ACETYLENE, DISSOLVED	1001	2		ADIPONITRILE	2205	6.1	
ACETYLENE, SOLVENT FREE	3374	2		Aeroplane flares, see	0093	1	
					0403	1	
					0404	1	
Acetylene tetrabromide, see	2504	6.1			0420	1	
					0421	1	
Acetylene tetrachloride, see	1702	6.1					
				AEROSOLS	1950	2	
ACETYL IODIDE	1898	8					
				AGENT, BLASTING, TYPE B	0331	1	

Name and description	UN No.	Class	Remarks	Name and description	UN No.	Class	Remarks
AGENT, BLASTING, TYPE E	0332	1		ALDEHYDES, FLAMMABLE, TOXIC, N.O.S.	1988	3	
AIR BAG INFLATORS	0503	1					
	3268	9		ALDOL	2839	6.1	
AIR BAG MODULES	0503	1		ALKALI METAL ALCOHOLATES, SELF-HEATING, CORROSIVE, N.O.S.	3206	4.2	
	3268	9					
AIR, COMPRESSED	1002	2					
Aircraft evacuation slides, see	2990	9		ALKALI METAL ALLOY, LIQUID, N.O.S.	1421	4.3	
AIRCRAFT HYDRAULIC POWER UNIT FUEL TANK (containing a mixture of anhydrous hydrazine and methylhydrazine) (M86 fuel)	3165	3		ALKALI METAL AMALGAM, LIQUID	1389	4.3	
				ALKALI METAL AMALGAM, SOLID	3401	4.3	
Aircraft survival kits, see	2990	9					
				ALKALI METAL AMIDES	1390	4.3	
AIR, REFRIGERATED LIQUID	1003	2		ALKALI METAL DISPERSION	1391	4.3	
ALCOHOLATES SOLUTION, N.O.S., in alcohol	3274	3		Alkaline corrosive battery fluid, see	2797	8	
Alcohol, denatured, see	1986	3		ALKALINE EARTH METAL ALCOHOLATES, N.O.S.	3205	4.2	
	1987	3					
Alcohol, industrial, see	1986	3		ALKALINE EARTH METAL ALLOY, N.O.S.	1393	4.3	
	1987	3					
ALCOHOLS, N.O.S.	1987	3		ALKALINE EARTH METAL AMALGAM, LIQUID	1392	4.3	
ALCOHOLS, FLAMMABLE, TOXIC, N.O.S.	1986	3		ALKALINE EARTH METAL AMALGAM, SOLID	3402	4.3	
ALCOHOLIC BEVERAGES, with more than 24% but not more than 70% alcohol by volume	3065	3		ALKALINE EARTH METAL DISPERSION	1391	4.3	
				ALKALOIDS, LIQUID, N.O.S.	3140	6.1	
ALCOHOLIC BEVERAGES, with more than 70% alcohol by volume	3065	3		ALKALOIDS, SOLID, N.O.S.	1544	6.1	
Aldehyde, see	1989	3		ALKALOID SALTS, LIQUID, N.O.S.	3140	6.1	
ALDEHYDES, N.O.S.	1989	3					

Name and description	UN No.	Class	Remarks	Name and description	UN No.	Class	Remarks
ALKALOID SALTS, SOLID, N.O.S.	1544	6.1		ALLYL GLYCIDYL ETHER	2219	3	
Alkyl aluminium halides, see	3394	4.2		ALLYL IODIDE	1723	3	
ALKYLPHENOLS, LIQUID, N.O.S. (including C_2-C_{12} homologues)	3145	8		ALLYL ISOTHIOCYANATE, STABILIZED	1545	6.1	
ALKYLPHENOLS, SOLID, N.O.S. (including C_2-C_{12} homologues)	2430	8		ALLYLTRICHLORO-SILANE, STABILIZED	1724	8	
				Aluminium alkyls, see	3394	4.2	
				Aluminium alkyl halides, liquid, see	3394	4.2	
ALKYLSULPHONIC ACIDS, LIQUID with more than 5% free sulphuric acid	2584	8		Aluminium alkyl halides, solid, see	3393	4.2	
				Aluminium alkyl hydrides, see	3394	4.2	
ALKYLSULPHONIC ACIDS, LIQUID with not more than 5% free sulphuric acid	2586	8		ALUMINIUM BOROHYDRIDE	2870	4.2	
				ALUMINIUM BOROHYDRIDE IN DEVICES	2870	4.2	
ALKYLSULPHONIC ACIDS, SOLID with more than 5% free sulphuric acid	2583	8		ALUMINIUM BROMIDE, ANHYDROUS	1725	8	
ALKYLSULPHONIC ACIDS, SOLID with not more than 5% free sulphuric acid	2585	8		ALUMINIUM BROMIDE SOLUTION	2580	8	
ALKYLSULPHURIC ACIDS	2571	8		ALUMINIUM CARBIDE	1394	4.3	
Allene, see	2200	2		ALUMINIUM CHLORIDE, ANHYDROUS	1726	8	
ALLYL ACETATE	2333	3		ALUMINIUM CHLORIDE SOLUTION	2581	8	
ALLYL ALCOHOL	1098	6.1					
ALLYLAMINE	2334	6.1		Aluminium dross, see	3170	4.3	
ALLYL BROMIDE	1099	3		ALUMINIUM FERROSILICON POWDER	1395	4.3	
ALLYL CHLORIDE	1100	3		ALUMINIUM HYDRIDE	2463	4.3	
Allyl chlorocarbonate, see	1722	6.1					
ALLYL CHLOROFORMATE	1722	6.1		ALUMINIUM NITRATE	1438	5.1	
				ALUMINIUM PHOSPHIDE	1397	4.3	
ALLYL ETHYL ETHER	2335	3		ALUMINIUM PHOSPHIDE PESTICIDE	3048	6.1	
ALLYL FORMATE	2336	3					

Name and description	UN No.	Class	Remarks	Name and description	UN No.	Class	Remarks
ALUMINIUM POWDER, COATED	1309	4.1		2-(2-AMINOETHOXY) ETHANOL	3055	8	
ALUMINIUM POWDER, UNCOATED	1396	4.3		N-AMINOETHYL-PIPERAZINE	2815	8	
ALUMINIUM REMELTING BY-PRODUCTS	3170	4.3		1-Amino-2-nitrobenzene, see	1661	6.1	
ALUMINIUM RESINATE	2715	4.1		1-Amino-3-nitrobenzene, see	1661	6.1	
ALUMINIUM SILICON POWDER, UNCOATED	1398	4.3		1-Amino-4-nitrobenzene, see	1661	6.1	
ALUMINIUM SMELTING BY-PRODUCTS	3170	4.3		AMINOPHENOLS (o-, m-, p-)	2512	6.1	
Amatols, see	0082	1		AMINOPYRIDINES (o-, m-, p-)	2671	6.1	
AMINES, FLAMMABLE, CORROSIVE, N.O.S.	2733	3		AMMONIA, ANHYDROUS	1005	2	
AMINES, LIQUID, CORROSIVE, N.O.S.	2735	8		AMMONIA, DEEPLY REFRIGERATED	9000	2	Admitted only for carriage in tank vessel
AMINES, LIQUID, CORROSIVE, FLAMMABLE, N.O.S.	2734	8		AMMONIA SOLUTION relative density between 0.880 and 0.957 at 15 °C in water, with more than 10% but not more than 35% ammonia	2672	8	
AMINES, SOLID, CORROSIVE, N.O.S.	3259	8		AMMONIA SOLUTION, relative density less than 0.880 at 15 °C in water, with more than 35% but not more than 50% ammonia	2073	2	
Aminobenzene, see	1547	6.1					
2-Aminobenzotrifluoruride, see	2942	6.1					
3-Aminobenzotrifluoruride, see	2948	6.1		AMMONIA SOLUTION, relative density less than 0.880 at 15 °C in water, with more than 50% ammonia	3318	2	
Aminobutane, see	1125	3		AMMONIUM ARSENATE	1546	6.1	
2-AMINO-4-CHLOROPHENOL	2673	6.1		Ammonium bichromate, see	1439	5.1	
2-AMINO-5-DIETHYLAMINOPENTANE	2946	6.1		Ammonium bifluoride solid, see	1727	8	
2-AMINO-4,6-DINITROPHENOL, WETTED with not less than 20% water, by mass	3317	4.1		Ammonium bifluoride solution, see	2817	8	
				Ammonium bisulphate, see	2506	8	

Name and description	UN No.	Class	Remarks	Name and description	UN No.	Class	Remarks
Ammonium bisulphite solution, see	2693	8		AMMONIUM NITRATE EMULSION, intermediate for blasting explosives, liquid	3375	5.1	
AMMONIUM DICHROMATE	1439	5.1		AMMONIUM NITRATE EMULSION, intermediate for blasting explosives, solid	3375	5.1	
AMMONIUM DINITRO-o-CRESOLATE, SOLID	1843	6.1		Ammonium nitrate explosive, see	0082	1	
					0331	1	
AMMONIUM DINITRO-o-CRESOLATE, SOLUTION	3424	6.1		AMMONIUM NITRATE BASED FERTILIZER	2067	5.1	
AMMONIUM FLUORIDE	2505	6.1		AMMONIUM NITRATE BASED FERTILIZER, uniform mixtures of the nitrogen/phosphate, nitrogen/potash or nitrogen/phosphate/potash type, containing not more than 70% ammonium nitrate and not more than 0.4% total combustible/organic material calculated as carbon or with not more than 45% ammonium nitrate and unrestricted combustible material	2071	9	
AMMONIUM FLUOROSILICATE	2854	6.1					
Ammonium hexafluorosilicate, see	2854	6.1					
AMMONIUM HYDROGENDIFLUORIDE, SOLID	1727	8					
AMMONIUM HYDROGENDIFLUORIDE SOLUTION	2817	8					
AMMONIUM HYDROGEN SULPHATE	2506	8					
Ammonium hydrosulphide solution (treat as ammonium sulphide solution), see	2683	8		AMMONIUM NITRATE GEL, intermediate for blasting explosives, liquid	3375	5.1	
AMMONIUM METAVANADATE	2859	6.1		AMMONIUM NITRATE GEL, intermediate for blasting explosives, solid	3375	5.1	
AMMONIUM NITRATE with more than 0.2% combustible substances, including any organic substance calculated as carbon, to the exclusion of any other added substance	0222	1		AMMONIUM NITRATE, LIQUID hot concentrated solution, in a concentration of more than 80% but not more than 93%	2426	5.1	
				AMMONIUM NITRATE SUSPENSION, intermediate for blasting explosives, liquid	3375	5.1	
AMMONIUM NITRATE with not more than 0.2% total combustible material, including any organic substance calculated as carbon, to the exclusion of any other added substance	1942	5.1		AMMONIUM NITRATE SUSPENSION, intermediate for blasting explosives, solid	3375	5.1	
				AMMONIUM PERCHLORATE	0402	1	
					1442	5.1	

Name and description	UN No.	Class	Remarks	Name and description	UN No.	Class	Remarks
Ammonium permanganate, see	1482	5.1		Ammunition, incendiary (water-activated contrivances) with burster, expelling charge or propelling charge, see	0248 0249	1 1	
AMMONIUM PERSULPHATE	1444	5.1					
AMMONIUM PICRATE dry or wetted with less than 10% water, by mass	0004	1		AMMUNITION, INCENDIARY, WHITE PHOSPHORUS with burster, expelling charge or propelling charge	0243 0244	1 1	
AMMONIUM PICRATE, WETTED with not less than 10% water, by mass	1310	4.1		Ammunition, industrial, see	0275 0276 0277 0278 0323 0381	1 1 1 1 1 1	
AMMONIUM POLYSULPHIDE SOLUTION	2818	8					
AMMONIUM POLYVANADATE	2861	6.1		Ammunition, lachrymatory, see	0018 0019 0301 2017	1 1 1 1	
Ammonium silicofluoride, see	2854	6.1					
AMMONIUM SULPHIDE SOLUTION	2683	8		AMMUNITION, PRACTICE	0362 0488	1 1	
Ammunition, blank, see	0014 0326 0327 0338 0413	1 1 1 1 1		AMMUNITION, PROOF	0363	1	
				AMMUNITION, SMOKE with or without burster, expelling charge or propelling charge	0015 0016 0303	1 1 1	
Ammunition, fixed	0005	1					
Ammunition, semi-fixed	0006	1		Ammunition, smoke (water-activated contrivances), white phosphorus with burster, expelling charge or propelling charge, see	0248	1	
Ammunition, separate loading, see	0007 0321 0348 0412	1 1 1 1					
AMMUNITION, ILLUMINATING with or without burster, expelling charge or propelling charge	0171 0254 0297	1 1 1		Ammunition, smoke (water-activated contrivances), without white phosphorus or phosphides with burster, expelling charge or propelling charge, see	0249	1	
AMMUNITION, INCENDIARY, liquid or gel, with burster, expelling charge or propelling charge	0247	1					
				AMMUNITION, SMOKE, WHITE PHOSPHORUS with burster, expelling charge or propelling charge	0245 0246	1 1	
AMMUNITION, INCENDIARY with or without burster, expelling charge or propelling charge	0009 0010 0300	1 1 1					

Name and description	UN No.	Class	Remarks	Name and description	UN No.	Class	Remarks
Ammunition, sporting, see	0012	1		AMYL MERCAPTAN	1111	3	
	0328	1		n-AMYL METHYL KETONE	1110	3	
	0339	1					
	0417	1					
AMMUNITION, TEAR-PRODUCING, NON-EXPLOSIVE without burster or expelling charge, non-fuzed	2017	6.1		AMYL NITRATE	1112	3	
				AMYL NITRITE	1113	3	
				AMYLTRICHLOROSILANE	1728	8	
AMMUNITION, TEAR-PRODUCING with burster, expelling charge or propelling charge	0018	1		Anaesthetic ether, see	1155	3	
	0019	1					
	0301	1		ANILINE	1547	6.1	
AMMUNITION, TOXIC with burster, expelling charge or propelling charge	0020	1	Carriage prohibited	Aniline chloride, see	1548	6.1	
				ANILINE HYDROCHLORIDE	1548	6.1	
AMMUNITION, TOXIC with burster, expelling charge or propelling charge	0021	1	Carriage prohibited	Aniline oil, see	1547	6.1	
				Aniline salt, see	1548	6.1	
Ammunition, toxic (water-activated contrivances) with burster, expelling charge or propelling charge, see	0248	1		ANISIDINES	2431	6.1	
	0249	1		ANISOLE	2222	3	
AMMUNITION, TOXIC, NON-EXPLOSIVE without burster or expelling charge, non-fuzed	2016	6.1		ANISOYL CHLORIDE	1729	8	
				Anthophyllite, see	2590	9	
				Antimonous chloride, see	1733	8	
Amosite, see	2212	9		ANTIMONY COMPOUND, INORGANIC, LIQUID, N.O.S.	3141	6.1	
AMYL ACETATES	1104	3					
AMYL ACID PHOSPHATE	2819	8		ANTIMONY COMPOUND, INORGANIC, SOLID, N.O.S.	1549	6.1	
Amyl aldehyde, see	2058	3					
AMYLAMINE	1106	3		Antimony hydride, see	2676	2	
n-Amylamine, see	1106	3		ANTIMONY LACTATE	1550	6.1	
AMYL BUTYRATES	2620	3		Antimony (III) lactate, see	1550	6.1	
AMYL CHLORIDE	1107	3		ANTIMONY PENTACHLORIDE, LIQUID	1730	8	
n-AMYLENE, see	1108	3					
AMYL FORMATES	1109	3		ANTIMONY PENTACHLORIDE SOLUTION	1731	8	

Name and description	UN No.	Class	Remarks	Name and description	UN No.	Class	Remarks
ANTIMONY PENTAFLUORIDE	1732	8		Arsenic chloride, see	1560	6.1	
Antimony perchloride, liquid, see	1730	8		ARSENIC COMPOUND, LIQUID, N.O.S., inorganic, including: Arsenates, n.o.s., Arsenites, n.o.s.; and Arsenic sulphides, n.o.s.	1556	6.1	
ANTIMONY POTASSIUM TARTRATE	1551	6.1		ARSENIC COMPOUND, SOLID, N.O.S., inorganic, including: Arsenates, n.o.s.; Arsenites, n.o.s.; and Arsenic sulphides, n.o.s.	1557	6.1	
ANTIMONY POWDER	2871	6.1					
ANTIMONY TRICHLORIDE	1733	8					
A.n.t.u., see	1651	6.1		Arsenic (III) oxide, see	1561	6.1	
ARGON, COMPRESSED	1006	2		Arsenic (V) oxide, see	1559	6.1	
ARGON, REFRIGERATED LIQUID	1951	2		ARSENIC PENTOXIDE	1559	6.1	
Arsenates, n.o.s., see	1556	6.1		Arsenic sulphides, see	1556	6.1	
	1557	6.1			1557	6.1	
ARSENIC	1558	6.1		ARSENIC TRICHLORIDE	1560	6.1	
ARSENIC ACID, LIQUID	1553	6.1		ARSENIC TRIOXIDE	1561	6.1	
ARSENIC ACID, SOLID	1554	6.1		Arsenious chloride, see	1560	6.1	
ARSENICAL DUST	1562	6.1		Arsenites, n.o.s., see	1556	6.1	
Arsenical flue dust, see	1562	6.1			1557	6.1	
ARSENICAL PESTICIDE, LIQUID, FLAMMABLE, TOXIC, flash-point less than 23 °C	2760	3		Arsenous chloride, see	1560	6.1	
				ARSINE	2188	2	
				ARTICLES, EEI, see	0486	1	
ARSENICAL PESTICIDE, LIQUID, TOXIC	2994	6.1		ARTICLES, EXPLOSIVE, EXTREMELY INSENSITIVE	0486	1	
ARSENICAL PESTICIDE, LIQUID, TOXIC, FLAMMABLE, flash-point not less than 23 °C	2993	6.1		ARTICLES, EXPLOSIVE, N.O.S.	0349	1	
					0350	1	
					0351	1	
					0352	1	
ARSENICAL PESTICIDE, SOLID, TOXIC	2759	6.1			0353	1	
					0354	1	
					0355	1	
ARSENIC BROMIDE	1555	6.1			0356	1	
					0462	1	
Arsenic (III) bromide, see	1555	6.1			0463	1	
					0464	1	
					0465	1	

Name and description	UN No.	Class	Remarks	Name and description	UN No.	Class	Remarks
	0466	1		Aviation regulated liquid, n.o.s.	3334	9	Not subject to ADN
	0467	1					
	0468	1					
	0469	1		Aviation regulated solid, n.o.s.	3335	9	Not subject to ADN
	0470	1					
	0471	1					
	0472	1					
ARTICLES, PRESSURIZED, HYDRAULIC (containing non-flammable gas)	3164	2		AZODICARBONAMIDE	3242	4.1	
				Bag charges, see	0242	1	
					0279	1	
					0414	1	
ARTICLES, PRESSURIZED, PNEUMATIC (containing non-flammable gas)	3164	2		Ballistite, see	0160	1	
					0161	1	
ARTICLES, PYROPHORIC	0380	1		Bangalore torpedoes, see	0136	1	
					0137	1	
ARTICLES, PYROTECHNIC for technical purposes	0428	1			0138	1	
	0429	1			0294	1	
	0430	1					
	0431	1		BARIUM	1400	4.3	
	0432	1					
ARYLSULPHONIC ACIDS, LIQUID with more than 5% free sulphuric acid	2584	8		BARIUM ALLOYS, PYROPHORIC	1854	4.2	
				BARIUM AZIDE, dry or wetted with less than 50% water, by mass	0224	1	
ARYLSULPHONIC ACIDS, LIQUID with not more than 5% free sulphuric acid	2586	8					
				BARIUM AZIDE, WETTED with not less than 50% water, by mass	1571	4.1	
ARYLSULPHONIC ACIDS, SOLID with more than 5% free sulphuric acid	2583	8					
				Barium binoxide, see	1449	5.1	
ARYLSULPHONIC ACIDS, SOLID with not more than 5% free sulphuric acid	2585	8		BARIUM BROMATE	2719	5.1	
				BARIUM CHLORATE, SOLID	1445	5.1	
Asbestos, blue or brown, see	2212	9					
Asbestos, white, see	2590	9		BARIUM CHLORATE, SOLUTION	3405	5.1	
Asphalt, with a flash-point not greater than 60 °C, see	1999	3		BARIUM COMPOUND, N.O.S.	1564	6.1	
Asphalt, with flash-point above 60 °C, at or above its flash-point, see	3256	9		BARIUM CYANIDE	1565	6.1	
Asphalt, at or above 100 °C and below its flash-point, see	3257	9		Barium dioxide, see	1449	5.1	

Name and description	UN No.	Class	Remarks	Name and description	UN No.	Class	Remarks
BARIUM HYPOCHLORITE with more than 22% available chlorine	2741	5.1					
				BENZENESULPHONYL CHLORIDE	2225	8	
BARIUM NITRATE	1446	5.1		Benzenethiol, see	2337	6.1	
BARIUM OXIDE	1884	6.1		BENZIDINE	1885	6.1	
BARIUM PERCHLORATE, SOLID	1447	5.1		Benzol, see	1114	3	
				Benzolene, see	1268	3	
BARIUM PERCHLORATE, SOLUTION	3406	5.1		BENZONITRILE	2224	6.1	
BARIUM PERMANGANATE	1448	5.1		BENZOQUINONE	2587	6.1	
				Benzosulphochloride, see	2225	8	
BARIUM PEROXIDE	1449	5.1		BENZOTRICHLORIDE	2226	8	
Barium selenate, see	2630	6.1		BENZOTRIFLUORIDE	2338	3	
Barium selenite, see	2630	6.1		BENZOYL CHLORIDE	1736	8	
Barium superoxide, see	1449	5.1		BENZYL BROMIDE	1737	6.1	
BATTERIES, CONTAINING SODIUM	3292	4.3		BENZYL CHLORIDE	1738	6.1	
BATTERIES, DRY, CONTAINING POTASSIUM HYDROXIDE SOLID, electric storage	3028	8		Benzyl chlorocarbonate, see	1739	8	
				BENZYL CHLOROFORMATE	1739	8	
BATTERIES, WET, FILLED WITH ACID, electric storage	2794	8		Benzyl cyanide, see	2470	6.1	
BATTERIES, WET, FILLED WITH ALKALI, electric storage	2795	8		BENZYLDIMETHYL-AMINE	2619	8	
BATTERIES, WET, NON-SPILLABLE, electric storage	2800	8		BENZYLIDENE CHLORIDE	1886	6.1	
				BENZYL IODIDE	2653	6.1	
BATTERY FLUID, ACID	2796	8		BERYLLIUM COMPOUND, N.O.S.	1566	6.1	
BATTERY FLUID, ALKALI	2797	8		BERYLLIUM NITRATE	2464	5.1	
Battery-powered vehicle or Battery-powered equipment	3171	9	Not subject to ADN	BERYLLIUM POWDER	1567	6.1	
				Bhusa	1327	4.1	Not subject to ADN
BENZALDEHYDE	1990	9					
BENZENE	1114	3					

Name and description	UN No.	Class	Remarks	Name and description	UN No.	Class	Remarks
BICYCLO[2.2.1]HEPTA-2,5-DIENE, STABILIZED	2251	3		Blasting cap assemblies, see	0360 0361	1 1	
Bifluorides, n.o.s., see	1740	8		Blasting caps, electric, see	0030 0255 0456	1 1 1	
BIOLOGICAL SUBSTANCE, CATEGORY B	3373	6.2					
				Bleaching powder, see	2208	5.1	
(BIO) MEDICAL WASTE, N.O.S.	3291	6.2		BLUE ASBESTOS (crocidolite)	2212	9	
BIPYRIDILIUM PESTICIDE, LIQUID, FLAMMABLE, TOXIC, flash-point less than 23 °C	2782	3		BOMBS with bursting charge	0033 0034 0035 0291	1 1 1 1	
BIPYRIDILIUM PESTICIDE, LIQUID, TOXIC	3016	6.1		Bombs, illuminating, see	0254	1	
BIPYRIDILIUM PESTICIDE, LIQUID, TOXIC, FLAMMABLE, flash-point not less than 23 °C	3015	6.1		BOMBS, PHOTO-FLASH	0037 0038 0039 0299	1 1 1 1	
BIPYRIDILIUM PESTICIDE, SOLID, TOXIC	2781	6.1		BOMBS, SMOKE, NON-EXPLOSIVE with corrosive liquid, without initiating device	2028	8	
BISULPHATES, AQUEOUS SOLUTION	2837	8					
BISULPHITES, AQUEOUS SOLUTION, N.O.S.	2693	8		Bombs, target identification, see	0171 0254 0297	1 1 1	
Bitumen, with a flash-point not greater than 60 °C,see	1999	3		BOMBS WITH FLAMMABLE LIQUID with bursting charge	0399 0400	1 1	
Bitumen, with a flash-point not greater than 60 °C, see	1999	3					
Bitumen, with flash-point above 60 °C, at or above its flash-point, see	3256	9		BOOSTERS WITH DETONATOR	0225 0268	1 1	
Bitumen, at or above 100 °C and below its flash-point, see	3257	9		BOOSTERS without detonator	0042 0283	1 1	
BLACK POWDER, COMPRESSED	0028	1		Borate and chlorate mixture, see	1458	5.1	
BLACK POWDER, granular or as a meal	0027	1		BORNEOL	1312	4.1	
				BORON TRIBROMIDE	2692	8	
BLACK POWDER, IN PELLETS	0028	1		BORON TRICHLORIDE	1741	2	

Name and description	UN No.	Class	Remarks	Name and description	UN No.	Class	Remarks
BORON TRIFLUORIDE ACETIC ACID COMPLEX, LIQUID	1742	8		omega-Bromoacetone, see	2645	6.4	
BORON TRIFLUORIDE ACETIC ACID COMPLEX, SOLID	3419	8		BROMOACETYL BROMIDE	2513	8	
BORON TRIFLUORIDE	1008	2		BROMOBENZENE	2514	3	
BORON TRIFLUORIDE DIETHYL ETHERATE	2604	8		BROMOBENZYL CYANIDES, LIQUID	1694	6.1	
BORON TRIFLUORIDE DIHYDRATE	2851	8		BROMOBENZYL CYANIDES, SOLID	3449	6.1	
BORON TRIFLUORIDE DIMETHYL ETHERATE	2965	4.3		1-BROMOBUTANE	1126	3	
				2-BROMOBUTANE	2339	3	
BORON TRIFLUORIDE PROPIONIC ACID COMPLEX, LIQUID	1743	8		BROMOCHLORO-METHANE	1887	6.1	
BORON TRIFLUORIDE PROPIONIC ACID COMPLEX, SOLID	3420	8		1-BROMO-3-CHLOROPROPANE	2688	6.1	
BROMATES, INORGANIC, N.O.S.	1450	5.1		1-Bromo-2,3-epoxypropane, see	2558	6.1	
				Bromoethane, see	1891	6.1	
BROMATES, INORGANIC, AQUEOUS SOLUTION, N.O.S	3213	5.1		2-BROMOETHYL ETHYL ETHER	2340	3	
BROMINE	1744	8		BROMOFORM	2515	6.1	
				Bromomethane, see	1062	2	
BROMINE CHLORIDE	2901	2		1-BROMO-3-METHYLBUTANE	2341	3	
BROMINE PENTAFLUORIDE	1745	5.1		BROMOMETHYL-PROPANES	2342	3	
BROMINE SOLUTION	1744	8		2-BROMO-2-NITROPROPANE-1,3-DIOL	3241	4.1	
BROMINE TRIFLUORIDE	1746	5.1		2-BROMOPENTANE	2343	3	
BROMOACETIC ACID, SOLID	3425	8		BROMOPROPANES	2344	3	
BROMOACETIC ACID, SOLUTION	1938	8		3-BROMOPROPYNE	2345	3	
BROMOACETONE	1569	6.1		BROMOTRIFLUORO-ETHYLENE	2419	2	

Name and description	UN No.	Class	Remarks	Name and description	UN No.	Class	Remarks
BROMOTRIFLUORO-METHANE	1009	2		BUTYL ACETATES	1123	3	
				Butyl acetate, secondary, see	1123	3	
BROWN ASBESTOS (amosite, mysorite)	2212	9		BUTYL ACID PHOSPHATE	1718	8	
BRUCINE	1570	6.1		BUTYL ACRYLATES, STABILIZED	2348	3	
BURSTERS, explosive	0043	1		n-Butyl alcohol, see	1120	3	
BUTADIENES AND HYDROCARBON MIXTURE, STABILIZED, having a vapour pressure at 70 °C not exceeding 1.1 MPa (11 bar) and a density at 50 °C not lower than 0.525 kg/l	1010	2		Butyl alcohols, see	1120	3	
				n-BUTYLAMINE	1125	3	
				N-BUTYLANILINE	2738	6.1	
				sec-Butyl benzene, see	2709	3	
BUTADIENES, STABILIZED, having a vapour pressure at 70 °C not exceeding 1.1 MPa (11 bar) and a density at 50 °C not lower than 0.525 kg/l	1010	2		BUTYLBENZENES	2709	3	
				n-Butyl bromide, see	1126	3	
				n-Butyl chloride, see	1127	3	
BUTANE	1011	2		n-BUTYL CHLOROFORMATE	2743	6.1	
BUTANEDIONE	2346	3		tert-BUTYLCYCLOHEXYL CHLOROFORMATE	2747	6.1	
Butane-1-thiol, see	2347	3					
BUTANOLS	1120	3		BUTYLENES MIXTURE or 1-BUTYLENE or CIS-2-BUTYLENE or TRANS-2-BUTYLENE	1012	2	
1-Butanol, see	1120	3					
Butan-2-ol, see	1120	3					
Butanol, secondary, see	1120	3		1,2-BUTYLENE OXIDE, STABILIZED	3022	3	
Butanol, tertiary, see	1120	3		Butyl ethers, see	1149	3	
Butanone, see	1193	3		Butyl ethyl ether, see	1179	3	
2-Butenal, see	1143	6.1		n-BUTYL FORMATE	1128	3	
Butene, see	1012	2		tert-BUTYL HYPOCHLORITE	3255	4.2	Carriage prohibited
Bute-1-ene-3-one, see	1251	3					
1,2-Buteneoxide, see	3022	3		N,n-BUTYLIMIDAZOLE	2690	6.1	
2-Buten-1-ol, see	2614	3		N,n-Butyliminazole, see	2690	6.1	
				n-BUTYL ISOCYANATE	2485	6.1	

Name and description	UN No.	Class	Remarks	Name and description	UN No.	Class	Remarks
tert-BUTYL ISOCYANATE	2484	6.1		Butyroyl chloride, see	2353	3	
Butyl lithium, see	3394	4.2		BUTYRYL CHLORIDE	2353	3	
BUTYL MERCAPTAN	2347	3		Cable cutters, explosive, see	0070	1	
n-BUTYL METHACRYLATE, STABILIZED	2227	3		CACODYLIC ACID	1572	6.1	
				CADMIUM COMPOUND	2570	6.1	
BUTYL METHYL ETHER	2350	3		CAESIUM	1407	4.3	
BUTYL NITRITES	2351	3		CAESIUM HYDROXIDE	2682	8	
Butylphenols, liquid, see	3145	8		CAESIUM HYDROXIDE SOLUTION	2681	8	
Butylphenols, solid, see	2430	8		CAESIUM NITRATE	1451	5.1	
BUTYL PROPIONATES	1914	3		Caffeine, see	1544	6.1	
p-tert-Butyltoluene, see	2667	6.1		Cajeputene, see	2052	3	
BUTYLTOLUENES	2667	6.1		CALCIUM	1401	4.3	
BUTYLTRICHLORO-SILANE	1747	8		CALCIUM ALLOYS, PYROPHORIC	1855	4.2	
5-tert-BUTYL-2,4,6-TRINITRO-m-XYLENE	2956	4.1		CALCIUM ARSENATE	1573	6.1	
BUTYL VINYL ETHER, STABILIZED	2352	3		CALCIUM ARSENATE AND CALCIUM ARSENITE MIXTURE, SOLID	1574	6.1	
But-1-yne, see	2452	2		Calcium bisulphite solution, see	2693	8	
1,4-BUTYNEDIOL	2716	6.1		CALCIUM CARBIDE	1402	4.3	
2-Butyne-1,4-diol, see	2716	6.1		CALCIUM CHLORATE	1452	5.1	
BUTYRALDEHYDE	1129	3		CALCIUM CHLORATE, AQUEOUS SOLUTION	2429	5.1	
n-Butyraldehyde, see	1129	3					
BUTYRALDOXIME	2840	3		CALCIUM CHLORITE	1453	5.1	
BUTYRIC ACID	2820	8		CALCIUM CYANAMIDE with more than 0.1% calcium carbide	1403	4.3	
BUTYRIC ANHYDRIDE	2739	8					
Butyrone, see	2710	3		CALCIUM CYANIDE	1575	6.1	
BUTYRONITRILE	2411	3		CALCIUM DITHIONITE	1923	4.2	

Name and description	UN No.	Class	Remarks	Name and description	UN No.	Class	Remarks
CALCIUM HYDRIDE	1404	4.3		CALCIUM RESINATE, FUSED	1314	4.1	
CALCIUM HYDROSULPHITE, see	1923	4.2		Calcium selenate, see	2630	6.1	
CALCIUM HYPOCHLORITE, DRY	1748	5.1		CALCIUM SILICIDE	1405	4.3	
				Calcium silicon, see	1405	4.3	
CALCIUM HYPOCHLORITE, HYDRATED with not less than 5.5% but not more than 16% water	2880	5.1		Calcium superoxide, see	1457	5.1	
				Camphanone, see	2717	4.1	
				CAMPHOR OIL	1130	3	
CALCIUM HYPOCHLORITE, HYDRATED MIXTURE with not less than 5.5% but not more than 16% water	2880	5.1		CAMPHOR, synthetic	2717	4.1	
				CAPROIC ACID	2829	8	
CALCIUM HYPOCHLORITE MIXTURE, DRY with more than 10% but not more than 39% available chlorine	2208	5.1		CARBAMATE PESTICIDE, LIQUID, FLAMMABLE, TOXIC, flash-point less than 23 °C	2758	3	
				CARBAMATE PESTICIDE, LIQUID, TOXIC	2992	6.1	
CALCIUM HYPOCHLORITE MIXTURE, DRY with more than 39% available chlorine (8.8% available oxygen)	1748	5.1		CARBAMATE PESTICIDE, LIQUID, TOXIC, FLAMMABLE, flash-point not less than 23 °C	2991	6.1	
CALCIUM MANGANESE SILICON	2844	4.3		CARBAMATE PESTICIDE, SOLID, TOXIC	2757	6.1	
CALCIUM NITRATE	1454	5.1		Carbolic acid, see	1671	6.1	
					2312	6.1	
Calcium oxide	1910	8	Not subject to ADN		2821	6.1	
				CARBON, animal or vegetable origin	1361	4.2	
CALCIUM PERCHLORATE	1455	5.1		CARBON, ACTIVATED	1362	4.2	
CALCIUM PERMANGANATE	1456	5.1		Carbon bisulphide, see	1131	3	
CALCIUM PEROXIDE	1457	5.1		Carbon black (animal or vegetable origin), see	1361	4.2	
CALCIUM PHOSPHIDE	1360	4.3					
CALCIUM, PYROPHORIC	1855	4.2		CARBON DIOXIDE	1013	2	
CALCIUM RESINATE	1313	4.1					

Name and description	UN No.	Class	Remarks	Name and description	UN No.	Class	Remarks
Carbon dioxide and ethylene oxide mixture, see	1041	2		CARTRIDGES FOR WEAPONS, BLANK	0014	1	
	1952	2			0326	1	
	3300	2			0327	1	
					0338	1	
CARBON DIOXIDE, REFRIGERATED LIQUID	2187	2			0413	1	
Carbon dioxide, solid	1845	9	Not subject to ADN	CARTRIDGES FOR WEAPONS, INERT PROJECTILE	0012	1	
					0328	1	
					0339	1	
					0417	1	
CARBON DISULPHIDE	1131	3		Cartridges, illuminating, see	0171	1	
					0254	1	
Carbonic anhydride, see	1013	2			0297	1	
	1845	9					
	2187	2		CARTRIDGES, OIL WELL	0277	1	
					0278	1	
CARBON MONOXIDE, COMPRESSED	1016	2		CARTRIDGES, POWER DEVICE	0275	1	
					0276	1	
Carbon oxysulphide, see	2204	2.3			0323	1	
					0381	1	
Carbon sulphide, see	1131	3		CARTRIDGES, SIGNAL	0054	1	
					0312	1	
CARBON TETRABROMIDE	2516	6.1			0405	1	
CARBON TETRACHLORIDE	1846	6.1		CARTRIDGES, SMALL ARMS	0012	1	
					0339	1	
					0417	1	
Carbonyl chloride, see	1076	2		CARTRIDGES, SMALL ARMS, BLANK	0014	1	
CARBONYL FLUORIDE	2417	2			0327	1	
					0338	1	
CARBONYL SULPHIDE	2204	2		Cartridges, starter, jet engine, see	0275	1	
Cartridge cases, empty, primed, see	0055	1			0276	1	
	0379	1			0323	1	
					0381	1	
Cartridges, actuating, for fire extinguisher or apparatus valve, see	0275	1					
	0276	1		CASES, CARTRIDGE, EMPTY, WITH PRIMER	0055	1	
	0323	1			0379	1	
	0381	1					
Cartridges, explosive, see	0048	1		CASES, COMBUSTIBLE, EMPTY, WITHOUT PRIMER	0446	1	
					0447	1	
CARTRIDGES, FLASH	0049	1					
	0050	1		Casinghead gasoline, see	1203	3	
CARTRIDGES FOR WEAPONS with bursting charge	0005	1		CASTOR BEANS	2969	9	
	0006	1					
	0007	1		CASTOR FLAKE	2969	9	
	0321	1					
	0348	1					
	0412	1					

Name and description	UN No.	Class	Remarks	Name and description	UN No.	Class	Remarks
CASTOR MEAL	2969	9		CHARGES, PROPELLING	0271	1	
					0272	1	
CASTOR POMACE	2969	9			0415	1	
					0491	1	
CAUSTIC ALKALI LIQUID, N.O.S.	1719	8		CHARGES, PROPELLING, FOR CANNON	0242	1	
					0279	1	
Caustic potash, see	1814	8			0414	1	
Caustic soda, see	1824	8		CHARGES, SHAPED, FLEXIBLE, LINEAR	0237	1	
					0288	1	
Caustic soda liquor, see	1824	8		CHARGES, SHAPED, without detonator	0059	1	
CELLS, CONTAINING SODIUM	3292	4.3			0439	1	
					0440	1	
					0441	1	
CELLULOID in block, rods, rolls, sheets, tubes, etc., except scrap	2000	4.1		CHARGES, SUPPLEMENTARY, EXPLOSIVE	0060	1	
CELLULOID, SCRAP	2002	4.2		CHEMICAL KIT	3316	9	
Cement, see	1133	3		CHEMICAL SAMPLE, TOXIC	3315	6.1	
CERIUM, slabs, ingots or rods	1333	4.1		Chile saltpetre, see	1498	5.1	
CERIUM, turnings or gritty powder	3078	4.3		CHLORAL, ANHYDROUS, STABILIZED	2075	6.1	
Cer mishmetall, see	1323	4.1		CHLORATE AND BORATE MIXTURE	1458	5.1	
Charcoal, activated, see	1362	4.1		CHLORATE AND MAGNESIUM CHLORIDE MIXTURE, SOLID	1459	5.1	
Charcoal, non-activated, see	1361	4.2					
CHARGES, BURSTING, PLASTICS BONDED	0457	1		CHLORATE AND MAGNESIUM CHLORIDE MIXTURE, SOLUTION	3407	5.1	
	0458	1					
	0459	1					
	0460	1					
CHARGES, DEMOLITION	0048	1		CHLORATES, INORGANIC, N.O.S.	1461	5.1	
CHARGES, DEPTH	0056	1		CHLORATES, INORGANIC, AQUEOUS SOLUTION, N.O.S.	3210	5.1	
Charges, expelling, explosive, for fire extinguishers, see	0275	1					
	0276	1					
	0323	1					
	0381	1		CHLORIC ACID, AQUEOUS SOLUTION with not more than 10% chloric acid	2626	5.1	
CHARGES, EXPLOSIVE, COMMERCIAL without detonator	0442	1					
	0443	1					
	0444	1					
	0445	1					

Name and description	UN No.	Class	Remarks	Name and description	UN No.	Class	Remarks
CHLORINE	1017	2		CHLOROBENZYL CHLORIDES, SOLID	3427	6.1	
CHLORINE PENTAFLUORIDE	2548	2		1-Chloro-3-bromopropane, see	2688	6.1	
CHLORINE TRIFLUORIDE	1749	2		1-Chlorobutane, see	1127	3	
CHLORITES, INORGANIC, N.O.S.	1462	5.1		2-Chlorobutane, see	1127	3	
				CHLOROBUTANES	1127	3	
CHLORITE SOLUTION	1908	8		CHLOROCRESOLS, SOLUTION	2669	6.1	
Chloroacetaldehyde, see	2232	6.1					
CHLOROACETIC ACID, MOLTEN	3250	6.1		CHLOROCRESOLS, SOLID	3437	6.1	
CHLOROACETIC ACID, SOLID	1751	6.1		CHLORODIFLUORO-BROMOMETHANE	1974	2	
CHLOROACETIC ACID SOLUTION	1750	6.1		1-CHLORO-1,1-DIFLUORO-ETHANE	2517	2	
CHLOROACETONE, STABILIZED	1695	6.1		CHLORODIFLUORO-METHANE	1018	2	
CHLOROACETONITRILE	2668	6.1		CHLORODIFLUORO-METHANE AND CHLORO-PENTAFLUOROETHANE MIXTURE with fixed boiling point, with approximately 49% chlorodifluoromethane	1973	2	
CHLOROACETOPHENONE, LIQUID	3416	6.1					
CHLOROACETOPHENONE, SOLID	1697	6.1		3-Chloro-1,2-dihydroxypropane, see	2689	6.1	
CHLOROACETYL CHLORIDE	1752	6.1		Chlorodimethyl ether, see	1239	6.1	
CHLOROANILINES, LIQUID	2019	6.1		1Chloro-2,2-dimethylpropane, see	1107	3	
CHLOROANILINES, SOLID	2018	6.1		CHLORODINITRO-BENZENES, LIQUID	1577	6.1	
CHLOROANISIDINES	2233	6.1		CHLORODINITRO-BENZENES, SOLID	3441	6.1	
CHLOROBENZENE	1134	3					
CHLOROBENZOTRIFLUO-RIDES	2234	3		2-CHLOROETHANAL	2232	6.1	
				Chloroethane, see	1037	2	
CHLOROBENZYL CHLORIDES, LIQUID	2235	6.1		Chloroethane nitrile, see	2668	6.1	
				2-Chloroethanol, see	1135	6.1	

Name and description	UN No.	Class	Remarks	Name and description	UN No.	Class	Remarks
CHLOROFORM	1888	6.1		CHLORONITROBENZENES SOLID	1578	6.1	
CHLOROFORMATES, TOXIC, CORROSIVE, N.O.S.	3277	6.1		CHLORONITRO-TOLUENES, LIQUID	2433	6.1	
CHLOROFORMATES, TOXIC, CORROSIVE, FLAMMABLE, N.O.S.	2742	6.1		CHLORONITRO-TOLUENES, SOLID	3457	6.1	
Chloromethane, see	1063	2		CHLOROPENTAFLUORO-ETHANE	1020	2	
1-Chloro-3-methylbutane, see	1107	3		1-Chloropentane	1107	3	
2-Chloro-2-methylbutane, see	1107	3		CHLOROPHENOLATES, LIQUID	2904	8	
CHLOROMETHYL CHLOROFORMATE	2745	6.1		CHLOROPHENOLATES, SOLID	2905	8	
Chloromethyl cyanide, see	2668	6.1		CHLOROPHENOLS, LIQUID	2021	6.1	
CHLOROMETHYL ETHYL ETHER	2354	3		CHLOROPHENOLS, SOLID	2020	6.1	
1-Chloro-3-methylbutane, see	1107	3		CHLOROPHENYL-TRICHLOROSILANE	1753	8	
1-Chloro-3-methylbutane, see	1107	3		CHLOROPICRIN	1580	6.1	
Chloromethyl methyl ether, see	1239	6.1		CHLOROPICRIN AND METHYL BROMIDE MIXTURE, with more than 2% chloropicrin	1581	2	
3-CHLORO-4-METHYL-PHENYL ISOCYANATE, LIQUID	2236	6.1		CHLOROPICRIN AND METHYL CHLORIDE MIXTURE	1582	2	
3-CHLORO-4-METHYL-PHENYL ISOCYANATE, SOLID	3428	6.1		CHLOROPICRIN MIXTURE, N.O.S.	1583	6.1	
1-Chloro-2-methylpropane, see	1127	3		CHLOROPLATINIC ACID, SOLID	2507	8	
2-Chloro-2-methylpropane, see	1127	3		CHLOROPRENE, STABILIZED	1991	3	
3-Chloro-2-methylprop-1-ene, see	2554	3		1-CHLOROPROPANE	1278	3	
CHLORONITROANILINES	2237	6.1		2-CHLOROPROPANE	2356	3	
CHLORONITROBENZENES LIQUID	3409	6.1		3-Chloro-propanediol-1,2, see	2689	6.1	

Name and description	UN No.	Class	Remarks	Name and description	UN No.	Class	Remarks
3-CHLOROPROPANOL-1	2849	6.1		CHLOROTOLUIDINES LIQUID	3429	6.1	
2-CHLOROPROPENE	2456	3		CHLOROTOLUIDINES SOLID	2239	6.1	
3-Chloropropene, see	1100	3					
3-Chloroprop-1-ene, see	1100	3		1-CHLORO-2,2,2-TRIFLUOROETHANE	1983	2	
2-CHLOROPROPIONIC ACID	2511	8		Chlorotrifluoroethylene, see	1082	2	
2-CHLOROPYRIDINE	2822	6.1		CHLOROTRIFLUORO-METHANE	1022	2	
CHLOROSILANES, CORROSIVE, N.O.S.	2987	8		CHLOROTRIFLUORO-METHANE AND TRIFLUOROMETHANE AZEOTROPIC MIXTURE with approximately 60% chlorotrifluoromethane	2599	2	
CHLOROSILANES, CORROSIVE, FLAMMABLE, N.O.S.	2986	8					
CHLOROSILANES, FLAMMABLE, CORROSIVE, N.O.S.	2985	3		Chromic acid, solid, see	1463	5.1	
				CHROMIC ACID SOLUTION	1755	8	
CHLOROSILANES, TOXIC, CORROSIVE, N.O.S.	3361	6.1		Chromic anhydride, solid, see	1463	5.1	
CHLOROSILANES, TOXIC, CORROSIVE, FLAMMABLE, N.O.S.	3362	6.1		CHROMIC FLUORIDE, SOLID	1756	8	
CHLOROSILANES, WATER-REACTIVE, FLAMMABLE, CORROSIVE, N.O.S.	2988	4.3		CHROMIC FLUORIDE SOLUTION	1757	8	
				Chromic nitrate, see	2720	5.1	
CHLOROSULPHONIC ACID (with or without sulphur trioxide)	1754	8		Chromium (VI) dichloride dioxide, see	1758	8	
				Chromium (III) fluoride, solid, see	1756	8	
1-CHLORO-1,2,2,2-TETRAFLUOROETHANE	1021	2		CHROMIUM NITRATE	2720	5.1	
CHLOROTOLUENES	2238	3		Chromium (III) nitrate, see	2720	5.1	
4-CHLORO-o-TOLUIDINE HYDROCHLORIDE, SOLID	1579	6.1		CHROMIUM OXYCHLORIDE	1758	8	
4-CHLORO-o-TOLUIDINE HYDROCHLORIDE, SOLUTION	3410	6.1		CHROMIUM TRIOXIDE, ANHYDROUS	1463	5.1	

Name and description	UN No.	Class	Remarks	Name and description	UN No.	Class	Remarks
CHROMOSULPHURIC ACID	2240	8		COMPRESSED GAS, N.O.S.	1956	2	
Chryosotile, see	2590	9		COMPRESSED GAS, FLAMMABLE, N.O.S.	1954	2	
Cinene, see	2052	3		COMPRESSED GAS, OXIDIZING, N.O.S.	3156	2	
Cinnamene, see	2055	3		COMPRESSED GAS, TOXIC, N.O.S.	1955	2	
Cinnamol, see	2055	3					
CLINICAL WASTE, UNSPECIFIED, N.O.S.	3291	6.2		COMPRESSED GAS, TOXIC, CORROSIVE, N.O.S.	3304	2	
COAL GAS, COMPRESSED	1023	2		COMPRESSED GAS, TOXIC, FLAMMABLE, N.O.S.	1953	2	
COAL TAR DISTILLATES, FLAMMABLE	1136	3					
Coal tar naphtha, see	1268	3		COMPRESSED GAS, TOXIC, FLAMMABLE, CORROSIVE, N.O.S.	3305	2	
Coal tar oil, see	1136	3		COMPRESSED GAS, TOXIC, OXIDIZING, N.O.S.	3303	2	
COATING SOLUTION (includes surface treatments or coatings used for industrial or other purposes such as vehicle under coating, drum or barrel lining)	1139	3		COMPRESSED GAS, TOXIC, OXIDIZING, CORROSIVE, N.O.S.	3306	2	
COBALT NAPHTHENATES, POWDER	2001	4.1		CONTRIVANCES, WATER-ACTIVATED with burster, expelling charge or propelling charge	0248 0249	1 1	
COBALT RESINATE, PRECIPITATED	1318	4.1		COPPER ACETOARSENITE	1585	6.1	
Cocculus, see	3172 3462	6.1 6.1		COPPER ARSENITE	1586	6.1	
Collodion cottons, see	0340 0341 0342 2059 2555 2556 2557	1 1 1 3 4.1 4.1 4.1		Copper (II) arsenite, see	1586	6.1	
				COPPER BASED PESTICIDE, LIQUID, FLAMMABLE, TOXIC, flash-point less than 23 °C	2776	3	
COMPONENTS, EXPLOSIVE TRAIN, N.O.S.	0382 0383 0384 0461	1 1 1 1		COPPER BASED PESTICIDE, LIQUID, TOXIC	3010	6.1	
Composition B, see	0118	1					

Name and description	UN No.	Class	Remarks	Name and description	UN No.	Class	Remarks
COPPER BASED PESTICIDE, LIQUID, TOXIC, FLAMMABLE, flash-point not less than 23 °C	3009	6.1		CORROSIVE LIQUID, FLAMMABLE, N.O.S.	2920	8	
COPPER BASED PESTICIDE, SOLID, TOXIC	2775	6.1		CORROSIVE LIQUID, OXIDIZING, N.O.S.	3093	8	
COPPER CHLORATE	2721	5.1		CORROSIVE LIQUID, SELF-HEATING, N.O.S.	3301	8	
Copper (II) chlorate, see	2721	5.1		CORROSIVE LIQUID, TOXIC, N.O.S.	2922	8	
COPPER CHLORIDE	2802	8		CORROSIVE LIQUID, WATER-REACTIVE, N.O.S.	3094	8	
COPPER CYANIDE	1587	6.1					
Copper selenate, see	2630	6.1		CORROSIVE SOLID, N.O.S.	1759	8	
Copper selenite, see	2630	6.1		CORROSIVE SOLID, ACIDIC, INORGANIC, N.O.S.	3260	8	
COPRA	1363	4.2					
CORD, DETONATING, flexible	0065	1		CORROSIVE SOLID, ACIDIC, ORGANIC, N.O.S.	3261	8	
	0289	1					
CORD, DETONATING, metal clad	0102	1		CORROSIVE SOLID, BASIC, INORGANIC, N.O.S.	3262	8	
	0290	1					
CORD, DETONATING, MILD EFFECT, metal clad	0104	1		CORROSIVE SOLID, BASIC, ORGANIC, N.O.S.	3263	8	
CORD, IGNITER	0066	1		CORROSIVE SOLID, FLAMMABLE, N.O.S.	2921	8	
Cordite, see	0160	1		CORROSIVE SOLID, OXIDIZING, N.O.S.	3084	8	
	0161	1					
CORROSIVE LIQUID, N.O.S.	1760	8		CORROSIVE SOLID, SELF-HEATING, N.O.S.	3095	8	
CORROSIVE LIQUID, ACIDIC, INORGANIC, N.O.S.	3264	8		CORROSIVE SOLID, TOXIC, N.O.S.	2923	8	
CORROSIVE LIQUID, ACIDIC, ORGANIC, N.O.S.	3265	8		CORROSIVE SOLID, WATER-REACTIVE, N.O.S.	3096	8	
				COTTON WASTE, OILY	1364	4.2	
CORROSIVE LIQUID, BASIC, INORGANIC, N.O.S.	3266	8		COTTON, WET	1365	4.2	
CORROSIVE LIQUID, BASIC, ORGANIC, N.O.S.	3267	8		COUMARIN DERIVATIVE PESTICIDE, LIQUID, FLAMMABLE, TOXIC, flash-point less than 23 °C	3024	3	

Name and description	UN No.	Class	Remarks	Name and description	UN No.	Class	Remarks
COUMARIN DERIVATIVE PESTICIDE, LIQUID, TOXIC	3026	6.1		Cut backs, at or above 100 °C and below its flash-point, see	3257	9	
COUMARIN DERIVATIVE PESTICIDE, LIQUID, TOXIC, FLAMMABLE, flash-point not less than 23 °C	3025	6.1		CUTTERS, CABLE, EXPLOSIVE	0070	1	
				CYANIDE SOLUTION, N.O.S.	1935	6.1	
COUMARIN DERIVATIVE PESTICIDE, SOLID, TOXIC	3027	6.1		CYANIDES, INORGANIC, SOLID, N.O.S.	1588	6.1	
Creosote, see	2810	6.1		Cyanides, organic, flammable, toxic, n.o.s., see	3273	3	
Creosote salts, see	1334	4.1		Cyanides, organic, toxic, n.o.s., see	3276	6.1	
CRESOLS, LIQUID	2076	6.1			3439	6.1	
CRESOLS, SOLID	3455	6.1		Cyanides, organic, toxic, flammable, n.o.s., see	3275	6.1	
CRESYLIC ACID	2022	6.1		Cyanoacetonitrile, see	2647	6.1	
Crocidolite, see	2212	9		CYANOGEN	1026	2	
CROTONALDEHYDE	1143	6.1		CYANOGEN BROMIDE	1889	6.1	
CROTONALDEHYDE, STABILIZED	1143	6.1		CYANOGEN CHLORIDE, STABILIZED	1589	2	
CROTONIC ACID, LIQUID	3472	8		CYANURIC CHLORIDE	2670	8	
CROTONIC ACID, SOLID	2823	8		CYCLOBUTANE	2601	2	
Crotonic aldehyde / Crotonic aldehyde, stabilized, see	1143	6.1		CYCLOBUTYL CHLOROFORMATE	2744	6.1	
CROTONYLENE	1144	3		1,5,9-CYCLODODECA-TRIENE	2518	6.1	
Crude naphtha, see	1268	3		CYCLOHEPTANE	2241	3	
Cumene, see	1918	3		CYCLOHEPTATRIENE	2603	3	
Cupric chlorate, see	2721	5.1		1,3,5-Cycloheptatriene, see	2603	3	
CUPRIETHYLENE-DIAMINE SOLUTION	1761	8		CYCLOHEPTENE	2242	3	
Cut backs, with a flash-point not greater than 60 °C, see	1999	3		1,4-Cyclohexadienedione, see	2587	6.1	
Cut backs, with flash-point above 60 °C, at or above its flash-point, see	3256	9		CYCLOHEXANE	1145	3	
				Cyclehexanethiol, see	3054	3	

Name and description	UN No.	Class	Remarks	Name and description	UN No.	Class	Remarks
CYCLOHEXANONE	1915	3		CYCLOTETRA-METHYLENE-TETRANITRAMINE, DESENSITIZED	0484	1	
CYCLOHEXENE	2256	3					
CYCLOHEXENYLTRI-CHLOROSILANE	1762	8		CYCLOTETRA-METHYLENE-TETRANITRAMINE, WETTED with not less than 15% water, by mass	0226	1	
CYCLOHEXYL ACETATE	2243	3					
CYCLOHEXYLAMINE	2357	8					
CYCLOHEXYL ISOCYANATE	2488	6.1		CYCLOTRIMETHYLENE-TRINITRAMINE AND CYCLOTETRA-METHYLENE-TETRANITRAMINE MIXTURE, DESENSITIZED with not less than 10% phlegmatiser by mass	0391	1	
CYCLOHEXYL MERCAPTAN	3054	3					
CYCLOHEXYLTRI-CHLOROSILANE	1763	8					
CYCLONITE AND CYCLOTETRA-METHYLENE-TETRANITRAMINE MIXTURE, WETTED with not less than 15% water, by mass or DESENSITIZED with not less than 10% phlegmatiser by mass, see	0391	1		CYCLOTRIMETHYLENE-TRINITRAMINE AND CYCLOTETRA-METHYLENE-TETRANITRAMINE MIXTURE, WETTED with not less than 15% water, by mass	0391	1	
CYCLONITE, DESENSITIZED, see	0483	1		CYCLOTRIMETHYLENE-TRINITRAMINE, DESENSITIZED	0483	1	
CYCLONITE, WETTED with not less than 15% water, by mass, see	0072	1		CYCLOTRIMETHYLENE-TRINITRAMINE, WETTED with not less than 15% water, by mass	0072	1	
CYCLOOCTADIENES	2520	3		CYMENES	2046	3	
CYCLOOCTADIENE PHOSPHINES, see	2940	4.2		Cymol, see	2046	3	
CYCLOOCTATETRAENE	2358	3		Deanol, see	2051	8	
CYCLOPENTANE	1146	3		Dangerous goods in machinery or dangerous goods in apparatus	3363	9	Not subject to ADN [see also 1.1.3.1 (b)]
CYCLOPENTANOL	2244	3					
CYCLOPENTANONE	2245	3					
CYCLOPENTENE	2246	3		DECABORANE	1868	4.1	
CYCLOPROPANE	1027	2		DECAHYDRO-NAPHTHALENE	1147	3	

Name and description	UN No.	Class	Remarks	Name and description	UN No.	Class	Remarks
Decalin, see	1147	3		DIALKYL-(C_{12}-C_{18})-DIMETHYL-AMMONIUM and 2-PROPANOL	3175	4.1	
n-DECANE	2247	3					
DEFLAGRATING METAL SALTS OF AROMATIC NITRODERIVATIVES, N.O.S.	0132	1		DIALLYLAMINE	2359	3	
				DIALLYL ETHER	2360	3	
				4,4'-DIAMINODIPHENYL-METHANE	2651	6.1	
Depth charge, see	0056	1					
DESENSITIZED EXPLOSIVE, LIQUID, N.O.S.	3379	3		1,2-Diaminoethane, see	1604	8	
				Diaminopropylamine, see	2269	8	
DESENSITIZED EXPLOSIVE, SOLID, N.O.S.	3380	4.1		DI-n-AMYLAMINE	2841	3	
				DIAZODINITROPHENOL, WETTED with not less than 40% water, or mixture of alcohol and water, by mass	0074	1	
Detonating relays, see	0029	1					
	0267	1					
	0360	1					
	0361	1					
	0455	1		Dibenzopyridine, see	2713	6.1	
	0500	1					
				DIBENZYLDICHLORO-SILANE	2434	8	
DETONATOR ASSEMBLIES, NON-ELECTRIC for blasting	0360	1					
	0361	1					
	0500	1		DIBORANE	1911	2	
DETONATORS FOR AMMUNITION	0073	1		1,2-DIBROMOBUTAN-3-ONE	2648	6.1	
	0364	1					
	0365	1		DIBROMOCHLORO-PROPANES	2872	6.1	
	0366	1					
DETONATORS, ELECTRIC for blasting	0030	1		1,2-Dibromo-3-chloropropane, see	2872	6.1	
	0255	1					
	0456	1					
DETONATORS, NON-ELECTRIC for blasting	0029	1		DIBROMODIFLUORO-METHANE	1941	9	
	0267	1					
	0455	1		DIBROMOMETHANE	2664	6.1	
DEUTERIUM, COMPRESSED	1957	2		DI-n-BUTYLAMINE	2248	8	
DEVICES, SMALL, HYDROCARBON GAS POWERED with release device	3150	2		DIBUTYLAMINO-ETHANOL	2873	6.1	
				2-Dibutylaminoethanol, see	2873	6.1	
DIACETONE ALCOHOL	1148	3		N,N-Di-n-butylaminoethanol, see	2873	6.1	

Name and description	UN No.	Class	Remarks	Name and description	UN No.	Class	Remarks
DIBUTYL ETHERS	1149	3		DICHLOROISOCYANURIC ACID SALTS	2465	5.1	
DICHLOROACETIC ACID	1764	8		DICHLOROISOPROPYL ETHER	2490	6.1	
1,3-DICHLOROACETONE	2649	6.1		DICHLOROMETHANE	1593	6.1	
DICHLOROACETYL CHLORIDE	1765	8		1,1-DICHLORO-1-NITROETHANE	2650	6.1	
DICHLOROANILINES, LIQUID	1590	6.1		DICHLOROPENTANES	1152	3	
DICHLOROANILINES, SOLID	3442	6.1		Dichlorophenol, see	2020 2021	6.1 6.1	
o-DICHLOROBENZENE	1591	6.1		DICHLOROPHENYL ISOCYANATES	2250	6.1	
2,2'-DICHLORODIETHYL ETHER	1916	6.1		DICHLOROPHENYLTRI-CHLOROSILANE	1766	8	
DICHLORODIFLUORO-METHANE	1028	2		1,2-DICHLOROPROPANE	1279	3	
DICHLORODIFLUORO-METHANE AND 1,1-DIFLUOROETHANE AZEOTROPIC MIXTURE with approximately 74% dichlorodifluoromethane	2602	2		1,3-DICHLORO-PROPANOL-2	2750	6.1	
				1,3-Dichloro-2-propanone, see	2649	6.1	
Dichlorodifluoromethane and ethylene oxide mixture, see	3070	2		DICHLOROPROPENES	2047	3	
				DICHLOROSILANE	2189	2	
DICHLORODIMETHYL ETHER, SYMMETRICAL	2249	6.1	Carriage prohibited	1,2-DICHLORO-1,1,2,2-TETRAFLUOROETHANE	1958	2	
1,1-DICHLOROETHANE	2362	3		Dichloro-s-triazine-2,4,6-trione, see	2465	5.1	
1,2-Dichloroethane, see	1184	3		1,4-Dicyanobutane, see	2205	6.1	
1,2-DICHLOROETHYLENE	1150	3		Dicycloheptadiene, see	2251	3	
Di(2-chloroethyl) ether, see	1916	6.1		DICYCLOHEXYLAMINE	2565	8	
DICHLOROFLUORO-METHANE	1029	2		Dicyclohexylamine nitrite, see	2687	4.1	
alpha-Dichlorohydrin, see	2750	6.1		DICYCLOHEXYL-AMMONIUM NITRITE	2687	4.1	
DICHLOROISOCYANURIC ACID, DRY	2465	5.1		DICYCLOPENTADIENE	2048	3	

Name and description	UN No.	Class	Remarks	Name and description	UN No.	Class	Remarks
1,2-DI-(DIMETHYLAMINO) ETHANE	2372	3		Di-(2-ethylhexyl) phosphoric acid, see	1902	8	
DIDYMIUM NITRATE	1465	5.1		DIETHYL KETONE	1156	3	
DIESEL FUEL	1202	3		DIETHYL SULPHATE	1594	6.1	
1,1-Diethoxyethane, see	1088	3		DIETHYL SULPHIDE	2375	3	
1,2-Diethoxyethane, see	1153	3		DIETHYLTHIO-PHOSPHORYL CHLORIDE	2751	8	
DIETHOXYMETHANE	2373	3					
3,3-DIETHOXYPROPENE	2374	3		Diethylzinc, see	3394	4.2	
DIETHYLAMINE	1154	3		2,4-Difluoroaniline, see	2941	6.1	
2-DIETHYLAMINO-ETHANOL	2686	8		Difluorochloroethane, see	2517	2	
				1,1-DIFLUOROETHANE	1030	2	
3-DIETHYL-AMINOPROPYLAMINE	2684	3		1,1-DIFLUOROETHYLENE	1959	2	
N,N-DIETHYLANILINE	2432	6.1		DIFLUOROMETHANE	3252	2	
DIETHYLBENZENE	2049	3		Difluoromethane, pentafluoroethane, and 1,1,1,2-tetrafluoroethane zeotropic mixture with approximately 10% difluoromethane and 70% pentafluoroethane, see	3339	2	
Diethylcarbinol, see	1105	3					
DIETHYL CARBONATE	2366	3					
DIETHYLDICHLORO-SILANE	1767	8		Difluoromethane, pentafluoroethane, and 1,1,1,2-tetrafluoroethane zeotropic mixture with approximately 20% difluoromethane and 40% pentafluoroethane, see	3338	2	
Diethylenediamine, see	2579	8					
DIETHYLENEGLYCOL DINITRATE, DESENSITIZED with not less than 25% non-volatile, water-insoluble phlegmatizer, by mass	0075	1					
				Difluoromethane, pentafluoroethane, and 1,1,1,2-tetrafluoroethane zeotropic mixture with approximately 23%difluoromethane and 25% pentafluoroethane, see	3340	2	
DIETHYLENETRIAMINE	2079	8					
N,N-Diethylethanolamine, see	2686	3					
DIETHYL ETHER	1155	3		DIFLUOROPHOSPHORIC ACID, ANHYDROUS	1768	8	
N,N-DIETHYLETHYLENE-DIAMINE	2685	8		2,3-DIHYDROPYRAN	2376	3	

Name and description	UN No.	Class	Remarks	Name and description	UN No.	Class	Remarks
DIISOBUTYLAMINE	2361	3		1,3-DIMETHYL-BUTYLAMINE	2379	3	
DIISOBUTYLENE, ISOMERIC COMPOUNDS	2050	3		DIMETHYLCARBAMOYL CHLORIDE	2262	8	
alpha-Diisobutylene, see	2050	3		DIMETHYL CARBONATE	1161	3	
beta-Diisobutylene, see	2050	3		DIMETHYL-CYCLOHEXANES	2263	3	
DIISOBUTYL KETONE	1157	3					
DIISOOCTYL ACID PHOSPHATE	1902	8		N,N-DIMETHYLCYCLO-HEXYLAMINE	2264	8	
DIISOPROPYLAMINE	1158	3		DIMETHYLDICHLORO-SILANE	1162	3	
DIISOPROPYL ETHER	1159	3		DIMETHYLDIETHOXY-SILANE	2380	3	
DIKETENE, STABILIZED	2521	6.1					
1,1-DIMETHOXYETHANE	2377	3		DIMETHYLDIOXANES	2707	3	
1,2-DIMETHOXYETHANE	2252	3		DIMETHYL DISULPHIDE	2381	3	
Dimethoxystrychnine, see	1570	6.1		Dimethylethanolamine, see	2051	8	
DIMETHYLAMINE, ANHYDROUS	1032	2		DIMETHYL ETHER	1033	2	
DIMETHYLAMINE AQUEOUS SOLUTION	1160	3		N,N-DIMETHYL-FORMAMIDE	2265	3	
2-DIMETHYLAMINO-ACETONITRILE	2378	3		DIMETHYLHYDRAZINE, SYMMETRICAL	2382	6.1	
2-DIMETHYLAMINO-ETHANOL	2051	8		DIMETHYLHYDRAZINE, UNSYMMETRICAL	1163	6.1	
2-DIMETHYL-AMINOETHYL ACRYLATE	3302	6.1		1,1-Dimethylhydrazine, see	1163	6.1	
2-DIMETHYL-AMINOETHYL METHACRYLATE	2522	6.1		N,N-Dimethyl-4-nitrosoaniline, see	1369	4.2	
				2,2-DIMETHYLPROPANE	2044	2	
N,N-DIMETHYLANILINE	2253	6.1		DIMETHYL-N-PROPYLAMINE	2266	3	
Dimethylarsenic acid, see	1572	6.1		DIMETHYL SULPHATE	1595	6.1	
N,N-Dimethylbenzylamine, see	2619	8		DIMETHYL SULPHIDE	1164	3	
2,3-DIMETHYLBUTANE	2457	3					

Name and description	UN No.	Class	Remarks	Name and description	UN No.	Class	Remarks
DIMETHYL THIOPHOSPHORYL CHLORIDE	2267	6.1		Dinitrotoluene mixed with sodium chlorate, see	0083	1	
Dimethylzinc, see	3394	4.2		DINITROTOLUENES, LIQUID	2038	6.1	
DINGU, see	0489	1					
DINITROANILINES	1596	6.1		DINITROTOLUENES, MOLTEN	1600	6.1	
DINITROBENZENES, LIQUID	1597	6.1		DINITROTOLUENES, SOLID	3454	6.1	
DINITROBENZENES, SOLID	3443	6.1		DIOXANE	1165	3	
				DIOXOLANE	1166	3	
Dinitrochlorobenzene, see	1577	6.1		DIPENTENE	2052	3	
	3441	6.1					
DINITRO-o-CRESOL	1598	6.1		DIPHENYLAMINE CHLOROARSINE	1698	6.1	
DINITROGEN TETROXIDE	1067	2		DIPHENYLCHLORO-ARSINE, LIQUID	1699	6.1	
DINITROGLYCOLURIL	0489	1					
DINITROPHENOL, dry or wetted with less than 15% water, by mass	0076	1		DIPHENYLCHLORO-ARSINE, SOLID	3450	6.1	
DINITROPHENOL SOLUTION	1599	6.1		DIPHENYLDICHLORO-SILANE	1769	8	
DINITROPHENOL, WETTED with not less than 15% water, by mass	1320	4.1		DIPHENYLMETHANE-4, 4'-DIISOCYANATE	9004	9	Dangerous in tank vessels only
DINITROPHENOLATES, alkali metals, dry or wetted with less than 15% water, by mass	0077	1		DIPHENYLMETHYL BROMIDE	1770	8	
				DIPICRYLAMINE, see	0079	1	
DINITROPHENOLATES, WETTED with not less than 15% water, by mass	1321	4.1		DIPICRYL SULPHIDE, dry or wetted with less than 10% water, by mass	0401	1	
DINITRORESORCINOL, dry or wetted with less than 15% water, by mass	0078	1		DIPICRYL SULPHIDE, WETTED with not less than 10% water, by mass	2852	4.1	
DINITRORESORCINOL, WETTED with not less than 15% water, by mass	1322	4.1		DIPROPYLAMINE	2383	3	
				Dipropylene triamine, see	2269	8	
DINITROSOBENZENE	0406	1		DI-n-PROPYL ETHER	2384	3	
				DIPROPYL KETONE	2710	3	

Name and description	UN No.	Class	Remarks	Name and description	UN No.	Class	Remarks
DISINFECTANT, LIQUID, CORROSIVE, N.O.S.	1903	8		Electrolyte (acid or alkaline) for batteries, see	2796 2797	8 8	
DISINFECTANT, LIQUID, TOXIC, N.O.S.	3142	6.1		ELEVATED TEMPERATURE LIQUID, N.O.S., at or above 100 °C and below its flash-point (including molten metals, molten salts, etc.)	3257	9	
DISINFECTANT, SOLID, TOXIC, N.O.S.	1601	6.1					
DISODIUM TRIOXOSILICATE	3253	8		ELEVATED TEMPERATURE LIQUID, FLAMMABLE, N.O.S. with flash-point above 60 °C, at or above its flash-point	3256	3	
DIVINYL ETHER, STABILIZED	1167	3					
DODECYLTRICHLORO-SILANE	1771	8		ELEVATED TEMPERATURE SOLID, N.O.S., at or above 240 °C	3258	9	
Dry ice, see	1845	9	Not subject to ADN				
				Empty battery-vehicle, uncleaned			See 4.3.2.4 of ADR, 5.1.3 and 5.4.1.1.6
DYE INTERMEDIATE, LIQUID, CORROSIVE, N.O.S.	2801	8					
DYE INTERMEDIATE, LIQUID, TOXIC, N.O.S.	1602	6.1		Empty IBC, uncleaned			See 4.1.1.11 of ADR, 5.1.3 and 5.4.1.1.6
DYE INTERMEDIATE, SOLID, CORROSIVE, N.O.S.	3147	8					
DYE INTERMEDIATE, SOLID, TOXIC, N.O.S.	3143	6.1		Empty large packaging, uncleaned			See 4.1.1.11 of ADR, 5.1.3 and 5.4.1.1.6
DYE, LIQUID, CORROSIVE, N.O.S.	2801	8					
DYE, LIQUID, TOXIC, N.O.S.	1602	6.1		Empty MEGC, uncleaned			See 4.3.2.4 of ADR, 5.1.3 and 5.4.1.1.6
DYE, SOLID, CORROSIVE, N.O.S.	3147	8		Empty packaging, uncleaned			See 4.1.1.11 of ADR, 5.1.3 and 5.4.1.1.6
DYE, SOLID, TOXIC, N.O.S.	3143	6.1					
Dynamite, see	0081	1		Empty receptacle, uncleaned			See 5.1.3 and 5.4.1.1.6
Electric storage batteries, see	2794 2795 2800 3028	8 8 8 8					

Name and description	UN No.	Class	Remarks	Name and description	UN No.	Class	Remarks
Empty tank, uncleaned			See 4.3.2.4 of ADR, 5.1.3 and 5.4.1.1.6	ETHANE, REFRIGERATED LIQUID	1961	2	
Empty vehicle, uncleaned			See 5.1.3 and 5.4.1.1.6	Ethanethiol, see	2363	3	
				ETHANOL	1170	3	
				ETHANOL SOLUTION	1170	3	
Enamel, see	1263	3		ETHANOLAMINE	2491	8	
	3066	8					
	3469	3		ETHANOLAMINE SOLUTION	2491	8	
	3470	8					
Engines, internal combustion	3166	9	Not subject to ADN	Ether, see	1155	3	
Engines, rocket, see	0250	1		ETHERS, N.O.S.	3271	3	
	0322	1					
				2-Ethoxyethanol, see	1171	3	
ENVIRONMENTALLY HAZARDOUS SUBSTANCE, LIQUID, N.O.S.	3082	9		2-Ethoxyethyl acetate, see	1172	3	
				Ethoxy propane-1, see	2615	3	
ENVIRONMENTALLY HAZARDOUS SUBSTANCE, SOLID, N.O.S.	3077	9		ETHYL ACETATE	1173	3	
				ETHYLACETYLENE, STABILIZED	2452	2	
EPIBROMOHYDRIN	2558	6.1		ETHYL ACRYLATE, STABILIZED	1917	3	
EPICHLOROHYDRIN	2023	6.1					
				ETHYL ALCOHOL, see	1170	3	
1,2-Epoxybutane, stabilized, see	3022	3		ETHYL ALCOHOL SOLUTION, see	1170	3	
Epoxyethane, see	1040	2					
1,2-EPOXY-3-ETHOXYPROPANE	2752	3		ETHYLAMINE	1036	2	
2,3-Epoxy-1-propanal, see	2622	3		ETHYLAMINE, AQUEOUS SOLUTION with not less than 50% but not more than 70% ethylamine	2270	3	
2,3-Epoxypropyl ethyl ether, see	2752	3					
				ETHYL AMYL KETONE	2271	3	
ESTERS, N.O.S.	3272	3		N-ETHYLANILINE	2272	6.1	
Ethanal, see	1089	3		2-ETHYLANILINE	2273	6.1	
ETHANE	1035	2		ETHYLBENZENE	1175	3	

Name and description	UN No.	Class	Remarks	Name and description	UN No.	Class	Remarks
N-ETHYL-N-BENZYLANILINE	2274	6.1		ETHYLENE, ACETYLENE AND PROPYLENE MIXTURE, REFRIGERATED LIQUID containing at least 71.5% ethylene with not more than 22.5% acetylene and not more than 6% propylene	3138	2	
N-ETHYLBENZYL-TOLUIDINES, LIQUID	2753	6.1					
N-ETHYLBENZYL-TOLUIDINES, SOLID	3460	6.1					
ETHYL BORATE	1176	3		ETHYLENE CHLOROHYDRIN	1135	6.1	
ETHYL BROMIDE	1891	6.1		ETHYLENE	1962	2	
ETHYL BROMOACETATE	1603	6.1		ETHYLENEDIAMINE	1604	8	
2-ETHYLBUTANOL	2275	3		ETHYLENE DIBROMIDE	1605	6.1	
2-ETHYLBUTYL ACETATE	1177	3					
ETHYL BUTYL ETHER	1179	3		Ethylene dibromide and methyl bromide, liquid mixture, see	1647	6.1	
2-ETHYLBUTYRALDEHYDE	1178	3					
ETHYL BUTYRATE	1180	3		ETHYLENE DICHLORIDE	1184	3	
ETHYL CHLORIDE	1037	2		ETHYLENE GLYCOL DIETHYL ETHER	1153	3	
ETHYL CHLOROACETATE	1181	6.1		ETHYLENE GLYCOL MONOETHYL ETHER	1171	3	
Ethyl chlorocarbonate, see	1182	6.1					
ETHYL CHLOROFORMATE	1182	6.1		ETHYLENE GLYCOL MONOETHYL ETHER ACETATE	1172	3	
ETHYL 2-CHLORO-PROPIONATE	2935	3		ETHYLENE GLYCOL MONOMETHYL ETHER	1188	3	
Ethyl-alpha-chloropropionate, see	2935	3		ETHYLENE GLYCOL MONOMETHYL ETHER ACETATE	1189	3	
ETHYL CHLORO-THIOFORMATE	2826	8		ETHYLENEIMINE, STABILIZED	1185	6.1	
ETHYL CROTONATE	1862	3		ETHYLENE OXIDE	1040	2	
ETHYLDICHLOROARSINE	1892	6.1		ETHYLENE OXIDE AND CARBON DIOXIDE MIXTURE with more than 87% ethylene oxide	3300	2	
ETHYLDICHLOROSILANE	1183	4.3					

Name and description	UN No.	Class	Remarks	Name and description	UN No.	Class	Remarks
ETHYLENE OXIDE AND CARBON DIOXIDE MIXTURE with more than 9% but not more than 87% ethylene oxide	1041	2		2-ETHYLHEXYL CHLOROFORMATE	2748	6.1	
				Ethylidene chloride, see	2362	3	
				ETHYL ISOBUTYRATE	2385	3	
ETHYLENE OXIDE AND CARBON DIOXIDE MIXTURE with not more than 9% ethylene oxide	1952	2		ETHYL ISOCYANATE	2481	3	
				ETHYL LACTATE	1192	3	
ETHYLENE OXIDE AND CHLOROTETRAFLUORO-ETHANE MIXTURE with not more than 8.8% ethylene oxide	3297	2		ETHYL MERCAPTAN	2363	3	
				ETHYL METHACRYLATE, STABILIZED	2277	3	
ETHYLENE OXIDE AND DICHLORODIFLUORO-METHANE MIXTURE with not more than 12.5% ethylene oxide	3070	2		ETHYL METHYL ETHER	1039	2	
				ETHYL METHYL KETONE	1193	3	
				ETHYL NITRITE SOLUTION	1194	3	
ETHYLENE OXIDE AND PENTAFLUOROETHANE MIXTURE with not more than 7.9% ethylene oxide	3298	2		ETHYL ORTHOFORMATE	2524	3	
				ETHYL OXALATE	2525	6.1	
ETHYLENE OXIDE AND PROPYLENE OXIDE MIXTURE, not more than 30% ethylene oxide	2983	3		ETHYLPHENYL-DICHLOROSILANE	2435	8	
				1-ETHYLPIPERIDINE	2386	3	
ETHYLENE OXIDE AND TETRAFLUOROETHANE MIXTURE with not more than 5.6% ethylene oxide	3299	2		ETHYL PROPIONATE	1195	3	
				ETHYL PROPYL ETHER	2615	3	
				Ethyl silicate, see	1292	3	
ETHYLENE OXIDE WITH NITROGEN up to a total pressure of 1 MPa (10 bar) at 50 °C	1040	2		Ethyl sulphate, see	1594	6.1	
				N-ETHYLTOLUIDINES	2754	6.1	
ETHYLENE, REFRIGERATED LIQUID	1038	2		ETHYLTRICHLORO-SILANE	1196	3	
ETHYL ETHER, see	1155	3		EXPLOSIVE, BLASTING, TYPE A	0081	1	
ETHYL FLUORIDE	2453	2		EXPLOSIVE, BLASTING, TYPE B	0082	1	
					0331	1	
ETHYL FORMATE	1190	3					
2-ETHYLHEXYLAMINE	2276	3		EXPLOSIVE, BLASTING, TYPE C	0083	1	

Name and description	UN No.	Class	Remarks	Name and description	UN No.	Class	Remarks
EXPLOSIVE, BLASTING, TYPE D	0084	1		FERROSILICON with 30% or more but less than 90% silicon	1408	4.3	
EXPLOSIVE, BLASTING, TYPE E	0241 0332	1 1		FERROUS ARSENATE	1608	6.1	
Explosives, emulsion, see	0241 0332	1 1		FERROUS METAL BORINGS in a form liable to self-heating	2793	4.2	
Explosive, seismic, see	0081 0082 0083 0331	1 1 1 1		FERROUS METAL CUTTINGS in a form liable to self-heating	2793	4.2	
Explosive, slurry, see	0241 0332	1 1		FERROUS METAL SHAVINGS in a form liable to self-heating	2793	4.2	
Explosive, water gel, see	0241 0332	1 1		FERROUS METAL TURNINGS in a form liable to self-heating	2793	4.2	
EXTRACTS, AROMATIC, LIQUID	1169	3		FERTILIZER AMMONIATING SOLUTION with free ammonia	1043	2	
EXTRACTS, FLAVOURING, LIQUID	1197	3		Fertilizer with ammonium nitrate, n.o.s., see	2067	5.1	
FABRICS, ANIMAL, N.O.S. with oil	1373	4.2		Fibres, animal, burnt wet or damp	1372	4.2	Not subject to ADN
FABRICS IMPREGNATED WITH WEAKLY NITRATED NITROCELLULOSE, N.O.S.	1353	4.1		FIBRES, ANIMAL, N.O.S. with oil	1373	4.2	
FABRICS, SYNTHETIC, N.O.S. with oil	1373	4.2		FIBRES IMPREGNATED WITH WEAKLY NITRATED NITROCELLULOSE, N.O.S.	1353	4.1	
FABRICS, VEGETABLE, N.O.S. with oil	1373	4.2		FIBRES, SYNTHETIC, N.O.S. with oil	1373	4.2	
FERRIC ARSENATE	1606	6.1		Fibres, vegetable, burnt wet or damp	1372	4.2	Not subject to ADN
FERRIC ARSENITE	1607	6.1					
FERRIC CHLORIDE, ANHYDROUS	1773	8		Fibres, vegetable, dry	3360	4.1	Not subject to ADN
FERRIC CHLORIDE SOLUTION	2582	8		FIBRES, VEGETABLE, N.O.S. with oil	1373	4.2	
FERRIC NITRATE	1466	5.1					
FERROCERIUM	1323	4.1					

Name and description	UN No.	Class	Remarks	Name and description	UN No.	Class	Remarks
Filler, liquid, see	1263	3		FLAMMABLE LIQUID, TOXIC, N.O.S.	1992	3	
	3066	8					
	3469	3					
	3470	8		FLAMMABLE LIQUID, TOXIC, CORROSIVE, N.O.S.	3286	3	
Films, nitrocellulose base, from which gelatin has been removed; film scrap, see	2002	4.2					
				FLAMMABLE SOLID, CORROSIVE, INORGANIC, N.O.S.	3180	4.1	
FILMS, NITROCELLULOSE BASE, gelatin coated, except scrap	1324	4.1					
				FLAMMABLE SOLID, CORROSIVE, ORGANIC, N.O.S.	2925	4.1	
FIRE EXTINGUISHER CHARGES, corrosive liquid	1774	8					
				FLAMMABLE SOLID, INORGANIC, N.O.S.	3178	4.1	
Fire extinguisher charges, expelling, explosive, see	0275	1					
	0276	1					
	0323	1		FLAMMABLE SOLID, ORGANIC, N.O.S.	1325	4.1	
	0381	1					
FIRE EXTINGUISHERS with compressed or liquefied gas	1044	2		FLAMMABLE SOLID, ORGANIC, MOLTEN, N.O.S.	3176	4.1	
FIRELIGHTERS, SOLID with flammable liquid	2623	4.1					
				FLAMMABLE SOLID, OXIDIZING, N.O.S.	3097	4.1	Carriage prohibited
FIREWORKS	0333	1	See 2.2.1.1.7				
	0334	1					
	0335	1		FLAMMABLE SOLID, TOXIC, INORGANIC, N.O.S.	3179	4.1	
	0336	1					
	0337	1					
FIRST AID KIT	3316	9		FLAMMABLE SOLID, TOXIC, ORGANIC, N.O.S.	2926	4.1	
FISH MEAL, STABILIZED	2216	9		FLARES, AERIAL	0093	1	
FISH MEAL, UNSTABILIZED	1374	4.2			0403	1	
					0404	1	
					0420	1	
FISH SCRAP, STABILIZED, see	2216	9			0421	1	
FISH SCRAP, UNSTABILIZED, see	1374	4.2		Flares, aeroplane, see	0093	1	
					0403	1	
					0404	1	
Flammable gas in lighters, see	1057	2			0420	1	
					0421	1	
FLAMMABLE LIQUID, N.O.S	1993	3					
				Flares, highway,	0191	1	
				Flares, distress, small,	0373	1	
FLAMMABLE LIQUID, CORROSIVE, N.O.S.	2924	3		Flares, railway or highway, see			

Name and description	UN No.	Class	Remarks	Name and description	UN No.	Class	Remarks
FLARES, SURFACE	0092	1		FORMALDEHYDE SOLUTION, FLAMMABLE	1198	3	
	0418	1					
	0419	1		Formalin, see	1198	3	
Flares, water-activated, see	0248	1			2209	8	
	0249	1		Formamidine sulphinic acid, see	3341	4.2	
FLASH POWDER	0094	1		FORMIC ACID with more than 85% acid by mass	1779	8	
	0305	1		FORMIC ACID with not more than 85% acid by mass	3412	8	
Flue dusts, toxic, see	1562	6.1					
				Formic aldehyde, see	1198	3	
Fluoric acid, see	1790	8			2209	8	
FLUORINE, COMPRESSED	1045	2		2-Formyl-3,4-dihydro-2H-pyran, see	2607	3	
FLUOROACETIC ACID	2642	6.1					
FLUOROANILINES	2941	6.1		FRACTURING DEVICES, EXPLOSIVE without detonator, for oil wells	0099	1	
2-Fluoroaniline, see	2941	6.1					
4-Fluoroaniline, see	2941	6.1		FUEL, AVIATION, TURBINE ENGINE	1863	3	
o-Fluoroaniline, see	2941	6.1					
p-Fluoroaniline, see	2941	6.1		FUEL CELL CARTRIDGES	3473	3	
FLUOROBENZENE	2387	3		Fumaroyl dichloride, see	1780	3	
FLUOROBORIC ACID	1775	8		FUMARYL CHLORIDE	1780	8	
Fluoroethane, see	2453	2		FUMIGATED UNIT	3359	9	
Fluoroform, see	1984	2		FURALDEHYDES	1199	6.1	
Fluoromethane, see	2454	2		FURAN	2389	3	
FLUOROPHOSPHORIC ACID, ANHYDROUS	1776	8		FURFURYL ALCOHOL	2874	6.1	
				FURFURYLAMINE	2526	3	
FLUOROSILICATES, N.O.S.	2856	6.1		Furyl carbinol, see	2874	6.1	
FLUOROSILICIC ACID	1778	8		FUSE, DETONATING, metal clad	0102	1	
FLUOROSULPHONIC ACID	1777	8			0290	1	
FLUOROTOLUENES	2388	3		FUSE, DETONATING, MILD EFFECT, metal clad	0104	1	
FORMALDEHYDE SOLUTION with not less than 25% formaldehyde	2209	8		FUSE, IGNITER, tubular, metal clad	0103	1	

Name and description	UN No.	Class	Remarks	Name and description	UN No.	Class	Remarks
FUSE, NON-DETONATING	0101	1		GAS SAMPLE, NON-PRESSURIZED, FLAMMABLE, N.O.S., not refrigerated liquid	3167	2	
FUSEL OIL	1201	3					
FUSE, SAFETY	0105	1		GAS SAMPLE, NON-PRESSURIZED, TOXIC, N.O.S., not refrigerated liquid	3169	2	
Fuze, combination, percussion or time, see	0106	1					
	0107	1					
	0257	1		GAS SAMPLE, NON-PRESSURIZED, TOXIC, FLAMMABLE, N.O.S., not refrigerated liquid	3168	2	
	0316	1					
	0317	1					
	0367	1					
	0368	1					
FUZES, DETONATING	0106	1		Gelatin, blasting, see	0081	1	
	0107	1					
	0257	1		Gelatin, dynamites, see	0081	1	
	0367	1					
FUZES, DETONATING with protective features	0408	1		GENETICALLY MODIFIED MICRO-ORGANISMS	3245	9	
	0409	1					
	0410	1		GENETICALLY MODIFIED ORGANISMS	3245	9	
FUZES, IGNITING	0316	1					
	0317	1		GERMANE	2192	2	
	0368	1					
				Germanium hydride, see	2192	2	
GALLIUM	2803	8					
				Glycer-1,3-dichlorohydrin, see	2750	6.1	
GAS CARTRIDGES without a release device, non-refillable, see	2037	2		GLYCEROL alpha-MONOCHLOROHYDRIN	2689	6.1	
Gas drips, hydrocarbon, see	3295	3		Glyceryl trinitrate, see	0143	1	
					0144	1	
GAS OIL	1202	3			1204	3	
					3064	3	
GASOLINE	1203	3					
				GLYCIDALDEHYDE	2622	3	
Gasoline, casinghead, see	1203	3					
				GRENADES, hand or rifle, with bursting charge	0284	1	
GAS, REFRIGERATED LIQUID, N.O.S.	3158	2			0285	1	
					0292	1	
					0293	1	
GAS, REFRIGERATED LIQUID, FLAMMABLE, N.O.S.	3312	2		Grenades, illuminating, see	0171	1	
					0254	1	
					0297	1	
GAS, REFRIGERATED LIQUID, OXIDIZING, N.O.S.	3311	2		GRENADES, PRACTICE, hand or rifle	0110	1	
					0318	1	
					0372	1	
					0452	1	

Name and description	UN No.	Class	Remarks	Name and description	UN No.	Class	Remarks
Grenades, smoke, see	0015	1		n-Heptanal, see	3056	3	
	0016	1		HEPTANES	1206	3	
	0245	1					
	0246	1		4-Heptanone, see	2710	3	
	0303	1					
GUANIDINE NITRATE	1467	5.1		n-HEPTENE	2278	3	
GUANYLNITROSAMINO-GUANYLIDENE HYDRAZINE, WETTED with not less than 30% water, by mass	0113	1		HEXACHLOROACETONE	2661	6.1	
				HEXACHLOROBENZENE	2729	6.1	
				HEXACHLORO-BUTADIENE	2279	6.1	
GUANYLNITROSAMINO-GUANYLTETRAZENE, WETTED with not less than 30% water, or mixture of alcohol and water, by mass	0114	1		Hexachloro-1,3-butadiene, see	2279	6.1	
				HEXACHLOROCYCLO-PENTADIENE	2646	6.1	
GUNPOWDER, COMPRESSED, see	0028	1		HEXACHLOROPHENE	2875	6.1	
				Hexachloro-2-propanone, see	2661	6.1	
GUNPOWDER, granular or as a meal, see	0027	1		HEXADECYLTRICHLORO-SILANE	1781	8	
GUNPOWDER, IN PELLETS, see	0028	1		HEXADIENES	2458	3	
Gutta percha solution, see	1287	3		HEXAETHYL TETRAPHOSPHATE	1611	6.1	
HAFNIUM POWDER, DRY	2545	4.2					
HAFNIUM POWDER, WETTED with not less than 25% water	1326	4.1		HEXAETHYL TETRAPHOSPHATE AND COMPRESSED GAS MIXTURE	1612	2	
Hay	1327	4.1	Not subject to ADN	HEXAFLUOROACETONE	2420	2	
				HEXAFLUOROACETONE HYDRATE, LIQUID	2552	6.1	
HEATING OIL, LIGHT	1202	3					
Heavy hydrogen, see	1957	2		HEXAFLUOROACETONE HYDRATE, SOLID	3436	6.1	
HELIUM, COMPRESSED	1046	2		HEXAFLUOROETHANE	2193	2	
HELIUM, REFRIGERATED LIQUID	1963	2		HEXAFLUORO-PHOSPHORIC ACID	1782	8	
HEPTAFLUOROPROPANE	3296	2		HEXAFLUORO-PROPYLENE	1858	2	
n-HEPTALDEHYDE	3056	3					

Name and description	UN No.	Class	Remarks	Name and description	UN No.	Class	Remarks
Hexahydrocresol, see	2617	3		HEXOLITE, dry or wetted with less than 15% water, by mass	0118	1	
Hexahydromethyl phenol, see	2617	3					
HEXALDEHYDE	1207	3		HEXOTOL, dry or wetted with less than 15% water, by mass, see	0118	1	
HEXAMETHYLENE-DIAMINE, SOLID	2280	8					
				HEXOTONAL	0393	1	
HEXAMETHYLENE-DIAMINE SOLUTION	1783	8		HEXOTONAL, cast, see	0393	1	
HEXAMETHYLENE DIISOCYANATE	2281	6.1		HEXYL, see	0079	1	
HEXAMETHYLENEIMINE	2493	3		HEXYLTRICHLORO-SILANE	1784	8	
HEXAMETHYLENE-TETRAMINE	1328	4.1		HMX, see	0391	1	
				HMX, DESENSITIZED, see	0484	1	
Hexamine, see	1328	4.1		HMX, WETTED with not less than 15% water, by mass, see	0226	1	
HEXANES	1208	3					
HEXANITRODIPHENYL-AMINE	0079	1		HYDRAZINE, ANHYDROUS	2029	8	
HEXANITROSTILBENE	0392	1		HYDRAZINE AQUEOUS SOLUTION, with more than 37% hydrazine by mass	2030	8	
Hexanoic acid, see	2829	8					
HEXANOLS	2282	3		HYDRAZINE, AQUEOUS SOLUTION with not more than 37% hydrazine, by mass	3293	6.1	
1-HEXENE	2370	3					
HEXOGEN AND CYCLOTETRA-METHYLENE-TETRANITRAMINE MIXTURE, WETTED with not less than 15% water, by mass or DESENSITIZED with not less than 10% phlegmatiser by mass, see	0391	1		Hydrides, metal, water-reactive, n.o.s., see	1409	4.3	
				Hydriodic acid, anhydrous, see	2197	2	
				HYDRIODIC ACID	1787	8	
				HYDROBROMIC ACID	1788	8	
HEXOGEN, DESENSITIZED, see	0483	1		HYDROCARBON GAS MIXTURE, COMPRESSED, N.O.S.	1964	2	
HEXOGEN, WETTED with not less than 15% water, by mass, see	0072	1					

Name and description	UN No.	Class	Remarks	Name and description	UN No.	Class	Remarks
HYDROCARBON GAS MIXTURE, LIQUEFIED, N.O.S. such as mixtures A, A01, A02, A0, A1, B1, B2, B or C	1965	2		HYDROGEN CHLORIDE, ANHYDROUS	1050	2	
				HYDROGEN CHLORIDE, REFRIGERATED LIQUID	2186	2	Carriage prohi-bited
HYDROCARBON GAS REFILLS FOR SMALL DEVICES with release device	3150	2		HYDROGEN, COMPRESSED	1049	2	
HYDROCARBONS, LIQUID, N.O.S.	3295	3		HYDROGEN CYANIDE, AQUEOUS SOLUTION with not more than 20% hydrogen cyanide, see	1613	6.1	
HYDROCHLORIC ACID	1789	8		HYDROGEN CYANIDE, SOLUTION IN ALCOHOL with not more than 45% hydrogen cyanide	3294	6.1	
HYDROCYANIC ACID, AQUEOUS SOLUTION with not more than 20% hydrogen cyanide	1613	6.1					
HYDROFLUORIC ACID with more than 60% but not more than 85% hydrogen fluoride	1790	8		HYDROGEN CYANIDE, STABILIZED containing less than 3% water	1051	6.1	
HYDROFLUORIC ACID with more than 85% hydrogen fluoride	1790	8		HYDROGEN CYANIDE, STABILIZED, containing less than 3% water and absorbed in a porous inert material	1614	6.1	
HYDROFLUORIC ACID with not more than 60% hydrogen fluoride	1790	8		HYDROGEN-DIFLUORIDES, SOLID, N.O.S.	1740	8	
HYDROFLUORIC ACID AND SULPHURIC ACID MIXTURE	1786	8		HYDROGEN-DIFLUORIDES SOLUTION, N.O.S.	3471	8	
Hydrofluoroboric acid, see	1775	8		HYDROGEN FLUORIDE, ANHYDROUS	1052	8	
Hydrofluorosilicic acid, see	1778	8		Hydrogen fluoride solution, see	1790	8	
HYDROGEN AND METHANE MIXTURE, COMPRESSED	2034	2		HYDROGEN IN A METAL HYDRIDE STORAGE SYSTEM	2	3468	
Hydrogen arsenide, see	2188	2					
HYDROGEN BROMIDE, ANHYDROUS	1048	2		HYDROGEN IODIDE, ANHYDROUS	2197	2	
Hydrogen bromide solution, see	1788	8		Hydrogen iodide solution, see	1787	8	

Name and description	UN No.	Class	Remarks	Name and description	UN No.	Class	Remarks
HYDROGEN PEROXIDE AND PEROXYACETIC ACID MIXTURE with acid(s), water and not more than 5% peroxyacetic acid, STABILIZED	3149	5.1		3-Hydroxyphenol, see	2876	6.1	
				HYPOCHLORITES, INORGANIC, N.O.S.	3212	5.1	
				HYPOCHLORITE SOLUTION	1791	8	
HYDROGEN PEROXIDE, AQUEOUS SOLUTION with not less than 8% but less than 20% hydrogen peroxide (stabilized as necessary)	2984	5.1		IGNITERS	0121	1	
					0314	1	
					0315	1	
					0325	1	
					0454	1	
HYDROGEN PEROXIDE, AQUEOUS SOLUTION with not less than 20% but not more than 60% hydrogen peroxide (stabilized as necessary)	2014	5.1		3,3'-IMINO-DIPROPYLAMINE	2269	8	
				Indiarubber, see	1287	3	
HYDROGEN PEROXIDE, AQUEOUS SOLUTION, STABILIZED with more than 60% hydrogen peroxide and not more than 70% hydrogen peroxide	2015	5.1		INFECTIOUS SUBSTANCE, AFFECTING ANIMALS only	2900	6.2	
				INFECTIOUS SUBSTANCE, AFFECTING HUMANS	2814	6.2	
HYDROGEN PEROXIDE, AQUEOUS SOLUTION, STABILIZED with more than 70% hydrogen peroxide	2015	5.1		Ink, printer's, flammable, see	1210	3	
				INSECTICIDE GAS, N.O.S.	1968	2	
HYDROGEN, REFRIGERATED LIQUID	1966	2		INSECTICIDE GAS, FLAMMABLE, N.O.S.	3354	2	
HYDROGEN SELENIDE, ANHYDROUS	2202	2		INSECTICIDE GAS, TOXIC, N.O.S.	1967	2	
Hydrogen silicide, see	2203	2		INSECTICIDE GAS, TOXIC, FLAMMABLE, N.O.S.	3355	2	
HYDROGEN SULPHIDE	1053	2		IODINE MONOCHLORIDE	1792	8	
Hydroselenic acid, see	2202	2		IODINE PENTAFLUORIDE	2495	5.1	
Hydrosilicofluoric acid, see	1778	8		2-IODOBUTANE	2390	3	
3-Hydroxybutan-2-one, see	2621	3		Iodomethane, see	2644	6.1	
HYDROXYLAMINE SULPHATE	2865	8		IODOMETHYLPROPANES	2391	3	
				IODOPROPANES	2392	3	
1-Hydroxy-3-methyl-2-penten-4-yne, see	2705	8		alpha-Iodotoluene, see	2653	6.1	

Name and description	UN No.	Class	Remarks	Name and description	UN No.	Class	Remarks
I.p.d.i., see	2290	6.1		ISOBUTYL ISOCYANATE	2486	3	
Iron chloride, anhydrous, see	1773	8		ISOBUTYL METHACRYLATE, STABILIZED	2283	3	
Iron (III) chloride, anhydrous, see	1773	8					
Iron chloride solution, see	2582	8		ISOBUTYL PROPIONATE	2394	3	
IRON OXIDE, SPENT obtained from coal gas purification	1376	4.2		ISOBUTYRALDEHYDE	2045	3	
				ISOBUTYRIC ACID	2529	3	
IRON PENTACARBONYL	1994	6.1		ISOBUTYRONITRILE	2284	3	
Iron perchloride, anhydrous, see	1773	8		ISOBUTYRYL CHLORIDE	2395	3	
Iron powder, pyrophoric, see	1383	4.2		ISOCYANATES, FLAMMABLE, TOXIC, N.O.S.	2478	3	
Iron sesquichloride, anhydrous, see	1773	8		ISOCYANATES, TOXIC, N.O.S.	2206	6.1	
IRON SPONGE, SPENT obtained from coal gas purification	1376	4.2		ISOCYANATES, TOXIC, FLAMMABLE, N.O.S.	3080	6.1	
Iron swarf, see	2793	4.2		ISOCYANATE SOLUTION, FLAMMABLE, TOXIC, N.O.S.	2478	3	
ISOBUTANE	1969	2					
ISOBUTANOL	1212	3		ISOCYANATE SOLUTION, TOXIC, N.O.S.	2206	6.1	
Isobutene, see	1055	2		ISOCYANATE SOLUTION, TOXIC, FLAMMABLE, N.O.S.	3080	6.1	
ISOBUTYL ACETATE	1213	3					
ISOBUTYL ACRYLATE, STABILIZED	2527	3		ISOCYANATO-BENZOTRIFLUORIDES	2285	6.1	
ISOBUTYL ALCOHOL, see	1212	3		3-Isocyanatomethyl-3,5,5-tri-methylcyclohexyl isocyanate, see	2290	6.1	
ISOBUTYL ALDEHYDE, see	2045	3					
				Isododecane, see	2286	3	
ISOBUTYLAMINE	1214	3		ISOHEPTENE	2287	3	
ISOBUTYLENE	1055	2		ISOHEXENE	2288	3	
ISOBUTYL FORMATE	2393	3		Isooctane, see	1262	3	
ISOBUTYL ISOBUTYRATE	2528	3		ISOOCTENE	1216	3	

Name and description	UN No.	Class	Remarks	Name and description	UN No.	Class	Remarks
Isopentane, see	1265	3		ISOPROPYL ISOBUTYRATE	2406	3	
ISOPENTENES	2371	3		ISOPROPYL ISOCYANATE	2483	3	
Isopentylamine, see	1106	3		Isopropyl mercaptan, see	2402	3	
Isopentyl nitrite, see	1113	3		ISOPROPYL NITRATE	1222	3	
ISOPHORONEDIAMINE	2289	8		ISOPROPYL PROPIONATE	2409	3	
ISOPHORONE DIISOCYANATE	2290	6.1		Isolpropyltoluene, see	2046	3	
ISOPRENE, STABILIZED	1218	3		Isopropyltoluol, see	2046	3	
ISOPROPANOL	1219	3		ISOSORBIDE DINITRATE MIXTURE with not less than 60% lactose, mannose, starch or calcium hydrogen phosphate	2907	4.1	
ISOPROPENYL ACETATE	2403	3					
ISOPROPENYLBENZENE	2303	3					
ISOPROPYL ACETATE	1220	3		ISOSORBIDE-5-MONONITRATE	3251	4.1	
ISOPROPYL ACID PHOSPHATE	1793	8		Isovaleraldehyde, see	2058	3	
ISOPROPYL ALCOHOL, see	1219	3		JET PERFORATING GUNS, CHARGED, oil well, without detonator	0124	1	
					0494	1	
ISOPROPYLAMINE	1221	3					
ISOPROPYLBENZENE	1918	3		Jet tappers, without detonator, see	0059	1	
ISOPROPYL BUTYRATE	2405	3		KEROSENE	1223	3	
Isopropyl chloride, see	2356	3		KETONES, LIQUID, N.O.S.	1224	3	
ISOPROPYL CHLOROACETATE	2947	3		KRYPTON, COMPRESSED	1056	2	
ISOPROPYL CHLOROFORMATE	2407	6.1		KRYPTON, REFRIGERATED LIQUID	1970	2	
ISOPROPYL 2-CHLORO-PROPIONATE	2934	3		Lacquer, see	1263	3	
					3066	8	
					3469	3	
Isopropyl-alpha-chloropropionate, see	2934	3			3470	8	
				Lacquer base, liquid, see	1263	3	
					3066	8	
Isopropyl ether, see	1159	3			3469	3	
					3470	8	
Isopropylethylene, see	2561	3		Lacquer base or lacquer chips, nitrocellulose, dry, see	2557	4.1	
Isopropyl formate, see	1281	3					

Name and description	UN No.	Class	Remarks	Name and description	UN No.	Class	Remarks
Lacquer base or lacquer chips, plastic, wet with alcohol or solvent, see	1263 2059 2555 2556	3 3 4.1 4.1		Lead tetraethyl, see	1649	6.1	
				Lead tetramethyl, see	1649	6.1	
LEAD ACETATE	1616	6.1		LEAD TRINITRO-RESORCINATE, WETTED with not less than 20% water, or mixture of alcohol and water, by mass, see	0130	1	
Lead (II) acetate, see	1616	6.1					
LEAD ARSENATES	1617	6.1					
LEAD ARSENITES	1618	6.1		LIFE-SAVING APPLIANCES NOT SELF-INFLATING containing dangerous goods as equipment	3072	9	
LEAD AZIDE, WETTED with not less than 20% water, or mixture of alcohol and water, by mass	0129	1					
				LIFE-SAVING APPLIANCES, SELF-INFLATING	2990	9	
Lead chloride, solid, see	2291	6.1					
LEAD COMPOUND, SOLUBLE, N.O.S.	2291	6.1		LIGHTER REFILLS containing flammable gas	1057	2	
LEAD CYANIDE	1620	6.1		LIGHTERS containing flammable gas	1057	2	
Lead (II) cyanide	1620	6.1		LIGHTERS, FUSE	0131	1	
LEAD DIOXIDE	1872	5.1		Limonene, inactive, see	2052	3	
LEAD NITRATE	1469	5.1		LIQUEFIED GAS, N.O.S.	3163	2	
Lead (II) nitrate	1469	5.1		LIQUEFIED GAS, FLAMMABLE, N.O.S.	3161	2	
LEAD PERCHLORATE, SOLID	1470	5.1		LIQUEFIED GASES, non-flammable, charged with nitrogen, carbon dioxide or air	1058	2	
LEAD PERCHLORATE, SOLUTION	3408	5.1					
Lead (II) perchlorate	1470 3408	5.1 5.1		LIQUEFIED GAS, OXIDIZING, N.O.S.	3157	2	
Lead peroxide, see	1872	5.1		LIQUEFIED GAS, TOXIC, N.O.S.	3162	2	
LEAD PHOSPHITE, DIBASIC	2989	4.1		LIQUEFIED GAS, TOXIC, CORROSIVE, N.O.S.	3308	2	
LEAD STYPHNATE, WETTED with not less than 20% water, or mixture of alcohol and water, by mass	0130	1		LIQUEFIED GAS, TOXIC, FLAMMABLE, N.O.S.	3160	2	
LEAD SULPHATE with more than 3% free acid	1794	8		LIQUEFIED GAS, TOXIC, FLAMMABLE, CORROSIVE, N.O.S.	3309	2	

Name and description	UN No.	Class	Remarks	Name and description	UN No.	Class	Remarks
LIQUEFIED GAS, TOXIC, OXIDIZING, N.O.S.	3307	2		LITHIUM HYPOCHLORITE, DRY	1471	5.1	
LIQUEFIED GAS, TOXIC, OXIDIZING, CORROSIVE, N.O.S.	3310	2		LITHIUM HYPOCHLORITE MIXTURE	1471	5.1	
Liquefied petroleum gas, see	1075	2		Lithium in cartouches, see	1415	4.3	
Liquid filler, see	1263	3		LITHIUM NITRATE	2722	5.1	
	3066	8		LITHIUM NITRIDE	2806	4.3	
	3469	3					
	3470	8		LITHIUM PEROXIDE	1472	5.1	
Liquid lacquer base, see	1263	3		Lithium silicide, see	1417	4.3	
	3066	8					
	3469	3		LITHIUM SILICON	1417	4.3	
	3470	8					
LITHIUM	1415	4.3		L.n.g., see	1972	2	
Lithium alkyls, liquid, see	3394	4.2		LONDON PURPLE	1621	6.1	
Lithium alkyls, solid, see	3393	4.2		L.p.g., see	1075	2	
LITHIUM ALUMINIUM HYDRIDE	1410	4.3		Lye, see	1823	8	
LITHIUM ALUMINIUM HYDRIDE, ETHEREAL	1411	4.3		Lythene, see	1268	3	
				MAGNESIUM in pellets, turnings or ribbons	1869	4.1	
LITHIUM BATTERIES	3090	9					
LITHIUM BATTERIES CONTAINED IN EQUIPMENT	3091	9		Magnesium alkyls, see	3394	4.2	
				MAGNESIUM ALLOYS with more than 50% magnesium in pellets, turnings or ribbons	1869	4.1	
LITHIUM BATTERIES PACKED WITH EQUIPMENT	3091	9		MAGNESIUM ALLOYS POWDER	1418	4.3	
LITHIUM BOROHYDRIDE	1413	4.3		MAGNESIUM ALUMINIUM PHOSPHIDE	1419	4.3	
LITHIUM FERROSILICON	2830	4.3					
LITHIUM HYDRIDE	1414	4.3		MAGNESIUM ARSENATE	1622	6.1	
LITHIUM HYDRIDE, FUSED SOLID	2805	4.3		Magnesium bisulphite solution, see	2693	8	
LITHIUM HYDROXIDE	2680	8		MAGNESIUM BROMATE	1473	5.1	
LITHIUM HYDROXIDE SOLUTION	2679	8		MAGNESIUM CHLORATE	2723	5.1	

Name and description	UN No.	Class	Remarks	Name and description	UN No.	Class	Remarks
Magnesium chloride and chlorate mixture, see	1459 3407	5.1 5.1		MANEB PREPARATION, STABILIZED against self-heating	2968	4.3	
MAGNESIUM DIAMIDE	2004	4.2		MANEB, STABILIZED against self-heating	2968	4.3	
Magnesium diphenyl, see	3393	4.2					
MAGNESIUM FLUOROSILICATE	2853	6.1		Manganese ethylene-di-dithiocarbamate, see	2210	4.2	
MAGNESIUM GRANULES, COATED, particle size not less than 149 microns	2950	4.3		Manganese ethylene-1,2-dithiocarbamate, see	2210	4.2	
				MANGANESE NITRATE	2724	5.1	
MAGNESIUM HYDRIDE	2010	4.3		Manganese (II) nitrate, see	2724	5.1	
MAGNESIUM NITRATE	1474	5.1		MANGANESE RESINATE	1330	4.1	
MAGNESIUM PERCHLORATE	1475	5.1		Manganous nitrate, see	2724	5.1	
MAGNESIUM PEROXIDE	1476	5.1		MANNITOL HEXANITRATE, WETTED with not less than 40% water, or mixture of alcohol and water, by mass	0133	1	
MAGNESIUM PHOSPHIDE	2011	4.3					
MAGNESIUM POWDER	1418	4.3					
Magnesium scrap, see	1869	4.1		MATCHES, FUSEE	2254	4.1	
MAGNESIUM SILICIDE	2624	4.3		MATCHES, SAFETY (book, card or strike on box)	1944	4.1	
Magnesium silicofluoride, see	2853	6.1		MATCHES, "STRIKE ANYWHERE"	1331	4.1	
Magnetized material	2807	9	Not subject to ADN	MATCHES, WAX "VESTA"	1945	4.1	
MALEIC ANHYDRIDE	2215	8		MEDICAL WASTE, N.O.S.	3291	6.2	
MALEIC ANHYDRIDE, MOLTEN	2215	8		MEDICINE, LIQUID, FLAMMABLE, TOXIC, N.O.S.	3248	3	
Malonic dinitrile, see	2647	6.1		MEDICINE, LIQUID, TOXIC, N.O.S.	1851	6.1	
Malonodinitrile, see	2647	6.1		MEDICINE, SOLID, TOXIC, N.O.S.	3249	6.1	
MALONONITRILE	2647	6.1					
MANEB	2210	4.2		p-Mentha-1,8-diene, see	2052	8	
MANEB PREPARATION with not less than 60% maneb	2210	4.2		MERCAPTANS, LIQUID, FLAMMABLE, N.O.S.	3336	3	

Name and description	UN No.	Class	Remarks	Name and description	UN No.	Class	Remarks
MERCAPTANS, LIQUID, FLAMMABLE, TOXIC, N.O.S.	1228	3		MERCURY BASED PESTICIDE, LIQUID, FLAMMABLE, TOXIC, flash-point less than 23 °C	2778	3	
MERCAPTANS, LIQUID, TOXIC, FLAMMABLE, N.O.S.	3071	6.1		MERCURY BASED PESTICIDE, LIQUID, TOXIC	3012	6.1	
MERCAPTAN MIXTURE, LIQUID, FLAMMABLE, N.O.S.	3336	3		MERCURY BASED PESTICIDE, LIQUID, TOXIC, FLAMMABLE, flash-point not less than 23 °C	3011	6.1	
MERCAPTAN MIXTURE, LIQUID, FLAMMABLE, TOXIC, N.O.S.	1228	3		MERCURY BASED PESTICIDE, SOLID, TOXIC	2777	6.1	
MERCAPTAN MIXTURE, LIQUID, TOXIC, FLAMMABLE, N.O.S.	3071	6.1		MERCURY BENZOATE	1631	6.1	
2-Mercaptoethanol, see	2966	6.1		Mercury bichloride, see	1624	6.1	
2-Mercaptopropionic acid, see	2936	6.1		MERCURY BROMIDES	1634	6.1	
5-MERCAPTOTETRAZOL-1-ACETIC ACID	0448	1		MERCURY COMPOUND, LIQUID, N.O.S.	2024	6.1	
MERCURIC ARSENATE	1623	6.1		MERCURY COMPOUND, SOLID, N.O.S.	2025	6.1	
MERCURIC CHLORIDE	1624	6.1		MERCURY CYANIDE	1636	6.1	
MERCURIC NITRATE	1625	6.1		MERCURY FULMINATE, WETTED with not less than 20% water, or mixture of alcohol and water, by mass	0135	1	
MERCURIC POTASSIUM CYANIDE	1626	6.1					
Mercuric sulphate, see	1645	6.1		MERCURY GLUCONATE	1637	6.1	
Mercurol, see	1639	6.1		MERCURY IODIDE	1638	6.1	
Mercurous bisulphate, see	1645	6.1		MERCURY NUCLEATE	1639	6.1	
MERCUROUS NITRATE	1627	6.1		MERCURY OLEATE	1640	6.1	
Mercurous sulphate, see	1645	6.1		MERCURY OXIDE	1641	6.1	
MERCURY	2809	8		MERCURY OXYCYANIDE, DESENSITIZED	1642	6.1	
MERCURY ACETATE	1629	6.1		MERCURY POTASSIUM IODIDE	1643	6.1	
MERCURY AMMONIUM CHLORIDE	1630	6.1		MERCURY SALICYLATE	1644	6.1	

Name and description	UN No.	Class	Remarks	Name and description	UN No.	Class	Remarks
MERCURY SULPHATE	1645	6.1		METAL SALTS OF ORGANIC COMPOUNDS, FLAMMABLE, N.O.S.	3181	4.1	
MERCURY THIOCYANATE	1646	6.1					
Metal alkyl halides, water-reactive, n.o.s. / Metal aryl halides, water-reactive, n.o.s., see	3394	4.2		METHACRYLALDEHYDE, STABILIZED	2396	3	
Metal alkyl hydrides, water-reactive, n.o.s. / Metal aryl hydrides, water-reactive, n.o.s., see	3394	4.2		METHACRYLIC ACID, STABILIZED	2531	8	
				METHACRYLONITRILE, STABILIZED	3079	3	
Metal alkyls, water-reactive, n.o.s. / Metal aryls, water-reactive, n.o.s., see	3393	4.2		METHALLYL ALCOHOL	2614	3	
Mesitylene, see	2325	3		Methanal, see	1198	3	
					2209	8	
MESITYL OXIDE	1229	3		Methane and hydrogen mixture, see	2034	2	
METAL CARBONYLS, LIQUID, N.O.S.	3281	6.1					
				METHANE, COMPRESSED	1971	2	
METAL CARBONYLS, SOLID, N.O.S.	3466	6.1		METHANE, REFRIGERATED LIQUID	1972	2	
METAL CATALYST, DRY	2881	4.2					
METAL CATALYST, WETTED with a visible excess of liquid	1378	4.2		METHANESULPHONYL CHLORIDE	3246	6.1	
				METHANOL	1230	3	
METALDEHYDE	1332	4.1		2-Methoxyethyl acetate, see	1189	3	
METAL HYDRIDES, FLAMMABLE, N.O.S.	3182	4.1		METHOXYMETHYL ISOCYANATE	2605	3	
METAL HYDRIDES, WATER-REACTIVE, N.O.S.	1409	4.3		4-METHOXY-4-METHYLPENTAN-2-ONE	2293	3	
METALLIC SUBSTANCE, WATER-REACTIVE, N.O.S.	3208	4.3		1-Methoxy-2-nitrobenzene, see	2730	6.1	
					3458	6.1	
METALLIC SUBSTANCE, WATER-REACTIVE, SELF-HEATING, N.O.S.	3209	4.3		1-Methoxy-3-nitrobenzene, see	2730	6.1	
					3458	6.1	
METAL POWDER, FLAMMABLE, N.O.S.	3089	4.1		1-Methoxy-4-nitrobenzene, see	2730	6.1	
					3458	6.1	
METAL POWDER, SELF-HEATING, N.O.S.	3189	4.2		1-METHOXY-2-PROPANOL	3092	3	
				METHYL ACETATE	1231	3	

Name and description	UN No.	Class	Remarks	Name and description	UN No.	Class	Remarks
METHYLACETYLENE AND PROPADIENE MIXTURE, STABILIZED such as mixture P1 or mixture P2	1060	2		METHYL BROMIDE AND ETHYLENE DIBROMIDE MIXTURE, LIQUID	1647	6.1	
beta-Methyl acrolein, see	1143	6.1		METHYL BROMOACETATE	2643	6.1	
METHYL ACRYLATE, STABILIZED	1919	3		2-METHYLBUTANAL	3371	3	
				3-METHYLBUTAN-2-ONE	2397	3	
METHYLAL	1234	3		2-METHYL-1-BUTENE	2459	3	
Methyl alcohol, see	1230	3		2-METHYL-2-BUTENE	2460	3	
Methyl allyl alcohol, see	2614	3		3-METHYL-1-BUTENE	2561	3	
METHYLALLYL CHLORIDE	2554	3		N-METHYLBUTYLAMINE	2945	3	
METHYLAMINE, ANHYDROUS	1061	2		METHYL tert-BUTYL ETHER	2398	3	
METHYLAMINE, AQUEOUS SOLUTION	1235	3		METHYL BUTYRATE	1237	3	
METHYLAMYL ACETATE	1233	3		METHYL CHLORIDE	1063	2	
Methyl amyl alcohol, see	2053	3		Methyl chloride and chloropicrin mixture, see	1582	2	
Methyl amyl ketone, see	1110	3		METHYL CHLORIDE AND METHYLENE CHLORIDE MIXTURE	1912	2	
N-METHYLANILINE	2294	6.1					
Methylated spirit, see	1986	3		METHYL CHLOROACETATE	2295	6.1	
	1987	3		Methyl chlorocarbonate, see	1238	6.1	
alpha-METHYLBENZYL ALCOHOL, LIQUID	2937	6.1		Methyl chloroform, see	2831	6.1	
alpha-METHYLBENZYL ALCOHOL, SOLID	3438	6.1		METHYL CHLOROFORMATE	1238	6.1	
METHYL BROMIDE with not more than 2% chloropicrin	1062	2		METHYL CHLOROMETHYL ETHER	1239	6.1	
Methyl bromide and chloropicrin mixture, with more than 2% chloropicrin, see	1581	2		METHYL 2-CHLORO-PROPIONATE	2933	3	
				Methyl alpha-chloropropionate, see	2933	3	
				METHYLCHLOROSILANE	2534	2	

Name and description	UN No.	Class	Remarks	Name and description	UN No.	Class	Remarks
Methyl cyanide, see	1648	3		2-METHYL-2-HEPTANETHIOL	3023	6.1	
METHYLCYCLOHEXANE	2296	3		5-METHYLHEXAN-2-ONE	2302	3	
METHYLCYCLO-HEXANOLS, flammable	2617	3		METHYLHYDRAZINE	1244	6.1	
METHYLCYCLO-HEXANONE	2297	3		METHYL IODIDE	2644	6.1	
METHYLCYCLOPENTANE	2298	3		METHYL ISOBUTYL CARBINOL	2053	3	
METHYL DICHLORO-ACETATE	2299	6.1		METHYL ISOBUTYL KETONE	1245	3	
METHYLDICHLORO-SILANE	1242	4.3		METHYL ISOCYANATE	2480	6.1	
Methylene bromide, see	2664	6.1		METHYL ISOPROPENYL KETONE, STABILIZED	1246	3	
Methylene chloride, see	1593	6.1		METHYL ISOTHIOCYANATE	2477	6.1	
Methylene chloride and methyl chloride mixture, see	1912	2		METHYL ISOVALERATE	2400	3	
Methylene cyanide, see	2647	6.1		METHYL MAGNESIUM BROMIDE IN ETHYL ETHER	1928	4.3	
p,p'-Methylene dianiline, see	2651	6.1					
Methylene dibromide, see	2664	6.1		METHYL MERCAPTAN	1064	2	
2,2'-Methylene-di-(3,4,6-trichlorophenol), see	2875	6.1		Methyl mercapto-propionaldehyde, see	2785	6.1	
Methyl ethyl ether, see	1039	2		METHYL METHACRYLATE MONOMER, STABILIZED	1247	3	
METHYL ETHYL KETONE, see	1193	3					
2-METHYL-5-ETHYLPYRIDINE	2300	6.1		4-METHYLMORPHOLINE	2535	3	
METHYL FLUORIDE	2454	2		N-METHYLMORPHOLINE, see	2535	3	
METHYL FORMATE	1243	3		METHYL NITRITE	2455	2	Carriage prohibited
2-METHYLFURAN	2301	3		METHYL ORTHOSILICATE	2606	6.1	
Methyl glycol, see	1188	3		METHYLPENTADIENE	2461	3	
Methyl glycol acetate, see	1189	3		Methylpentanes, see	1208	3	
				2-METHYLPENTAN-2-OL	2560	3	

Name and description	UN No.	Class	Remarks	Name and description	UN No.	Class	Remarks
4-Methylpentan-2-ol, see	2053	3		MINES with bursting charge	0136	1	
					0137	1	
3-Methyl-2-penten-4ynol, see	2705	8			0138	1	
					0294	1	
METHYLPHENYL-DICHLOROSILANE	2437	8		Mirbane oil, see	1662	6.1	
2-Methyl-2-phenylpropane, see	2709	3		Missiles, guided, see	0180	1	
					0181	1	
					0182	1	
1-METHYLPIPERIDINE	2399	3			0183	1	
					0295	1	
METHYL PROPIONATE	1248	3			0397	1	
					0398	1	
Methylpropylbenzene, see	2046	3			0436	1	
					0437	1	
METHYL PROPYL ETHER	2612	3			0438	1	
METHYL PROPYL KETONE	1249	3		Mixtures A, A01, A02, A0, A1, B1, B2, B or C, see	1965	2	
Methyl pyridines, see	2313	3		Mixture F1, mixture F2 or mixture F3, see	1078	2	
Methylstyrene, inhibited, see	2618	3		MIXTURES OF 1,3-BUTADIENE AND HYDROCARBONS, STABILIZED, having a vapour pressure at 70 °C not exceeding 1.1 MPa (11 bar) and a density at 50 °C not lower than 0.525 kg/l	1010	2	
alpha-Methylstyrene, see	2303	3					
Methyl sulphate, see	1595	6.1					
Methyl sulphide, see	1164	3					
METHYLTETRAHYDRO-FURAN	2536	3					
METHYL TRICHLOROACETATE	2533	6.1		Mixture P1 or mixture P2, see	1060	2	
METHYLTRICHLORO-SILANE	1250	3		MOLYBDENUM PENTACHLORIDE	2508	8	
alpha-METHYLVALERAL-DEHYDE	2367	3		Monochloroacetic acid, see	1750	6.1	
					1751	6.1	
Methyl vinyl benzene, inhibited, see	2618	3		Monochlorobenzene, see	1134	3	
METHYL VINYL KETONE, STABILIZED	1251	6.1		Monochlorodifluoromethane, see	1018	2	
				Monochlorodifluoromethane and monochloro-pentafluoroethane mixture, see	1973	2	
M.i.b.c., see	2053	3					
				Monochlorodifluoromono-bromomethane, see	1974	2	

Name and description	UN No.	Class	Remarks	Name and description	UN No.	Class	Remarks
Monochloropentafluoroethane and monochloro-difluoromethane mixture, see	1973	2		NATURAL GAS, REFRIGERATED LIQUID with high methane content	1972	2	
Monoethylamine, see	1036	2		Natural gasoline, see	1203	3	
MONONITROTOLUIDINES, see	2660	6.1		Neohexane, see	1208	3	
Monopropylamine, see	1277	3		NEON, COMPRESSED	1065	2	
MORPHOLINE	2054	8		NEON, REFRIGERATED LIQUID	1913	2	
MOTOR FUEL ANTI-KNOCK MIXTURE	1649	6.1		Neothyl, see	2612	3	
MOTOR SPIRIT	1203	3		NICKEL CARBONYL	1259	6.1	
Muriatic acid, see	1789	8		NICKEL CYANIDE	1653	6.1	
MUSK XYLENE, see	2956	4.1		Nickel (II) cyanide, see	1653	6.1	
Mysorite, see	2212	9		NICKEL NITRATE	2725	5.1	
Naphta, see	1268	3		Nickel (II) nitrate, see	2725	5.1	
Naphta, petroleum, see	1268	3		NICKEL NITRITE	2726	5.1	
Naphta, solvent, see	1268	3		Nickel (II) nitrite, see	2726	5.1	
NAPHTHALENE, CRUDE	1334	4.1		Nickelous nitrate, see	2725	5.1	
NAPHTHALENE, MOLTEN	2304	4.1		Nickelous nitrite, see	2726	5.1	
NAPHTHALENE, REFINED	1334	4.1		Nickel tetracarbonyl, see	1259	6.1	
alpha-NAPHTHYLAMINE	2077	6.1		NICOTINE	1654	6.1	
beta-NAPHTHYLAMINE, SOLID	1650	6.1		NICOTINE COMPOUND, LIQUID, N.O.S	3144	6.1	
beta-NAPHTHYLAMINE, SOLUTION	3411	6.1		NICOTINE COMPOUND, SOLID, N.O.S	1655	6.1	
NAPHTHYLTHIOUREA	1651	6.1		NICOTINE HYDROCHLORIDE, LIQUID	1656	6.1	
1-Naphthylthiourea, see	1651	6.1		NICOTINE HYDROCHLORIDE, SOLID	3444	6.1	
NAPHTHYLUREA	1652	6.1					
NATURAL GAS, COMPRESSED with high methane content	1971	2		NICOTINE HYDROCHLORIDE SOLUTION	1656	6.1	

Name and description	UN No.	Class	Remarks	Name and description	UN No.	Class	Remarks
NICOTINE PREPARATION, LIQUID, N.O.S.	3144	6.1		NITRIC OXIDE AND DINITROGEN TETROXIDE MIXTURE	1975	2	
NICOTINE PREPARATION, SOLID, N.O.S.	1655	6.1		NITRIC OXIDE AND NITROGEN DIOXIDE MIXTURE, see	1975	2	
NICOTINE SALICYLATE	1657	6.1					
NICOTINE SULPHATE, SOLID	3445	6.1		NITRILES, FLAMMABLE, TOXIC, N.O.S.	3273	3	
NICOTINE SULPHATE, SOLUTION	1658	6.1		NITRILES, TOXIC, LIQUID, N.O.S.	3276	6.1	
NICOTINE TARTRATE	1659	6.1		NITRILES, TOXIC, SOLID, N.O.S.	3439	6.1	
NITRATES, INORGANIC, N.O.S.	1477	5.1		NITRILES, TOXIC, FLAMMABLE, N.O.S.	3275	6.1	
NITRATES, INORGANIC, AQUEOUS SOLUTION, N.O.S.	3218	5.1		NITRITES, INORGANIC, N.O.S.	2627	5.1	
NITRATING ACID MIXTURE with more than 50% nitric acid	1796	8		NITRITES, INORGANIC, AQUEOUS SOLUTION, N.O.S.	3219	5.1	
NITRATING ACID MIXTURE with not more than 50% nitric acid	1796	8		NITROANILINES (o-, m-, p-)	1661	6.1	
NITRATING ACID MIXTURE, SPENT, with more than 50% nitric acid	1826	8		NITROANISOLES, LIQUID	2730	6.1	
				NITROANISOLES, SOLID	3458	6.1	
NITRATING ACID MIXTURE, SPENT, with not more than 50% nitric acid	1826	8		NITROBENZENE	1662	6.1	
				Nitrobenzene bromide, see	2732	6.1	
NITRIC ACID, other than red fuming, with more than 70% nitric acid	2031	8		NITROBENZENE-SULPHONIC ACID	2305	8	
				Nitrobenzol, see	1662	6.1	
NITRIC ACID, other than red fuming, with not more than 70% nitric acid	2031	8		5-NITROBENZOTRIAZOL	0385	1	
				NITROBENZOTRIFLUO-RIDES, LIQUID	2306	6.1	
NITRIC ACID, RED FUMING	2032	8		NITROBENZOTRIFLUO-RIDES, SOLID	3431	6.1	
NITRIC OXIDE, COMPRESSED	1660	2		NITROBROMOBENZENES, LIQUID	2732	6.1	

Name and description	UN No.	Class	Remarks	Name and description	UN No.	Class	Remarks
NITROBROMOBENZENES, SOLID	3459	6.1		NITROCELLULOSE, WETTED with not less than 25% alcohol, by mass	0342	1	
NITROCELLULOSE, dry or wetted with less than 25% water (or alcohol), by mass	0340	1		NITROCELLULOSE WITH ALCOHOL (not less than 25% alcohol, by mass, and not more than 12.6% nitrogen, by dry mass)	2556	4.1	
NITROCELLULOSE, unmodified or plasticized with less than 18% plasticizing substance, by mass	0341	1		NITROCELLULOSE WITH WATER (not less than 25% water, by mass)	2555	4.1	
NITROCELLULOSE MEMBRANE FILTERS, with not more than 12.6% nitrogen, by dry mass	3270	4.1		Nitrochlorobenzenes, see	1578 3409	6.1 6.1	
NITROCELLULOSE, with not more than 12.6% nitrogen, by dry mass, MIXTURE WITH PLASTICIZER, WITH PIGMENT	2557	4.1		3-NITRO-4-CHLOROBENZO-TRIFLUORIDE	2307	6.1	
NITROCELLULOSE, with not more than 12.6% nitrogen, by dry mass, MIXTURE WITH PLASTICIZER, WITHOUT PIGMENT	2557	4.1		NITROCRESOLS, LIQUID	3434	6.1	
				NITROCRESOLS, SOLID	2446	6.1	
NITROCELLULOSE, with not more than 12.6% nitrogen, by dry mass, MIXTURE WITHOUT PLASTICIZER, WITH PIGMENT	2557	4.1		NITROETHANE	2842	3	
				NITROGEN, COMPRESSED	1066	2	
				NITROGEN DIOXIDE, see	1067	2	
NITROCELLULOSE, with not more than 12.6% nitrogen, by dry mass, MIXTURE WITHOUT PLASTICIZER, WITHOUT PIGMENT	2557	4.1		NITROGEN, REFRIGERATED LIQUID	1977	2	
				NITROGEN TRIFLUORIDE	2451	2	
NITROCELLULOSE, PLASTICIZED with not less than 18% plasticizing substance, by mass	0343	1		NITROGEN TRIOXIDE	2421	2	Carriage prohibited
				NITROGLYCERIN, DESENSITIZED with not less than 40% non-volatile water-insoluble phlegmatizer, by mass	0143	1	
NITROCELLULOSE SOLUTION, FLAMMABLE with not more than 12.6% nitrogen, by dry mass, and not more than 55% nitrocellulose	2059	3		NITROGLYCERIN MIXTURE, DESENSITIZED, LIQUID, N.O.S. with not more than 30% nitroglycerin, by mass	3357	3	

Name and description	UN No.	Class	Remarks	Name and description	UN No.	Class	Remarks
NITROGLYCERIN MIXTURE, DESENSITIZED, LIQUID, FLAMMABLE, N.O.S. with not more than 30% nitroglycerin, by mass	3343	3		NITROPROPANES	2608	3	
				p-NITROSODIMETHYL-ANILINE	1369	4.2	
NITROGLYCERIN MIXTURE, DESENSITIZED, SOLID, N.O.S. with more than 2% but not more than 10% nitroglycerin, by mass	3319	4.1		NITROSTARCH, dry or wetted with less than 20% water, by mass	0146	1	
				NITROSTARCH, WETTED with not less than 20% water, by mass	1337	4.1	
NITROGLYCERIN, SOLUTION IN ALCOHOL with more than 1% but not more than 5% nitroglycerin	3064	3		NITROSYL CHLORIDE	1069	2	
				NITROSYLSULPHURIC ACID, LIQUID	2308	8	
NITROGLYCERIN SOLUTION IN ALCOHOL with more than 1% but not more than 10% nitroglycerin	0144	1		NITROSYLSULPHURIC ACID, SOLID	3456	8	
				NITROTOLUENES, LIQUID	1664	6.1	
NITROGLYCERIN SOLUTION IN ALCOHOL with not more than 1% nitroglycerin	1204	3		NITROTOLUENES, SOLID	3446	6.1	
				NITROTOLUIDINES	2660	6.1	
NITROGUANIDINE, dry or wetted with less than 20% water, by mass	0282	1		NITROTRIAZOLONE	0490	1	
				NITRO UREA	0147	1	
NITROGUANIDINE, WETTED with not less than 20% water, by mass	1336	4.1		NITROUS OXIDE	1070	2	
				NITROUS OXIDE, REFRIGERATED LIQUID	2201	2	
NITROHYDROCHLORIC ACID	1798	8	Carriage prohibited	NITROXYLENES, LIQUID	1665	6.1	
				NITROXYLENES, SOLID	3447	6.1	
NITROMANNITE, WETTED, see	0133	1		Non-activated carbon, see	1361	4.2	
				Non-activated charcoal, see	1361	4.2	
NITROMETHANE	1261	3		NONANES	1920	3	
Nitromuriatic acid, see	1798	8		NONYLTRICHLORO-SILANE	1799	8	
NITRONAPHTHALENE	2538	4.1					
NITROPHENOLS (o-, m-, p-)	1663	6.1		2,5-NORBORNADIENE, STABILIZED, see	2251	3	
4-NITROPHENYL-HYDRAZINE, with not less than 30% water, by mass	3376	4.1		Normal propyl alcohol, see	1274	3	

Name and description	UN No.	Class	Remarks	Name and description	UN No.	Class	Remarks
NTO, see	0490	1		ORGANIC PEROXIDE TYPE B, SOLID, TEMPERATURE CONTROLLED	3112	5.2	
OCTADECYLTRICHLORO-SILANE	1800	8					
OCTADIENE	2309	3		ORGANIC PEROXIDE TYPE C, LIQUID	3103	5.2	
OCTAFLUOROBUT-2-ENE	2422	2					
OCTAFLUOROCYCLO-BUTANE	1976	2		ORGANIC PEROXIDE TYPE C, LIQUID, TEMPERATURE CONTROLLED	3113	5.2	
OCTAFLUOROPROPANE	2424	2					
OCTANES	1262	3		ORGANIC PEROXIDE TYPE C, SOLID	3104	5.2	
OCTOGEN, see	0226	1		ORGANIC PEROXIDE TYPE C, SOLID, TEMPERATURE CONTROLLED	3114	5.2	
	0391	1					
	0484	1					
OCTOL, dry or wetted with less than 15% water, by mass, see	0266	1		ORGANIC PEROXIDE TYPE D, LIQUID	3105	5.2	
OCTOLITE, dry or wetted with less than 15% water, by mass	0266	1		ORGANIC PEROXIDE TYPE D, LIQUID, TEMPERATURE CONTROLLED	3115	5.2	
OCTONAL	0496	1		ORGANIC PEROXIDE TYPE D, SOLID	3106	5.2	
OCTYL ALDEHYDES	1191	3					
tert-Octyl mercaptan, see	3023	6.1		ORGANIC PEROXIDE TYPE D, SOLID, TEMPERATURE CONTROLLED	3116	5.2	
OCTYLTRICHLORO-SILANE	1801	8					
Oenanthol, see	3056	3		ORGANIC PEROXIDE TYPE E, LIQUID	3107	5.2	
OIL GAS, COMPRESSED	1071	2		ORGANIC PEROXIDE TYPE E, LIQUID, TEMPERATURE CONTROLLED	3117	5.2	
Oleum, see	1831	8					
ORGANIC PEROXIDE TYPE B, LIQUID	3101	5.2		ORGANIC PEROXIDE TYPE E, SOLID	3108	5.2	
ORGANIC PEROXIDE TYPE B, LIQUID, TEMPERATURE CONTROLLED	3111	5.2		ORGANIC PEROXIDE TYPE E, SOLID, TEMPERATURE CONTROLLED	3118	5.2	
ORGANIC PEROXIDE TYPE B, SOLID	3102	5.2					

Name and description	UN No.	Class	Remarks	Name and description	UN No.	Class	Remarks
ORGANIC PEROXIDE TYPE F, LIQUID	3109	5.2		ORGANOMETALLIC COMPOUND, TOXIC, SOLID, N.O.S.	3467	6.1	
ORGANIC PEROXIDE TYPE F, LIQUID, TEMPERATURE CONTROLLED	3119	5.2		Organometallic compound, solid, water-reactive, flammable, n.o.s., see	3396	4.3	
ORGANIC PEROXIDE TYPE F, SOLID	3110	5.2		Organometallic compound or Organometallic compound solution or Organometallic compound dispersion, water-reactive, flammable, n.o.s., see	3399	4.3	
ORGANIC PEROXIDE TYPE F, SOLID, TEMPERATURE CONTROLLED	3120	5.2		ORGANOMETALLIC SUBSTANCE, LIQUID, PYROPHORIC	3392	4.2	
Organic peroxides, see 2.2.52.4 for an alphabetic list of currently assigned organic peroxides and see	3101 to 3120	5.2		ORGANOMETALLIC SUBSTANCE, SOLID, PYROPHORIC	3391	4.2	
ORGANIC PIGMENTS, SELF-HEATING	3313	4.2		ORGANOMETALLIC SUBSTANCE, SOLID, SELF-HEATING	3400	4.2	
ORGANOARSENIC COMPOUND, LIQUID, N.O.S.	3280	6.1		ORGANOMETALLIC SUBSTANCE, LIQUID, PYROPHORIC, WATER-REACTIVE	3394	4.2	
ORGANOARSENIC COMPOUND, SOLID, N.O.S.	3465	6.1		ORGANOMETALLIC SUBSTANCE, SOLID, PYROPHORIC, WATER-REACTIVE	3393	4.2	
ORGANOCHLORINE PESTICIDE, LIQUID, FLAMMABLE, TOXIC, flash-point less than 23 °C	2762	3		ORGANOMETALLIC SUBSTANCE, LIQUID, WATER-REACTIVE	3398	4.3	
ORGANOCHLORINE PESTICIDE, LIQUID, TOXIC	2996	6.1		ORGANOMETALLIC SUBSTANCE, SOLID, WATER-REACTIVE	3395	4.3	
ORGANOCHLORINE PESTICIDE, LIQUID, TOXIC, FLAMMABLE, flash-point not less than 23 °C	2995	6.1		ORGANOMETALLIC SUBSTANCE, LIQUID, WATER-REACTIVE, FLAMMABLE	3399	4.3	
ORGANOCHLORINE PESTICIDE, SOLID, TOXIC	2761	6.1		ORGANOMETALLIC SUBSTANCE, SOLID, WATER-REACTIVE, FLAMMABLE	3396	4.3	
ORGANOMETALLIC COMPOUND, TOXIC, LIQUID, N.O.S.	3282	6.1					

Name and description	UN No.	Class	Remarks	Name and description	UN No.	Class	Remarks
ORGANOMETALLIC SUBSTANCE, SOLID, WATER-REACTIVE, SELF-HEATING	3397	4.3		ORGANOTIN PESTICIDE, SOLID, TOXIC	2786	6.1	
				Orthophospohoric acid, see	1805	8	
ORGANOPHOSPHORUS COMPOUND, TOXIC, LIQUID, N.O.S.	3278	6.1		OSMIUM TETROXIDE	2471	6.1	
				OXIDIZING LIQUID, N.O.S.	3139	5.1	
ORGANOPHOSPHORUS COMPOUND, TOXIC, SOLID, N.O.S.	3464	6.1		OXIDIZING LIQUID, CORROSIVE, N.O.S.	3098	5.1	
ORGANOPHOSPHORUS COMPOUND, TOXIC, FLAMMABLE, N.O.S.	3279	6.1		OXIDIZING LIQUID, TOXIC, N.O.S.	3099	5.1	
				OXIDIZING SOLID, N.O.S.	1479	5.1	
ORGANOPHOSPHORUS PESTICIDE, LIQUID, FLAMMABLE, TOXIC, flash-point less than 23 °C	2784	3		OXIDIZING SOLID, CORROSIVE, N.O.S.	3085	5.1	
				OXIDIZING SOLID, FLAMMABLE, N.O.S.	3137	5.1	Carriage prohibited
ORGANOPHOSPHORUS PESTICIDE, LIQUID, TOXIC	3018	6.1		OXIDIZING SOLID, SELF-HEATING, N.O.S.	3100	5.1	Carriage prohibited
ORGANOPHOSPHORUS PESTICIDE, LIQUID, TOXIC, FLAMMABLE, flash-point not less than 23 °C	3017	6.1		OXIDIZING SOLID, TOXIC, N.O.S.	3087	5.1	
				OXIDIZING SOLID, WATER-REACTIVE, N.O.S.	3121	5.1	Carriage prohibited
ORGANOPHOSPHORUS PESTICIDE, SOLID, TOXIC	2783	6.1		Oxirane, see	1040	2	
ORGANOTIN COMPOUND, LIQUID, N.O.S.	2788	6.1		OXYGEN, COMPRESSED	1072	2	
ORGANOTIN COMPOUND, SOLID, N.O.S.	3146	6.1		OXYGEN DIFLUORIDE, COMPRESSED	2190	2	
ORGANOTIN PESTICIDE, LIQUID, FLAMMABLE, TOXIC, flash-point less than 23 °C	2787	3		OXYGEN GENERATOR, CHEMICAL	3356	5.1	
ORGANOTIN PESTICIDE, LIQUID, TOXIC	3020	6.1		OXYGEN, REFRIGERATED LIQUID	1073	2	
				1-Oxy-4-nitrobenzene, see	1663	6.1	
ORGANOTIN PESTICIDE, LIQUID, TOXIC, FLAMMABLE, flash-point not less than 23 °C	3019	6.1		PAINT (including paint, lacquer, enamel, stain, shellac, varnish, polish, liquid filler and liquid lacquer base)	1263 3066 3469 3470	3 8 3 8	

Name and description	UN No.	Class	Remarks	Name and description	UN No.	Class	Remarks
PAINT RELATED MATERIAL (including paint thinning and reducing compound)	1263 3066 3469 3470	3 8 3 8		Pentafluoroethane, 1,1,1-trifluoroethane, and 1,1,1,2-tetrafluoroethane zeotropic mixture with approximately 44% pentafluoroethane and 52% 1,1,1-trifluoroethane, see	3337	2	
Paint thinning and reducing compound, see	1263 3066 3469 3470	3 8 3 8		PENTAMETHYLHEPTANE	2286	3	
PAPER, UNSATURATED OIL TREATED, incompletely dried (including carbon paper)	1379	4.2		Pentanal, see	2058	3	
				PENTANE-2,4-DIONE	2310	3	
Paraffin, see	1223	3		PENTANES, liquid	1265	3	
PARAFORMALDEHYDE	2213	4.1		n-Pentane, see	1265	3	
PARALDEHYDE	1264	3		PENTANOLS	1105	3	
PCBs, see	2315 3432	9 9		n-Pentanol, see	1105	3	
PENTABORANE	1380	4.2		3-Pentanol, see	1105	3	
PENTACHLOROETHANE	1669	6.1		1-PENTENE	1108	3	
PENTACHLOROPHENOL	3155	6.1		1-PENTOL	2705	8	
PENTAERYTHRITE TETRANITRATE with not less than 7% wax, by mass	0411	1		PENTOLITE, dry or wetted with less than 15% water, by mass	0151	1	
PENTAERYTHRITE TETRANITRATE, DESENSITIZED with not less than 15% phlegmatizer, by mass	0150	1		Pentyl nitrite, see	1113	3	
				PERCHLORATES, INORGANIC, N.O.S.	1481	5.1	
PENTAERYTHRITE TETRANITRATE MIXTURE, DESENSITIZED, SOLID, N.O.S. with more than 10% but not more than 20% PETN, by mass	3344	4.1		PERCHLORATES, INORGANIC, AQUEOUS SOLUTION, N.O.S.	3211	5.1	
				PERCHLORIC ACID with more than 50% but not more than 72% acid, by mass	1873	5.1	
PENTAERYTHRITE TETRANITRATE, WETTED with not less than 25% water, by mass	0150	1		PERCHLORIC ACID with not more than 50% acid, by mass	1802	8	
				Perchlorobenzene, see	2729	6.1	
PENTAERYTHRITOL TETRANITRATE, see	0150 0411	1 1		Perchlorocyclopentadiene, see	2646	6.1	
PENTAFLUOROETHANE	3220	2		Perchloroethylene, see	1897	6.1	

Name and description	UN No.	Class	Remarks	Name and description	UN No.	Class	Remarks
PERCHLOROMETHYL MERCAPTAN	1670	6.1		PETN, see	0150	1	
					0411	1	
PERCHLORYL FLUORIDE	3083	2		PETN/TNT, see	0151	1	
Perfluoroacetylchloride, see	3057	2		PETROL	1203	3	
PERFLUORO(ETHYL VINYL ETHER)	3154	2		PETROLEUM CRUDE OIL	1267	3	
PERFLUORO(METHYL VINYL ETHER)	3153	2		PETROLEUM DISTILLATES, N.O.S.	1268	3	
Perfluoropropane, see	2424	2		Petroleum ether, see	1268	3	
PERFUMERY PRODUCTS with flammable solvents	1266	3		PETROLEUM GASES, LIQUEFIED	1075	2	
PERMANGANATES, INORGANIC, N.O.S.	1482	5.1		Petroleum naphtha, see	1268	3	
				Petroleum oil, see	1268	3	
PERMANGANATES, INORGANIC, AQUEOUS SOLUTION, N.O.S.	3214	5.1		PETROLEUM PRODUCTS, N.O.S.	1268	3	
				Petroleum raffinate, see	1268	3	
PEROXIDES, INORGANIC, N.O.S.	1483	5.1		Petroleum spirit, see	1268	3	
PERSULPHATES, INORGANIC, N.O.S.	3215	5.1		PHENACYL BROMIDE	2645	6.1	
				PHENETIDINES	2311	6.1	
PERSULPHATES, INORGANIC, AQUEOUS SOLUTION, N.O.S.	3216	5.1		PHENOLATES, LIQUID	2904	8	
				PHENOLATES, SOLID	2905	8	
PESTICIDE, LIQUID, FLAMMABLE, TOXIC, N.O.S., flash-point less than 23 °C	3021	3		PHENOL, MOLTEN	2312	6.1	
				PHENOL, SOLID	1671	6.1	
PESTICIDE, LIQUID, TOXIC, N.O.S.	2902	6.1		PHENOL SOLUTION	2821	6.1	
PESTICIDE, LIQUID, TOXIC, FLAMMABLE, N.O.S., flash-point not less than 23 °C	2903	6.1		PHENOLSULPHONIC ACID, LIQUID	1803	8	
PESTICIDE, SOLID, TOXIC, N.O.S.	2588	6.1		PHENOXYACETIC ACID DERIVATIVE PESTICIDE, LIQUID, FLAMMABLE, TOXIC, flash-point less than 23 °C	3346	3	
Pesticide, toxic, under compressed gas, n.o.s, see	1950	2					

Name and description	UN No.	Class	Remarks	Name and description	UN No.	Class	Remarks
PHENOXYACETIC ACID DERIVATIVE PESTICIDE, LIQUID, TOXIC	3348	6.1		PHENYLMERCURIC HYDROXIDE	1894	6.1	
PHENOXYACETIC ACID DERIVATIVE PESTICIDE, LIQUID, TOXIC, FLAMMABLE, flash-point not less than 23 °C	3347	6.1		PHENYLMERCURIC NITRATE	1895	6.1	
				PHENYLPHOSPHORUS DICHLORIDE	2798	8	
				PHENYLPHOSPHORUS THIODICHLORIDE	2799	8	
PHENOXYACETIC ACID DERIVATIVE PESTICIDE, SOLID, TOXIC	3345	6.1		2-Phenylpropene, see	2303	3	
PHENYLACETONITRILE, LIQUID	2470	6.1		PHENYLTRICHLORO-SILANE	1804	8	
PHENYLACETYL CHLORIDE	2577	8		PHOSGENE	1076	2	
				9-PHOSPHABICYCLO-NONANES	2940	4.2	
Phenylamine, see	1547	6.1					
1-Phenylbutane, see	2709	3		PHOSPHINE	2199	2	
2-Phenylbutane, see	2709	3		Phosphoretted hydrogen, see	2199	2	
PHENYLCARBYLAMINE CHLORIDE	1672	6.1		PHOSPHORIC ACID, SOLUTION	1805	8	
PHENYL CHLOROFORMATE	2746	6.1		PHOSPHORIC ACID, SOLID	3453	8	
Phenyl cyanide, see	2224	6.1		Phosphoric acid, anhydrous, see	1807	8	
PHENYLENEDIAMINES (o-, m-, p-)	1673	6.1		PHOSPHOROUS ACID	2834	8	
Phenylethylene, see	2055	3		PHOSPHORUS, AMORPHOUS	1338	4.1	
PHENYLHYDRAZINE	2572	6.1		Phosphorus bromide, see	1808	8	
PHENYL ISOCYANATE	2487	6.1		Phosphorus chloride, see	1809	6.1	
Phenylisocyanodichloride, see	1672	6.1		PHOSPHORUS HEPTASULPHIDE, free from yellow and white phosphorus	1339	4.1	
PHENYL MERCAPTAN	2337	6.1					
PHENYLMERCURIC ACETATE	1674	6.1		PHOSPHORUS OXYBROMIDE	1939	8	
PHENYLMERCURIC COMPOUND, N.O.S.	2026	6.1		PHOSPHORUS OXYBROMIDE, MOLTEN	2576	8	

Name and description	UN No.	Class	Remarks	Name and description	UN No.	Class	Remarks
PHOSPHORUS OXYCHLORIDE	1810	8		PHOSPHORUS, YELLOW, DRY	1381	4.2	
PHOSPHORUS PENTABROMIDE	2691	8		PHOSPHORUS, YELLOW, IN SOLUTION	1381	4.2	
PHOSPHORUS PENTACHLORIDE	1806	8		PHOSPHORUS, YELLOW, UNDER WATER	1381	4.2	
PHOSPHORUS PENTAFLUORIDE	2198	2		Phosphoryl chloride, see	1810	8	
PHOSPHORUS PENTASULPHIDE, free from yellow and white phosphorus	1340	4.3		PHTHALIC ANHYDRIDE with more than 0.05% of maleic anhydride	2214	8	
PHOSPHORUS PENTOXIDE	1807	8		PICOLINES	2313	3	
PHOSPHORUS SESQUISULPHIDE, free from yellow and white phosphorus	1341	4.1		PICRAMIDE, see	0153	1	
				PICRIC ACID WETTED, see	3364	4.1	
Phosphorus (V) sulphide, free from yellow and white phosphorus, see	1340	4.3		PICRITE, see	0282	1	
				PICRITE, WETTED, see	1336	4.1	
Phosphorus sulphochloride, see	1837	8		Picrotoxin, see	3172	6.1	
					3462	6.1	
PHOSPHORUS TRIBROMIDE	1808	8		PICRYL CHLORIDE, see	0155	1	
PHOSPHORUS TRICHLORIDE	1809	6.1		PICRYL CHLORIDE, WETTED, see	3365	4.1	
				alpha-PINENE	2368	3	
PHOSPHORUS TRIOXIDE	2578	8		PINE OIL	1272	3	
PHOSPHORUS TRISULPHIDE, free from yellow and white phosphorus	1343	4.1		PIPERAZINE	2579	8	
				PIPERIDINE	2401	8	
PHOSPHORUS, WHITE, DRY	1381	4.2		Pivaloyl chloride, see	2438	6.1	
				Plastic explosives , see	0084	1	
PHOSPHORUS, WHITE IN SOLUTION	1381	4.2		PLASTICS MOULDING COMPOUND in dough, sheet or extruded rope form evolving flammable vapour	3314	9	
PHOSPHORUS, WHITE, MOLTEN	2447	4.2		PLASTICS, NITROCELLULOSE-BASED, SELF-HEATING, N.O.S.	2006	4.2	
PHOSPHORUS, WHITE, UNDER WATER	1381	4.2					

Name and description	UN No.	Class	Remarks	Name and description	UN No.	Class	Remarks
Polish, see	1263	3		Potassium bisulphate, see	2509	8	
	3066	8		Potassium bisulphite solution, see	2693	8	
	3469	3					
	3470	8					
POLYAMINES, FLAMMABLE, CORROSIVE, N.O.S.	2733	3		POTASSIUM BOROHYDRIDE	1870	4.3	
POLYAMINES, LIQUID, CORROSIVE, N.O.S.	2735	8		POTASSIUM BROMATE	1484	5.1	
				POTASSIUM CHLORATE	1485	5.1	
POLYAMINES, LIQUID, CORROSIVE, FLAMMABLE, N.O.S.	2734	8		POTASSIUM CHLORATE, AQUEOUS SOLUTION	2427	5.1	
POLYAMINES, SOLID, CORROSIVE, N.O.S.	3259	8		Potassium chlorate mixed with mineral oil, see	0083	1	
POLYCHLORINATED BIPHENYLS, LIQUID	2315	9		POTASSIUM CUPROCYANIDE	1679	6.1	
POLYCHLORINATED BIPHENYLS, SOLID	3432	9		POTASSIUM CYANIDE, SOLID	1680	6.1	
POLYESTER RESIN KIT	3269	3		POTASSIUM CYANIDE, SOLUTION	3413	6.1	
POLYHALOGENATED BIPHENYLS, LIQUID	3151	9		Potassium dicyanocuprate (I), see	1679	6.1	
POLYHALOGENATED BIPHENYLS, SOLID	3152	9		POTASSIUM DITHIONITE	1929	4.2	
POLYHALOGENATED TERPHENYLS, LIQUID	3151	9		POTASSIUM FLUORIDE, SOLID	1812	6.1	
POLYHALOGENATED TERPHENYLS, SOLID	3152	9		POTASSIUM FLUORIDE, SOLUTION	3422	6.1	
POLYMERIC BEADS, EXPANDABLE, evolving flammable vapour	2211	9		POTASSIUM FLUOROACETATE	2628	6.1	
Polystyrene beads, expandable, see	2211	9		POTASSIUM FLUOROSILICATE	2655	6.1	
				Potassium hexafluorosilicate, see	2655	6.1	
POTASSIUM	2257	4.3					
POTASSIUM ARSENATE	1677	6.1		Potassium hydrate, see	1814	8	
POTASSIUM ARSENITE	1678	6.1		POTASSIUM HYDROGENDIFLUORIDE, SOLID	1811	8	
Potassium bifluoride, see	1811	8					

Name and description	UN No.	Class	Remarks	Name and description	UN No.	Class	Remarks
POTASSIUM HYDROGENDIFLUORIDE, SOLUTION	3421	8		Potassium selenate, see	2630	6.1	
				Potassium selenite, see	2630	6.1	
POTASSIUM HYDROGEN SULPHATE	2509	8		Potassium silicofluoride, see	2655	6.1	
POTASSIUM HYDROSULPHITE, see	1929	4.2		POTASSIUM SODIUM ALLOYS, LIQUID	1422	4.3	
Potassium hydroxide, liquid, see	1814	8		POTASSIUM SODIUM ALLOYS, SOLID	3404	4.3	
POTASSIUM HYDROXIDE, SOLID	1813	8		POTASSIUM SULPHIDE with less than 30% water of crystallization	1382	4.2	
POTASSIUM HYDROXIDE SOLUTION	1814	8		POTASSIUM SULPHIDE, ANHYDROUS	1382	4.2	
POTASSIUM METAL ALLOYS, LIQUID	1420	4.3		POTASSIUM SULPHIDE, HYDRATED with not less than 30% water of crystallization	1847	8	
POTASSIUM METAL ALLOYS, SOLID	3403	4.3					
POTASSIUM METAVANADATE	2864	6.1		POTASSIUM SUPEROXIDE	2466	5.1	
				Potassium tetracyano-mercurate (II), see	1626	6.1	
POTASSIUM MONOXIDE	2033	8					
POTASSIUM NITRATE	1486	5.1		POWDER CAKE, WETTED with not less than 17% alcohol, by mass	0433	1	
Potassium nitrate and sodium nitrate mixture, see	1499	5.1					
POTASSIUM NITRATE AND SODIUM NITRITE MIXTURE	1487	5.1		POWDER CAKE, WETTED with not less than 25% water, by mass	0159	1	
POTASSIUM NITRITE	1488	5.1		POWDER PASTE, see	0159	1	
					0433	1	
POTASSIUM PERCHLORATE	1489	5.1		POWDER, SMOKELESS	0160	1	
					0161	1	
POTASSIUM PERMANGANATE	1490	5.1		Power devices, explosive, see	0275	1	
					0276	1	
					0323	1	
POTASSIUM PEROXIDE	1491	5.1			0381	1	
				PRIMERS, CAP TYPE	0044	1	
POTASSIUM PERSULPHATE	1492	5.1			0377	1	
					0378	1	
POTASSIUM PHOSPHIDE	2012	4.3		Primers, small arms, see	0044	1	

Name and description	UN No.	Class	Remarks	Name and description	UN No.	Class	Remarks
PRIMERS, TUBULAR	0319	1		Propellant with a single base,	0160	1	
	0320	1		Propellant with a double base,	0161	1	
	0376	1		Propellant with a triple base, see			
PRINTING INK, flammable or PRINTING INK RELATED MATERIAL (including printing ink thinning or reducing compound), flammable	1210	3					
				Propene, see	1077	2	
				PROPIONALDEHYDE	1275	3	
				PROPIONIC ACID with not less than 10% and less than 90% acid by mass	1848	8	
Projectiles, illuminating, see	0171	1					
	0254	1					
	0297	1		PROPIONIC ACID with not less than 90% acid by mass	3463	8	
PROJECTILES, inert with tracer	0345	1					
	0424	1					
	0425	1		PROPIONIC ANHYDRIDE	2496	8	
				PROPIONITRILE	2404	3	
PROJECTILES with burster or expelling charge	0346	1					
	0347	1		PROPIONYL CHLORIDE	1815	3	
	0426	1					
	0427	1		n-PROPYL ACETATE	1276	3	
	0434	1					
	0435	1		PROPYL ALCOHOL, NORMAL, see	1274	3	
PROJECTILES with bursting charge	0167	1					
	0168	1		PROPYLAMINE	1277	3	
	0169	1					
	0324	1		n-PROPYLBENZENE	2364	3	
	0344	1					
				Propyl chloride, see	1278	3	
PROPADIENE, STABILIZED	2200	2					
				n-PROPYL CHLOROFORMATE	2740	6.1	
Propadiene and methyl acetylene mixture, stabilized, see	1060	2					
				PROPYLENE	1077	2	
				PROPYLENE CHLOROHYDRIN	2611	6.1	
PROPANE	1978	2					
PROPANETHIOLS	2402	3		1,2-PROPYLENEDIAMINE	2258	8	
n-PROPANOL	1274	3		Propylene dichloride, see	1279	3	
PROPELLANT, LIQUID	0495	1		PROPYLENEIMINE, STABILIZED	1921	3	
	0497	1					
PROPELLANT, SOLID	0498	1		PROPYLENE OXIDE	1280	3	
	0499	1					
	0501	1		PROPYLENE TETRAMER	2850	3	
				Propylene trimer, see	2057	3	

Name and description	UN No.	Class	Remarks	Name and description	UN No.	Class	Remarks
PROPYL FORMATES	1281	3		PYROPHORIC SOLID, ORGANIC, N.O.S.	2846	4.2	
n-PROPYL ISOCYANATE	2482	6.1		PYROSULPHURYL CHLORIDE	1817	8	
Propyl mercaptan, see	2402	3					
n-PROPYL NITRATE	1865	3		Pyroxylin solution, see	2059	3	
PROPYLTRICHLOROSILANE	1816	8		PYRROLIDINE	1922	3	
Pyrazine hexahydride, see	2579	8		QUINOLINE	2656	6.1	
PYRETHROID PESTICIDE, LIQUID, FLAMMABLE, TOXIC, flash-point less than 23 °C	3350	3		Quinone, see	2587	6.1	
				RADIOACTIVE MATERIAL, EXCEPTED PACKAGE - ARTICLES MANUFACTURED FROM NATURAL URANIUM or DEPLETED URANIUM or NATURAL THORIUM	2909	7	
PYRETHROID PESTICIDE, LIQUID, TOXIC	3352	6.1					
PYRETHROID PESTICIDE, LIQUID, TOXIC, FLAMMABLE, flash-point not less than 23 °C	3351	6.1		RADIOACTIVE MATERIAL, EXCEPTED PACKAGE - EMPTY PACKAGING	2908	7	
PYRETHROID PESTICIDE, SOLID, TOXIC	3349	6.1		RADIOACTIVE MATERIAL, EXCEPTED PACKAGE - INSTRUMENTS or ARTICLES	2911	7	
PYRIDINE	1282	3					
Pyrophoric organometallic compound, water-reactive, n.o.s., liquid, see	3394	4.2		RADIOACTIVE MATERIAL, EXCEPTED PACKAGE - LIMITED QUANTITY OF MATERIAL	2910	7	
Pyrophoric organometallic compound, water-reactive, n.o.s., solid, see	3393	4.2					
PYROPHORIC ALLOY, N.O.S.	1383	4.2		RADIOACTIVE MATERIAL, LOW SPECIFIC ACTIVITY (LSA-I), non fissile or fissile-excepted	2912	7	
PYROPHORIC LIQUID, INORGANIC, N.O.S.	3194	4.2					
PYROPHORIC LIQUID, ORGANIC, N.O.S.	2845	4.2		RADIOACTIVE MATERIAL, LOW SPECIFIC ACTIVITY (LSA-II), FISSILE	3324	7	
PYROPHORIC METAL, N.O.S.	1383	4.2					
PYROPHORIC SOLID, INORGANIC, N.O.S.	3200	4.2					

Name and description	UN No.	Class	Remarks	Name and description	UN No.	Class	Remarks
RADIOACTIVE MATERIAL, LOW SPECIFIC ACTIVITY (LSA-II), non fissile or fissile-excepted	3321	7		RADIOACTIVE MATERIAL, TYPE A PACKAGE, SPECIAL FORM, FISSILE	3333	7	
RADIOACTIVE MATERIAL, LOW SPECIFIC ACTIVITY, (LSA-III), FISSILE	3325	7		RADIOACTIVE MATERIAL, TYPE A PACKAGE, SPECIAL FORM, non fissile or fissile-excepted	3332	7	
RADIOACTIVE MATERIAL, LOW SPECIFIC ACTIVITY (LSA-III), non fissile or fissile-excepted	3322	7		RADIOACTIVE MATERIAL, TYPE B(M) PACKAGE, FISSILE	3329	7	
RADIOACTIVE MATERIAL, SURFACE CONTAMINATED OBJECTS (SCO-I or SCO-II), FISSILE	3326	7		RADIOACTIVE MATERIAL, TYPE B(M) PACKAGE, non fissile or fissile-excepted	2917	7	
RADIOACTIVE MATERIAL, SURFACE CONTAMINATED OBJECTS (SCO-I or SCO-II), non fissile or fissile-excepted	2913	7		RADIOACTIVE MATERIAL, TYPE B(U) PACKAGE, FISSILE	3328	7	
				RADIOACTIVE MATERIAL, TYPE B(U) PACKAGE, non fissile or fissile-excepted	2916	7	
RADIOACTIVE MATERIAL, TRANSPORTED UNDER SPECIAL ARRANGEMENT, FISSILE	3331	7		RADIOACTIVE MATERIAL, TYPE C PACKAGE, FISSILE	3330	7	
RADIOACTIVE MATERIAL, TRANSPORTED UNDER SPECIAL ARRANGEMENT, non fissile or fissile-excepted	2919	7		RADIOACTIVE MATERIAL, TYPE C PACKAGE, non fissile or fissile-excepted	3323	7	
				RADIOACTIVE MATERIAL, URANIUM HEXAFLUORIDE, FISSILE	2977	7	
RADIOACTIVE MATERIAL, TYPE A PACKAGE, FISSILE, non-special form	3327	7		RADIOACTIVE MATERIAL, URANIUM HEXAFLUORIDE, non fissile or fissile-excepted	2978	7	
RADIOACTIVE MATERIAL, TYPE A PACKAGE, non-special form, non fissile or fissile-excepted	2915	7		Rags, oily	1856	4.2	Not subject to ADN
				RDX, see	0072	1	
					0391	1	
					0483	1	

Name and description	UN No.	Class	Remarks	Name and description	UN No.	Class	Remarks
RECEPTACLES, SMALL, CONTAINING GAS without a release device, non-refillable	2037	2		REFRIGERANT GAS R 124, see	1021	2	
Red phosphorus, see	1338	4.1		REFRIGERANT GAS R 125, see	3220	2	
REFRIGERANT GAS, N.O.S., such as mixture F1, mixture F2 or mixture P2	1078	2		REFRIGERANT GAS R 133a, see	1983	2	
REFRIGERANT GAS R 12, see	1028	2		REFRIGERANT GAS R 134a, see	3159	2	
REFRIGERANT GAS R 12B1, see	1974	2		REFRIGERANT GAS R 142b, see	2517	2	
REFRIGERANT GAS R 13, see	1022	2		REFRIGERANT GAS R 143a, see	2035	2	
REFRIGERANT GAS R 13B1, see	1009	2		REFRIGERANT GASR 152a, see	1030	2	
REFRIGERANT GAS R 14, see	1982	2		REFRIGERANT GAS R 161, see	2453	2	
REFRIGERANT GAS R 21, see	1029	2		REFRIGERANT GAS R 218, see	2424	2	
REFRIGERANT GAS R 22, see	1018	2		REFRIGERANT GAS R 227, see	3296	2	
REFRIGERANT GAS R 23, see	1984	2		REFRIGERANT GAS R 404A	3337	2	
REFRIGERANT GAS R 32, see	3252	2		REFRIGERANT GAS R 407A	3338	2	
REFRIGERANT GAS R 40, see	1063	2		REFRIGERANT GAS R 407B	3339	2	
REFRIGERANT GAS R 41, see	2454	2		REFRIGERANT GAS R 407C	3340	2	
REFRIGERANT GAS R 114, see	1958	2		REFRIGERANT GAS R 500, see	2602	2	
REFRIGERANT GAS R 115, see	1020	2		REFRIGERANT GAS R 502, see	1973	2	
REFRIGERANT GAS R 116, see	2193	2		REFRIGERANT GAS R 503, see	2599	2	
				REFRIGERANT GAS R 1132a, see	1959	2	

Name and description	UN No.	Class	Remarks	Name and description	UN No.	Class	Remarks
REFRIGERANT GAS R 1216, see	1858	2		ROCKETS with expelling charge	0436 0437 0438	1 1 1	
REFRIGERANT GAS R 1318, see	2422	2		ROCKETS with inert head	0183 0502	1 1	
REFRIGERANT GAS RC 318, see	1976	2		ROCKETS, LINE-THROWING	0238 0240 0453	1 1 1	
REFRIGERATING MACHINES containing flammable, non-toxic, liquefied gas	3358	2		ROCKETS, LIQUID FUELLED with bursting charge	0397 0398	1 1	
REFRIGERATING MACHINES containing non-flammable, non-toxic, gases or ammonia solutions (UN 2672)	2857	2		ROSIN OIL	1286	3	
				RUBBER SCRAP, powdered or granulated	1345	4.1	
REGULATED MEDICAL WASTE, N.O.S.	3291	6.2		RUBBER SHODDY, powdered or granulated	1345	4.1	
RELEASE DEVICES, EXPLOSIVE	0173	1		RUBBER SOLUTION	1287	3	
RESIN SOLUTION, flammable	1866	3		RUBIDIUM	1423	4.3	
				RUBIDIUM HYDROXIDE	2678	8	
Resorcin, see	2876	6.1		RUBIDIUM HYDROXIDE SOLUTION	2677	8	
RESORCINOL	2876	6.1					
RIVETS, EXPLOSIVE	0174	1		Saltpetre, see	1486	5.1	
ROCKET MOTORS	0186 0280 0281	1 1 1		SAMPLES, EXPLOSIVE, other than initiating explosive	0190	1	
				Sand acid, see	1778	8	
ROCKET MOTORS, LIQUID FUELLED	0395 0396	1 1		SEAT-BELT PRETENSIONERS	0503 3268	1 9	
ROCKET MOTORS WITH HYPERGOLIC LIQUIDS with or without expelling charge	0250 0322	1 1		SEED CAKE with more than 1.5% oil and not more than 11% moisture	1386	4.2	
ROCKETS with bursting charge	0180 0181 0182 0295	1 1 1 1		SEED CAKE with not more than 1.5% oil and not more than 11% moisture	2217	4.2	
				Seed expellers, see	1386 2217	4.2 4.2	

Name and description	UN No.	Class	Remarks	Name and description	UN No.	Class	Remarks
SELENATES	2630	6.1		SELF-HEATING SOLID, ORGANIC, N.O.S.	3088	4.2	
SELENIC ACID	1905	8					
SELENITES	2630	6.1		SELF-HEATING SOLID, OXIDIZING, N.O.S	3127	4.2	Carriage prohi-bited
SELENIUM COMPOUND, LIQUID, N.O.S.	3440	6.1		SELF-HEATING SOLID, TOXIC, INORGANIC, N.O.S.	3191	4.2	
SELENIUM COMPOUND, SOLID, N.O.S.	3283	6.1		SELF-HEATING SOLID, TOXIC, ORGANIC, N.O.S.	3128	4.2	
SELENIUM DISULPHIDE	2657	6.1		SELF-REACTIVE LIQUID TYPE B	3221	4.1	
SELENIUM HEXAFLUORIDE	2194	2		SELF-REACTIVE LIQUID TYPE B, TEMPERATURE CONTROLLED	3231	4.1	
SELENIUM OXYCHLORIDE	2879	8		SELF-REACTIVE LIQUID TYPE C	3223	4.1	
SELF-HEATING LIQUID, CORROSIVE, INORGANIC, N.O.S.	3188	4.2		SELF-REACTIVE LIQUID TYPE C, TEMPERATURE CONTROLLED	3233	4.1	
SELF-HEATING LIQUID, CORROSIVE, ORGANIC, N.O.S.	3185	4.2		SELF-REACTIVE LIQUID TYPE D	3225	4.1	
SELF-HEATING LIQUID, INORGANIC, N.O.S.	3186	4.2		SELF-REACTIVE LIQUID TYPE D, TEMPERATURE CONTROLLED	3235	4.1	
SELF-HEATING LIQUID, ORGANIC, N.O.S.	3183	4.2		SELF-REACTIVE LIQUID TYPE E	3227	4.1	
SELF-HEATING LIQUID, TOXIC, INORGANIC, N.O.S.	3187	4.2		SELF-REACTIVE LIQUID TYPE E, TEMPERATURE CONTROLLED	3237	4.1	
SELF-HEATING LIQUID, TOXIC, ORGANIC, N.O.S.	3184	4.2		SELF-REACTIVE LIQUID TYPE F	3229	4.1	
SELF-HEATING SOLID, CORROSIVE, INORGANIC, N.O.S.	3192	4.2		SELF-REACTIVE LIQUID TYPE F, TEMPERATURE CONTROLLED	3239	4.1	
SELF-HEATING SOLID, CORROSIVE, ORGANIC, N.O.S.	3126	4.2		SELF-REACTIVE SOLID TYPE B	3222	4.1	
SELF-HEATING SOLID, INORGANIC, N.O.S.	3190	4.2					

Name and description	UN No.	Class	Remarks	Name and description	UN No.	Class	Remarks
SELF-REACTIVE SOLID TYPE B, TEMPERATURE CONTROLLED	3232	4.1		SIGNALS, RAILWAY TRACK, EXPLOSIVE	0192	1	
					0193	1	
					0492	1	
					0493	1	
SELF-REACTIVE SOLID TYPE C	3224	4.1		SIGNALS, SMOKE	0196	1	
					0197	1	
SELF-REACTIVE SOLID TYPE C, TEMPERATURE CONTROLLED	3234	4.1			0313	1	
					0487	1	
				SILANE	2203	2	
SELF-REACTIVE SOLID TYPE D	3226	4.1		Silicofluoric acid, see	1778	8	
SELF-REACTIVE SOLID TYPE D, TEMPERATURE CONTROLLED	3236	4.1		Silicofluorides, n.o.s., see	2856	6.1	
				Silicon chloride, see	1818	8	
SELF-REACTIVE SOLID TYPE E	3228	4.1		SILICON POWDER, AMORPHOUS	1346	4.1	
SELF-REACTIVE SOLID TYPE E, TEMPERATURE CONTROLLED	3238	4.1		SILICON TETRACHLORIDE	1818	8	
				SILICON TETRAFLUORIDE	1859	2	
SELF-REACTIVE SOLID TYPE F	3230	4.1		SILVER ARSENITE	1683	6.1	
SELF-REACTIVE SOLID TYPE F, TEMPERATURE CONTROLLED	3240	4.1		SILVER CYANIDE	1684	6.1	
				SILVER NITRATE	1493	5.1	
SHALE OIL	1288	3		SILVER PICRATE, WETTED with not less than 30% water, by mass	1347	4.1	
Shaped charges, see	0059	1					
	0439	1					
	0440	1		SLUDGE ACID	1906	8	
	0441	1					
				SODA LIME with more than 4% sodium hydroxide	1907	8	
Shellac, see	1263	3					
	3066	8					
	3469	3		SODIUM	1428	4.3	
	3470	8					
SIGNAL DEVICES, HAND	0191	1		Sodium aluminate, solid	2812	8	Not subject to ADN
	0373	1					
SIGNALS, DISTRESS, ship	0194	1		SODIUM ALUMINATE SOLUTION	1819	8	
	0195	1					
Signals, distress, ship, water-activated, see	0249	1		SODIUM ALUMINIUM HYDRIDE	2835	4.3	

Name and description	UN No.	Class	Remarks	Name and description	UN No.	Class	Remarks
SODIUM AMMONIUM VANADATE	2863	6.1		SODIUM CUPROCYANIDE, SOLID	2316	6.1	
SODIUM ARSANILATE	2473	6.1		SODIUM CUPROCYANIDE SOLUTION	2317	6.1	
SODIUM ARSENATE	1685	6.1		SODIUM CYANIDE, SOLID	1689	6.1	
SODIUM ARSENITE, AQUEOUS SOLUTION	1686	6.1		SODIUM CYANIDE, SOLUTION	3414	6.1	
SODIUM ARSENITE, SOLID	2027	6.1		Sodium dicyanocuprate (I), solid, see	2316	6.1	
SODIUM AZIDE	1687	6.1		Sodium dicyanocuprate (I) solution, see	2317	6.1	
Sodium bifluoride, see	2439	8		Sodium dimethylarsenate, see	1688	6.1	
Sodium binoxide, see	1504	5.1		SODIUM DINITRO-o-CRESOLATE, dry or wetted with less than 15% water, by mass	0234	1	
Sodium bisulphite solution, see	2693	8					
SODIUM BOROHYDRIDE	1426	4.3		SODIUM DINITRO-o-CRESOLATE, WETTED with not less than 10% water, by mass	3369	4.1	
SODIUM BOROHYDRIDE AND SODIUM HYDROXIDE SOLUTION, with not more than 12% sodium borohydride and not more than 40% sodium hydroxide by mass	3320	8					
				SODIUM DINITRO-o-CRESOLATE, WETTED with not less than 15% water, by mass	1348	4.1	
SODIUM BROMATE	1494	5.1					
SODIUM CACODYLATE	1688	6.1		Sodium dioxide, see	1504	5.1	
SODIUM CARBONATE PEROXYHYDRATE	3378	5.1		SODIUM DITHIONITE	1384	4.2	
SODIUM CHLORATE	1495	5.1		SODIUM FLUORIDE, SOLID	1690	6.1	
SODIUM CHLORATE, AQUEOUS SOLUTION	2428	5.1		SODIUM FLUORIDE, SOLUTION	3415	6.1	
Sodium chlorate mixed with dinitrotoluene, see	0083	1		SODIUM FLUOROACETATE	2629	6.1	
SODIUM CHLORITE	1496	5.1		SODIUM FLUOROSILICATE	2674	6.1	
SODIUM CHLOROACETATE	2659	6.1		Sodium hexafluorosilicate, see	2674	6.1	
				Sodium hydrate, see	1824	8	

Name and description	UN No.	Class	Remarks	Name and description	UN No.	Class	Remarks
SODIUM HYDRIDE	1427	4.3		SODIUM PERBORATE MONOHYDRATE	3377	5.1	
Sodium hydrogen 4-amino-phenylarsenate, see	2473	6.1		SODIUM PERCHLORATE	1502	5.1	
SODIUM HYDROGENDIFLUORIDE	2439	8		SODIUM PERMANGANATE	1503	5.1	
SODIUM HYDROSULPHIDE with less than 25% water of crystallization	2318	4.2		SODIUM PEROXIDE	1504	5.1	
SODIUM HYDROSULPHIDE, HYDRATED with not less than 25% water of crystallization	2949	8		SODIUM PEROXOBORATE, ANHYDROUS	3247	5.1	
				SODIUM PERSULPHATE	1505	5.1	
SODIUM HYDROSULPHITE, see	1384	4.2		SODIUM PHOSPHIDE	1432	4.3	
				SODIUM PICRAMATE, dry or wetted with less than 20% water, by mass	0235	1	
SODIUM HYDROXIDE, SOLID	1823	8		SODIUM PICRAMATE, WETTED with not less than 20% water, by mass	1349	4.1	
SODIUM HYDROXIDE SOLUTION	1824	8					
Sodium metasilicate pentahydrate, see	3253	8		Sodium potassium alloys, liquid, see	1422	4.3	
				Sodium selenate, see	2630	6.1	
SODIUM METHYLATE	1431	4.2		Sodium selenite, see	2630	6.1	
SODIUM METHYLATE SOLUTION in alcohol	1289	3		Sodium silicofluoride, see	2674	6.1	
				SODIUM SULPHIDE, ANHYDROUS	1385	4.2	
SODIUM MONOXIDE	1825	8					
SODIUM NITRATE	1498	5.1		SODIUM SULPHIDE with less than 30% water of crystallization	1385	4.2	
SODIUM NITRATE AND POTASSIUM NITRATE MIXTURE	1499	5.1					
				SODIUM SULPHIDE, HYDRATED with not less than 30% water	1849	8	
SODIUM NITRITE	1500	5.1					
Sodium nitrite and potassium nitrate mixture, see	1487	5.1		SODIUM SUPEROXIDE	2547	5.1	
SODIUM PENTACHLOROPHENATE	2567	6.1		SOLIDS CONTAINING CORROSIVE LIQUID, N.O.S.	3244	8	

Name and description	UN No.	Class	Remarks	Name and description	UN No.	Class	Remarks
SOLIDS or mixtures of solids (such as preparations and wastes) CONTAINING FLAMMABLE LIQUID, N.O.S. having a flash-point up to 60°C	3175	4.1		STRONTIUM PERCHLORATE	1508	5.1	
				STRONTIUM PEROXIDE	1509	5.1	
				STRONTIUM PHOSPHIDE	2013	4.3	
SOLIDS CONTAINING TOXIC LIQUID, N.O.S.	3243	6.1		STRYCHNINE	1692	6.1	
				STRYCHNINE SALTS	1692	6.1	
Solvents, flammable, n.o.s., see	1993	3		STYPHNIC ACID, see	0219	1	
					0394	1	
Solvents, flammable, toxic, n.o.s., see	1992	3		STYRENE MONOMER, STABILIZED	2055	3	
SOUNDING DEVICES, EXPLOSIVE	0204	1					
	0296	1		SUBSTANCES, EVI, N.O.S., see	0482	1	
	0374	1					
	0375	1		SUBSTANCES, EXPLOSIVE, N.O.S.	0357	1	
Squibs, see	0325	1			0358	1	
	0454	1			0359	1	
					0473	1	
Stain, see	1263	3			0474	1	
	3066	8			0475	1	
	3469	3			0476	1	
	3470	8			0477	1	
STANNIC CHLORIDE, ANHYDROUS	1827	8			0478	1	
					0479	1	
					0480	1	
STANNIC CHLORIDE PENTAHYDRATE	2440	8			0481	1	
					0485	1	
STANNIC PHOSPHIDES	1433	4.3		SUBSTANCES, EXPLOSIVE, VERY INSENSITIVE, N.O.S.	0482	1	
Steel swarf, see	2793	4.2					
STIBINE	2676	2		Substances liable to spontaneous combustion, n.o.s., see	2845	4.2	
					2846	4.2	
					3194	4.2	
Straw	1327	4.1	Not subject to ADN		3200	4.2	
				SUBSTANCES WITH A FLASH-POINT ABOVE 60 °C which are carried heated within a limiting range of 15K below their flash-point	9001	3	Dangerous in tank vessels only
Strontium alloys, pyrophoric, see	1383	4.2					
STRONTIUM ARSENITE	1691	6.1		SUBSTANCES WITH A FLASH-POINT ABOVE 60 °C AND NOT MORE THAN 100 °C, which do not belong to another Class	9003	9	Dangerous in tank vessels only
STRONTIUM CHLORATE	1506	5.1					
Strontium dioxide, see	1509	5.1					
STRONTIUM NITRATE	1507	5.1					

Name and description	UN No.	Class	Remarks	Name and description	UN No.	Class	Remarks
SUBSTANCES WITH AN AUTO-IGNITION TEMPERATURE OF 200 °C AND BELOW, n.o.s.	9002	3	Dangerous in tank vessels only	SULPHUR, MOLTEN	2448	4.1	
SUBSTITUTED NITROPHENOL PESTICIDE, LIQUID, FLAMMABLE, TOXIC, flash-point less than 23 °C	2780	3		Sulphur monochloride, see	1828	8	
				SULPHUROUS ACID	1833	8	
				SULPHUR TETRAFLUORIDE	2418	2	
SUBSTITUTED NITROPHENOL PESTICIDE, LIQUID, TOXIC	3014	6.1		SULPHUR TRIOXIDE, STABILIZED	1829	8	
				SULPHURYL CHLORIDE	1834	8	
				SULPHURYL FLUORIDE	2191	2	
SUBSTITUTED NITROPHENOL PESTICIDE, LIQUID, TOXIC, FLAMMABLE, flash-point not less than 23 °C	3013	6.1		Talcum with tremolite and/or actinolite, see	2590	9	
				TARS, LIQUID, including road asphalt and oils, bitumen and cut backs, with a flash-point not greater than 60 °C	1999	3	
SUBSTITUTED NITROPHENOL PESTICIDE, SOLID, TOXIC	2779	6.1					
SULPHAMIC ACID	2967	8		Tars, liquid, with flash-point above 60 °C, at or above its flash-point, see	3256	9	
SULPHUR	1350	4.1		Tars, liquid, at or above 100 °C and below its flash-point, see	3257	9	
SULPHUR CHLORIDES	1828	8					
Sulphur dichloride, see	1828	8		Tartar emetic, see	1551	6.1	
SULPHUR DIOXIDE	1079	2		TEAR GAS CANDLES	1700	6.1	
Sulphuretted hydrogen, see	1053	2		TEAR GAS SUBSTANCE, LIQUID, N.O.S.	1693	6.1	
SULPHUR HEXAFLUORIDE	1080	2		TEAR GAS SUBSTANCE, SOLID, N.O.S.	3448	6.1	
SULPHURIC ACID with more than 51% acid	1830	8		TELLURIUM COMPOUND, N.O.S.	3284	6.1	
SULPHURIC ACID with not more than 51% acid	2796	8		TELLURIUM HEXAFLUORIDE	2195	2	
SULPHURIC ACID, FUMING	1831	8		TERPENE HYDROCARBONS, N.O.S.	2319	3	
SULPHURIC ACID, SPENT	1832	8		TERPINOLENE	2541	3	
Sulphuric and hydrofluoric acid mixture, see	1786	8		TETRABROMOETHANE	2504	6.1	

Name and description	UN No.	Class	Remarks	Name and description	UN No.	Class	Remarks
1,1,2,2-TETRACHLORO-ETHANE	1702	6.1		TETRAMETHYL-AMMONIUM HYDROXIDE, SOLUTION	1835	8	
TETRACHLORO-ETHYLENE	1897	6.1		Tetramethylene, see	2601	2	
TETRAETHYL DITHIO-PYROPHOSPHATE	1704	6.1		Tetramethylene cyanide, see	2205	6.1	
TETRAETHYLENE-PENTAMINE	2320	8		Tetramethyl lead, see	1649	6.1	
Tetraethyl lead, see	1649	6.1		TETRAMETHYLSILANE	2749	3	
TETRAETHYL SILICATE	1292	3		TETRANITROANILINE	0207	1	
Tetraethyoxysilane, see	1292	3		TETRANITROMETHANE	1510	5.1	
Tetrafluorodichloroethane, see	1958	2		TETRAPROPYL ORTHOTITANATE	2413	3	
1,1,1,2-TETRA-FLUOROETHANE	3159	2		TETRAZENE, WETTED with not less than 30% water, or mixture of alcohol and water, by mass, see	0114	1	
TETRAFLUORO-ETHYLENE, STABILIZED	1081	2		TETRAZOL-1-ACETIC ACID	0407	1	
TETRAFLUOROMETHANE	1982	2		1H-TETRAZOLE	0504	1	
1,2,3,6-TETRAHYDRO-BENZALDEHYDE	2498	3		TETRYL, see	0208	1	
TETRAHYDROFURAN	2056	3		Textile waste, wet	1857	4.2	Not subject to ADN
TETRAHYDRO-FURFURYLAMINE	2943	3		THALLIUM CHLORATE	2573	5.1	
Tetrahydro-1,4-oxazine, see	2054	3		Thallium (I) chlorate, see	2573	5.1	
TETRAHYDROPHTHALIC ANHYDRIDES with more than 0.05% of maleic anhydride	2698	8		THALLIUM COMPOUND, N.O.S.	1707	6.1	
				THALLIUM NITRATE	2727	6.1	
1,2,3,6-TETRAHYDRO-PYRIDINE	2410	3		Thallium (I) nitrate, see	2727	6.1	
TETRAHYDROTHIOPHENE	2412	3		Thallous chlorate, see	2573	5.1	
Tetramethoxysilane, see	2606	6.1		4-THIAPENTANAL	2785	6.1	
				Thia-4-pentanal, see	2785	6.1	
TETRAMETHYL-AMMONIUM HYDROXIDE, SOLID	3423	8		THIOACETIC ACID	2436	3	

Name and description	UN No.	Class	Remarks	Name and description	UN No.	Class	Remarks
THIOCARBAMATE PESTICIDE, LIQUID, FLAMMABLE, TOXIC, flash-point less than 23 °C	2772	3		TITANIUM POWDER, WETTED with not less than 25% water	1352	4.1	
THIOCARBAMATE PESTICIDE, LIQUID, TOXIC	3006	6.1		TITANIUM SPONGE GRANULES	2878	4.1	
THIOCARBAMATE PESTICIDE, LIQUID, TOXIC, FLAMMABLE, flash-point not less than 23 °C	3005	6.1		TITANIUM SPONGE POWDERS	2878	4.1	
				TITANIUM TETRACHLORIDE	1838	8	
THIOCARBAMATE PESTICIDE, SOLID, TOXIC	2771	6.1		TITANIUM TRICHLORIDE MIXTURE	2869	8	
THIOGLYCOL	2966	6.1		TITANIUM TRICHLORIDE MIXTURE, PYROPHORIC	2441	4.2	
THIOGLYCOLIC ACID	1940	8		TITANIUM TRICHLORIDE, PYROPHORIC	2441	4.2	
THIOLACTIC ACID	2936	6.1		TNT, see	0209	1	
THIONYL CHLORIDE	1836	8			0388	1	
					0389	1	
THIOPHENE	2414	3		TNT mixed with aluminium, see	0390	1	
Thiophenol, see	2337	6.1					
THIOPHOSGENE	2474	6.1		TNT, WETTED with not less than 30% water, by mass, see	1356	4.1	
THIOPHOSPHORYL CHLORIDE	1837	8		TNT, WETTED with not less than 10% water, by mass, see	3366	4.1	
THIOUREA DIOXIDE	3341	4.2		Toe puffs, nitrocellulose base, see	1353	4.1	
Tin (IV) chloride, anhydrous, see	1827	8					
Tin (IV) chloride pentahydrate, see	2440	8		TOLUENE	1294	3	
TINCTURES, MEDICINAL	1293	3		TOLUENE DIISOCYANATE	2078	6.1	
Tin tetrachloride, see	1827	8		TOLUIDINES, LIQUID	1708	6.1	
				TOLUIDINES, SOLID	3451	6.1	
TITANIUM DISULPHIDE	3174	4.2		Toluol, see	1294	3	
TITANIUM HYDRIDE	1871	4.1		2,4-TOLUYLENEDIAMINE, SOLID	1709	6.1	
TITANIUM POWDER, DRY	2546	4.2		2,4-TOLUYLENEDIAMINE, SOLUTION	3418	6.1	

Name and description	UN No.	Class	Remarks	Name and description	UN No.	Class	Remarks
Toluylene diisocyanate, see	2078	6.1		TOXIC BY INHALATION LIQUID, FLAMMABLE, N.O.S. with an inhalation toxicity lower than or equal to 200 ml/m^3 and saturated vapour concentration greater than or equal to 500 LC_{50}	3383	6.1	
Tolylene diisocyanate, see	2078	6.1					
Tolylethylene, inhibited, see	2618	3					
TORPEDOES with bursting charge	0329	1		TOXIC BY INHALATION LIQUID, FLAMMABLE, N.O.S. with an inhalation toxicity lower than or equal to 1000 ml/m^3 and saturated vapour concentration greater than or equal to 10 LC_{50}	3384	6.1	
	0330	1					
	0451	1					
TORPEDOES, LIQUID FUELLED with inert head	0450	1					
TORPEDOES, LIQUID FUELLED with or without bursting charge	0449	1		TOXIC BY INHALATION LIQUID, OXIDIZING, N.O.S. with an inhalation toxicity lower than or equal to 200 ml/m^3 and saturated vapour concentration greater than or equal to 500 LC_{50}	3387	6.1	
TOXIC BY INHALATION LIQUID, N.O.S. with an inhalation toxicity lower than or equal to 200 ml/m^3 and saturated vapour concentration greater than or equal to 500 LC_{50}	3381	6.1					
TOXIC BY INHALATION LIQUID, N.O.S. with an inhalation toxicity lower than or equal to 1000 ml/m^3 and saturated vapour concentration greater than or equal to 10 LC_{50}	3382	6.1		TOXIC BY INHALATION LIQUID, OXIDIZING, N.O.S. with an inhalation toxicity lower than or equal to 1000 ml/m^3 and saturated vapour concentration greater than or equal to 10 LC_{50}	3387	6.1	
TOXIC BY INHALATION LIQUID, CORROSIVE, N.O.S. with an inhalation toxicity lower than or equal to 200 ml/m^3 and saturated vapour concentration greater than or equal to 500 LC_{50}	3389	6.1		TOXIC BY INHALATION LIQUID, OXIDIZING, N.O.S. with an inhalation toxicity lower than or equal to 1000 ml/m^3 and saturated vapour concentration greater than or equal to 10 LC_{50}	3388	6.1	
TOXIC BY INHALATION LIQUID, CORROSIVE, N.O.S. with an inhalation toxicity lower than or equal to 1000 ml/m^3 and saturated vapour concentration greater than or equal to 10 LC_{50}	3390	6.1		TOXIC BY INHALATION LIQUID, WATER-REACTIVE, N.O.S. with an inhalation toxicity lower than or equal to 200 ml/m^3 and saturated vapour concentration greater than or equal to 500 LC_{50}	3385	6.1	

Name and description	UN No.	Class	Remarks	Name and description	UN No.	Class	Remarks
TOXIC BY INHALATION LIQUID, WATER-REACTIVE, N.O.S. with an inhalation toxicity lower than or equal to 1000 ml/m^3 and saturated vapour concentration greater than or equal to 10 LC$_{50}$	3386	6.1		TOXIC SOLID, SELF-HEATING, N.O.S.	3124	6.1	
				TOXIC SOLID, WATER-REACTIVE, N.O.S.	3125	6.1	
				TOXINS, EXTRACTED FROM LIVING SOURCES, LIQUID, N.O.S.	3172	6.1	
TOXIC LIQUID, CORROSIVE, INORGANIC, N.O.S.	3289	6.1		TOXINS, EXTRACTED FROM LIVING SOURCES, SOLID, N.O.S.	3462	6.1	
TOXIC LIQUID, CORROSIVE, ORGANIC, N.O.S.	2927	6.1		TRACERS FOR AMMUNITION	0212 0306	1 1	
TOXIC LIQUID, FLAMMABLE, ORGANIC, N.O.S.	2929	6.1		Tremolite, see	2590	9	
				TRIALLYLAMINE	2610	3	
TOXIC LIQUID, INORGANIC, N.O.S.	3287	6.1		TRIALLYL BORATE	2609	6.1	
TOXIC LIQUID, ORGANIC, N.O.S.	2810	6.1		TRIAZINE PESTICIDE, LIQUID, FLAMMABLE, TOXIC, flash-point less than 23 °C	2764	3	
TOXIC LIQUID, OXIDIZING, N.O.S.	3122	6.1		TRIAZINE PESTICIDE, LIQUID, TOXIC	2998	6.1	
TOXIC LIQUID, WATER-REACTIVE, N.O.S.	3123	6.1		TRIAZINE PESTICIDE, LIQUID, TOXIC, FLAMMABLE, flash-point not less than 23 °C	2997	6.1	
TOXIC SOLID, CORROSIVE, INORGANIC, N.O.S.	3290	6.1					
TOXIC SOLID, CORROSIVE, ORGANIC, N.O.S.	2928	6.1		TRIAZINE PESTICIDE, SOLID, TOXIC	2763	6.1	
				Tribromoborane, see	2692	8	
TOXIC SOLID, FLAMMABLE, ORGANIC, N.O.S.	2930	6.1		TRIBUTYLAMINE	2542	6.1	
				TRIBUTYLPHOSPHANE	3254	4.2	
TOXIC SOLID, INORGANIC, N.O.S.	3288	6.1		Trichloroacetaldehyde, see	2075	6.1	
				TRICHLOROACETIC ACID	1839	8	
TOXIC SOLID, ORGANIC, N.O.S.	2811	6.1		TRICHLOROACETIC ACID SOLUTION	2564	8	
TOXIC SOLID, OXIDIZING, N.O.S.	3086	6.1		Trichlororaceticaldehyde, see	2075	6.1	

Name and description	UN No.	Class	Remarks	Name and description	UN No.	Class	Remarks
TRICHLOROACETYL CHLORIDE	2442	8		1,1,1-TRIFLUOROETHANE	2035	2	
TRICHLOROBENZENES, LIQUID	2321	6.1		TRIFLUOROMETHANE	1984	2	
				TRIFLUOROMETHANE, REFRIGERATED LIQUID	3136	2	
TRICHLOROBUTENE	2322	6.1		2-TRIFLUOROMETHYL-ANILINE	2942	6.1	
1,1,1-TRICHLOROETHANE	2831	6.1					
TRICHLOROETHYLENE	1710	6.1		3-TRIFLUOROMETHYL-ANILINE	2948	6.1	
TRICHLOROISO-CYANURIC ACID, DRY	2468	5.1		TRIISOBUTYLENE	2324	3	
Trichloronitromethane, see	1580	6.1		TRIISOPROPYL BORATE	2616	3	
TRICHLOROSILANE	1295	4.3		TRIMETHYLACETYL CHLORIDE	2438	6.1	
1,3,5-Trichloro-s-triazine-2,4,6-trione, see	2468	5.1		TRIMETHYLAMINE, ANHYDROUS	1083	2	
2,4,6-Trichloro-1,3,5- triazine, see	2670	8		TRIMETHYLAMINE, AQUEOUS SOLUTION, not more than 50% trimethylamine, by mass	1297	3	
TRICRESYL PHOSPHATE with more than 3% ortho isomer	2574	6.1					
				1,3,5-TRIMETHYL-BENZENE	2325	3	
TRIETHYLAMINE	1296	3					
Triethyl borate, see	1176	3		TRIMETHYL BORATE	2416	3	
TRIETHYLENE-TETRAMINE	2259	8		TRIMETHYLCHLORO-SILANE	1298	3	
Triethyl orthoformate, see	2524	3		TRIMETHYLCYCLO-HEXYLAMINE	2326	8	
TRIETHYL PHOSPHITE	2323	3		Trimethylene chlorobromide, see	2688	6.1	
TRIFLUOROACETIC ACID	2699	8					
TRIFLUOROACETYL CHLORIDE	3057	2		TRIMETHYLHEXA-METHYLENEDIAMINES	2327	8	
Trifluorobromomethane, see	1009	2		TRIMETHYLHEXA-METHYLENE DIISOCYANATE	2328	6.1	
Trifluorochloroethane, see	1983	2					
TRIFLUOROCHLORO-ETHYLENE, STABILIZED	1082	2		2,4,4-Trimethylpentene-1, see	2050	3	
				2,4,4-Trimethylpentene-2, see	2050	3	
Trifluorochloromethane, see	1022	2					

Name and description	UN No.	Class	Remarks	Name and description	UN No.	Class	Remarks
TRIMETHYL PHOSPHITE	2329	3		TRINITROPHENOL, WETTED with not less than 30% water, by mass	1344	4.1	
TRINITROANILINE	0153	1					
TRINITROANISOLE	0213	1		TRINITROPHENOL WETTED with not less than 10% water, by mass	3364	4.1	
TRINITROBENZENE, dry or wetted with less than 30% water, by mass	0214	1		TRINITROPHENYL-METHYLNITRAMINE	0208	1	
TRINITROBENZENE, WETTED with not less than 10% water, by mass	3367	4.1		TRINITRORESORCINOL, dry or wetted with less than 20% water, or mixture of alcohol and water, by mass	0219	1	
TRINITROBENZENE, WETTED with not less than 30% water, by mass	1354	4.1		TRINITRORESORCINOL, WETTED with not less than 20% water, or mixture of alcohol and water, by mass	0394	1	
TRINITROBENZENE-SULPHONIC ACID	0386	1					
TRINITROBENZOIC ACID, dry or wetted with less than 30% water, by mass	0215	1		TRINITROTOLUENE (TNT), dry or wetted with less than 30% water, by mass	0209	1	
TRINITROBENZOIC ACID, WETTED with not less than 10% water, by mass	3368	4.1		TRINITROTOLUENE AND HEXANITROSTILBENE MIXTURE	0388	1	
TRINITROBENZOIC ACID, WETTED with not less than 30% water, by mass	1355	4.1		TRINITROTOLUENE MIXTURE CONTAINING TRINITROBENZENE AND HEXANITROSTILBENE	0389	1	
TRINITROCHLORO-BENZENE	0155	1		TRINITROTOLUENE AND TRINITROBENZENE MIXTURE	0388	1	
TRINITROCHLOROBENZENE WETTED with not less than 10% water, by mass	3365	4.1					
TRINITRO-m-CRESOL	0216	1		TRINITROTOLUENE, WETTED with not less than 10% water, by mass	3366	4.1	
TRINITROFLUORENONE	0387	1		TRINITROTOLUENE, WETTED with not less than 30% water, by mass	1356	4.1	
TRINITRONAPHTHALENE	0217	1					
TRINITROPHENETOLE	0218	1		TRIPROPYLAMINE	2260	3	
TRINITROPHENOL, dry or wetted with less than 30% water, by mass	0154	1		TRIPROPYLENE	2057	3	
				TRIS-(1-AZIRIDINYL) PHOSPHINE OXIDE SOLUTION	2501	6.1	

Name and description	UN No.	Class	Remarks	Name and description	UN No.	Class	Remarks
TRITONAL	0390	1		VANADIUM TETRACHLORIDE	2444	8	
Tropilidene, see	2603	3		VANADIUM TRICHLORIDE	2475	8	
TUNGSTEN HEXAFLUORIDE	2196	2		VANADYL SULPHATE	2931	6.1	
TURPENTINE	1299	3		Varnish, see	1263	3	
					3066	8	
TURPENTINE SUBSTITUTE	1300	3			3469	3	
					3470	8	
UNDECANE	2330	3		Vehicle, flammable gas powered or vehicle, flammable liquid powered	3166	9	Not subject to ADN
UREA HYDROGEN PEROXIDE	1511	5.1		Villiaumite, see	1690	6.1	
UREA NITRATE, dry or wetted with less than 20% water, by mass	0220	1		VINYL ACETATE, STABILIZED	1301	3	
UREA NITRATE, WETTED with not less than 10% water, by mass	3370	4.1		Vinylbenzene, see	2055	3	
				VINYL BROMIDE, STABILIZED	1085	2	
UREA NITRATE, WETTED with not less than 20% water, by mass	1357	4.1		VINYL BUTYRATE, STABILIZED	2838	3	
Valeral, see	2058	3		VINYL CHLORIDE, STABILIZED	1086	2	
VALERALDEHYDE	2058	3		VINYL CHLOROACETATE	2589	6.1	
n-Valeraldehyde, see	2058	3		VINYL ETHYL ETHER, STABILIZED	1302	3	
Valeric aldehyde, see	2058	3					
VALERYL CHLORIDE	2502	8		VINYL FLUORIDE, STABILIZED	1860	2	
VANADIUM COMPOUND, N.O.S.	3285	6.1		VINYLIDENE CHLORIDE, STABILIZED	1303	3	
Vanadium (IV) oxide sulphate, see	2931	6.1		VINYL ISOBUTYL ETHER, STABILIZED	1304	3	
Vanadium oxysulphate, see	2931	6.1		VINYL METHYL ETHER, STABILIZED	1087	2	
VANADIUM OXYTRICHLORIDE	2443	8					
VANADIUM PENTOXIDE, non-fused form	2862	6.1		VINYLPYRIDINES, STABILIZED	3073	6.1	

Name and description	UN No.	Class	Remarks	Name and description	UN No.	Class	Remarks
VINYLTOLUENES, STABILIZED	2618	3		WHITE ASBESTOS (chrysotile, actinolite, anthophyllite, tremolite)	2590	9	
VINYLTRICHLORO-SILANE	1305	3		White spirit, see	1300	3	
Warheads for guided missiles, see	0286 0287 0369 0370 0371	1 1 1 1 1		WOOD PRESERVATIVES, LIQUID	1306	3	
				Wool waste, wet	1387	4.2	Not subject to ADN
WARHEADS, ROCKET with burster or expelling charge	0370 0371	1 1		XANTHATES	3342	4.2	
				XENON	2036	2	
WARHEADS, ROCKET with bursting charge	0286 0287 0369	1 1 1		XENON, REFRIGERATED LIQUID	2591	2	
WARHEADS, TORPEDO with bursting charge	0221	1		XYLENES	1307	3	
				XYLENOLS, LIQUID	3430	6.1	
WATER-REACTIVE LIQUID, N.O.S.	3148	4.3		XYLENOLS, SOLID	2261	6.1	
				XYLIDINES, LIQUID	1711	6.1	
WATER-REACTIVE LIQUID, CORROSIVE, N.O.S.	3129	4.3		XYLIDINES, SOLID	3452	6.1	
				Xylols, see	1307	3	
WATER-REACTIVE LIQUID, TOXIC, N.O.S.	3130	4.3		XYLYL BROMIDE, LIQUID	1701	6.1	
WATER-REACTIVE SOLID, N.O.S.	2813	4.3		XYLYL BROMIDE, SOLID	3417	6.1	
WATER-REACTIVE SOLID, CORROSIVE, N.O.S.	3131	4.3		ZINC AMMONIUM NITRITE	1512	5.1	
				ZINC ARSENATE	1712	6.1	
WATER-REACTIVE SOLID, FLAMMABLE, N.O.S.	3132	4.3	Carriage prohibited	ZINC ARSENATE AND ZINC ARSENITE MIXTURE	1712	6.1	
WATER-REACTIVE SOLID, OXIDIZING, N.O.S.	3133	4.3	Carriage prohibited	ZINC ARSENITE	1712	6.1	
WATER-REACTIVE SOLID, SELF-HEATING, N.O.S.	3135	4.3	Carriage prohibited	ZINC ASHES	1435	4.3	
				Zinc bisulphite solution, see	2693	8	
WATER-REACTIVE SOLID, TOXIC, N.O.S.	3134	4.3		ZINC BROMATE	2469	5.1	
White arsenic, see	1561	6.1		ZINC CHLORATE	1513	5.1	

Name and description	UN No.	Class	Remarks	Name and description	UN No.	Class	Remarks
ZINC CHLORIDE, ANHYDROUS	2331	8		ZIRCONIUM, DRY, coiled wire, finished metal sheets, strip (thinner than 254 microns but not thinner than 18 microns)	2858	4.1	
ZINC CHLORIDE SOLUTION	1840	8					
ZINC CYANIDE	1713	6.1		ZIRCONIUM, DRY, finished sheets, strip or coiled wire	2009	4.2	
ZINC DITHIONITE	1931	9		ZIRCONIUM HYDRIDE	1437	4.1	
ZINC DUST	1436	4.3		ZIRCONIUM NITRATE	2728	5.1	
ZINC FLUOROSILICATE	2855	6.1		ZIRCONIUM PICRAMATE, dry or wetted with less than 20% water, by mass	0236	1	
Zinc hexafluorosilicate, see	2855	6.1					
ZINC HYDROSULPHITE, see	1931	9		ZIRCONIUM PICRAMATE, WETTED with not less than 20% water, by mass	1517	4.1	
ZINC NITRATE	1514	5.1					
ZINC PERMANGANATE	1515	5.1		ZIRCONIUM POWDER, DRY	2008	4.2	
ZINC PEROXIDE	1516	5.1		ZIRCONIUM POWDER, WETTED with not less than 25% water	1358	4.1	
ZINC PHOSPHIDE	1714	4.3					
ZINC POWDER	1436	4.3		ZIRCONIUM SCRAP	1932	4.2	
ZINC RESINATE	2714	4.1		ZIRCONIUM SUSPENDED IN A FLAMMABLE LIQUID	1308	3	
Zinc selenate, see	2630	4.1					
Zinc selenite, see	2630	4.1		ZIRCONIUM TETRACHLORIDE	2503	8	
Zinc silicofluoride, see	2855	6.1					

CHAPTER 3.3

SPECIAL PROVISIONS APPLICABLE TO CERTAIN ARTICLES OR SUBSTANCES

3.3.1   When Column (6) of Table A of Chapter 3.2 indicates that a special provision is relevant to a substance or article, the meaning and requirements of that special provision are as set forth below.

16   Samples of new or existing explosive substances or articles may be carried as directed by the competent authorities (see 2.2.1.1.3) for purposes including: testing, classification, research and development, quality control, or as a commercial sample. Explosive samples which are not wetted or desensitised shall be limited to 10 kg in small packages as specified by the competent authorities. Explosive samples which are wetted or desensitised shall be limited to 25 kg.

23   Even though this substance has a flammability hazard, it only exhibits such hazard under extreme fire conditions in confined areas.

32   This substance is not subject to the requirements of ADN when in any other form.

37   This substance is not subject to the requirements of ADN when coated.

38   This substance is not subject to the requirements of ADN when it contains not more than 0.1% calcium carbide.

39   This substance is not subject to the requirements of ADN when it contains less than 30% or not less than 90% silicon.

43   When offered for carriage as pesticides, these substances shall be carried under the relevant pesticide entry and in accordance with the relevant pesticide provisions (see 2.2.61.1.10 to 2.2.61.1.11.2).

45   Antimony sulphides and oxides which contain not more than 0.5% of arsenic calculated on the total mass are not subject to the requirements of ADN.

47   Ferricyanides and ferrocyanides are not subject to the requirements of ADN.

48   The carriage of this substance, when it contains more than 20% hydrocyanic acid, is prohibited.

59   These substances are not subject to the requirements of ADN when they contain not more than 50% magnesium.

60   If the concentration is more than 72%, the carriage of this substance is prohibited.

61   The technical name which shall supplement the proper shipping name shall be the ISO common name (see also ISO 1750:1981 "*Pesticides and other agrochemicals - common names*", as amended), other name listed in the WHO "*Recommended Classification of Pesticides by Hazard and Guidelines to Classification*" or the name of the active substance (see also 3.1.2.8.1 and 3.1.2.8.1.1).

62   This substance is not subject to the requirements of ADN when it contains not more than 4% sodium hydroxide.

65   Hydrogen peroxide aqueous solutions with less than 8% hydrogen peroxide are not subject to the requirements of ADN.

103 The carriage of ammonium nitrites and mixtures of an inorganic nitrite with an ammonium salt is prohibited.

105 Nitrocellulose meeting the descriptions of UN No. 2556 or UN No. 2557 may be classified in Class 4.1.

113 The carriage of chemically unstable mixtures is prohibited.

119 Refrigerating machines include machines or other appliances which have been designed for the specific purpose of keeping food or other items at a low temperature in an internal compartment, and air conditioning units. Refrigerating machines and refrigerating machine components are not subject to the provisions of ADN if they contain less than 12 kg of gas in Class 2, group A or O according to 2.2.2.1.3, or if they contain less than 12 litres ammonia solution (UN No. 2672).

122 The subsidiary risks, control and emergency temperatures if any, and the UN number (generic entry) for each of the currently assigned organic peroxide formulations are given in 2.2.52.4.

127 Other inert material or inert material mixture may be used, provided this inert material has identical phlegmatizing properties.

131 The phlegmatized substance shall be significantly less sensitive than dry PETN.

135 The dihydrated sodium salt of dichloroisocyanuric acid is not subject to the requirements of ADN.

138 p-Bromobenzyl cyanide is not subject to the requirements of ADN.

141 Products which have undergone sufficient heat treatment so that they present no hazard during carriage are not subject to the requirements of ADN.

142 Solvent extracted soya bean meal containing not more than 1.5% oil and 11% moisture, which is substantially free of flammable solvent, is not subject to the requirements of ADN.

144 An aqueous solution containing not more than 24% alcohol by volume is not subject to the requirements of ADN.

145 Alcoholic beverages of packing group III, when carried in receptacles of 250 litres or less, are not subject to the requirements of ADN.

152 The classification of this substance will vary with particle size and packaging, but borderlines have not been experimentally determined. Appropriate classifications shall be made in accordance with 2.2.1.

153 This entry applies only if it is demonstrated, on the basis of tests, that the substances when in contact with water are not combustible nor show a tendency to auto-ignition and that the mixture of gases evolved is not flammable.

163 A substance mentioned by name in Table A of Chapter 3.2 shall not be carried under this entry. Substances carried under this entry may contain 20% or less nitrocellulose provided the nitrocellulose contains not more than 12.6% nitrogen (by dry mass).

168 Asbestos which is immersed or fixed in a natural or artificial binder (such as cement, plastics, asphalt, resins or mineral ore) in such a way that no escape of hazardous quantities of respirable asbestos fibres can occur during carriage is not subject to the requirements of ADN. Manufactured articles containing asbestos and not meeting this provision are nevertheless not subject to the requirements of ADN when packed so that no escape of hazardous quantities of respirable asbestos fibres can occur during carriage.

169 Phthalic anhydride in the solid state and tetrahydrophthalic anhydrides, with not more than 0.05% maleic anhydride, are not subject to the requirements of ADN. Phthalic anhydride molten at a temperature above its flash-point, with not more than 0.05% maleic anhydride, shall be classified under UN No. 3256.

172 For radioactive material with a subsidiary risk:

(a)   The packages shall be labelled with a label corresponding to each subsidiary risk exhibited by the material; corresponding placards shall be affixed to vehicles or containers in accordance with the relevant provisions of 5.3.1;

(b)   The radioactive material shall be allocated to packing groups I, II or III, as and if appropriate, by application of the grouping criteria provided in Part 2 corresponding to the nature of the predominant subsidiary risk.

The description required in 5.4.1.2.5.1 (b) shall include a description of these subsidiary risks (e.g. "Subsidiary risk: 3, 6.1"), the name of the constituents which most predominantly contribute to this (these) subsidiary risk(s), and where applicable, the packing group.

177 Barium sulphate is not subject to the requirements of ADN.

178 This designation shall be used only when no other appropriate designation exists in Table A of Chapter 3.2, and only with the approval of the competent authority of the country of origin (see 2.2.1.1.3).

181 Packages containing this type of substance shall bear a label conforming to model No. 1 (see 5.2.2.2.2) unless the competent authority of the country of origin has permitted this label to be dispensed with for the specific packaging employed because test data have proved that the substance in this packaging does not exhibit explosive behaviour (see 5.2.2.1.9).

182 The group of alkali metals includes lithium, sodium, potassium, rubidium and caesium.

183 The group of alkaline earth metals includes magnesium, calcium, strontium and barium.

186 In determining the ammonium nitrate content, all nitrate ions for which a molecular equivalent of ammonium ions is present in the mixture shall be calculated as ammonium nitrate.

188   Lithium cells and batteries offered for carriage are not subject to other provisions of ADN if they meet the following:

(a)   For a lithium metal or lithium alloy cell, the lithium content is not more than 1 g, and for a lithium-ion cell, the lithium-equivalent content is not more than 1.5 g;

(b)   For a lithium metal or lithium alloy battery the aggregate lithium content is not more than 2 g, and for a lithium-ion battery, the aggregate lithium-equivalent content is not more than 8 g;

(c)   Each cell or battery is of the type proved to meet the requirements of each test in the *Manual of Tests and Criteria*, Part III, sub-section 38.3;

(d)   Cells and batteries are separated so as to prevent short circuits and are packed in strong packagings, except when installed in equipment; and

(e)   Except when installed in equipment, each package containing more than 24 lithium cells or 12 lithium batteries shall in addition meet the following requirements:

(i)   Each package shall be marked indicating that it contains lithium batteries and that special procedures should be followed in the event that the package is damaged;

(ii)   Each shipment shall be accompanied with a document indicating that packages contain lithium batteries and that special procedures should be followed in the event a package is damaged;

(iii)   Each package is capable of withstanding a 1.2 m drop test in any orientation without damage to cells or batteries contained therein, without shifting of the contents so as to allow battery to battery (or cell to cell) contact and without release of contents; and

(iv)   Except in the case of lithium batteries packed with equipment, packages may not exceed 30 kg gross mass.

As used above and elsewhere in ADN, "lithium content" means the mass of lithium in the anode of a lithium metal or lithium alloy cell, except in the case of a lithium-ion cell the "lithium-equivalent content" in grams is calculated to be 0.3 times the rated capacity in ampere-hours.

190   Aerosol dispensers shall be provided with protection against inadvertent discharge. Aerosols with a capacity not exceeding 50 ml containing only non-toxic constituents are not subject to the requirements of ADN.

191   Receptacles, small, with a capacity not exceeding 50 ml, containing only non-toxic constituents are not subject to the requirements of ADN.

193 This entry may only be used for uniform ammonium nitrate based fertilizer mixtures of the nitrogen, phosphate or potash type, containing not more than 70% ammonium nitrate and not more than 0.4% total combustible/organic material calculated as carbon or with not more than 45% ammonium nitrate and unrestricted combustible material. Fertilizers within these composition limits are not subject to the requirements of ADN if shown by a Trough Test (see *Manual of Tests and Criteria*, Part III, sub-section 38.2) not to be liable to self-sustaining decomposition.

194 The control and emergency temperatures, if any, and the UN number (generic entry) for each of the currently assigned self-reactive substances are given in 2.2.41.4.

196 Formulations which in laboratory testing neither detonate in the cavitated state nor deflagrate, which show no effect when heated under confinement and which exhibit no explosive power may be carried under this entry. The formulation must also be thermally stable (i.e. the SADT is 60 °C or higher for a 50 kg package). Formulations not meeting these criteria shall be carried under the provisions of Class 5.2, (see 2.2.52.4).

198 Nitrocellulose solutions containing not more than 20 % nitrocellulose may be carried as paint or printing ink, as applicable (see UN Nos. 1210, 1263 and 3066).

199 Lead compounds which, when mixed in a ratio of 1:1000 with 0.07M hydrochloric acid and stirred for one hour at a temperature of 23 °C ± 2 °C, exhibit a solubility of 5 % or less are considered insoluble. See ISO 3711:1990 "*Lead chromate pigments and lead chromate - molybdate pigments - Specifications and methods of test*".

201 Lighters and lighter refills shall comply with the provisions of the country in which they were filled. They shall be provided with protection against inadvertent discharge. The liquid portion of the gas shall not exceed 85% of the capacity of the receptacle at 15 °C. The receptacles, including the closures, shall be capable of withstanding an internal pressure of twice the pressure of the liquefied petroleum gas at 55 °C. The valve mechanisms and ignition devices shall be securely sealed, taped or otherwise fastened or designed to prevent operation or leakage of the contents during carriage. Lighters shall not contain more than 10 g of liquefied petroleum gas. Lighter refills shall not contain more than 65 g of liquefied petroleum gas.

203 This entry shall not be used for polychlorinated biphenyls, liquid, UN No. 2315 and polychlorinated biphenyls, solid, UN No. 3432.

205 This entry shall not be used for UN No. 3155 PENTACHLOROPHENOL.

207 Polymeric beads and moulding compounds may be made from polystyrene, poly(methyl methacrylate) or other polymeric material.

208 The commercial grade of calcium nitrate fertilizer, when consisting mainly of a double salt (calcium nitrate and ammonium nitrate) containing not more than 10% ammonium nitrate and at least 12% water of crystallization, is not subject to the requirements of ADN.

210 Toxins from plant, animal or bacterial sources which contain infectious substances, or toxins that are contained in infectious substances, shall be classified in Class 6.2.

215 This entry only applies to the technically pure substance or to formulations derived from it having an SADT higher than 75 °C and therefore does not apply to formulations which are self-reactive substances (for self-reactive substances, see 2.2.41.4). Homogeneous mixtures containing not more than 35 % by mass of

azodicarbonamide and at least 65 % of inert substance are not subject to the requirements of ADN unless criteria of other classes are met.

216   Mixtures of solids which are not subject to the requirements of ADN and flammable liquids may be carried under this entry without first applying the classification criteria of Class 4.1, provided there is no free liquid visible at the time the substance is loaded or at the time the packaging, vehicle or container is closed. Sealed packets and articles containing less than 10 ml of a packing group II or III flammable liquid absorbed into a solid material are not subject to ADN provided there is no free liquid in the packet or article.

217   Mixtures of solids which are not subject to the requirements of ADN and toxic liquids may be carried under this entry without first applying the classification criteria of Class 6.1, provided there is no free liquid visible at the time the substance is loaded or at the time the packaging, vehicle or container is closed. This entry shall not be used for solids containing a packing group I liquid.

218   Mixtures of solids which are not subject to the requirements of ADN and corrosive liquids may be carried under this entry without first applying the classification criteria of Class 8, provided there is no free liquid visible at the time the substance is loaded or at the time the packaging, vehicle or container is closed.

219   Genetically modified micro-organisms and genetically modified organisms which meet the definition of an infectious substance and the criteria for inclusion in Class 6.2 in accordance with section 2.2.62 shall be carried as UN No. 2814, UN No. 2900 or UN No. 3373, as appropriate.

220   Only the technical name of the flammable liquid component of this solution or mixture shall be shown in parentheses immediately following the proper shipping name.

221   Substances included under this entry shall not be of packing group I.

224   Unless it can be demonstrated by testing that the sensitivity of the substance in its frozen state is no greater than in its liquid state, the substance shall remain liquid during normal transport conditions. It shall not freeze at temperatures above -15 °C.

225   Fire extinguishers under this entry may include installed actuating cartridges (cartridges, power device of classification code 1.4C or 1.4S), without changing the classification of Class 2, group A or O according to 2.2.2.1.3 provided the total quantity of deflagrating (propellant) explosives does not exceed 3.2 g per extinguishing unit.

226   Formulations of this substance containing not less than 30% non-volatile, non-flammable phlegmatizer are not subject to the requirements of ADN.

227   When phlegmatized with water and inorganic inert material the content of urea nitrate may not exceed 75% by mass and the mixture shall not be capable of being detonated by the Series 1, type (a), test in the *Manual of Tests and Criteria*, Part 1.

228   Mixtures not meeting the criteria for flammable gases (see 2.2.2.1.5) shall be carried under UN No. 3163.

230　This entry applies to cells and batteries containing lithium in any form, including lithium polymer and lithium ion cells and batteries.

Lithium cells and batteries may be carried under this entry if they meet the following provisions:

(a)　Each cell or battery is of the type proved to meet the requirements of each test of the *Manual of Tests and Criteria*, Part III, sub-section 38.3;

(b)　Each cell and battery incorporates a safety venting device or is designed to preclude a violent rupture under normal conditions of carriage;

(c)　Each cell and battery is equipped with an effective means of preventing external short circuits;

(d)　Each battery containing cells or series of cells connected in parallel is equipped with effective means as necessary to prevent dangerous reverse current flow (e.g. diodes, fuses, etc.).

235　This entry applies to articles which contain Class 1 explosive substances and which may also contain dangerous goods of other classes. These articles are used as life-saving vehicle air bag inflators or air bag modules or seat-belt pretensioners.

236　Polyester resin kits consist of two components: a base material (Class 3, packing group II or III) and an activator (organic peroxide). The organic peroxide shall be type D, E or F, not requiring temperature control. Packing group shall be II or III, according to the criteria for Class 3, applied to the base material. The quantity limit referred to in Column (7) of Table A of Chapter 3.2 applies to the base material.

237　The membrane filters, including paper separators, coating or backing materials, etc., that are present in carriage, shall not be liable to propagate a detonation as tested by one of the tests described in the *Manual of Tests and Criteria*, Part I, Test series 1 (a).

In addition the competent authority may determine, on the basis of the results of suitable burning rate tests taking account of the standard tests in the *Manual of Tests and Criteria*, Part III, sub-section 33.2.1, that nitrocellulose membrane filters in the form in which they are to be carried are not subject to the requirements applicable to flammable solids in Class 4.1.

238　(a)　Batteries can be considered as non-spillable provided that they are capable of withstanding the vibration and pressure differential tests given below, without leakage of battery fluid.

Vibration test: The battery is rigidly clamped to the platform of a vibration machine and a simple harmonic motion having an amplitude of 0.8 mm (1.6 mm maximum total excursion) is applied. The frequency is varied at the rate of 1 Hz/min between the limits of 10 Hz and 55 Hz. The entire range of frequencies and return is traversed in 95 ± 5 minutes for each mounting position (direction of vibration) of the battery. The battery is tested in three mutually perpendicular positions (to include testing with fill openings and vents, if any, in an inverted position) for equal time periods.

Pressure differential test: Following the vibration test, the battery is stored for six hours at 24 °C ± 4 °C while subjected to a pressure differential of at least 88 kPa. The battery is tested in three mutually perpendicular positions (to include testing with fill openings and vents, if any, in an inverted position) for at least six hours in each position.

(b)   Non-spillable batteries are not subject to the requirements of ADN if, at a temperature of 55 °C, the electrolyte will not flow from a ruptured or cracked case and there is no free liquid to flow and if, as packaged for carriage, the terminals are protected from short circuit.

239   Batteries or cells shall not contain dangerous substances other than sodium, sulphur and/or polysulphides. Batteries or cells shall not be offered for carriage at a temperature such that liquid elemental sodium is present in the battery or cell unless approved and under the conditions established by the competent authority of the country of origin. If the country of origin is not a Contracting Party to ADN, the approval and conditions of carriage shall be recognized by the competent authority of the first country Contracting Party to ADN reached by the consignment.

Cells shall consist of hermetically sealed metal casings which fully enclose the dangerous substances and which are so constructed and closed as to prevent the release of the dangerous substances under normal conditions of carriage.

Batteries shall consist of cells secured within and fully enclosed by a metal casing so constructed and closed as to prevent the release of the dangerous substances under normal conditions of carriage.

241   The formulation shall be prepared so that it remains homogeneous and does not separate during carriage. Formulations with low nitrocellulose contents and not showing dangerous properties when tested for their liability to detonate, deflagrate or explode when heated under defined confinement by tests of Test series 1 (a), 2 (b) and 2 (c) respectively in the *Manual of Tests and Criteria*, Part I and not being a flammable solid when tested in accordance with test No. 1 in the *Manual of Tests and Criteria*, Part III, sub-section 33.2.1.4 (chips, if necessary, crushed and sieved to a particle size of less than 1.25 mm) are not subject to the requirements of ADN.

242   Sulphur is not subject to the requirements of ADN when it has been formed to a specific shape (e.g. prills, granules, pellets, pastilles or flakes).

243   Gasoline, motor spirit and petrol for use in spark-ignition engines (e.g. in automobiles, stationary engines and other engines) shall be assigned to this entry regardless of variations in volatility.

244   This entry includes e.g. aluminium dross, aluminium skimmings, spent cathodes, spent potliner, and aluminium salt slags.

247   Alcoholic beverages containing more than 24% alcohol but not more than 70% by volume, when carried as part of the manufacturing process, may be carried in wooden barrels with a capacity of more than 250 litres and not more than 500 litres meeting the general requirements of 4.1.1 of ADR, as appropriate, on the following conditions:

(a)   The wooden barrels shall be checked and tightened before filling;

(b)   Sufficient ullage (not less than 3%) shall be left to allow for the expansion of the liquid;

(c)   The wooden barrels shall be carried with the bungholes pointing upwards;

(d)   The wooden barrels shall be carried in containers meeting the requirements of the CSC. Each wooden barrel shall be secured in custom-made cradles and be wedged by appropriate means to prevent it from being displaced in any way during carriage.

249   Ferrocerium, stabilized against corrosion, with a minimum iron content of 10% is not subject to the requirements of ADN.

250   This entry may only be used for samples of chemicals taken for analysis in connection with the implementation of the Convention on the Prohibition of the Development, Production, Stockpiling and Use of Chemical Weapons and on their Destruction. The carriage of substances under this entry shall be in accordance with the chain of custody and security procedures specified by the Organisation for the Prohibition of Chemical Weapons.

The chemical sample may only be carried providing prior approval has been granted by the competent authority or the Director General of the Organisation for the Prohibition of Chemical Weapons and providing the sample complies with the following provisions:

(a)   It shall be packed according to packing instruction 623 in the ICAO Technical Instructions (see S-3-8 of the Supplement); and

(b)   During carriage, a copy of the document of approval for transport, showing the quantity limitations and the packing provisions shall be attached to the transport document.

251   The entry CHEMICAL KIT or FIRST AID KIT is intended to apply to boxes, cases etc. containing small quantities of various dangerous goods which are used for example for medical, analytical or testing or repair purposes. Such kits may not contain dangerous goods for which the code "LQ0" has been indicated in Column (7) of Table A of Chapter 3.2.

Components shall not react dangerously (see "dangerous reaction" in 1.2.1). The total quantity of dangerous goods in any one kit shall not exceed either 1 *l* or 1 kg. The packing group assigned to the kit as a whole shall be the most stringent packing group assigned to any individual substance in the kit.

Kits which are carried on board vessels for first-aid or operating purposes are not subject to the requirements of ADN.

Chemical kits and first aid kits containing dangerous goods in inner packagings which do not exceed the quantity limits applicable to individual substances as specified in Column (7) of Table A of Chapter 3.2 in accordance with the LQ code defined in 3.4.6 may be carried in accordance with Chapter 3.4.

252   Provided the ammonium nitrate remains in solution under all conditions of carriage, aqueous solutions of ammonium nitrate, with not more than 0.2% combustible material, in a concentration not exceeding 80%, are not subject to the requirements of ADN.

266   This substance, when containing less alcohol, water or phlegmatizer than specified, shall not be carried unless specifically authorized by the competent authority (see 2.2.1.1).

267   Any explosives, blasting, type C containing chlorates shall be segregated from explosives containing ammonium nitrate or other ammonium salts.

270   Aqueous solutions of Class 5.1 inorganic solid nitrate substances are considered as not meeting the criteria of Class 5.1 if the concentration of the substances in solution at the minimum temperature encountered during carriage is not greater than 80% of the saturation limit.

271   Lactose or glucose or similar materials, may be used as a phlegmatizer provided that the substance contains not less than 90%, by mass, of phlegmatizer. The competent authority may authorize these mixtures to be classified in Class 4.1 on the basis of a test Series 6 (c) of Section 16 of Part I of the *Manual of Tests and Criteria* on at least three packages as prepared for carriage. Mixtures containing at least 98%, by mass, of phlegmatizer are not subject to the requirements of ADN. Packages containing mixtures with not less than 90%, by mass, of phlegmatizer need not bear a label conforming to model No. 6.1.

272   This substance shall not be carried under the provisions of Class 4.1 unless specifically authorized by the competent authority (see UN No. 0143).

273   Maneb and maneb preparations stabilized against self-heating need not be classified in Class 4.2 when it can be demonstrated by testing that a cubic volume of 1 m^3 of substance does not self-ignite and that the temperature at the centre of the sample does not exceed 200 °C, when the sample is maintained at a temperature of not less than 75 °C ± 2 °C for a period of 24 hours.

274   The provisions of 3.1.2.8 apply.

278   These substances shall not be classified and carried unless authorized by the competent authority on the basis of results from Series 2 tests and a Series 6(c) test of Part I of the *Manual of Tests and Criteria* on packages as prepared for carriage (see 2.2.1.1). The competent authority shall assign the packing group on the basis of 2.2.3 criteria and the package type used for the Series 6(c) test.

279   The substance is assigned to this classification or packing group based on human experience rather than the strict application of classification criteria set out in ADN.

280   This entry applies to articles which are used as life-saving vehicle air bag inflators, or air bag modules or seat-belt pretensioners and which contain dangerous goods of Class 1 or dangerous goods of other classes and when carried as component parts and when these articles as presented for carriage have been tested in accordance with Test series 6 (c) of Part I of the *Manual of Tests and Criteria*, with no explosion of the device, no fragmentation of device casing or pressure vessel, and no projection hazard nor thermal effect which would significantly hinder fire-fighting or other emergency response efforts in the immediate vicinity.

283   Articles, containing gas, intended to function as shock absorbers, including impact energy-absorbing devices, or pneumatic springs are not subject to the requirements of ADN provided:

       (a)   Each article has a gas space capacity not exceeding 1.6 litres and a charge pressure not exceeding 280 bar where the product of the capacity (litres) and charge pressure (bars) does not exceed 80 (i.e. 0.5 litres gas space and 160 bar charge pressure, 1 litre gas space and 80 bar charge pressure, 1.6 litres gas space and 50 bar charge pressure, 0.28 litres gas space and 280 bar charge pressure);

(b)   Each article has a minimum burst pressure of 4 times the charge pressure at 20 °C for products not exceeding 0.5 litres gas space capacity and 5 times charge pressure for products greater than 0.5 litres gas space capacity;

(c)   Each article is manufactured from material which will not fragment upon rupture;

(d)   Each article is manufactured in accordance with a quality assurance standard acceptable to the competent authority; and

(e)   The design type has been subjected to a fire test demonstrating that the article relieves its pressure by means of a fire degradable seal or other pressure relief device, such that the article will not fragment and that the article does not rocket.

See also 1.1.3.2 (d) of ADR for equipment used for the operation of the vehicle.

284   An oxygen generator, chemical, containing oxidizing substances shall meet the following conditions:

(a)   The generator when containing an explosive actuating device shall only be carried under this entry when excluded from Class 1 in accordance with the NOTE under paragraph 2.2.1.1.1 (b);

(b)   The generator, without its packaging, shall be capable of withstanding a 1.8 m drop test onto a rigid, non-resilient, flat and horizontal surface, in the position most likely to cause damage, without loss of its contents and without actuation;

(c)   When a generator is equipped with an actuating device, it shall have at least two positive means of preventing unintentional actuation.

286   Nitrocellulose membrane filters covered by this entry, each with a mass not exceeding 0.5 g, are not subject to the requirements of ADN when contained individually in an article or a sealed packet.

288   These substances shall not be classified and carried unless authorized by the competent authority on the basis of results from Series 2 tests and a Series 6 (c) test of Part I of the *Manual of tests and Criteria* on packages as prepared for carriage (see 2.2.1.1).

289   Air bags or seat-belts installed in conveyances or in completed conveyance components such as steering columns, door panels, seats, etc. are not subject to the requirements of ADN.

290   When this material meets the definitions and criteria of other classes as defined in Part 2, it shall be classified in accordance with the predominant subsidiary risk. Such material shall be declared under the proper shipping name and UN number appropriate for the material in that predominant Class, with the addition of the name applicable to this material according to Column (2) of Table A of Chapter 3.2, and shall be carried in accordance with the provisions applicable to that UN number. In addition, all other requirements specified in 2.2.7.9.1 shall apply, except 5.2.1.7.2.

291   Flammable liquefied gases shall be contained within refrigerating machine components. These components shall be designed and tested to at least three times the working pressure of the machinery. The refrigerating machines shall be designed and constructed to contain the liquefied gas and preclude the risk of bursting or cracking of

the pressure retaining components during normal conditions of carriage. Refrigerating machines and refrigerating-machine components are not subject to the requirements of ADN if they contain less than 12 kg of gas.

292   Mixtures containing not more than 23.5% oxygen by volume may be carried under this entry when no other oxidizing gases are present. A label conforming to model 5.1 is not required for any concentrations within this limit.

293   The following definitions apply to matches:

(a)   Fusee matches are matches the heads of which are prepared with a friction-sensitive igniter composition and a pyrotechnic composition which burns with little or no flame, but with intense heat;

(b)   Safety matches are matches which are combined with or attached to the box, book or card that can be ignited by friction only on a prepared surface;

(c)   Strike anywhere matches are matches that can be ignited by friction on a solid surface;

(d)   Wax Vesta matches are matches that can be ignited by friction either on a prepared surface or on a solid surface.

295   Batteries need not be individually marked and labelled if the pallet bears the appropriate mark and label.

296   These entries apply for life-saving appliances such as life rafts, personal flotation devices and self-inflating slides. UN No. 2990 applies to self-inflating appliances and UN No. 3072 applies to life-saving appliances that are not self-inflating. Life-saving appliances may contain:

(a)   Signal devices (Class 1) which may include smoke and illumination signal flares packed in packagings that prevent them from being inadvertently activated;

(b)   For UN No. 2990 only, cartridges, power device of Division 1.4, compatibility group S, may be contained for purposes of the self-inflating mechanism and provided that the quantity of explosives per appliance does not exceed 3.2 g;

(c)   Class 2 compressed gases, group A or O, according to 2.2.2.1.3;

(d)   Electric storage batteries (Class 8) and lithium batteries (Class 9);

(e)   First aid kits or repair kits containing small quantities of dangerous goods (e.g.: substances of Class 3, 4.1, 5.2, 8 or 9); or

(f)   "Strike anywhere" matches packed in packagings that prevent them from being inadvertently activated.

300   Fish meal or fish scrap shall not be loaded if the temperature at the time of loading exceeds 35 °C or 5 °C above the ambient temperature whichever is higher.

302   In the proper shipping name, the word "UNIT" means: a vehicle, a wagon, a container or a tank.

Fumigated vehicles, containers and tanks are only subject to the provisions of 5.5.2.

303   Receptacles shall be assigned to the classification code of the gas or mixture of gases contained therein determined in accordance with the provisions of section 2.2.2.

304   Batteries, dry, containing corrosive electrolyte which will not flow out of the battery if the battery case is cracked are not subject to the requirements of ADN provided the batteries are securely packed and protected against short-circuits. Examples of such batteries are: alkali-manganese, zinc-carbon, nickel-metal hydride and nickel-cadmium batteries.

305   These substances are not subject to the requirements of ADN when in concentrations of not more than 50 mg/kg.

306   This entry may only be used for substances that do not exhibit explosive properties of Class 1 when tested in accordance to Test Series 1 and 2 of Class 1 (see *Manual of Tests and Criteria*, Part I).

307   This entry may only be used for uniform mixtures containing ammonium nitrate as the main ingredient within the following composition limits:

(a)   Not less than 90% ammonium nitrate with not more than 0.2% total combustible/organic material calculated as carbon and with added matter, if any, which is inorganic and inert towards ammonium nitrate; or

(b)   Less than 90% but more than 70% ammonium nitrate with other inorganic materials or more than 80% but less than 90% ammonium nitrate mixed with calcium carbonate and/or dolomite and not more than 0.4% total combustible/organic material calculated as carbon; or

(c)   Nitrogen type ammonium nitrate based fertilizers containing mixtures of ammonium nitrate and ammonium sulphate with more than 45% but less than 70% ammonium nitrate and not more than 0.4% total combustible/organic material calculated as carbon such that the sum of the percentage compositions of ammonium nitrate and ammonium sulphate exceeds 70%.

309   This entry applies to non sensitized emulsions, suspensions and gels consisting primarily of a mixture of ammonium nitrate and fuel, intended to produce a Type E blasting explosive only after further processing prior to use.

The mixture for emulsions typically has the following composition: 60-85% ammonium nitrate, 5-30% water, 2-8% fuel, 0.5-4% emulsifier agent, 0-10% soluble flame suppressants, and trace additives. Other inorganic nitrate salts may replace part of the ammonium nitrate.

The mixture for suspensions and gels typically has the following composition: 60-85% ammonium nitrate, 0-5% sodium or potassium perchlorate, 0-17% hexamine nitrate or monomethylamine nitrate, 5-30% water, 2-15% fuel, 0.5-4% thickening agent, 0-10% soluble flame suppressants, and trace additives. Other inorganic nitrate salts may replace part of the ammonium nitrate.

Substances shall satisfactorily pass Test Series 8 of the Manual of Tests and Criteria, Part I, Section 18 and be approved by the competent authority.

310   The testing requirements in sub-section 38.3 of the *Manual of Tests and Criteria* do not apply to production runs consisting of not more than 100 lithium cells and

batteries, or to pre-production prototypes of lithium cells and batteries when these prototypes are carried for testing, if:

(a)   the cells and batteries are carried in an outer packaging that is a metal, plastics or plywood drum or a metal, plastics or wooden box and that meets the criteria for packing group I; and

(b)   each cell and battery is individually packed in an inner packaging inside an outer packaging and is surrounded by cushioning material that is non-combustible, and non-conductive.

311   Substances shall not be carried under this entry unless approved by the competent authority on the basis of the results of appropriate tests according to Part I of the *Manual of Tests and Criteria*. Packaging shall ensure that the percentage of diluent does not fall below that stated in the competent authority approval, at any time during carriage.

312   *(Reserved)*.

313   Substances and mixtures meeting the criteria for Class 8 shall bear a subsidiary risk label conforming to model No. 8 (see 5.2.2.2.2).

314   (a)   These substances are liable to exothermic decomposition at elevated temperatures. Decomposition can be initiated by heat or by impurities (e.g. powdered metals (iron, manganese, cobalt, magnesium) and their compounds);

        (b)   During the course of carriage, these substances shall be shaded from direct sunlight and all sources of heat and be placed in adequately ventilated areas.

315   This entry shall not be used for Class 6.1 substances which meet the inhalation toxicity criteria for packing group I described in 2.2.61.1.8.

316   This entry applies only to calcium hypochlorite, dry, when carried in non friable tablet form.

317   "Fissile-excepted" applies only to those packages complying with 6.4.11.2 of ADR.

318   For the purposes of documentation, the proper shipping name shall be supplemented with the technical name (see 3.1.2.8). When the infectious substances to be carried are unknown, but suspected of meeting the criteria for inclusion in category A and assignment to UN No. 2814 or 2900, the words "suspected category A infectious substance" shall be shown, in parentheses, following the proper shipping name on the transport document.

319   Substances packed and packages marked in accordance with packing instruction P650 are not subject to any other requirements of ADN.

321   These storage systems shall always be considered as containing hydrogen.

322   When carried in non-friable tablet form, these goods are assigned to packing group III.

323   *(Reserved)*.

324   This substance needs to be stabilized when in concentrations of not more than 99%.

325 In the case of non-fissile or fissile excepted uranium hexafluoride, the material shall be classified under UN No 2978.

326 In the case of fissile uranium hexafluoride, the material shall be classified under UN No 2977.

327 Waste aerosols consigned in accordance with 5.4.1.1.3 may be carried under this entry for the purposes of reprocessing or disposal. They need not be protected against inadvertent discharge provided that measures to prevent dangerous build up of pressure and dangerous atmospheres are addressed. Waste aerosols, other than those leaking or severely deformed, shall be packed in accordance with packing instruction P003 of ADR and special provision PP87 of ADR, or packing instruction LP02 of ADR and special packing provision L2 of ADR. Leaking or severely deformed aerosols shall be carried in salvage packagings provided appropriate measures are taken to ensure there is no dangerous build up of pressure.

NOTE: For maritime carriage, waste aerosols shall not be carried in closed containers.

328 This entry applies to fuel cell cartridges containing flammable liquids including methanol or methanol/water solutions. Fuel cell cartridge means a container that stores fuel for discharge into fuel cell powered equipment through a valve(s) that controls the discharge of fuel into such equipment and is free of electric charge generating components. The cartridge shall be designed and constructed to prevent the fuel from leaking during normal conditions of carriage.
This entry applies to fuel cell cartridge design types shown without their packaging to pass an internal pressure test at a pressure of 100 kPa (gauge).

329 *(Reserved).*

330 Alcohols containing petroleum products (e.g. gasoline) up to 5% shall be carried under the entry UN 1987 ALCOHOLS, N.O.S.

331-499 *(Reserved).*

500 UN No. 3064 nitroglycerin, solution in alcohol with more than 1% but not more than 5% nitroglycerin, packed in accordance with packing instruction P300 of 4.1.4.1 of ADR, is a substance of Class 3.

501 For naphthalene, molten, see UN No. 2304.

502 UN No. 2006 plastics, nitrocellulose-based, self-heating, n.o.s., and 2002 celluloid scrap are substances of Class 4.2.

503 For phosphorus, white or yellow, molten, see UN No. 2447.

504 UN No. 1847 potassium sulphide, hydrated with not less than 30% water of crystallization, UN No. 1849 sodium sulphide, hydrated with not less than 30% water of crystallization and UN No. 2949 sodium hydrosulphide with not less than 25% water of crystallization are substances of Class 8.

505 UN No. 2004 magnesium diamide is a substance of Class 4.2.

506 Alkaline earth metals and alkaline earth metal alloys in pyrophoric form are substances of Class 4.2.

UN No. 1869 magnesium or magnesium alloys containing more than 50% magnesium as pellets, turnings or ribbons, are substances of Class 4.1.

507 UN No. 3048 aluminium phosphide pesticides, with additives inhibiting the emission of toxic flammable gases are substances of Class 6.1.

508 UN No. 1871 titanium hydride and UN No. 1437 zirconium hydride are substances of Class 4.1. UN No. 2870 aluminium borohydride is a substance of Class 4.2.

509 UN No. 1908 chlorite solution is a substance of Class 8.

510 UN No. 1755 chromic acid solution is a substance of Class 8.

511 UN No. 1625 mercuric nitrate, UN No. 1627 mercurous nitrate and UN No. 2727 thallium nitrate are substances of Class 6.1. Thorium nitrate, solid, uranyl nitrate hexahydrate solution and uranyl nitrate, solid are substances of Class 7.

512 UN No. 1730 antimony pentachloride, liquid, UN No. 1731 antimony pentachloride solution, UN No. 1732 antimony pentafluoride and UN No. 1733 antimony trichloride are substances of Class 8.

513 UN No. 0224 barium azide, dry or wetted with less than 50% water, by mass, is a substance of Class 1. UN No. 1571 barium azide, wetted with not less than 50% water, by mass, is a substance of Class 4.1. UN No. 1854 barium alloys, pyrophoric, are substances of Class 4.2. UN No. 1445 barium chlorate, solid, UN No. 1446 barium nitrate, UN No. 1447 barium perchlorate, solid, UN No. 1448 barium permanganate, UN No. 1449 barium peroxide, UN No. 2719 barium bromate, UN No. 2741 barium hypochlorite with more than 22% available chlorine, UN No. 3405 barium chlorate, solution and UN No. 3406 barium perchlorate, solution, are substances of Class 5.1. UN No. 1565 barium cyanide and UN No. 1884 barium oxide are substances of Class 6.1.

514 UN No. 2464 beryllium nitrate is a substance of Class 5.1.

515 UN No. 1581 chloropicrin and methyl bromide mixture and UN No. 1582 chloropicrin and methyl chloride mixture are substances of Class 2.

516 UN No. 1912 methyl chloride and methylene chloride mixture is a substance of Class 2.

517 UN No. 1690 sodium fluoride, solid, UN No. 1812 potassium fluoride, solid, UN No. 2505 ammonium fluoride, UN No. 2674 sodium fluorosilicate, UN No. 2856 fluorosilicates, n.o.s., UN No. 3415 sodium fluoride, solution and UN No. 3422 potassium fluoride, solution, are substances of Class 6.1.

518 UN No. 1463 chromium trioxide, anhydrous (chromic acid, solid) is a substance of Class 5.1.

519 UN No. 1048 hydrogen bromide, anhydrous, is a substance of Class 2.

520 UN No. 1050 hydrogen chloride, anhydrous, is a substance of Class 2.

521 Solid chlorites and hypochlorites are substances of Class 5.1.

522 UN No. 1873 perchloric acid aqueous solution with more than 50% but not more than 72% pure acid, by mass are substances of Class 5.1. Perchloric acid solutions containing more than 72% pure acid, by mass, or mixtures of perchloric acid with any liquid other than water, are not to be accepted for carriage.

523 UN No. 1382 anhydrous potassium sulphide and UN No. 1385 anhydrous sodium sulphide and their hydrates with less than 30% water of crystallization, and UN No. 2318 sodium hydrosulphide with less than 25% water of crystallization are substances of Class 4.2.

524 UN No. 2858 finished zirconium products of a thickness of 18 μm or more are substances of Class 4.1.

525 Solutions of inorganic cyanides with a total cyanide ion content of more than 30% shall be classified in packing group I, solutions with a total cyanide ion content of more than 3% and not more than 30% in packing group II and solutions with a cyanide ion content of more than 0.3% and not more than 3% in packing group III.

526 UN No. 2000 celluloid is assigned to Class 4.1.

528 UN No. 1353 fibres or fabrics impregnated with weakly nitrated cellulose, non-self heating are articles of Class 4.1.

529 UN No. 0135 mercury fulminate, wetted with not less than 20% water, or mixture of alcohol and water, by mass, is a substance of Class 1. Mercurous chloride (calomel) is a substance of Class 9 (UN No. 3077).

530 UN No. 3293 hydrazine, aqueous solution with not more than 37% hydrazine, by mass, is a substance of Class 6.1.

531 Mixtures having a flash-point below 23 °C and containing more than 55% nitrocellulose, whatever its nitrogen content or containing not more than 55% nitrocellulose with a nitrogen content above 12.6% (by dry mass), are substances of Class 1 (see UN Nos. 0340 or 0342) or of Class 4.1.

532 UN No. 2672 ammonia solution containing not less than 10% but not more than 35% ammonia is a substance of Class 8.

533 UN No. 1198 formaldehyde solutions, flammable are substances of Class 3. Formaldehyde solutions, non-flammable, with less than 25% formaldehyde are not subject to the requirements of ADN.

534 While in some climatic conditions, petrol (gasoline) may have a vapour pressure at 50 °C of more than 110 kPa (1.10 bar) but not more than 150 kPa (1.50 bar) it is to continue to be considered as a substance having a vapour pressure at 50 °C of not more than 110 kPa (1.10 bar).

535 UN No. 1469 lead nitrate, UN No. 1470 lead perchlorate, solid and UN No. 3408 lead perchlorate, solution are substances of Class 5.1.

536 For naphthalene, solid, see UN No. 1334.

537 UN No. 2869 titanium trichloride mixture, not pyrophoric, is a substance of Class 8.

538 For sulphur (in the solid state), see UN No. 1350.

539   Solutions of isocyanates having a flash-point of not less than 23 °C are substances of Class 6.1.

540   UN No. 1326 hafnium powder, wetted, UN No. 1352 titanium powder, wetted or UN No. 1358 zirconium powder, wetted, with not less than 25% water, are substances of Class 4.1.

541   Nitrocellulose mixtures with a water content, alcohol content or plasticizer content lower than the stated limits are substances of Class 1.

542   Talc containing tremolite and/or actinolite is covered by this entry.

543   UN No. 1005 ammonia, anhydrous, UN No. 3318 ammonia solution with more than 50% ammonia and UN No. 2073 ammonia solution, with more than 35% but not more than 50% ammonia, are substances of Class 2. Ammonia solutions with not more than 10% ammonia are not subject to the requirements of ADN.

544   UN No. 1032 dimethylamine, anhydrous, UN No. 1036 ethylamine, UN No. 1061 methylamine, anhydrous and UN No. 1083 trimethylamine, anhydrous, are substances of Class 2.

545   UN No. 0401 dipicryl sulphide, wetted with less than 10% water by mass is a substance of Class 1.

546   UN No. 2009 zirconium, dry, finished sheets, strip or coiled wire, in thicknesses of less than 18 μm, is a substance of Class 4.2. Zirconium, dry, finished sheets, strip or coiled wire, in thicknesses of 254 μm or more, is not subject to the requirements of ADN.

547   UN No. 2210 maneb or UN No. 2210 maneb preparations in self-heating form are substances of Class 4.2.

548   Chlorosilanes which, in contact with water, emit flammable gases, are substances of Class 4.3.

549   Chlorosilanes having a flash-point of less than 23 °C and which, in contact with water, do not emit flammable gases are substances of Class 3. Chlorosilanes having a flash-point equal to or greater than 23 °C and which, in contact with water, do not emit flammable gases are substances of Class 8.

550   UN No. 1333 cerium in slabs, rods or ingots is a substance of Class 4.1.

551   Solutions of these isocyanates having a flash-point below 23 °C are substances of Class 3.

552   Metals and metal alloys in powdered or other flammable form, liable to spontaneous combustion, are substances of Class 4.2. Metals and metal alloys in powdered or other flammable form which, in contact with water, emit flammable gases are substances of Class 4.3.

553   This mixture of hydrogen peroxide and peroxyacetic acid shall, in laboratory testing (see *Manual of Tests and Criteria*, Part II, section 20), neither detonate in the cavitated state nor deflagrate at all and shall show no effect when heated under confinement nor any explosive power. The formulation shall be thermally stable (self-accelerating decomposition temperature 60 °C or higher for a 50 kg package), and a liquid

compatible with peroxyacetic acid shall be used for desensitization. Formulations not meeting these criteria are to be regarded as substances of Class 5.2 (see *Manual of Tests and Criteria*, Part II, paragraph 20.4.3 (g)).

554   Metal hydrides which, in contact with water, emit flammable gases are substances of Class 4.3. UN No. 2870 aluminium borohydride or UN No. 2870 aluminium borohydride in devices is a substance of Class 4.2.

555   Dust and powder of metals in non-spontaneously combustible form, non-toxic which nevertheless, in contact with water, emit flammable gases, are substances of Class 4.3.

556   Organometallic compounds and their solutions which ignite spontaneously are substances of Class 4.2. Flammable solutions with organometallic compounds in concentrations which, in contact with water, neither emit flammable gases in dangerous quantities nor ignite spontaneously are substances of Class 3.

557   Dust and powder of metals in pyrophoric form are substances of Class 4.2.

558   Metals and metal alloys in pyrophoric form are substances of Class 4.2. Metals and metal alloys which, in contact with water, do not emit flammable gases and are not pyrophoric or self-heating, but which are easily ignited, are substances of Class 4.1.

559   Mixtures of a hypochlorite with an ammonium salt are not to be accepted for carriage. UN No. 1791 hypochlorite solution is a substance of Class 8.

560   UN No. 3257 elevated temperature liquid, n.o.s., at or above 100 °C and, for a substance with a flash-point, below its flash-point (including molten metals and molten salts) is a substance of Class 9.

561   Chloroformates having predominantly corrosive properties are substances of Class 8.

562   Spontaneously combustible organometallic compounds are substances of Class 4.2. Water-reactive organometallic compounds, flammable, are substances of Class 4.3.

563   UN No. 1905 selenic acid is a substance of Class 8.

564   UN No. 2443 vanadium oxytrichloride, UN No. 2444 vanadium tetrachloride and UN No. 2475 vanadium trichloride are substances of Class 8.

565   Unspecified wastes resulting from medical/veterinary treatment of humans/animals or from biological research, and which are unlikely to contain substances of Class 6.2 shall be assigned to this entry. Decontaminated clinical wastes or wastes resulting from biological research which previously contained infectious substances are not subject to the requirements of Class 6.2.

566   UN No. 2030 hydrazine aqueous solution, with more than 37% hydrazine, by mass, is a substance of Class 8.

567   Mixtures containing more than 21% oxygen by volume shall be classified as oxidizing.

568   Barium azide with a water content lower than the stated limit is a substance of Class 1, UN No. 0224.

569-579 *(Reserved).*

580 Tank-vehicles, specialized vehicles and specially equipped vehicles for carriage in bulk shall bear on both sides and at the rear the mark referred to in 5.3.3. Tank-containers, portable tanks, special containers and specially equipped containers for carriage in bulk shall bear this mark on both sides and at each end.

581 This entry covers mixtures of methylacetylene and propadiene with hydrocarbons, which as:

Mixture P1, contain not more than 63% methylacetylene and propadiene by volume and not more than 24% propane and propylene by volume, the percentage of C_4-saturated hydrocarbons being not less than 14% by volume; and as

Mixture P2, contain not more than 48% methylacetylene and propadiene by volume and not more than 50% propane and propylene by volume, the percentage of C_4-saturated hydrocarbons being not less than 5% by volume,

as well as mixtures of propadiene with 1 to 4% methylacetylene.

When relevant, in order to meet the requirements for the transport document (5.4.1.1), the term "Mixture P1" or "Mixture P2" may be used as technical name.

582 This entry covers, <u>inter alia</u>, mixtures of gases indicated by the letter R ..., which as

Mixture F1, have a vapour pressure at 70° C not exceeding 1.3 MPa (13 bar) and a density at 50 °C not lower than that of dichlorofluoromethane (1.30 kg/l);

Mixture F2, have a vapour pressure at 70 °C not exceeding 1.9 MPa (19 bar) and a density at 50 °C not lower than that of dichloridifluoromethane (1.21 kg/l);

Mixture F3, have a vapour pressure at 70 °C not exceeding 3 MPa (30 bar) and a density at 50 °C not lower than that of chlorodifluoromethane (1.09 kg/l).

NOTE: *Trichlorofluoromethane (refrigerant R 11), 1,1,2-trichloro-1,2,2-trifluoroethane (refrigerant R 113), 1,1,1-trichloro-2,2,2-trifluoroethane (refrigerant R 113a), 1-chloro-1,2,2-trifluoroethane (refrigerant R 133) and 1-chloro-1,1,2-trifluoroethane (refrigerant R 133 b) are not substances of Class 2. They may, however, enter into the composition of mixtures F1 to F3.*

When relevant, in order to meet the requirements for the transport document (5.4.1.1), the term "Mixture F1", "Mixture F2" or "Mixture F3" may be used as technical name.

583 This entry covers, <u>inter alia</u>, mixtures which as:

Mixture A, have a vapour pressure at 70 °C not exceeding 1.1 MPa (11 bar) and a density at 50 °C not lower than 0.525 kg/l;

Mixture A01, have a vapour pressure at 70 °C not exceeding 1.6 MPa (16 bar) and a density at 50 °C not lower than 0.516 kg/l;

Mixture A02, have a vapour pressure at 70 °C not exceeding 1.6 MPa (16 bar) and a density at 50 °C not lower than 0.505 kg/l;

Mixture A0, have a vapour pressure at 70 °C not exceeding 1.6 MPa (16 bar) and a density at 50 °C not lower than 0.495 kg/l;

Mixture A1, have a vapour pressure at 70 °C not exceeding 2.1 MPa (21 bar) and a density at 50 °C not lower than 0.485 kg/l;

Mixture B1, have a vapour pressure at 70 °C not exceeding 2.6 MPa (26 bar) and a density at 50 °C not lower than 0.474 kg/l;

Mixture B2, have a vapour pressure at 70 °C not exceeding 2.6 MPa (26 bar) and a density at 50 °C not lower than 0.463 kg/l;

Mixture B, have a vapour pressure at 70 °C not exceeding 2.6 MPa (26 bar) and a density at 50 °C not lower than 0.450 kg/l;

Mixture C, have a vapour pressure at 70 °C not exceeding 3.1 MPa (31 bar) and a density at 50 °C not lower than 0.440 kg/l;

When relevant, in order to meet the requirements for the transport document (5.4.1.1), the following terms may be used as technical name:

–   "Mixture A" or "Butane";

–   "Mixture A01" or "Butane";

–   "Mixture A02" or "Butane";

–   "Mixture A0" or "Butane";

–   "Mixture A1";

–   "Mixture B1";

–   "Mixture B2";

–   "Mixture B";

–   "Mixture C" or "Propane".

For carriage in tanks, the trade names "butane" or "propane" may be used only as a complement.

584   This gas is not subject to the requirements of ADN when:

–   it is in the gaseous state;

–   it contains not more than 0.5% air;

–   it is contained in metal capsules (sodors, sparklets) free from defects which may impair their strength;

–   the leakproofness of the closure of the capsule is ensured;

–   a capsule contains not more than 25 g of this gas;

–   a capsule contains not more than 0.75 g of this gas per cm^3 of capacity.

585   Cinnabar is not subject to the requirements of ADN.

586 Hafnium, titanium and zirconium powders shall contain a visible excess of water. Hafnium, titanium and zirconium powders, wetted, mechanically produced, of a particle size of 53 μm and over, or chemically produced, of a particle size of 840 μm and over, are not subject to the requirements of ADN.

587 Barium stearate and barium titanate are not subject to the requirements of ADN.

588 Solid hydrated forms of aluminium bromide and aluminium chloride are not subject to the requirements of ADN.

589 Calcium hypochlorite mixtures, dry, containing not more than 10% available chlorine are not subject to the requirements of ADN.

590 Ferric chloride hexahydrate is not subject to the requirements of ADN.

591 Lead sulphate with not more than 3% free acid is not subject to the requirements of ADN.

592 Uncleaned empty packagings (including empty IBCs and large packagings), empty tank-vehicles, empty demountable tanks, empty portable tanks, empty tank-containers and empty small containers which have contained this substance are not subject to the requirements of ADN.

593 This gas, intended for the cooling of e.g. medical or biological specimens, if contained in double wall receptacles which comply with the provisions of packing instruction P203 (12) of 4.1.4.1 of ADR is not subject to the requirements of ADN.

594 The following articles, manufactured and filled according to the regulations of the manufacturing State and packaged in strong outer packagings, are not subject to the requirements of ADN:

    – UN No. 1044 fire extinguishers provided with protection against inadvertent discharge;

    – UN No. 3164 articles, pressurized pneumatic or hydraulic, designed to withstand stresses greater than the internal gas pressure by virtue of transmission of force, intrinsic strength or construction.

596 Cadmium pigments, such as cadmium sulphides, cadmium sulphoselenides and cadmium salts of higher fatty acids (e.g. cadmium stearate), are not subject to the requirements of ADN.

597 Acetic acid solutions with not more than 10% pure acid by mass, are not subject to the requirements of ADN.

598 The following are not subject to the requirements of ADN:

    (a) New storage batteries when:

        – they are secured in such a way that they cannot slip, fall or be damaged;

        – they are provided with carrying devices, unless they are suitably stacked, e.g. on pallets;

        – there are no dangerous traces of alkalis or acids on the outside;

	–	they are protected against short circuits;

(b)   Used storage batteries when:

- their cases are undamaged;

- they are secured in such a way that they cannot leak, slip, fall or be damaged, e.g. by stacking on pallets;

- there are no dangerous traces of alkalis or acids on the outside of the articles;

- they are protected against short circuits.

"Used storage batteries" means storage batteries carried for recycling at the end of their normal service life.

599   Manufactured articles or instruments containing not more than 1 kg of mercury are not subject to the requirements of ADN.

600   Vanadium pentoxide, fused and solidified, is not subject to the requirements of ADN.

601   Pharmaceutical products (medicines) ready for use, which are substances manufactured and packaged for retail sale or distribution for personal or household consumption are not subject to the requirements of ADN.

602   Phosphorus sulphides which are not free from yellow and white phosphorus are not to be accepted for carriage.

603   Anhydrous hydrogen cyanide not meeting the description for UN No. 1051 or UN No. 1614 is not to be accepted for carriage. Hydrogen cyanide (hydrocyanic acid) containing less than 3% water is stable, if the pH-value is 2.5 ± 0.5 and the liquid is clear and colourless.

604   Ammonium bromate and its aqueous solutions and mixtures of a bromate with an ammonium salt are not to be accepted for carriage.

605   Ammonium chlorate and its aqueous solutions and mixtures of a chlorate with an ammonium salt are not to be accepted for carriage.

606   Ammonium chlorite and its aqueous solutions and mixtures of a chlorite with an ammonium salt are not to be accepted for carriage.

607   Mixtures of potassium nitrate and sodium nitrite with an ammonium salt are not to be accepted for carriage.

608   Ammonium permanganate and its aqueous solutions and mixtures of a permanganate with an ammonium salt are not to be accepted for carriage.

609   Tetranitromethane not free from combustible impurities is not to be accepted for carriage.

610   The carriage of this substance, when it contains more than 45% hydrogen cyanide is prohibited.

611   Ammonium nitrate containing more than 0.2% combustible substances (including any organic substance calculated as carbon) is not to be accepted for carriage unless it is a constituent of a substance or article of Class 1.

612   *(Reserved).*

613   Chloric acid solution containing more than 10% chloric acid and mixtures of chloric acid with any liquid other than water is not to be accepted for carriage.

614   2,3,7,8-tetrachlorodibenzo-p-dioxin (TCDD) in concentrations considered highly toxic according to the criteria in 2.2.61.1 is not to be accepted for carriage.

615   *(Reserved).*

616   Substances containing more than 40% liquid nitric esters shall satisfy the exudation test specified in 2.3.1.

617   In addition to the type of explosive, the commercial name of the particular explosive shall be marked on the package.

618   In receptacles containing 1,2-butadiene, the oxygen concentration in the gaseous phase shall not exceed 50 ml/m^3.

619-622 *(Reserved).*

623   UN No. 1829 sulphur trioxide shall be inhibited. Sulphur trioxide, 99.95% pure or above, may be carried without inhibitor in tanks provided that its temperature is maintained at or above 32.5 °C. For the carriage of this substance without inhibitor in tanks at a minimum temperature of 32.5 °C, the specification **"Transport under minimum temperature of the product of 32.5 °C"** shall appear in the transport document.

625   Packages containing these articles shall be clearly marked as follows:
"**UN 1950 AEROSOLS**"

626-627 *(Reserved).*

632   Considered to be spontaneously flammable (pyrophoric).

633   Packages and small containers containing this substance shall bear the following marking: "**Keep away from any source of ignition**". This marking shall be in an official language of the forwarding country, and also, if that language is not English, French or German, in English, French or German, unless any agreements concluded between the countries concerned in the transport operation provide otherwise.

635   Packages containing these articles need not bear a label conforming to model No. 9 unless the article is fully enclosed by packaging, crates or other means that prevent the ready identification of the article.

636   (a)   Used lithium cells and batteries collected and presented for carriage for disposal between the consumer collecting point and the intermediate processing facility, together with other non-lithium cells or batteries or alone, are not subject to the other provisions of ADN if they meet the following conditions:

              (i)   The gross mass of each lithium cell or battery does not exceed 250 g;

(ii) The provisions of packing instruction P903b (2) of ADR are complied with.

(b) Cells contained in equipment shall not be capable of being discharged during carriage to the extent that the open circuit voltage falls below 2 volts or two thirds of the voltage of the undischarged cell, whichever is the lower.

(c) Packages containing used cells or batteries in unmarked packagings shall bear the inscription: "**Used lithium cells**".

637 Genetically modified micro-organisms are those which are not dangerous for humans and animals, but which could alter animals, plants, microbiological substances and ecosystems in such a way as cannot occur naturally. Genetically modified micro-organisms which have received a consent for deliberate release into the environment[1] are not subject to the requirements of Class 9. Live vertebrate or invertebrate animals shall not be used to carry these substances classified under this UN number unless the substance can be carried in no other way. For the carriage of easily perishable substances under this UN number appropriate information shall be given, e.g.: "**Cool at +2 °/+4 °C**" or "**Carry in frozen state**" or "**Do not freeze**".

638 Substances related to self-reactive substances (see 2.2.41.1.19).

639 See 2.2.2.3, classification code 2F, UN No. 1965, Note 2.

640 The physical and technical characteristics mentioned in column (2) of Table A of Chapter 3.2 determine different tank codes for the carriage of substances of the same packing group in tanks conforming to Chapter 6.8 of RID or ADR.

In order to identify these physical and technical characteristics of the product carried in the tank, the following shall be added, to the particulars required in the transport document/consignment note, only in case of carriage in tanks conforming to Chapter 6.8 of ADR or RID:

"Special provision 640X" where "X" is the applicable capital letter appearing after the reference to special provision 640 in column (6) of Table A of Chapter 3.2.

These particulars may, however, be dispensed with in the case of carriage in the type of tank which, for substances of a specific packing group of a specific UN number, meets at least the most stringent requirements.

643 Stone or aggregate asphalt mixture is not subject to the requirements for Class 9.

644 This substance is admitted for carriage provided that:

– The pH is between 5 and 7 measured in an aqueous solution of 10% of the substance carried;

– The solution does not contain more than 0.2% combustible material or chlorine compounds in quantities such that the chlorine level exceeds 0.02%.

---

[1] *See in particular Part C of Directive 2001/18/EC of the European Parliament and of the Council on the deliberate release into the environment of genetically modified organisms and repealing Council Directive 90/220/EEC (Official Journal of the European Communities, No. L 106, of 17 April 2001, pp. 8-14), which sets out the authorization procedures for the European Community.*

645 The classification code as mentioned in Column (3b) of Table A of Chapter 3.2 shall be used only with the approval of the competent authority of a Contracting Party to ADN prior to carriage. When assignment to a division is made in accordance with the procedure in 2.2.1.1.7.2, the competent authority may require the default classification to be verified on the basis of test data derived from Test Series 6 of the Manual of Tests and Criteria.

646 Carbon made by steam activation process is not subject to the requirements of ADN.

647 Except for carriage in tank vessels, the carriage of vinegar and acetic acid with not more than 25 % pure acid by mass is subject only to the following requirements:

(a) Packagings, including IBCs and large packagings, and tanks shall be manufactured from stainless steel or plastic material which is permanently resistant to corrosion of vinegar/acetic acid food grade;

(b) Packagings, including IBCs and large packagings, and tanks shall be subjected to a visual inspection by the owner at least once a year. The results of the inspections shall be recorded and the records kept for at least one year. Damaged packagings, including IBCs and large packagings, and tanks shall not be filled;

(c) Packagings, including IBCs and large packagings, and tanks shall be filled in a way that no product is spilled or adheres to the outer surface;

(d) Seals and closures shall be resistant to vinegar/acetic acid food grade. Packagings, including IBCs and large packagings, and tanks shall be hermetically sealed by the person in charge of packaging and/or filling so that under normal conditions of carriage there will be no leakage;

(e) Combination packagings with inner packaging made of glass or plastic (see packing instruction P001 in 4.1.4.1 of ADR) which fulfil the general packing requirements of 4.1.1.1, 4.1.1.2, 4.1.1.4, 4.1.1.5, 4.1.1.6, 4.1.1.7 and 4.1.1.8 of ADR may be used;

The other provisions of ADN do not apply.

648 Articles impregnated with this pesticide, such as fibreboard plates, paper strips, cotton-wool balls, sheets of plastics material, in hermetically closed wrappings, are not subject to the provisions of ADN.

649 To determine the initial boiling point, as mentioned under 2.2.3.1.3, packing group I, the test method according to standard ASTM D86-01[2] is suitable.

Substances which have an initial boiling point above 35 °C determined with this method are substances of packing group II and shall be classified in accordance with the applicable entry of this packing group.

650 Waste consisting of packaging residues, solidified residues and liquid residues of paint may be carried under the conditions of packing group II. In addition to the provisions of UN No. 1263, packing group II, the waste may also be packed and carried as follows:

---

[2] *Standard Test Method for Distillation of Petroleum Products at Atmospheric Pressure, published September 2001 by ASTM International.*

(a) The waste may be packed in accordance with packing instruction P002 of 4.1.4.1 of ADR or to packing instruction IBC006 of 4.1.4.2 of ADR;

(b) The waste may be packed in flexible IBCs of types 13H3, 13H4 and 13H5 in overpacks with complete walls;

(c) Testing of packagings and IBCs indicated under (a) or (b) may be carried out in accordance with the requirements of Chapters 6.1 or 6.5 of ADR, as appropriate, in relation to solids, at the packing group II performance level.

The tests shall be carried out on packagings and IBCs, filled with a representative sample of the waste, as prepared for carriage;

(d) Carriage in bulk in sheeted wagons, movable roof wagons/sheeted vehicles, closed containers or sheeted large containers, all with complete walls is allowed. The wagons, containers or body of vehicles shall be leakproof or rendered leakproof, for example by means of a suitable and sufficiently stout inner lining;

(e) If the waste is carried under the conditions of this special provision, the goods shall be declared in accordance with 5.4.1.1.3 in the transport document, as follows: "WASTE, UN 1263 PAINT, 3, II.

651 Special provision V2 (1) of ADR is only applicable for a net explosive content of more than 3,000 kg (4,000 kg with trailer).

653 The carriage of this gas in cylinders with a maximum capacity of 0.5 litres, is not subject to the other provisions of ADN if the following conditions are met:

- The provisions for construction and testing of cylinders are observed;

- The cylinders are contained in outer packagings which at least meet the requirements of Part 4 for combination packagings. The general provisions of packing of 4.1.1.1, 4.1.1.2 and 4.1.1.5 to 4.1.1.7 of ADR shall be observed;

- The cylinders are not packed together with other dangerous goods;

- The total gross mass of a package does not exceed 30 kg; and

Each package is clearly and durably marked with "UN 1013". This marking is displayed within a diamond-shaped area surrounded by a line that measures at least 100 mm by 100 mm.

800 Oil seeds, crushed seeds and seedcake containing vegetable oil, treated with solvents, not subject to spontaneous combustion, are allocated to UN No. 3175. These substances are not subject to ADN when they have been prepared or treated to ensure that they cannot give off dangerous gases in dangerous quantities (no risk of explosion) during carriage and when this is mentioned in the transport document.

801 Ferrosilicon with between 25 and 30% or more than 90% silicon content by mass is a dangerous substance of Class 4.3 for carriage in bulk or without packaging by inland navigation vessel.

802 See 7.1.4.10.

CHAPTER 3.4

EXEMPTIONS RELATED TO DANGEROUS GOODS PACKED IN LIMITED QUANTITIES

3.4.1 General requirements

3.4.1.1   Packagings used in accordance with 3.4.3 to 3.4.6 below, need only to conform to the general provisions of 4.1.1.1, 4.1.1.2 and 4.1.1.4 to 4.1.1.8 of ADR.

3.4.1.2   The maximum gross mass of a combination packaging shall not exceed 30 kg and for shrink and stretched wrapped trays shall not exceed 20 kg.

NOTE: The limit for combination packagings does not apply when LQ5 is issued.

3.4.1.3   Subject to the maximum limits in 3.4.1.2 and individual limits in table 3.4.6, dangerous goods may be packed together with other articles or substances, provided they will not react dangerously in the event of leakage.

3.4.2   When the code "LQ0" is shown in Column (7) of Table A in Chapter 3.2 for a given substance or article, that substance or article is not exempted from any of the applicable provisions of ADN when it is packed in limited quantities, unless otherwise specified in these annexed Regulations.

3.4.3   Unless otherwise provided in this Chapter, when one of the codes "LQ1" or "LQ2" is shown in Column (7) of Table A in Chapter 3.2 for a given substance or article, the provisions of other Chapters of ADN do not apply to the carriage of that substance or article, provided:

(a)   the provisions of 3.4.5 (a) to (c) are observed; with respect to these provisions, articles are considered to be inner packagings;

(b)   inner packagings meet the conditions of 6.2.1.2, 6.2.4.1, 6.2.4.3 of ADR.

3.4.4   Unless otherwise provided in this Chapter, when the code "LQ3" is shown in Column (7) of Table A in Chapter 3.2 for a given substance, the provisions of other Chapters of ADN do not apply to the carriage of that substance, provided:

(a)   The substance is carried in combination packagings, the following outer packagings being allowed:

- steel or aluminium drums with removable head;

- steel or aluminium jerricans with removable head;

- plywood or fibre drums;

- plastics drums or jerricans with removable head;

- boxes of natural wood, plywood, reconstituted wood, fibreboard, plastics, steel or aluminium;

and be so designed that they meet the relevant construction requirements of 6.1.4 of ADR;

(b)   The maximum net quantities per inner packaging shown in columns (2) or (4) and per package in columns (3) or (5), where indicated, of table 3.4.6 are not exceeded;

(c)   Each package is clearly and durably marked with :

    (i)   the UN number of the goods contained therein, as given in Column (1) of Table A in Chapter 3.2, preceded by the letters "UN";

    (ii)   in the case of different goods with different UN numbers within a single package:

        –   the UN numbers of the goods contained therein, preceded by the letters "UN", or

        –   the letters "LQ"[1].

These markings shall be displayed within a diamond-shaped area surrounded by a line that measures at least 100 mm × 100 mm. The width of line forming the diamond shall be at least 2 mm; the number shall be at least 6 mm high. Where more than one substance assigned to different UN numbers are included in the package, the diamond shall be large enough to include each relevant UN number. If the size of the package so requires, the dimension may be reduced, provided the markings remain clearly visible.

3.4.5   Unless otherwise provided in this Chapter, when one of the codes "LQ4" to "LQ19" and "LQ22" to "LQ28" is shown in Column (7) of Table A in Chapter 3.2 for a given substance, the provisions of other Chapters of ADN do not apply to the carriage of that substance, provided:

(a)   The substance is carried:

    –   in combination packagings, corresponding to the prescriptions of 3.4.4 (a), or

    –   in metal or plastics inner packagings which are not liable to break or be easily punctured, placed in shrink-wrapped or stretch-wrapped trays;

(b)   The maximum net quantities per inner packaging shown in columns (2) or (4) and per package in columns (3) or (5), where indicated, of table 3.4.6 are not exceeded;

(c)   Each package is clearly and durably marked as indicated in 3.4.4 (c).

---

[1]   *The letters "LQ" are an abbreviation of the English words "Limited Quantities". The letters "LQ" are not permitted by the IMDG Code or the ICAO Technical Instructions.*

3.4.6 Table

Code	Combination packagings[a] Maximum net quantity		Inner packagings placed in shrink-wrapped or stretch-wrapped trays[a] Maximum net quantity	
	per inner packaging	per package[b]	per inner packaging	per package[b]
(1)	(2)	(3)	(4)	(5)
LQ0	No exemption under the conditions of 3.4.2.			
LQ1	120 ml		120 ml	
LQ2	1 *l*		1 *l*	
LQ3 [c]	500 ml	1 *l*	Not allowed	Not allowed
LQ4 [c]	3 *l*		1 *l*	
LQ5 [c]	5 *l*	Unlimited	1 *l*	
LQ6 [c]	5 *l*		1 *l*	
LQ7 [c]	5 *l*		5 *l*	
LQ8	3 kg		500 g	
LQ9	6 kg		3 kg	
LQ10	500 ml		500 ml	
LQ11	500 g		500 g	
LQ12	1 kg		1 kg	
LQ13	1 *l*		1 *l*	
LQ14	25 ml		25 ml	
LQ15	100 g		100 g	
LQ16	125 ml		125 ml	
LQ17	500 ml	2 *l*	100 ml	2 *l*
LQ18	1 kg	4kg	500 g	4 kg
LQ19	5 kg		5 kg	
LQ20	Reserved	Reserved	Reserved	Reserved
LQ21	Reserved	Reserved	Reserved	Reserved
LQ22	1 *l*		500 ml	
LQ23	3 kg		1 kg	
LQ24	6 kg		2 kg	
LQ25 [d]	1 kg		1 kg	
LQ26 [d]	500 ml	2 *l*	500 ml	2 *l*
LQ27	6 kg		6 kg	
LQ28	3 *l*		3 *l*	

[a] *See 3.4.1.2.*

[b] *See 3.4.1.3.*

[c] *In the case of homogenous mixtures of Class 3 containing water, the quantities specified relate only to the substance of Class 3 contained in those mixtures.*

[d] *For UN Nos. 2315, 3151, 3152 and 3432 when carried in apparatus, the inner packaging quantities shall not be exceeded per piece of apparatus. The apparatus shall be carried in a leakproof packaging and the complete package shall conform to 3.4.4 (c). Shrink-wrapped and stretch-wrapped trays shall not be used for apparatus.*

3.4.7   Overpacks containing packages conforming to 3.4.3, 3.4.4 or 3.4.5 shall be marked, as required by 3.4.4 (c) for each item of dangerous goods contained in the overpack, unless markings representative of all dangerous goods contained in the overpack are visible.